Javier García de Jalón   Eduardo Bayo

Kinematic and Dynamic Simulation of Multibody Systems

The Real-Time Challenge

With 268 Figures

Springer-Verlag

New York Berlin Heidelberg London Paris
Tokyo Hong Kong Barcelona Budapest

Javier García de Jalón
Department of Applied Mechanics
University of Navarra and CEIT
Paseo de Manuel Lardizabal 13
20009 Donostia—San Sebastian
Spain

Eduardo Bayo
Department of Mechanical Engineering
University of California, Santa Barbara
Santa Barbara, CA 93106
USA

Series Editor
Frederick F. Ling
Ernest F. Gloyna Regents Chair in Engineering
Department of Mechanical Engineering
The University of Texas at Austin
Austin, TX 78712-1063 USA
and
Distinguished William Howard Hart Professor Emeritus
Department of Mechanical Engineering,
  Aeronautical Engineering, and Mechanics
Rensselaer Polytechnic Institute
Troy, NY 12180-3590
USA

Library of Congress Cataloging-in-Publication Data
García de Jalón, Javier
    Kinematic and dynamic simulation of multibody systems : the real
    time challenge/Javier García de Jalón and Eduardo Bayo.
        p. cm.
    Includes bibliographical references and index.
    ISBN-13: 978-1-4612-7601-2   e-ISBN-13: 978-1-4612-2600-0
    DOI: 10.1007/978-1-4612-2600-0
    1. Dynamics—Computer simulation.   2. Kinematics—Computer
simulation.   I. Bayo, Eduardo.   II. Title.
    QA845.G37   1993   93-29030

Printed on acid-free paper.

Production managed by Natalie Johnson; manufacturing supervised by Vincent Scelta.
Camera-ready copy prepared by the author.

9 8 7 6 5 4 3 2 1

Series Preface

Mechanical engineering, an engineering discipline born of the needs of the industrial revolution, is once again asked to do its substantial share in the call for industrial renewal. The general call is urgent as we face profound issues of productivity and competitiveness that require engineering solutions, among others. The Mechanical Engineering Series features graduate texts and research monographs intended to address the need for information in contemporary areas of mechanical engineering.

The series is conceived as a comprehensive one that will cover a broad range of concentrations important to mechanical engineering graduate education and research. We are fortunate to have a distinguished roster of consulting editors, each an expert in one of the areas of concentration. The names of the consulting editors are listed on the front page of the volume. The areas of concentration are applied mechanics, biomechanics, computational mechanics, dynamic systems and control, energetics, mechanics of material, processing, thermal science, and tribology.

Professor Leckie, the consulting editor for applied mechanics, and I are pleased to present this volume of the series: *Kinematic and Dynamic Simulation of Multibody Systems: The Real-Time Challenge* by Professors García de Jalón and Bayo. The selection of this volume underscores again the interest of the Mechanical Engineering Series to provide our readers with topical monographs as well as graduate texts.

Austin Texas   Frederick F. Ling

*The first author dedicates this book to the memory of
Prof. F. Tegerizo († 1988), who introduced him to kinematics.
The second author dedicates this book
to his wife Elisabeth Anne and children:
Carolina, Eduardo, Joseph, Annelise, and Christopher.*

Preface

The name *multibody* stands as a general term that encompasses a wide range of systems such as: mechanisms, automobiles and trucks (including steering systems, suspensions, etc.), robots, trains, industrial machinery (textile, packaging, etc.), space structures, antennas, satellites, the human body, and others.

The use of computer aided kinematic and dynamic simulation has emerged as a powerful tool for the analysis and design of multibody systems in fields such as automobile industry, aerospace, robotics, machinery, biomechanics, and others. The attention that it has received recently can be measured by the amount of computer-aided analysis programs proliferating in the market for engineering software, a phenomena similar to that produced by the finite element method in the early seventies for structural design. Efficient formulations for dynamics and reliable computational methods play a key role in achieving good simulation tools.

The purpose of this book is to describe not only the commonly used methods for multibody kinematic and dynamic simulation, but also the advanced topics and the state of the art techniques. These include numerical methods and improved dynamic formulations that allow *real-time simulation* response. The real time response in multibody simulation is a characteristic that the engineering profession is demanding more and more for analysis and design purposes. The analyst and designer are interested in visualizing a whole set of successive responses of a multibody in real time under different conditions, so as to get a clear picture of the actual performance of the system that will help them to optimize the design process.

The main features that characterize this book and distinguish it from other texts are:

a) The use of the *natural or fully Cartesian coordinates* which allow for a simple representation of multibodies, and lead to important advantages for kinematic and dynamic simulation.

b) The consideration of advanced topics such as: friction, backlash, forward and inverse dynamics of flexible multibodies, sensitivity analysis, and others.

c) The detailed description of numerical methods and improved dynamic formulations that allow *real time simulation* response.

Contents

The first part of the book contains a description of the basic approaches and methods for kinematic and dynamic analyses. Chapter 1 serves as an introduction where the basic concepts and definitions are explained, the different types of problems identified, and the general ways they may be solved are outlined. Chapter 2 describes the types of coordinates commonly used for the analysis of multibody systems. Emphasis is placed on the fully Cartesian coordinates for 2- and 3-dimensional systems which are treated thoroughly along with the types of constraint conditions that they generate for different kinematic pairs. Chapter 3 deals with kinematic analysis. The solution of problems such as initial position, finite displacements, finding of the velocities and accelerations, treatment of redundant constraints (over constrained systems), and the study of the Jacobian nullspace that contains the possible motions are thoroughly exposed in this chapter. Dynamic analysis starts in Chapter 4 with the formulation of the inertial forces (mass matrices) generated by the different kind of bodies, and the external and gravitational forces. Chapter 5 continues with a detailed description of the different methods currently available for the dynamic analysis. Special attention is given to the description of the methods in both dependent and independent coordinates, and those based on velocity transformations. Chapter 6 deals with the analysis static equilibrium position and the inverse dynamic problem.

The more advanced topics are dealt with in the second part initiated in Chapter 7 which describes the numerical integration of the resultant equations of motion. Attention is given to the methods available for the solution of nonlinear ordinary differential equations and differential-algebraic equations, and emphasis is placed not only on accuracy but on stability for real time simulation. Improved dynamic formulations of order $O(N)$ and $O(N^3)$ such as recursive formalisms, improved use of velocity transformations, and some particular implementations of the penalty formulations in dependent coordinates are dealt with in Chapter 8. Emphasis is placed on the real time simulation from the viewpoint of versatility, generality, ease of implementation and possibilities of parallelization. The linearized dynamic analysis is treated in Chapter 9. Chapter 10 deals with further topics such as backlash, Coulomb friction, impacts, singular positions, kinematic synthesis, and sensitivity analysis. Some of these topics offer open areas for further research. Chapter 11 covers the forward dynamic analysis (simulation problem) of multibodies with flexible elements. The formulations that arise from the use of moving reference frames as well as inertial frames that require large displacements and rotation elastic theories are explained in detail. Chapter 12 deals with the newly developed inverse dynamics of flexible multibodies that leads to the time anticipatory joint efforts capable of reproducing a specified endpoint trajectory.

Audience for the Book

The aims of the book are twofold: educational and tutorial on the one hand, and a state of the art compilation of techniques and results on the other. The basic ideas are presented in sufficient detail in the first part of the book which can be used either as a textbook for an undergraduate elective course or basic graduate class on computer aided kinematic and dynamic analysis of mechanical systems, or as a self-study tool for the newcomer. The more advanced topics presented in the second part of the book are aimed at both the graduate student and researcher, who will find a compendium of state of the art information and a number of areas identified for further research. This part may be of interest not only to the mechanical, biomedical, or aerospace engineer, but also to people in other fields such as the numerical analyst, computer scientist, and even the software developer with interest in the computational aspects on the analysis of multibody systems.

Emphasis has been put on techniques which are basic to understanding the subject, and results included are felt to be of essentially permanent value. The authors believe that a very important feature of this book is the simplicity and easiness (rather than the sophistication) of the methods explained therein. All chapters include solved examples. Problem assignments can be found in addition at the end of each chapter of the first part of the book. Finally, the book requires only a minimal amount of background in physics and mathematics that does not exceed the basics known by undergraduate junior students in science and engineering in both American and European universities. We have purposely tried to avoid more advanced mathematics such as tensorial calculus, dyads, and quaternions, which are used in this field but are not part of that basic background.

Acknowledgments

We would like to thank all the colleagues and students who have helped us with their support and contributions in the preparation of this book. A thank-you is given in particular to Dr. Leckie for his appreciation of the ideas and methods presented, Dr. A. Avello, Dr. J.M. Jimenez and Dr. R. Ledesma for their contributions and critical comments. We would also like to thank M. Rodríguez, J. Gómez and Ana Leiza for the preparation of the manuscript.

Special thanks are due to the CICYT of the Spanish Ministry of Education for its economical support that allowed the second author to spend his sabbatical leave at the University of Navarra, and to the NATO Scientific Affairs Division which provided the travel grants that have allowed team cooperation.

Finally, we would like to thank our families, friends and colleagues for the patience and encouragement they have provided in the completion of this book.

San Sebastián, Spain   Javier García de Jalón
Santa Barbara, California   Eduardo Bayo

Contents

1

Introduction and Basic Concepts

The kinematics and dynamics of multibody systems is an important part of what is referred to as CAD (*Computer Aided Design*) and MCAE (*Mechanical Computer Aided Engineering*). Figures 1.1 to 1.6 illustrate some practical examples of computer generated models for the simulation of real multibody systems. The mechanical systems included under the definition of *multibodies* comprise robots, heavy machinery, spacecraft, automobile suspensions and steering systems, graphic arts and textile machinery, packaging machinery, machine tools, and others. Normally, the mechanisms used in all these applications are subjected to large displacements, hence, their geometric configuration undergoes large variations under normal service conditions. Moreover, in recent years operating speeds have been increased, and consequently, there has been an increase in accelerations and inertial forces. These large forces inevitably lead to the appearance of dynamic problems that one must be able to predict and control.

The advantage of computer simulations performed by CAD and MCAE tools is that they allow one to predict the kinematic and dynamic behavior of all types of multibody systems in great detail during all the design stages from the first design concepts to the final prototypes. At any design stage, computer-aided analysis is an auxiliary tool of great value, providing a sufficient amount of data for the engineer to study the influence of the different design parameters, since it allows him to carry out a large number of simulations quickly and economically.

The *analysis* programs simulate the behavior of a multibody system once all of its geometric and dynamic characteristics have been defined. The analysis programs are certainly very useful. At the present time they are the only general purpose tools available for the largest number of applications. We are also witnessing the advent of the *design* programs that will not only perform system analyses, but also modify automatically its parameters so as to obtain an optimal behavior. An intermediate step between the analysis and optimal design programs are the *parametric* analyses, which determine the different responses of a multibody system with respect to the variation of one of the design variables. In any case, the analysis programs constitute the basis of the design programs. This book is particularly oriented towards the study of the analytical methods and numerical algorithms that are necessary to build such simulation tools. Nevertheless, we will also pay attention to some important design issues.

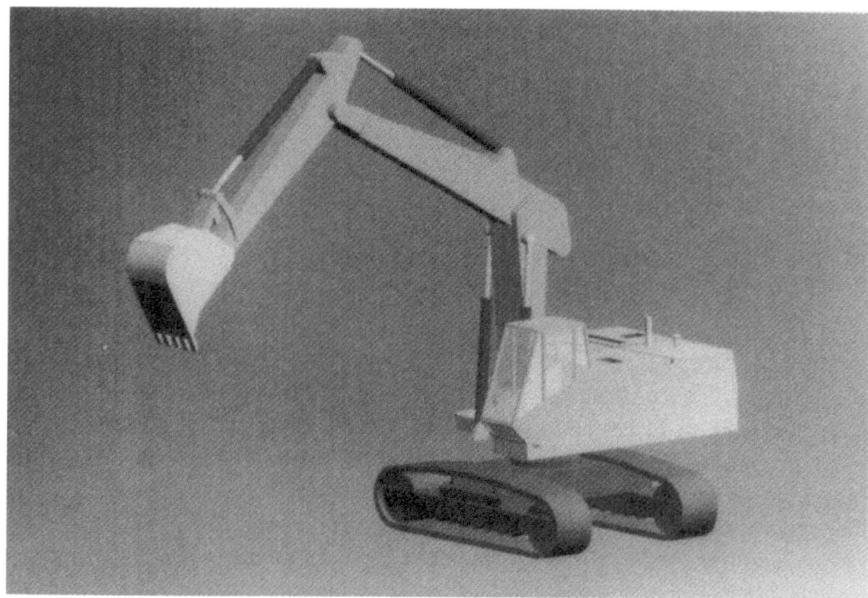

Figure 1.1. Computer model of heavy machinery.

Figure 1.2. Computer model of the steering and suspension system of a car.

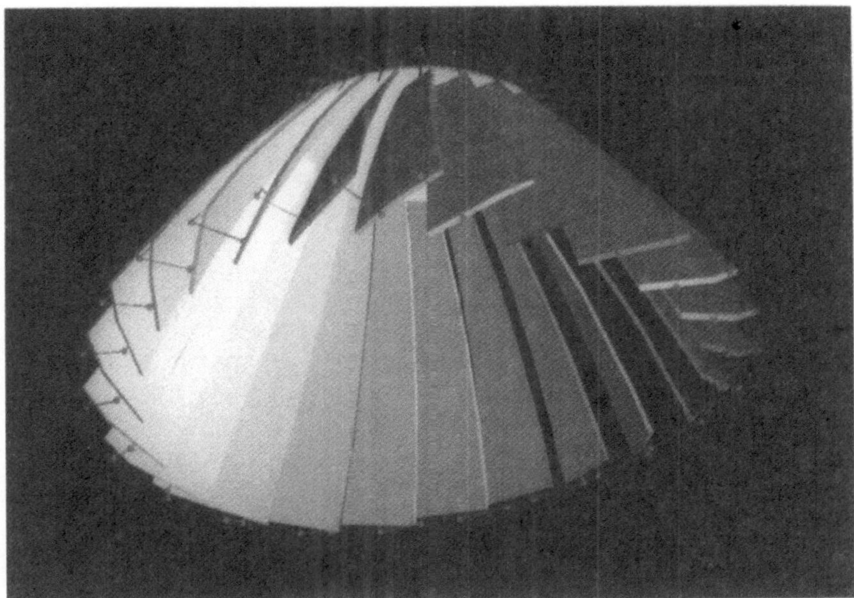

Figure 1.3. Computer model of a space deployable antenna.

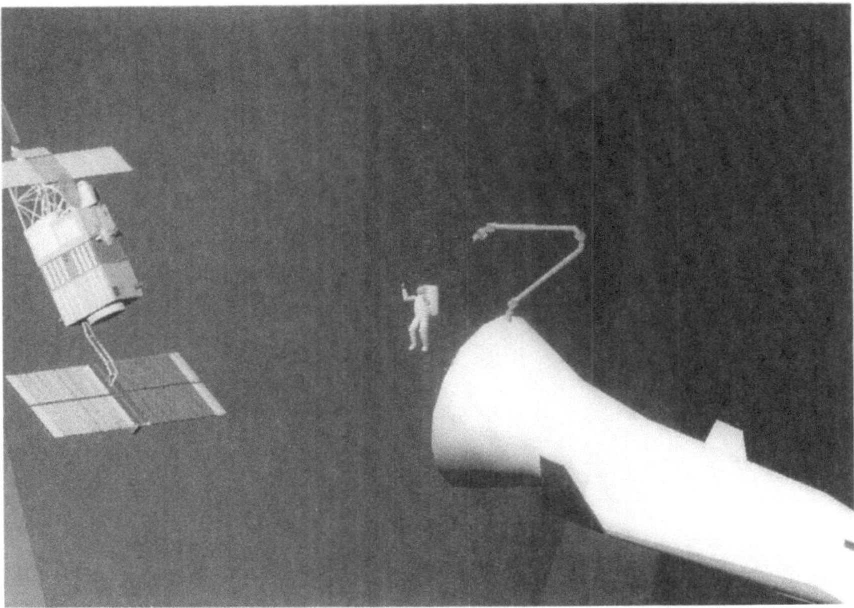

Figure 1.4. Computer model of a complex space scenario, including a satellite, the earth, a shuttle with a robot, and an astronaut.

Figure 1.5. Computer model of a bicycle and a cyclist.

Figure 1.6. Computer animation of two long jump finalists in the 1992 Barcelona Olympic Games.

1.1 Computer Methods for Multibody Systems

Computer systems, while increasing tremendously in power in recent years, are so affordable nowadays, that their use has become widely spread in many different fields and for an immense amount of applications. Today we consider the computer as a necessary tool, whose availability is taken for granted by engineers, scientists, businessmen, writers, and others. We can take the PC as an example of a system currently used by students in the classroom, laboratory and at home, whose power exceeds that of the mainframes used in the sixties and seventies, and which only the largest corporations could afford. Engineers working for consulting firms or large corporations in the analysis or design of new products, perform their work using personal workstations. These workstations have a capability for number crunching that vastly exceeds that of minicomputers, which only a few years ago used to be considered powerful enough to satisfy the needs of a whole engineering department. The field of mechanical engineering has not been an exception to this trend. There is an increasing demand for faster executions and better graphical interfaces that will facilitate and improve the tedious tasks of data entry and interpretation of the results. The help of the computer is sought in the decision-making process for optimal designs.

The two authors of this book started their professional careers in the late seventies working on the finite element method in Spain and the United States. At that time, the analysis of a medium size finite element model with several thousands of degrees of freedom, or the complete dynamic analysis of a multibody system, could last for over twelve hours in a mainframe computer. The analyst would spend a long time preparing the data entry consisting of large numbers of punched cards and interpreting the results shown on the endless pages of computer output. This process has changed quite a bit up to the present method. The analyst can prepare the input data in an interactive manner with the help of a preprocessor running in sophisticated graphic terminals. The execution time has been reduced to a fraction of an hour of CPU of modern workstations. For larger problems the analyst considers the access to super computers or parallel architectures remotely connected to his personal system. The tedious work of interpreting the pages and pages of computer output has been alleviated and even made pleasant through the use of graphic terminals which can show an animated picture of the results.

Although *finite element analysis* and *multibody simulation* are part of the MCAE family, they are substantially different not only in their respective aims but in their *modus operandi*, namely, in the way they work. Finite element analysis must be fast. It is essentially a *batch* process, in which the user does not usually interact with the computer analysis from the beginning to the end of that process. On the other hand, the kinematic and dynamic analyses of multibody systems are processes which are most appropriately performed using *interactive analysis*. The analyst is interested in visualizing a whole set of successive responses of the multibody system, with a simulation of its behavior and operation over all the workspace and over a certain period of time. In certain cases it may be necessary to obtain a *real-time* response, and introduce the analyst as an additional element in the simulation, called *man-in-the-loop*, who may act by introducing external forces or control over specific degrees of freedom. This obviously imposes constraints on the computer hardware and software, which exceed those imposed by the finite element analysis. Real-time analysis now requires the use of mostly top of the line workstations, and is not yet possible for the very large problems. Interactive and real-time analysis will help the engineer optimize productivity and the use of his own time, which is really the most expensive part of the simulation process. Obviously the class and size of problems that may be solved in real time will increase as the computer hardware and numerical algorithms improve in the ensuing years. In any case, readers will find that the methods described in this book will always help to speed up and improve the interactive analysis of multibody systems.

The advent of powerful workstations in the computer market is making this interactive analysis now possible for the engineering profession in general and for the multibody system analysis in particular. These workstations can currently reach 100 Mips and 20 Mflops of processing power, and draw hundreds of thousands of three-dimensional vectors and polygons per second. They run under standard operating systems and graphic interfaces such as UNIX, X-Windows, MOTIF, PHIGS, etc., and may be obtained at very affordable prices. Given the rate at which the computer hardware has been improving in the past, we can only expect better and faster hardware platforms in years to come. As a consequence, it is foreseeable that the use of general purpose computer programs for the interactive three-dimensional analysis of multibody systems will be considered by the engineering profession not only as a necessary tool but also as something to be taken for granted in the design process. We intend to describe in this book formulations and numerical methods aimed at this end.

Traditional methods of analysis, such as *graphical* and *analytical*, may be limited when they are applied to complicated problems. Graphic methods, although they provide a good understanding of the kinematics, lack accuracy and tend to be time-consuming. These are the reasons why they are not used for repetitive or three-dimensional analyses. Analytical or closed-form methods can be extremely efficient, although they are application-dependent, and may suffer from an excessive complexity in a multitude of practical problems.

An alternative to overcome these limitations is to resort to *numerical analysis* and the fast processing of alphanumeric data available in current digital computers. Several books have recently appeared (Nikravesh (1988), Roberson and Schwertassek (1988), Haug (1989), Shabana (1989), Huston (1990), and Amirouche (1992)) that emphasize the use of formulations and computational methods for multibody dynamic simulation. Various general purpose programs for multibody kinematics and dynamics (Schiehlen, (1990)) have been described simultaneously in the literature or made available in the market.

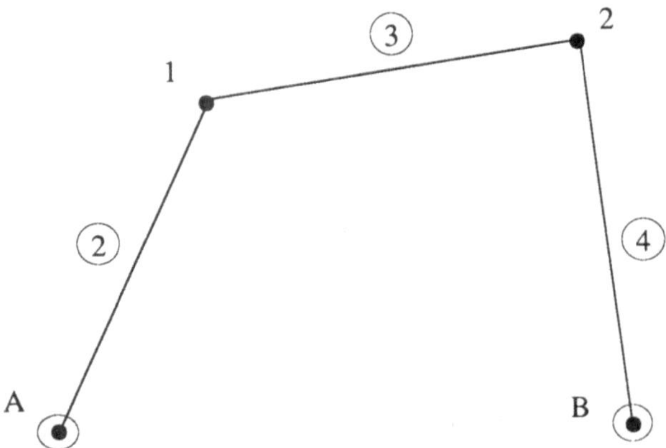

Figure 1.7. Four-bar articulated quadrilateral

1.2 Basic Concepts

1.2.1 Multibody Systems and Joints

We define a *multibody system* as an assembly of two or more rigid bodies (also called elements) imperfectly joined together, having the possibility of relative movement between them. This imperfect joining of the two rigid bodies that makes up a multibody system is called a *kinematic pair* or *joint*, or simply a *joint*. A joint permits certain degrees of freedom of relative motion and prevents or restricts others. A *class I* joint allows one degree of freedom, a *class II* allows two degrees of freedom and so forth. For example, a revolute joint (R) is a *class I* joint that only allows one relative rotation. In planar multibodies, the most used joints are *revolute* (R) and *prismatic* (P), which allow one relative rotation and translation, respectively. In three-dimensional multibodies, *cylindrical* (C), *spherical* (S), *universal* (U), and *helical* (H) joints are also used. Other joints such as *gears* (G) and the *track-wheel rolling contact* (W) are sometimes used. All these will be seen in detail in Chapters 2 and 3.

Usually the elements of a multibody system are linked by means of joints, as shown in the articulated quadrilateral of Figure 1.7. At times, the elements do not have direct contact with one another but rather are interrelated via force transmission elements, such as springs, shock absorbers or dampers.

Multibody systems are classified as *open-chain* or *closed-chain* systems. If a system is composed of bodies without closed branches (or *loops*), then it is called an open-chain system; otherwise, it is called a closed-chain multibody system. A double pendulum and a tree-type of system are good examples of an open-chain configuration. The four-bar mechanism of Figure 1.7 is an example of a closed-chain system.

1.2.2 Dependent and Independent Coordinates

In order to describe a multibody system, the first important point to consider is that of choosing a mathematical way or model that will describe its position and motion. In other words, select a set of parameters or coordinates that will allow one to unequivocally define the position, velocity, and acceleration of the multibody system at all times. There are several ways to go about solving this problem, and different authors have opted for one way or another depending on their preferences or the peculiarities of their own formulation.

Even though the same multibody system can be described with different types of coordinates, this does not mean that they are all equivalent in the sense that they will allow for formulations that are just as efficient or as easy to implement. In fact, it will be shown in Chapters 2 and 3 that there are differences in computational efficiency and simplicity of implementation when using different sets of coordinates. The different dynamic formulations may also benefit from the characteristics of a particular set of coordinates.

Consequently, the first problem encountered at the time of modeling the motion of a multibody system is that of finding an appropriate system of coordinates. A first choice is that of using a system of *independent coordinates*, whose number coincides with the number of *degrees of freedom* of motion of the multibody system and is thereby minimal. The second choice is to adopt an expanded system of *dependent coordinates* in a number larger than that of the degrees of freedom, which can describe the multibody system much more easily but which are not independent, but interrelated through certain equations known as *constraint equations*. The number of constraints is equal to the difference between the number of dependent coordinates and the number of degrees of freedom. Constraint equations are generally nonlinear, and play a main role in the kinematics and dynamics of multibody systems.

Studies on this subject tend to conclude that independent coordinates are not a suitable solution for a general purpose analysis, because they do not meet one of the most important requirements: that the coordinate system should unequivocally define the position of the multibody system. Independent coordinates directly determine the position of the input bodies or the value of the externally driven coordinates, but not the position of the entire system. Therefore additional non-trivial analysis (seen in Chapter 3) need be performed to this end. For some particular applications, independent coordinates can be very useful to describe with a minimum data set the actual velocities or accelerations and small variations in the position. In addition, they may lead to the highest computational efficiency.

For general cases, the alternative choice to the independent set of coordinates is a system of *dependent coordinates*, which uniquely determine the position of all the bodies. Three major types of coordinates have been proposed to solve this problem: *relative* coordinates, *reference point* or *Cartesian* coordinates, and *natural* or *fully Cartesian* coordinates. The latter are the ones most frequently used in this book. These types of coordinates are described in detail in Chapter

2, both for planar and three-dimensional multibody systems. Although this book deals with these three types of coordinates, it emphasizes the use of the later ones. By means of these coordinates, the position of a three-dimensional object is defined using the Cartesian coordinates of two or more points and the components of one or more unit vectors rigidly attached to the body. Chapters 2 and 3 describe these coordinates in detail, along with other sets of coordinates.

1.2.3 Symbolic vs. Numerical Formulations

Among the computer programs for kinematic and dynamic analysis of multibody systems, there are two groups with very different approaches and capabilities: *symbolic* programs and strictly *numerical* programs.

Symbolic programs do not process numbers from the outset, but variable names and analytical expressions. Their outcome is a list of statements in FORTRAN, C, Pascal, or any other scientific programming language containing the mathematical equations that model the kinematics and/or dynamics of the system in question. If the problem is of large complexity, the formulae can occupy dozens or even hundreds of pages of listings. The advantages of symbolic methods are mainly that they eliminate those operations with variables having zero values, and also allow one to explicitly see the influence of each variable in the equations that control the behavior of the assembly. However, in order for a symbolic code to achieve maximum efficiency it also must be able to compact and simplify the equations by extracting common factors and by compacting trigonometric expressions. Such operations are alleviated through the use of symbolic tools such as MACSYMA, MAPLE, and MATHEMATICA.

Symbolic formulations can be advantageous when the generation of the equations is performed only once and is valid for the entire range of motions that the multibody may undergo. However, one of their major problems stems from the fact that a multibody system may undergo a qualitative change in its kinematic configuration during its motion, thus, demanding a substantial change in the equations of motion. Such situations occur with changes in generalized coordinates, the appearance and disappearance of kinematic constraints, impacts and shocks, backlash, Coulomb friction, etc. Special provisions need to be made in these cases to avoid the complete reformulating of the symbolic equations of motion.

Numerical programs, on the other hand, provide a real general purpose solution to the kinematic and dynamic analysis of all types of multibody systems. These programs formulate the equations of motion numerically without generating analytical equations suited to the specific problem. In many cases, numerical methods are less efficient than the symbolic counterparts. However, their generality and the fact that they are easy to use is a definite advantage. In addition, recent advances in numerical methods have allowed a substantial improvement in the efficiency of numerical approaches and have made them more competitive for many types of applications. These advances include the use of sparse matrix techniques that eliminate operations involving zero terms, and the possibility of using improved dynamic formulations (See Chapter 8).

1.3 Types of Problems

We briefly describe in this section the most important types of kinematic and dynamic problems that occur in real everyday situations.

1.3.1 Kinematic Problems

Kinematic problems are those in which the position or motion of the multibody system are studied, irrespective of the forces and reactions that generate it. Kinematic problems are of a purely geometrical nature and can be solved, irrespective not only of the forces but also of the inertial characteristics of elements such as mass, moments of inertial, and the position of the center of gravity.

We define *input elements* of a multibody system as those whose position or motion is known or specified. The position and motion of the other elements of the system are found in accordance with the position and motion of the input elements. There are as many input elements as there are degrees of freedom for the multibody system. As an example, let us consider the four-bar mechanism of Figure 1.7 in which the crank A-1 (body or element 2) is the input element. Sometimes the kinematic problem is based not on an input element, but on an input coordinate or degree of freedom, such as an angle or a distance.

Below is a brief description of the different kinematic problems that occur in practice and which will be discussed in detail in Chapter 3.

Initial Position Problem. The initial position or *assembly* problem consists of finding the position of all the elements of the multibody system once that of the input elements is known. In general, the position problem is difficult to solve, since it leads to a system of nonlinear algebraic equations which has in general several solutions. The more complicated the system is, the larger the number of possible solutions.

Finite Displacement Problem. This problem is a variation of the initial position problem, both from a conceptual point of view as well as from the mathematical methods that can be used to solve it. Given a fixed position on the multibody system and a known finite displacement (not infinitesimal) for the input bodies (or elements), the problem of finite displacements consists of finding the final position of the system's remaining bodies.

In practice, the finite displacement problem ends up being easier to solve than the initial position problem, mainly because one starts from a known position of the system, which can be used as a starting point for the iterative process needed for the solution of the resulting nonlinear equations. The problem of having multiple solutions is not as critical in this case, because usually one is only interested in the solution nearest to the previous position.

Velocity and Acceleration Analysis. Given the position of the multibody system and the velocity of the input elements, velocity analysis consists of determining the velocities of all the other elements and all the points of interest. This problem is much easier to solve than the position problems discussed earlier, mainly because it is linear and has a unique solution. This means in mathematical terms that it is modeled by a system of linear equations. Consequently, the principle of superposition holds that the velocity of an element is the sum of the velocities produced by each one of the input elements. If all the input velocities are zero, then the velocities of all the elements will also be null.

Given the position and velocity of all the elements in the system, and the acceleration of the input elements, the acceleration analysis consists of finding the acceleration of all the remaining elements and points of interest. Just as the velocity problem is linear, so also is the acceleration problem linear. Moreover, the matrix of the system of linear equations that models this problem is the same as the one in the velocity problem.

Kinematic Simulation. The kinematic simulation provides a view of the entire range of a multibody's motion. The solution of the kinematic simulation encompasses all the previous problems with emphasis on the finite displacement problem. It permits one to detect collisions, study the trajectories of points, sequences of the positions of an element of the multibody system, and the rotation angles of rocker levers and connecting rods.

1.3.2 Dynamic Problems

In general, dynamic problems are much more complicated to solve than kinematics ones. The kinematic problems need to be solved before the dynamic problems. Henceforth, it will be assumed that the velocity and acceleration problems can be solved without any difficulty. The outstanding characteristic about dynamic problems is that they involve the forces that act on the multibody system and its inertial characteristics as follows: mass, inertial tensor, and the position of its center of gravity (seen in detail in Chapter 4). We will describe briefly the most important dynamic problems encountered in practice.

Static Equilibrium Position Problem. The static equilibrium position problem (dealt with extensively in Chapter 6) consists of determining the position of the system in which all the gravitational and external forces, elastic forces in the springs, and external reactions are balanced. This problem is not really a dynamic problem, but a static one, that depends on the weight and the position of the center of gravity of the multibody system and not on its inertial properties.

The problem of determining the static equilibrium position takes place very frequently in vehicles with spring suspension systems. It is not always easy (for example, when the loads are not centered) to determine the static equilibrium position at a glance or by means of simple calculations. The general solution to this problem also leads to a system of nonlinear equations which need to be solved iteratively. Even though there might also be several solutions for this case, there are reliable initial estimates normally available that lead to the right solution.

Linearized Dynamics. A problem closely related to the previous one is that of determining the natural vibration modes and frequencies of the small oscillations that take place about the static (or dynamic) equilibrium position. This problem is solved by first linearizing the equations of motion at a particular position, and then performing a step-by-step time history or an eigenvalue analysis. A knowledge of the natural vibration modes and frequencies gives an idea of the system's dynamic stiffness, and it also allows one to design different control systems. This problem is discussed in Chapter 9.

Inverse Dynamic Problem. The inverse dynamic problem aims at determining the motor or driving forces that produce a specific motion, as well as the reactions that appear at each one of the multibody system's joints. It is necessary to know the velocities and accelerations to be able to estimate the inertia forces which, together with the weight, the forces in the springs and dampers and all the other known external forces, will provide the basis to calculate the required actuating forces.

The solution to the inverse dynamics (See Chapter 6) has different applications. In the first place, it determines the forces to which the multibody system is subjected, for both dynamic and kinematic simulation problems. Extremely important is the fact that the inverse dynamics yields the driving forces necessary to control a system so that it follows a desired trajectory.

Forward Dynamic Problem (Dynamic Simulation). The forward dynamic problem yields the motion of a multibody system over a given time interval, as a consequence of the applied forces and given initial conditions. The importance of the direct dynamic problem lies in the fact that it allows one to simulate and predict the system's actual behavior; the motion is always the result of the forces that produce it.

The forward dynamics implies the solution of a system of nonlinear ordinary differential equations (initial value problem). These differential equations are numerically integrated starting from the initial conditions. An important characteristic of this mathematical problem is that it is computationally intensive. Because of this, it is very important to choose the most efficient method for dealing with and solving this problem. The results of the dynamic simulation problem can be displayed numerically, or they can be depicted graphically by means of a plotter of a graphics terminal in the same way as with the kinematic simulation results. Chapter 5 deals with the most common formulations used for dynamic analysis. Chapter 8 presents the most recent ones, including those most suited for real time analysis.

Forward and Inverse Dynamics of Elastic Multibodies. So far we have assumed that all the bodies in a multibody system satisfy the *rigid body condition*. A body is assumed to be rigid if any pair of its material points does not present relative displacements. In practice, bodies suffer some degree of deformation. This tends, however, to be so small that it does affect the system's behavior, and therefore, it can be neglected without committing an appreciable error.

There are some important cases in which deformation plays an important role in the dynamic analysis. It happens, for instance, in lightweight spatial structures and manipulators, or in high-speed machinery. The complexity and size of the equations of motion considering deformation grow considerably, since all the variables defining the deformation must also be considered. Chapter 11 deals with the forward dynamics of elastic multibodies and Chapter 12 with the inverse dynamics. This case is of particular interest since the driving forces are now non-causal, which means that there is a time delay between actuation and response and the solution goes to negative time and future time.

Percussions and Impacts. Mechanically, a percussion is a force with a large value that occurs in a very short period of time. It is convenient to distinguish between *percussion* and *impact* problems. In the case of percussions, it is assumed that a very large force of known value acts during an infinitesimal amount of time. Bear in mind that the percussion is the value of a mechanical impulse (the integral of the force in relation to time). A typical characteristic of a percussion is that it produces discontinuities (finite jumps) in the distribution of velocities, which are determined from the value of the applied percussion. This problem is of limited practical importance because in practice the percussion value is seldom ever known. The impact problem is more important.

The impact involves the collision of bodies in which at least one of them experiences a sudden change in velocities. The point of contact undergoes a percussion which is generally unknown. In order to be able to calculate the effect of the impact on the system's velocity distribution, it is necessary to introduce an additional equation of experimental nature which measures the nature of the surfaces in contact and the type of impact.

The study of the effects of percussions and impacts in the distribution of velocities of a multibody system can be carried out separately or within a dynamic simulation program. Chapter 10 deals with this problem as well as the design issues outlined in the next section.

1.3.3 Other Problems: Synthesis or Design

We have outlined in the previous sections the most important analysis problems that can occur in the kinematics and dynamics of multibody systems. In all these problems, it is assumed that the geometry and physical properties of the system are known (either because it is an existing multibody system or it has been previously designed). When wishing to design a new system to comply with certain specifications, with only analysis tools available, one must proceed in an iterative manner by means of rough calculations. A preliminary design is carried out and the system is analyzed. Once the results of the analysis have been obtained, the design is then modified if they are not entirely satisfactory, and another analysis is performed. The same procedure is followed until the desired effect is attained. This process may be slow and rather dependent upon the experience of the designer.

Synthesis or design methods help to overcome this difficulty, or at least lessen it. These methods directly lead, without the intervention of an analyst, to a design which complies with the given specifications, or which is the optimal one from a certain design point of view. The design of a multibody system can also be carried out from a more general perspective by taking dynamic factors into account. Two different problems can be considered: pure kinematic design also called *synthesis*, and the more general dynamic sensitivity analysis.

Kinematic Synthesis of Multibody Systems. Kinematic synthesis entails the finding of the best possible dimensions for a given type of multibody system. This is mainly a geometric problem, about which much has been written in the last half of the past century and in the first half of the present one. During this time, many methods were developed, almost all of them graphic and containing a notable amount of ingenuousness and originality. The majority of the methods were focused on the planar four-bar mechanism. However, graphic methods of kinematic synthesis are limited, too specific, and at times difficult to use. In recent years, more general programs based on numerical methods have been developed, and they are applicable to many different types of planar and three-dimensional multibody systems.

Sensitivity Analysis and Optimal Design. The optimal design of a multibody system is started by defining an *objective function* which will optimize the system performance. The solution to the problem will be the configuration that minimizes the objective function. This function is minimized in relation to certain variables which depend on the design of the multibody system and are referred to as *design variables*. It may or may not have *design constraint equations*, that is, equalities or inequalities that should comply with certain specific functions of the design variables. The constraint equations mathematically introduce certain physical design limitations into the problem. An example of a design limitation is that there cannot be any elements with negative mass or length.

The objective functions are defined depending on the application of the multibody system. Since multibody dynamics is a process that takes place over a period of time, it often turns out that the objective function is defined as the time integral of a specific function or as a series of conditions that the multibody must satisfy within certain intervals of time or at specific moments.

There are several optimization methods or ways to minimize the objective function, which are applicable to this problem. Almost all of them are based on the knowledge of the *derivatives* of the objective function with respect to the design variables. The determination of these derivatives is known as *sensitivity* analysis and is the first phase in the optimization process which can also be considered separately. Sensitivity analysis determines the tendencies of the objective function with respect to design variations, and is very useful in a non-automatic interactive design process. Sensitivity analysis is considered in Chapter 10.

1.4 Summary

In this introductory chapter we have tried to outline the different problems that arise in the analysis and design of multibody systems. We have also succinctly commented on the different ways these can be tackled in a very general form, and in this respect it is important to mention that at the present time there is not a formulation or method that can solve all the problems in the best possible manner. Some methods are preferable over others depending on the type of configuration and motion. In the chapters that follow we will deal with all these different problems and try to offer methods for their solution.

References

Amirouche, F.M.L, *Computational Methods for Multibody Dynamics*, Prentice-Hall, (1992).

Haug, E.J., *Computer-Aided Kinematics and Dynamics of Mechanical Systems, Volume I: Basic Methods*, Allyn and Bacon, (1989).

Huston, R.L., *Multibody Dynamics*, Butterworth-Heinemann, (1990).

Nikravesh, P.E., *Computer-Aided Analysis of Mechanical Systems*, Prentice-Hall, (1988).

Roberson, R.E. and Schwertassek, R., *Dynamics of Multibody Systems*, Springer-Verlag, (1988).

Schiehlen, W.O., *Multibody System Handbook*, Springer-Verlag, (1990).

Shabana, A.A., *Dynamics of Multibody Systems*, Wiley, (1989).

2
Dependent Coordinates and Related Constraint Equations

In either the kinematic or dynamic analysis of multibody systems described in Chapter 1, the first issue to consider is that of modeling the system, which involves the selection of a set of parameters or coordinates that will allow one to define unequivocally at all times the position, velocity and acceleration of the multibody system. There are several ways to solve this problem, and different authors have opted for one way or another depending on their preferences or the peculiarities of their own formulation.

Even though the same multibody system can be described with different types of dependent coordinates, their definition is not a trivial problem. They are all not equivalent in the sense that they will lead to formulations that are just as efficient or as easy to implement. In fact, there are in practical applications large differences both in efficiency and simplicity among the different sets of coordinates. We will provide some examples that corroborate this fact.

The most important types of coordinates currently used to define the motion of planar and three-dimensional multibody systems are *relative* coordinates, *reference point* coordinates (also called *Cartesian* coordinates), and *natural* coordinates (also called *fully Cartesian* coordinates). These will be described in detail in this chapter. A qualitative comparison among them will also be provided. No general quantitative comparison is yet available, although some preliminary results for 2-D systems have been already published by Unda et al. (1987). We will also deal extensively with the constraint equations that the dependent coordinates generate. A combination of the ideas and concepts arising from the different types of coordinates (relative, reference point, and natural) explained in this chapter, are the basis for very efficient dynamic formulations that will be seen in Chapter 8.

2.1 Planar Multibody Systems

Different sets of dependent coordinates for *planar multibody systems* and the related constraint equations are described below. These systems are a simpler alternative to the three-dimensional ones and make it easier to understand the impor-

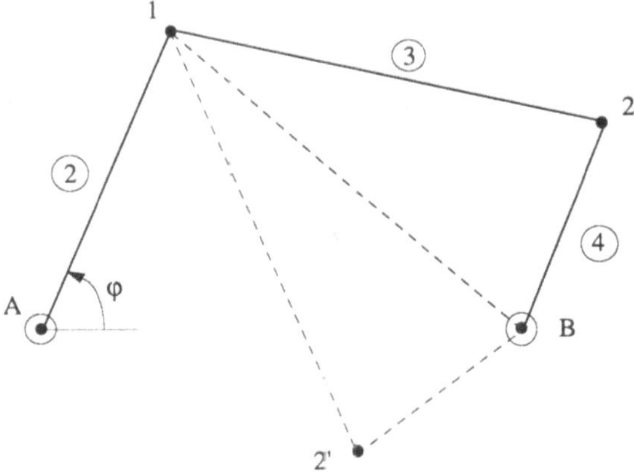

Figure 2.1. Solutions of the position problem in a four-bar mechanism.

tant concepts and the differences between the various types of dependent coordinates.

The first dilemma encountered when choosing a system of coordinates which may describe the motion by position, velocity and acceleration is the problem of either adopting a set of *independent coordinates*, whose number coincides with the number of degrees of freedom and is thereby minimal, or adopting an expanded system of *dependent coordinates*. The latter can describe the system much more easily, but they are not independent but instead related through certain *constraint equations*.

Studies on this subject tend to conclude that generally a system of independent coordinates is not an acceptable solution, because it does not meet one of the most important conditions: the system of coordinates should be capable of *unequivocally* describing the position of the multibody system. Independent coordinates directly determine the position of the input elements or the value of the driven degrees of freedom but not the position of the other elements. In order to determine the position of the entire system, the position problem must first be solved. As was already explained in Chapter 1, there are multiple solutions to this problem. For example, the four-bar mechanism of Figure 2.1 has one degree of freedom and one independent coordinate, the angle φ. It may be seen that there are two possible solutions for the position of the elements 3 and 4. The same thing generally occurs with other multibody systems.

Once the independent coordinates have been ruled out for the description of the position, a system of dependent coordinates larger than the number of degrees of freedom must be adopted to determine the position of *each and every one* of the bodies. Three major types of coordinates have been described in the literature: *relative* coordinates, *reference point* coordinates, and *natural* coordinates. These types of coordinates will be described in detail in the following sections, both for planar and three-dimensional multibody systems.

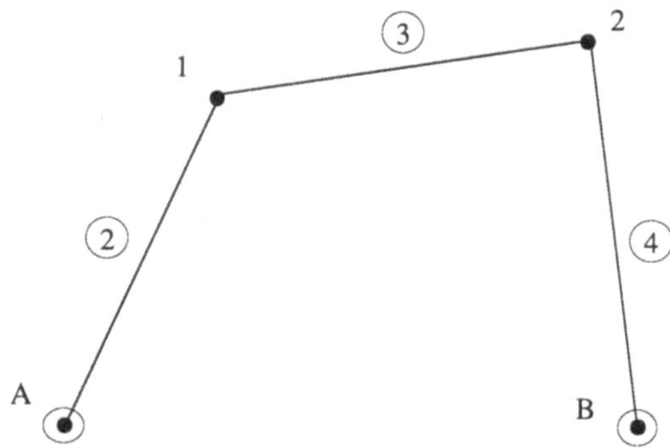

Figure 2.2. Representation of a four-bar mechanism using natural coordinates.

An important aspect of the dependent coordinates is precisely their *dependent* nature, or in other words, the fact that they are related by algebraic *constraint equations* in a number equal to the difference between the number of dependent coordinates and the number of degrees of freedom. Constraint equations are generally nonlinear and play a main role in the kinematic and dynamic analysis of multibody systems. Therefore, the description of the dependent coordinates included below and their comparative study will be completed with the study of the specific constraint equations generated by each one of the types of dependent coordinates. The concept of constraint equation is not complicated and neither is its mathematical formulation. A very simple example will be presented next.

Example   2.1

Figure 2.2 illustrates a four-bar mechanism modeled with natural coordinates, i.e. with the Cartesian coordinates of points 1 and 2. There are four dependent coordinates (x_1, y_1, x_2, y_2) and the mechanism has one degree of freedom. Hence, there should be three constraint equations relating the four dependent coordinates.

The constraint equations shall guarantee that points 1 and 2 move in accordance with the limitations imposed on them by the three moving bars of the four-bar mechanism. It is precisely from there that the three constraint equations arise: from the fact of imposing the *rigid body condition* (a constant distance between points) on the three elements of the mechanism. These conditions can be formulated mathematically as follows:

$$(x_1 - x_A)^2 + (y_1 - y_A)^2 - L_2^2 = 0$$

$$(x_2 - x_1)^2 + (y_2 - y_1)^2 - L_3^2 = 0$$

$$(x_2 - x_B)^2 + (y_2 - y_B)^2 - L_4^2 = 0$$

These are the three constraint equations that correspond to the mechanism of Figure 2.2. It may be seen that they are nonlinear equations (quadratic in this case). A similar system of equations can be established for any other type of coordinates and for any other multibody system.

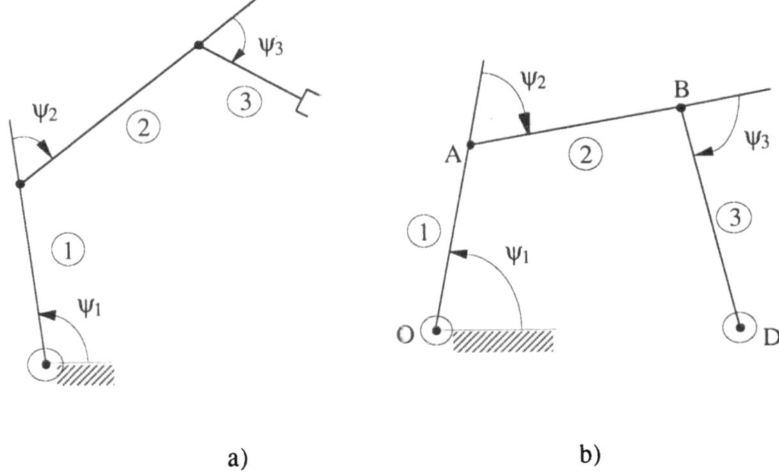

Figure 2.3. Representation of a four-bar mechanism using relative coordinates
(a) Open chain, (b) Closed chain.

In the following sections, three types of dependent coordinates will be discussed: relative, reference point, and natural, both for planar and three-dimensional systems. The generation of constraint equations will be studied in detail. For the case of planar multibody systems, the explanations will be illustrated with simple completely developed examples.

2.1.1 Relative Coordinates

Relative coordinates were the first ones used in the general purpose planar and three-dimensional analysis programs of Paul and Krajcinovic (1970), Sheth and Uicker (1972), and Smith et al. (1973).

Relative coordinates define the position of each element in relation to the previous element in the kinematic chain by using the parameters or coordinates corresponding to the relative degrees of freedom allowed by the joint linking these elements. In the case of planar multibody systems, if two elements are linked by means of a joint R (revolute), their relative position is defined by means of an angle. If they are linked by a joint P (prismatic), their relative position is defined by means of a distance. Figure 2.3 shows two examples of mechanisms with four bars that are described with relative coordinates.

Relative coordinates make up a system with a *minimum number* of dependent coordinates. In fact, in the particular case of open kinematic chain systems, as in Figure 2.3a, the number of relative coordinates coincides with the number of degrees of freedom; therefore there will not be constraint equations. Likewise, Figure 2.4 shows a more complicated mechanism modeled by means of relative coordinates.

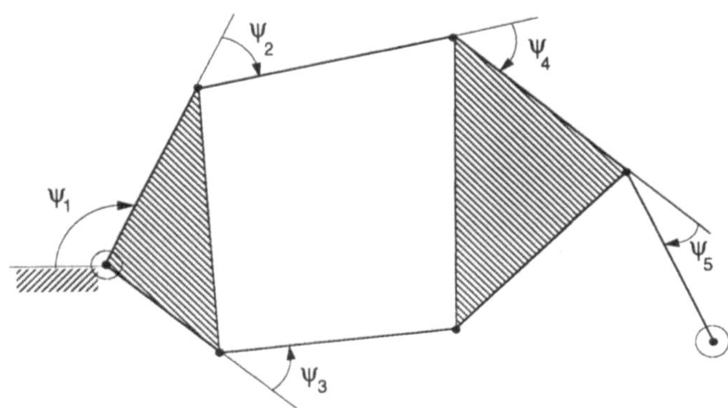

Figure 2.4. Multi-loop mechanism modeled with relative coordinates.

The advantages of relative coordinates can be summarized as follows:

1. Reduced number of coordinates, hence good numerical efficiency.
2. Relative coordinates are specially suited for open-chain configurations.
3. The consideration of the corresponding degree of freedom at each joint. This has an important advantage when the joint has a motor or actuator attached to it, since it allows to control the motion of the corresponding degree of freedom directly.

The following are considered to be the most important difficulties of the relative coordinates:

1. The mathematical formulation can be more involved, because the absolute position of an element depends on the positions of the previous elements in the kinematic chain.
2. They lead to equations of motion with matrices that, although small, are full and sometimes expensive to evaluate.
3. They require some preprocessing work (to determine the independent constraint equations) and postprocessing (to determine the absolute motion of each point and element).

In the case of planar multibody systems formulated with relative coordinates, the constraint equations arise from the condition of the *vector closure of the kinematic loops*.

Example 2.2

If a mechanism has an open kinematic chain type (See Figure 2.3a), then there will not be any constraint equation. In the case of the four-bar mechanism shown in Figure 2.3b, there are three relative coordinates and one degree of freedom; therefore there should be two constraint equations.

Vectorially, the condition of closed loop for the four-bar mechanism of Figure 2.3b can be expressed as follows:

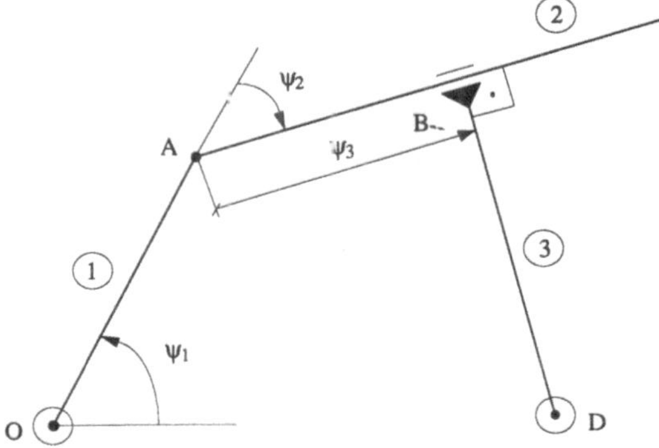

Figure 2.5. Relative coordinates in a four-bar mechanism with a prismatic joint.

$$\overrightarrow{OA} + \overrightarrow{AB} + \overrightarrow{BD} - \overrightarrow{OD} = \vec{0}$$

This vector equation is equivalent to two algebraic equations which correspond to the components x and y of the previous equation

$$L_1 \cos \Psi_1 + L_2 \cos (\Psi_1 + \Psi_2) + L_3 \cos (\Psi_1 + \Psi_2 + \Psi_3) - OD = 0$$

$$L_1 \sin \Psi_1 + L_2 \sin (\Psi_1 + \Psi_2) + L_3 \sin (\Psi_1 + \Psi_2 + \Psi_3) = 0$$

It may be seen that these equations are nonlinear and contain transcendental functions. This is a common characteristic of all the multibody systems formulated with relative coordinates.

Example 2.3

Figure 2.5 shows a four-bar mechanism with a prismatic joint. The number of constraint equations is the same as before (two), and they also come from the vector closure of the only loop that the mechanism has

$$\overrightarrow{OA} + \overrightarrow{AB} + \overrightarrow{BD} - \overrightarrow{OD} = \vec{0}$$

This is equivalent to the following algebraic expressions

$$L_1 \cos \Psi_1 + \Psi_3 \cos (\Psi_1 + \Psi_2) + L_3 \cos (\Psi_1 + \Psi_2 - \pi/2) - OD = 0$$

$$L_1 \sin \Psi_1 + \Psi_3 \sin (\Psi_1 + \Psi_2) + L_3 \sin (\Psi_1 + \Psi_2 - \pi/2) = 0$$

It may be seen in these equations that since Ψ_3 is not an angular coordinate, it is not affected by the sine and cosine functions.

Example 2.4

As a last example, let us consider the mechanism of Figure 2.6, which has six elements and one degree of freedom. This mechanism can be modeled with five relative coordinates. Then four constraint equations must be found. By examining the mechanism, it may be seen that there are three closed loops which satisfy the following vector equations:

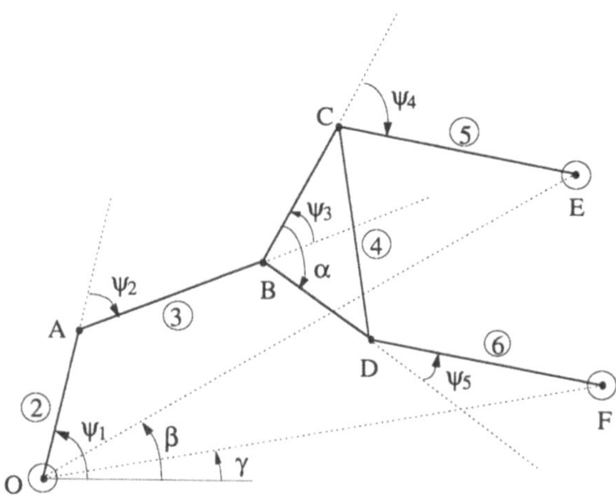

Figure 2.6. Geometrical representation to generate the constraint equations with relative coordinates.

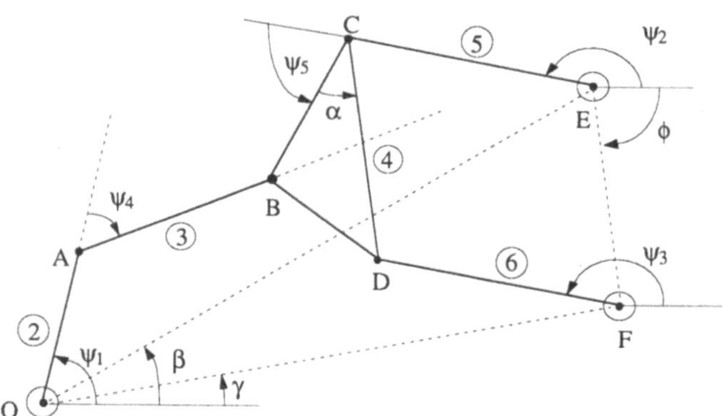

Figure 2.7. Alternative representation to generate the constraint equations with relative coordinates.

$$\overrightarrow{OA} + \overrightarrow{AB} + \overrightarrow{BC} + \overrightarrow{CE} - \overrightarrow{OE} = \vec{0}$$

$$\overrightarrow{OA} + \overrightarrow{AB} + \overrightarrow{BD} + \overrightarrow{DF} - \overrightarrow{OF} = \vec{0}$$

$$\overrightarrow{FD} + \overrightarrow{DC} + \overrightarrow{CE} - \overrightarrow{FE} = \vec{0}$$

These three vector equations give rise to six algebraic equations; however, only four of them are independent. Four equations can be chosen corresponding to any two of the three loops. For example, by using the first two loops,

$$L_2 \cos \Psi_1 + L_3 \cos (\Psi_1 + \Psi_2) + \overline{BC} \cos (\Psi_1 + \Psi_2 + \Psi_3)$$
$$+ L_5 \cos (\Psi_1 + \Psi_2 + \Psi_3 + \Psi_4) - \overline{OE} \cos \beta = 0$$

$$L_2 \sin \Psi_1 + L_3 \sin (\Psi_1 + \Psi_2) + \overline{BC} \sin (\Psi_1 + \Psi_2 + \Psi_3)$$
$$+ L_5 \sin (\Psi_1 + \Psi_2 + \Psi_3 + \Psi_4) - \overline{OE} \sin \beta = 0$$

$$L_2 \cos \Psi_1 + L_3 \cos (\Psi_1 + \Psi_2) + \overline{BD} \cos (\Psi_1 + \Psi_2 + \Psi_3 + \alpha)$$
$$+ L_6 \cos (\Psi_1 + \Psi_2 + \Psi_3 + \alpha + \Psi_5) - \overline{OF} \cos \gamma = 0$$

$$L_2 \sin \Psi_1 + L_3 \sin (\Psi_1 + \Psi_2) + \overline{BD} \sin (\Psi_1 + \Psi_2 + \Psi_3 + \alpha)$$
$$+ L_6 \sin (\Psi_1 + \Psi_2 + \Psi_3 + \alpha + \Psi_5) - \overline{OF} \sin \gamma = 0$$

These equations make up a nonlinear system of four equations with five unknown variables also involving transcendental functions.

Relative coordinates may be chosen in many different ways; and the one that was used in this example—although perfectly valid—may not necessarily be the best. Instead of beginning to establish relative coordinates from one of the fixed points to other fixed points, passing through all the joints on the mechanism, one can define relative coordinates from the three fixed points. Simultaneously one can advance and pass through all the joints, and in this way make the constraint equations simpler. Figure 2.7 shows the mechanism of Figure 2.6 modeled according to this new criteria.

Now the loop closure vector equations are:

$$\vec{OA} + \vec{AB} - \vec{OE} - \vec{EC} - \vec{CB} = 0$$

$$\vec{EC} + \vec{CD} - \vec{EF} - \vec{FD} = 0$$

thus, resulting in the following algebraic equations:

$$L_2 \cos \Psi_1 + L_3 \cos (\Psi_1 + \Psi_4) - \overline{OE} \cos \beta - L_5 \cos \Psi_2 - \overline{CB} \cos (\Psi_2 + \Psi_5) = 0$$

$$L_2 \sin \Psi_1 + L_3 \sin (\Psi_1 + \Psi_4) - \overline{OE} \sin \beta - L_5 \sin \Psi_2 - \overline{CB} \sin (\Psi_2 + \Psi_5) = 0$$

$$L_5 \cos \Psi_2 + \overline{CD} \cos (\Psi_2 + \Psi_5 + \alpha) - \overline{EF} \cos \phi - L_6 \cos \Psi_3 = 0$$

$$L_5 \sin \Psi_2 + \overline{CD} \cos (\Psi_2 + \Psi_5 + \alpha) - \overline{EF} \sin \phi - L_6 \sin \Psi_3 = 0$$

These expressions are more manageable and easier to evaluate than their counterparts as shown in Figure 2.6.

From the previous example it may be concluded that relative coordinates are particularly *suitable for open-Chains* or with few closed loops. When relative coordinates are used in multibody systems with many closed loops, it is very important to correctly choose the independent loops with which the constraint equations will be formulated and at what point the loop is going to be intersected or broken to establish the loop closure vector equation. In the previous example (Figure 2.7) the loops were intersected at B and D. This task is referred to as system *preprocessing*. In some computer implementations this can be automatically carried out using *graph theory*, but in others it is assumed that the preprocessing is carried out by the analyst.

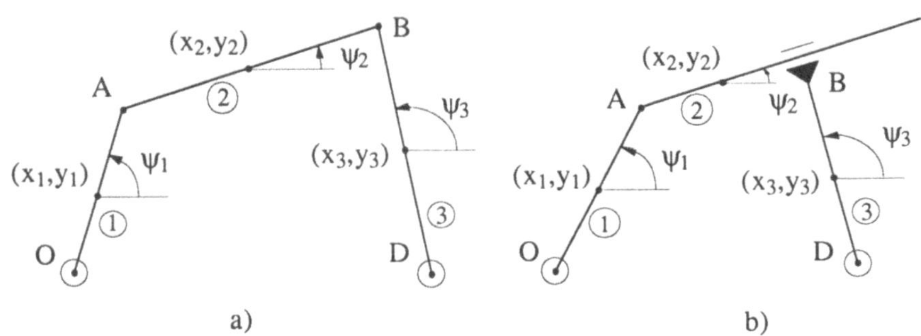

Figure 2.8. Representation of two four-bar mechanisms using reference point coordinates: a) Revolute joint, b) Prismatic joint.

2.1.2 Reference Point Coordinates

The reference point coordinates try to remedy the disadvantages of the relative coordinates by directly defining, using three coordinates or parameters, the absolute position of each one of the elements of the system. This is done by determining the position of a point of the element (the *reference point*, which often is the center of gravity) with two Cartesian coordinates, and by determining with an angle the orientation of the body in relation to a system of inertial axes. Figure 2.8 shows two four-bar mechanisms represented with reference point coordinates. Note that although the two mechanisms are different, they both are modeled with the same coordinates.

The reference point coordinates require a much larger number of variables than the relative coordinates (nine as compared to three in the four-bar mechanism, 15 as compared to five in the mechanism of Figure 2.6) and do not take into account at all if it is an open chain configuration or not. This means that for some particular cases, and from the numerical efficiency point of view, reference point coordinates may not be the most suitable ones.

The advantages of reference point coordinates can be listed as follows:

1. The position of each element is directly determined; hence the formulation is easier with less preprocessing and postprocessing requirements.
2. The matrices appearing in the equations of motion are sparse, meaning that they have very few non-zero elements. If one takes advantage of this condition and uses special techniques for this type of matrices, then one may make the formulation numerically efficient.

As mentioned earlier, the apparent disadvantages are their large number and the difficulty to be adapted for particular topologies such as open kinematic chains.

Using reference point coordinates, one can develop the constraint equations by considering the constraints that the joints introduce in the relative motion of contiguous elements. With these coordinates the motion of each element is defined regardless of the motion of the rest of them. However, the motion of the adjacent elements, linked by a kinematic joint, cannot be arbitrary. Rather it must generate a relative motion between these elements according to the nature of the joint. For instance, a class I kinematic joint (allowing one degree of freedom of relative motion) will constrain two degrees of freedom for a planar system. This means that it must also generate two constraint equations.

Example   2.5

Let us consider the four-bar mechanism of Figure 2.8a. This mechanism has nine dependent coordinates and one degree of freedom, which means that there should be eight constraint equations. These eight equations will originate from the four kinematic joints (points O, A, B and D), with two equations per joint.

The eight algebraic constraint equations are as follows (for the sake of simplicity, it will be assumed that the reference points are located at the middle points on the bars):

$$(x_1 - z_0) - L_1/2 \, cos \, \Psi_1 = 0$$

$$(y_1 - y_0) - L_1/2 \, sin \, \Psi_1 = 0$$

$$(x_2 - x_1) - L_1/2 \, cos \, \Psi_1 - L_2/2 \, cos \, \Psi_2 = 0$$

$$(y_2 - y_1) - L_1/2 \, sin \, \Psi_1 - L_2/2 \, sin \, \Psi_2 = 0$$

$$(x_3 - x_2) - L_2/2 \, cos \, \Psi_2 + L_3/2 \, cos \, \Psi_3 = 0$$

$$(y_3 - y_2) - L_2/2 \, sin \, \Psi_2 + L_3/2 \, sin \, \Psi_3 = 0$$

$$(x_3 - x_D) - L_3/2 \, cos \, \Psi_3 = 0$$

$$(y_3 - y_D) - L_3/2 \, sin \, \Psi_3 = 0$$

It can be seen that these equations are more sparse (a lower number of variables intervenes in each one of them) than the ones corresponding to relative coordinates (see Example 2.2). Likewise, it is evident that the constraint equations are nonlinear and cause transcendental functions to come into play.

Figure 2.8b shows a mechanism with four bars, three revolute joints, and one prismatic joint. The reference point coordinates are identical to the ones in the articulated quadrilateral of Figure 2.8a, as are the constraint equations corresponding to joints O, A and D. However, the constraint equations corresponding to joint B change as follows: the first equation directly indicates the constant relationship existing between angles Ψ_2 and Ψ_3.

$$\Psi_2 - \Psi_3 - \pi/2 = 0$$

The second equation is more complicated. In order to understand it better, one should carry out the graphic construction of Figure 2.9. Bear in mind that the G_3-B segment is equal to the sum of the projections coming from segments M-G_2 and M-G_3 which result in the following equation:

$$(y_2 - y_3) \, cos \, \Psi_2 + (x_3 - x_2) \, sin \, \Psi_2 - L_3/2 = 0$$

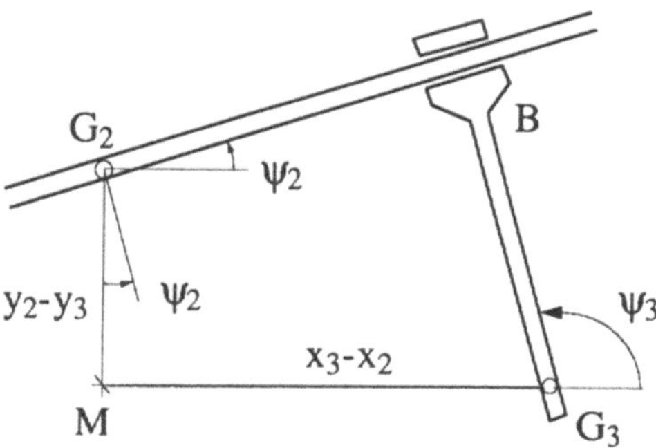

Figure 2.9. Detail of the representation of a prismatic joint.

It is very easy to generalize these results for any planar mechanism and for those cases where the reference point occupies an arbitrary position on the element.

A very important characteristic of the reference point coordinates is that for the constraint equations corresponding to a particular joint, the only coordinates that intervene are the ones of the elements related to this joint. This means that, unlike relative coordinates, constraint equations are established at a *local level*; therefore a particular joint will always have the same ones regardless of the system's complexity. Thus the reference point coordinates do not require preprocessing like the relative ones do, and it becomes much easier to generate the constraint equations automatically on a computer program.

The simplicity of reference point coordinates and the related constraint equations in 2-D cannot be extrapolated to 3-D directly, because in the latter case there are many more types of joints and the definition of orientation is more complicated.

2.1.3 Natural Coordinates

Natural coordinates represent an interesting alternative to relative coordinates and reference point coordinates. These coordinates were originally introduced by García de Jalón et al. (1981) and Serna et al. (1982) for planar cases, and García de Jalón et al. (1986 and 1987) for spatial systems.

In the case of planar multibody systems, natural coordinates can be considered as an *evolution of the reference point coordinates* in which the points are moved to the joints or to other important points of the elements, so that each element has at least two points (See Figure 2.10).

It is important to point out that since each body has at least two points, its position and angular orientation are determined by the Cartesian coordinates of these points, and the angular variables used by reference point coordinates are no longer necessary. It will be seen later on that this simplifies the formulation of

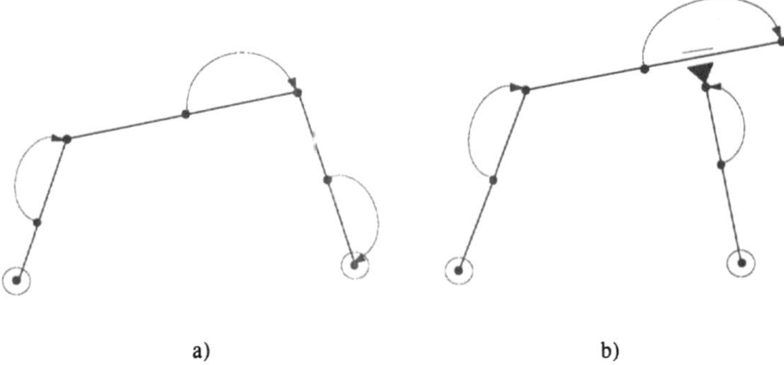

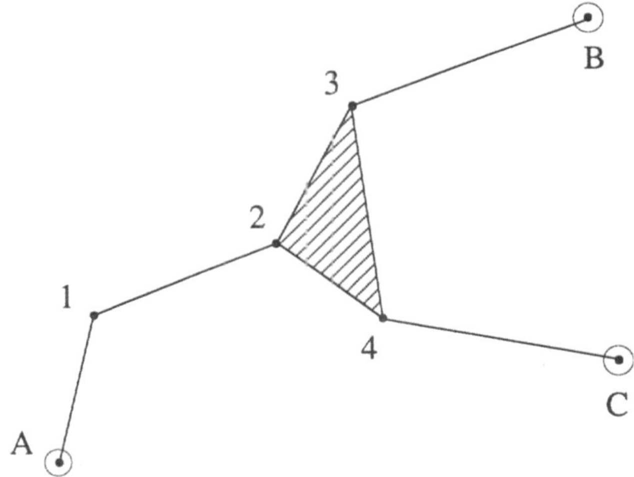

Figure 2.10. Evolution of the reference point coordinates to the natural coordinates.

Figure 2.11. Representation of the mechanism of Figure 2.6 in natural coordinates.

the constraint equations along with the fact that points can be shared at the joints.

Thus the natural coordinates in the case of planar multibody systems are made up of Cartesian coordinates of a series of points. We will call these the *basic points*, and they are distributed throughout the entire mechanism. These points should be chosen according to the following rules or criteria:

1. Each element should have at least two basic points for the motion to be defined.
2. There should be a basic point in each revolute joint R. This point is shared by the two elements linked at this joint.
3. Each prismatic joint P links two bodies, and the two basic points at one of these determine the direction of the relative motion. Although one of the ba-

sic points of the other body can be located on the segment determined by the two basic points of the first one, this is not absolutely necessary.

4. In addition to the basic points that model the body, any other important point of any body can be selected as a basic point, and its coordinates would then automatically become part of the set of unknown variables.

The number of natural coordinates tends to be an average between the number of relative coordinates and the number of reference point coordinates. For example, in the mechanisms of Figure 2.10, the number of natural coordinates is four and six respectively, as opposed to three and three relative coordinates, and nine and nine reference point coordinates. The reason for the decrease in the number of coordinates is due, on one hand, to the elimination of the angular coordinates and, on the other hand, to the sharing of the basic points (located at the joints R) by two or more bodies. Thus, they have the advantage of describing the position of bodies with a reduced number of unknowns.

Figure 2.11 shows the six-body mechanism of Figure 2.6 modeled with natural coordinates. In this example and in the previous ones, another characteristic of the natural coordinates can be observed: preprocessing and postprocessing are practically not required. In fact, once the basic point coordinates are known, drawing the position of the mechanism on a plotter or on a terminal is absolutely trivial. Drawing the velocity and acceleration vectors of the different points is just as easy.

Finally, it should be pointed out that perhaps the most important advantage of natural coordinates is their easy formulation and implementation from a programming standpoint. As may be seen in the next paragraphs, the constraint equations and their Jacobian matrix are very easy to evaluate. Some numerical tests performed by Unda et al. (1987) have shown that for some 2-D multibody systems the advantages mentioned for natural coordinates versus reference point coordinates can be translated into some reductions in calculation times.

Even though natural coordinates can be explained as an evolution of reference point coordinates, in reality their history is quite different. In fact, natural coordinates historically came about as an adaptation of the *displacement method* for matrix analysis of structures to the analysis of multibody systems. A multibody system can be considered as an underconstrained structure that lacks bars or elements, therefore becoming unstable. Van der Werf (1979) and Van der Werf and Jonker (1985) developed a multibody analysis method entirely based on the finite element method. The difference between the method proposed by these authors and the one described in this book is that while the former remained entirely based on the principles of the finite element method, the method based on natural coordinates has been entirely reformulated mathematically; so it can be introduced and considered as a new method expressly developed for analysis of multibody systems.

It has been shown in the previous sections how the relative coordinates lead to constraint equations that are originated from closed loops of the system. The constraints with reference point coordinates originate in the kinematic joints.

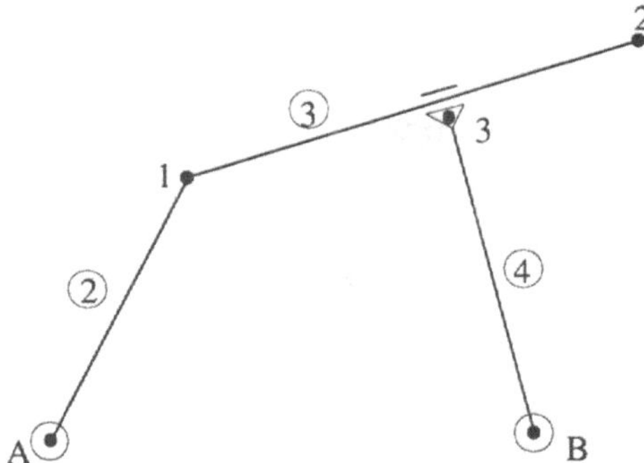

Figure 2.12. Four-bar mechanism with a prismatic joint modeled with natural coordinates.

Further on, it will be seen how the constraint equations originate for two sources when using natural coordinates:

1. Rigid body condition of each element.
2. Constraints corresponding to some kinematic joints.

A four-bar mechanism modeled with natural coordinates was shown in Example 2.1. Its three constraint equations come from the constant length condition for each one of the moving bars. No joint constraint was present for this case. A second example in which joint constraints appear is given below:

Example 2.6

Figure 2.12 shows a four-bar mechanism with one prismatic joint modeled with natural coordinates. There are six coordinates and one degree of freedom, which means that there should be five constraint equations. These equations can be obtained as follows. In the first place, the mechanism in question has three elements, each of which contains two points. It must be guaranteed that these elements move as rigid bodies. To do this, the following three constant length conditions must be imposed:

$$(x_1 - x_A)^2 + (y_1 - y_A)^2 - L_2^2 = 0$$

$$(x_2 - x_1)^2 + (y_2 - y_1)^2 - L_3^2 = 0$$

$$(x_2 - x_B)^2 + (y_2 - y_B)^2 - L_4^2 = 0$$

In the case of the fixed points A and B, the constraint equations are automatically taken into account when considering that their coordinates do not vary. Note that the revolute joints do not generate any constraint equation. In the case of the revolute joint located at point 1 the joint constraints also have been automatically taken into account, when considering that 1 is a point common to or shared by elements 2 and 3. As long as point 1 belongs and contributes to the definition of both elements, the only possibility of relative motion that these elements have is that of relative rotation.

The two remaining equations will originate in the prismatic joint. In this kinematic pair, the adjacent elements do not share anything; therefore two equations are required to constrain the two relative degrees of freedom eliminated by the prismatic joint.

The first equation originates by imposing point 3 be permanently aligned with points 1 and 2. This can be done in two ways. The first way is by imposing the following condition of proportionality:

$$\frac{x_3 - x_1}{x_2 - x_1} = \frac{y_3 - y_1}{y_2 - y_1}$$

This equation can also be expressed as follows:

$$(x_3 - x_1)(y_2 - y_1) - (x_2 - x_1)(y_3 - y_1) = 0$$

An equivalent result can be obtained by imposing the constant area condition (zero area, in this case) of the triangle determined by points 1, 2 and 3. Using the formula of the determinant, whose value is equal to twice the area of the triangle,

$$det \begin{vmatrix} 1 & x_1 & y_1 \\ 1 & x_2 & y_2 \\ 1 & x_3 & y_3 \end{vmatrix} = 2\,A_{123} = 0$$

By expanding the determinant, the following equation is obtained:

$$(x_2 - x_1)(y_3 - y_1) - (x_3 - x_1)(y_2 - y_1) = 0$$

which coincides with the equation obtained previously. The advantage of the area formula is that it can be applied to non-aligned points, forming a triangle with a constant area. For example, in the mechanism of Figure 2.12, the area of the triangle (1-2-B) is constant, because segment (B-3) moves perpendicular to segment (1-2). This means that the previous equation could be substituted by the equation:

$$det \begin{vmatrix} 1 & x_1 & y_1 \\ 1 & x_B & y_B \\ 1 & x_2 & y_2 \end{vmatrix} - 2\,A_{12B} = 0$$

and by expanding the determinant

$$(x_B - x_1)(y_2 - y_1) - (x_2 - x_1)(y_B - y_1) - 2\,A_{12B} = 0$$

One more equation still remains to be obtained, the one corresponding to the condition that the angle between elements 3 and 4 is maintained constant. This condition, when the angle does not have a value close to $0°$, can be imposed by means of the scalar or dot product of vectors (1-2) and (B-3),

$$(x_2 - x_1)(x_3 - x_B) + (y_2 - y_1)(y_3 - y_B) - L_3 L_4 \cos\phi = 0$$

where ϕ is the angle formed by both elements.

Let us now consider the constraint equations corresponding to angular quantities. When the angle is close to $0°$, the scalar product is not valid for imposing the constant angle condition, and the cross product of vectors must be used instead (more specifically, in the case of planar systems, the z component of the cross product). In turn, the cross product is not valid when the angle has a value close to $\pm 90°$. The reason for this can be understood by observing the two parts

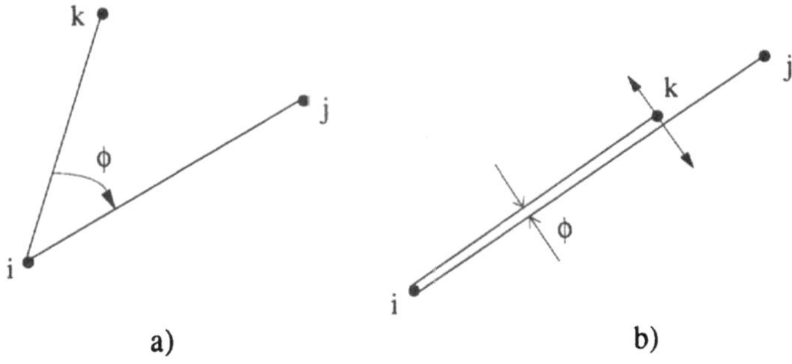

Figure 2.13. Variation of the scalar product in terms of the angle ϕ.

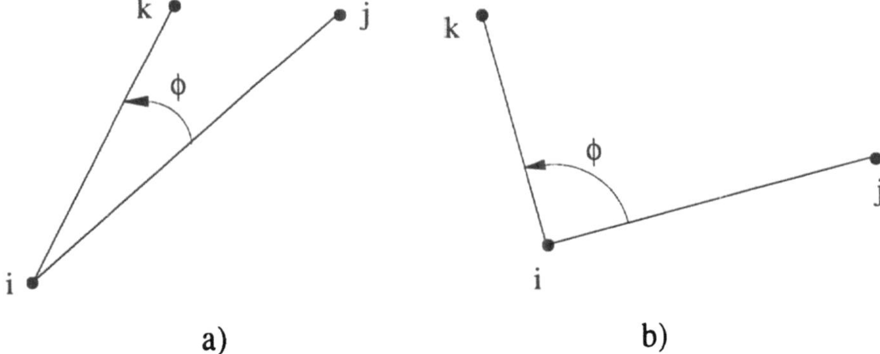

Figure 2.14. Variation of the triangular area in terms of the angle ϕ.

of Figure 2.13. The scalar product keeps the angle between the segments constant by controlling the projection of one over the other. In Figure 2.13a, the angle ϕ cannot be changed without changing the projection of segment $(i\text{-}k)$ on $(i\text{-}j)$. However in Figure 2.13b, where ϕ is zero, the angle can vary infinitesimally without varying the projection of $(i\text{-}k)$ on $(i\text{-}j)$. Therefore the scalar product is not a good method for driving an angle when it is zero or close to zero. The same statement is valid when the angle is near 180°.

On the other hand, the module of the cross product is related to the area of the triangle determined by the three points, which means that the cross product uses the area of the triangle to control the angle. It may be seen in Figure 2.14a that small variations in the angle ϕ produce significant variations in the area of the triangle; while in Figure 2.14b it is observed that when ϕ is equal to or close to 90°, small variations of ϕ do not produce any variation in the value of the area. From this it can be concluded that the cross product is not valid for driving the value of the angle when the latter has a value close to 90°. The $\phi = 180°$ case is similar to the $\phi = 0°$ case, and the $\phi = -90°$ case is similar to the $\phi = 90°$ case.

The cross product can be formulated by means of the well known determinant formula. In the case of Figure 2.14:

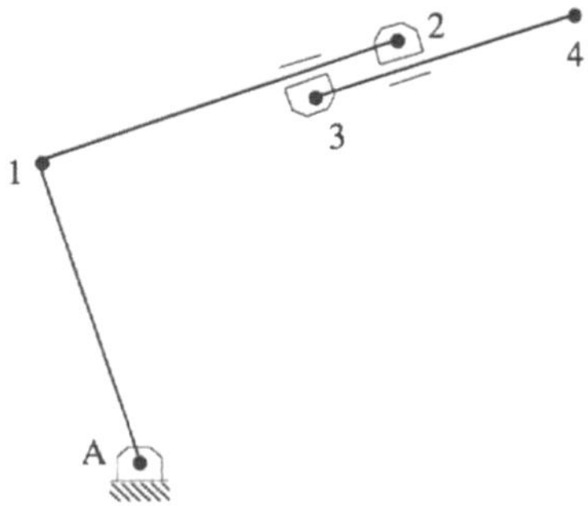

Figure 2.15. Mechanism with a telescopic joint.

$$det \begin{vmatrix} u_x & u_y & u_z \\ x_k - x_i & y_k - y_i & 0 \\ x_j - x_i & y_j - y_i & 0 \end{vmatrix} = (x_k - x_i)(y_j - y_i) - (x_j - x_i)(y_k - y_i) \quad (2.1)$$

and equating this value to twice the area of the triangle $(i\text{-}j\text{-}k)$, we obtain

$$(x_k - x_i)(y_j - y_i) - (x_j - x_i)(y_k - y_i) - 2\,A_{ijk} = 0 \quad (2.2)$$

Example 2.7

Accordingly, it is not difficult to establish the constraint equations corresponding to the prismatic pair of the mechanism shown in Figure 2.15, which is formed by a telescopic element. The two constraint equations required can originate from the zero area (alignment) condition of triangles (1-2-3) and (2-3-4).

$$(x_2 - x_1)(y_3 - y_1) - (x_3 - x_1)(y_2 - y_1) - 2\,A_{123} = 0$$

$$(x_3 - x_2)(y_4 - y_2) - (x_4 - x_2)(y_3 - y_2) - 2\,A_{234} = 0$$

It should be clear at this point how to establish the constraint equations for prismatic planar joints. In regard to the rigid body constraints of elements defined by more than two basic points, there are a few particular cases that should be analyzed in detail.

Example 2.8

Figure 2.16 shows three elements with more than two basic points in each one of them. The element in Figure 2.16a is standard, and its rigid body condition can be established by imposing the constant length condition on the three sides of the triangle,

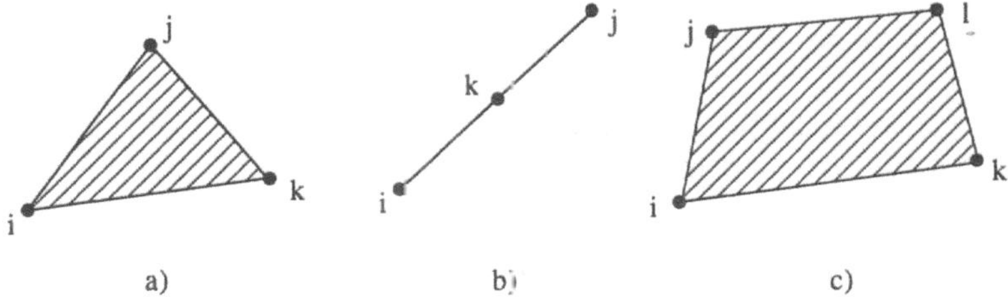

Figure 2.16. Different types of elements with more than two basic points.

$$(x_i - x_j)^2 + (y_i - y_j)^2 - L_{ij}^2 = 0 \qquad\qquad \text{(i)}$$

$$(x_j - x_k)^2 + (y_j - y_k)^2 - L_{jk}^2 = 0 \qquad\qquad \text{(ii)}$$

$$(x_k - x_i)^2 + (y_k - y_i)^2 - L_{ki}^2 = 0 \qquad\qquad \text{(iii)}$$

These three equations are not valid, however, for the element in Figure 2.16b where the three points are aligned. In principle, the three equations (i)-(iii) could also be valid in this case. It can be realized that they are not independent and therefore are not capable of guaranteeing that the element will move as a rigid body. The reason for this is that point k can be separated infinitesimally from segment (i-j) without varying the distances to points i and j.

The solution to the problem of the element in Figure 2.16b is to substitute one of the equations (i)-(iii) for the zero area condition of the triangle (i-j-k) which guarantees that k will be aligned with i and j. This equation is as follows:

$$(x_k - x_j)\,(y_j - y_i) - (x_j - x_i)\,(y_k - y_i) = 0 \qquad\qquad \text{(iv)}$$

However, this is not the only possible solution. An easier solution can be obtained by adopting the following constraint equations for the element in Figure 2.16b:

$$(x_i - x_j)^2 + (y_i - y_j)^2 - L_{ij}^2 = 0 \qquad\qquad \text{(v)}$$

$$(x_j - x_i) - C\,(x_k - x_i) = 0 \qquad\qquad \text{(vi)}$$

$$(y_j - y_i) - C\,(y_k - y_i) = 0 \qquad\qquad \text{(vii)}$$

where C is a constant whose value is (L_{ij}/L_{ik}). Equation (v) is a constant distance condition. Equations (vi) and (vii) indicate that the segment (i-j) is equal to segment (i-k) scaled by the constant C. Note that both segments always have the same direction.

Equation (iv) is a little more complicated than equation (iii). Even though both of them are quadratic, equation (iv) involves three points, while equation (iii) only involves two. On the other hand, equations (v)-(vii) are one quadratic and two linear; whereas equations (i)-(iii) are all quadratic.

Regarding the body in Figure 2.16c, bear in mind that it has four basic points, therefore eight coordinates, and that it has three degrees of freedom as a rigid body moving on the plane. Thus, it is necessary to establish five constraint equations. An immediate solution is to establish the constant length condition for the four sides of the body and for one of the two diagonals. This guarantees that the element will be non-deformable. These equations are as follows:

$$(x_k - x_i)^2 + (y_k - y_i)^2 - L_{ki}^2 = 0$$

$$(x_l - x_k)^2 + (y_l - y_k)^2 - L_{lk}^2 = 0$$

$$(x_j - x_l)^2 + (y_j - y_l)^2 - L_{jl}^2 = 0$$

$$(x_i - x_j)^2 + (y_i - y_j)^2 - L_{ij}^2 = 0$$

$$(x_l - x_i)^2 + (y_l - y_i)^2 - L_{li}^2 = 0$$

However, another possibility exists which results in simpler equations. This solution will insure the non-deformability of the triangle $(i\text{-}j\text{-}k)$ by means of the following equations:

$$(x_i - x_j)^2 + (y_i - y_j)^2 - L_{ij}^2 = 0$$

$$(x_k - x_i)^2 + (y_k - y_i)^2 - L_{ki}^2 = 0$$

$$(x_k - x_j)^2 + (y_k - y_j)^2 - L_{kj}^2 = 0$$

Then the condition is imposed that the vector $(i\text{-}l)$ be expressed as a linear combination of vectors $(i\text{-}k)$ and $(i\text{-}j)$, with coefficients α and β properly determined:

$$(x_l - x_i) - \alpha\,(x_k - x_i) - \beta\,(x_j - x_i) = 0$$

$$(y_l - y_i) - \alpha\,(y_k - y_i) - \beta\,(y_j - y_i) = 0$$

The advantage of these last two equations lies in the fact that they are linear instead of quadratic.

As before, constraint equations can also be generated for planar elements with any number of basic points. The generation of constraint equations with natural coordinates can be easily automated on a computer program. No preprocessing is required and the resulting equations are sparse. In addition, the natural coordinates generate quadratic or linear constraint equations. These equations are easier to evaluate than the transcendental equations obtained with both relative and reference point coordinates.

2.1.4 Mixed and Two-Stage Coordinates

It was mentioned previously that one of the advantages of the relative coordinates is the possibility of directly accounting for the relative degrees of freedom permitted by the joints. This type of coordinates allows the direct inclusion of motors or actuators at the joints with no further difficulties. On the other hand, neither natural coordinates nor reference point coordinates have this advantage. However, *mixed coordinates* can solve this problem. Mixed coordinates are obtained by adding, to natural coordinates or to reference point coordinates, angular or linear variables corresponding to the degrees of freedom of the system joints.

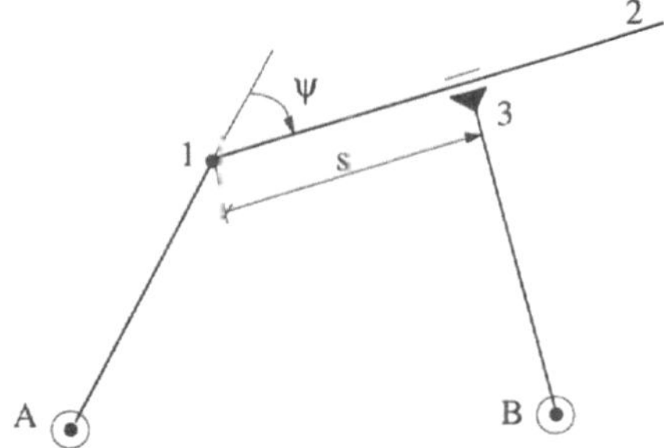

Figure 2.17. Four-bar mechanism with a prismatic pair modeled with mixed coordinates.

It is very easy to add relative coordinates to natural coordinates, as can be seen in the next example.

Example 2.9

A mechanism with six Cartesian (natural) and two relative coordinates can be seen in Figure 2.17.

The constraint equations that must be added, because of the introduction of angle Ψ and distance s, come from the scalar product of vectors and are, respectively

$$(x_1 - x_A)(x_2 - x_1) + (y_1 - y_A)(y_2 - y_1) - L_{1A} L_{12} \cos\Psi = 0$$

$$(x_3 - x_1)^2 + (y_3 - y_1)^2 - s^2 = 0$$

An angle Ψ different from $0°$ or $180°$ has been assumed in order for the scalar product to define the relative angle.

When considering mixed coordinates, joint variables do not replace the other coordinates; rather they are simply added to them. When increasing the number of dependent coordinates without modifying the number of degrees of freedom, one should increase the number of constraint equations by the same amount.

Some authors as Jerkovski (1978) and Kim and Vanderploeg (1986) use two different coordinate systems in two stages of the analysis. First they describe the mechanism using reference point coordinates, and then they perform the analysis using relative coordinates, hoping this will be more effective. This successive use of two different types of coordinates is also called *velocity transformations*, and should be distinguished from the use of mixed coordinates. Velocity transformations can improve the efficiency significantly and will be considered in more detail in Chapters 5 and 8.

2.2 Spatial Multibody Systems

The same types of coordinates discussed in the previous section for planar multibody systems also apply to three-dimensional ones. Although the formulation is at times substantially more complicated, the basic concepts hardly differ, therefore the explanations tend to be quite straightforward.

The general principles and guidelines for constraint equations in three-dimensional multibody systems will be developed next. In this case, the constraint equations corresponding to relative and reference point coordinates will not be developed in detail as previously done for the case of planar systems, because they are much more involved. Following this is a general description of the basic guidelines used for obtaining them. Natural coordinates and mixed coordinates will be explained in detail.

2.2.1 Relative Coordinates

The main difference regarding the use of relative coordinates, between three-dimensional multibody systems, and the planar ones, is the great variety of joints that appear in the three-dimensional case and the fact that many of these joints allow more than one degree of freedom of relative motion. It is also necessary to introduce a relative coordinate for each one of the degrees of freedom permitted by the joint. Thus a ball-joint or spherical joint (S) introduces three rotations, a cylindrical joint (C) one rotation and one translation with coincident axes, and so forth.

Some authors propose to simplify the problem and minimize the number of possibilities that may arise with all the types of joints by combining the revolute joints (R) and/or the prismatic joints (P), thus creating a combination of joints with a single degree of freedom. This substitution is carried out by introducing fictitious elements with zero mass and dimensions. For example, a cylindrical joint can be substituted by a prismatic joint and a revolute joint by simply introducing an intermediate element. A spherical joint can be substituted by three revolute joints with concurrent axes forming 90° angles between them. Consequently any multibody system can be transformed by this substitution method into an equivalent one with more elements containing only R and P joints. The analysis should be thus considerably simplified.

Robots constitute a specific case of three-dimensional mechanisms in which the relative coordinates are especially effective and suitable. Usually, robots are open chain systems with revolute and/or prismatic joints and with an actuator controlling each one of these joints. These are the ideal conditions for relative coordinates. In fact they are the ones used almost exclusively in robotic applications. Relative coordinates in three-dimensional multibody systems and the corresponding matrix formulation for the constraint equations were introduced by Hartenberg and Denavit (1963). Other authors (Sheth and Uicker (1972), Wittenburg and Wolz (1985)) developed three-dimensional computer codes based on this type of coordinates.

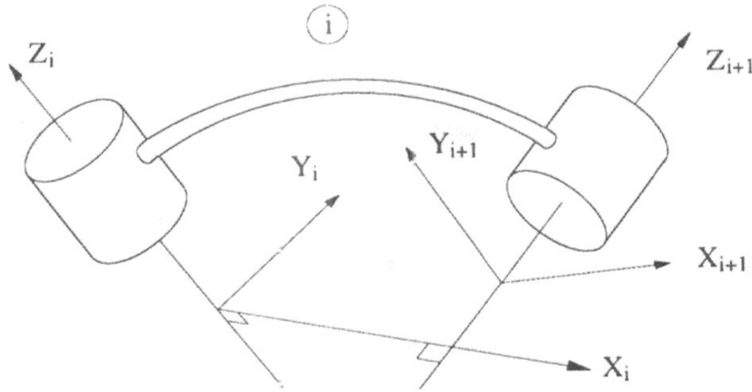

Figure 2.18. Hartenberg and Denavit representation of a binary element.

The advantages and disadvantages of relative coordinates in three-dimensional multibody systems are similar to those described for planar ones. The degree of involvement of the resulting formulation grows at a higher order with the addition of the spatial dimension.

Similar to the planar case, constraint equations are generated, with three-dimensional relative coordinates, by closing the kinematic loops. In the case of three-dimensional configurations, the constraint equations are usually formulated in matrix form instead of vectorially using the Hartenberg and Denavit (1963) method and notation. This technique usually starts by reducing all the class II joints, III etc., to class I joints. As explained previously this is done by introducing as many fictitious elements as required.

For binary links, a system of Cartesian coordinates rigidly attached to each one of the moving elements is defined next (See Figure 2.18). In this way, axis Z_i coincides with the axis of the pair (R or P) that joins the elements (i–1) and (i). Axis X_i is drawn along the common normal line to axes Z_i and Z_{i+1}. Axis Y_i is the normal common to axes X_i and Z_i. The key to the Hartenberg and Denavit method is that it is possible to find a (4×4) transformation matrix ${}^{i}T_{i+1}(\psi_{i+1})$ that permits passing from the frame $(X_{i+1}, Y_{i+1}, Z_{i+1})$ to the frame (X_i, Y_i, Z_i).

This matrix depends on a series of constant lengths and angles that are characteristic of element (i), and of the joint variable ψ_{i+1} which will be an angle or a distance depending on whether it is a revolute or prismatic joint. If element (i) has more than 2 joints, it will be necessary to define as many local frames as needed for it (all of them will have axis Z_i in common) and the corresponding (4×4) transformation matrices between the reference frames.

The constraint equations with relative coordinates are established by carrying out all the coordinate transformations along a closed loop of the multibody system and by imposing the condition that the product of all those transformations be the unit matrix (one will end up at the original axis). A loop can also be in-

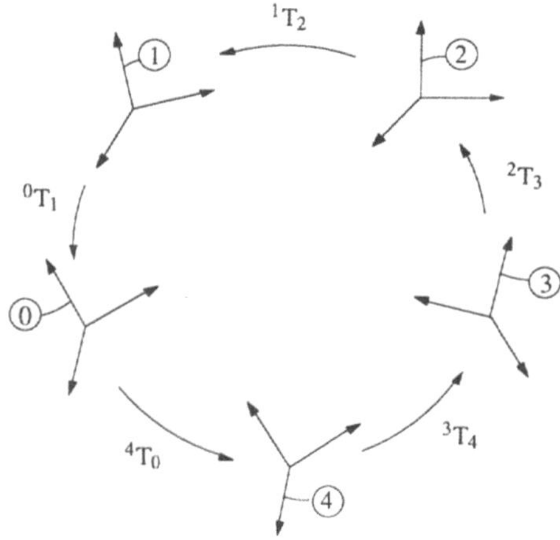

Figure 2.19. Series of Hartenberg and Denavit transformations.

tersected by a specific element to obtain the transformation matrix between the fixed element and that element in two ways, equating the corresponding results.

In the loop shown in Figure 2.19, the closure matrix equation would be as follows:

$$^0\mathbf{T}_1(\psi_1)\,^1\mathbf{T}_2(\psi_2)\,^2\mathbf{T}_3(\psi_3)\,^3\mathbf{T}_4(\psi_4)\,^4\mathbf{T}_0(\psi_0) = \mathbf{I} \tag{2.3}$$

where $^i\mathbf{T}_{i+1}(\psi_{i+1})$ is the (4×4) matrix that permits the transformation from frame (i+1) to frame (i), which depends on the coordinate ψ_{i+1}. Starting from the matrix equation (2.3), one can obtain the corresponding algebraic equations by formulating the matrix product and equating the sufficient number of elements of the left hand side to the corresponding elements of the unit matrix. When closing a loop, there are many more equations available than needed. The problem lies in correctly choosing the equations so that they will be independent and that the solution they provide will suit all the other equations. This second condition is hard to meet. What is usually done is to gather more constraint equations than required and to solve at the time of analysis an over determined system of equations using, for instance, a least square method. This enables one to solve the problem, even at the price of a greater computational effort.

2.2.2 *Reference Point Coordinates*

In the case of three-dimensional multibody systems, reference point coordinates define the position of an element by means of the Cartesian coordinates of one of its points and by means of the angular orientation of a reference frame, rigidly attached to the element, in relation to an inertial or fixed reference frame. As is well known, the definition of the angular orientation of two frames is a classical

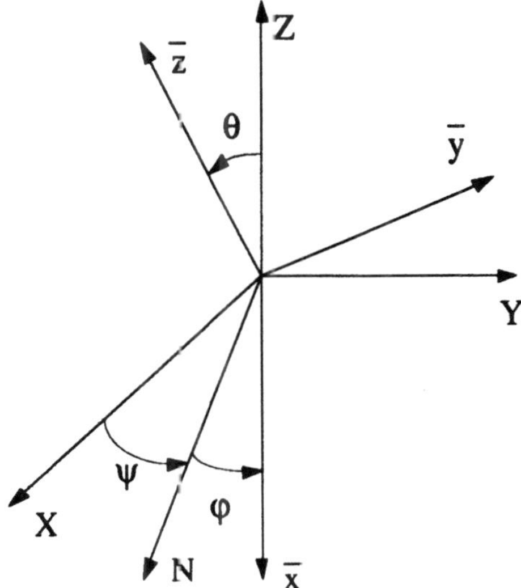

Figure 2.20. Euler angles.

problem of mechanics that has received different solutions (Argyris (1982)). One way of unequivocally defining this orientation is by means of the nine elements of the rotation matrix **A**, whose columns contain the three direction cosines for each one of the moving axes in relation to the fixed frame. The nine elements of the matrix **A** are not independent, but are related by means of six equations (a unit module for each column and orthogonality between the three columns). The main drawback is that no subset of three elements of the matrix **A** is capable by itself of unequivocally representing the orientation of the moving frame at any possible position.

Various three-parameter systems have been developed to solve this problem that define the relative orientation between the two reference frames. The best known ones are the *Roll*, *Pitch*, and *Yaw* rotations or *Tait-Bryant* angles (a succession of three rotations which carry the moving frame from the position of the fixed frame to its final position: α around the Z axis, β around the Y axis, and γ around the X axis; and the well-known *Euler angles* (See Figure 2.20). The problem is that all the existing three-parameter systems have singular positions in those locations where these parameters are not defined unequivocally. For example, for Euler angles when the nutation angle θ is zero, the node line N is not defined (neither are the precession angle ψ and the rotation angle φ, although their sum is still defined). This shortcoming can be corrected by changing one of the coordinate frames every time the movable body approaches a singular position. For instance, the moving frame can be rotated in relation to the element on the mechanism to which it is linked. This approach will solve the problem but tends to complicate the implementation on a computer program. Reference point coordinates with Euler angles were used, for instance, by Orlandea et al. (1977).

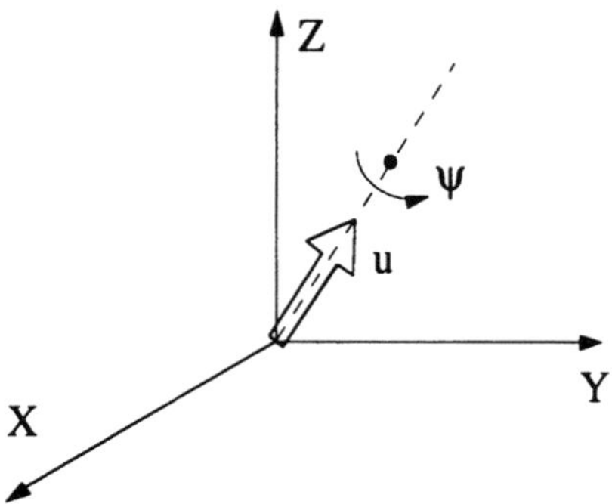

Figure 2.21. Description of the orientation and motion of a rotating axis.

Some other programs based on reference point coordinates use sets of four non-independent parameter systems to describe the angular orientation of the elements. These systems do not have the drawbacks of the three-parameter systems. However, they pay the price as the number of coordinates is increased (seven instead of six for each element). There is also the additional problem of having to take into account the constraint equation that relates the four parameters .

The simplest and easiest to understand of the four-parameter systems is that which defines the orientation of the moving frame by means of a rotation angle ψ around an axis defined by the three direction cosines of the unit vector **u**. This axis and angle represent the rotation that must be transmitted to the fixed frame to make it coincide with the moving one (See Figure 2.21).

The relation that exists between these four parameters is that the module of the axis direction vector **u** should be the unit value, thus

$$u_x^2 + u_y^2 + u_z^2 = 1 \tag{2.4}$$

In practice, the four-parameter system used the most is not the one suggested, but one made up of the so-called *Euler parameters*, that is closely related to the former. The Euler parameters are defined as follows:

$$p_1 = u_x \, cos\left(\psi/2\right) \tag{2.5}$$

$$p_2 = u_y \, cos\left(\psi/2\right) \tag{2.6}$$

$$p_3 = u_z \, cos\left(\psi/2\right) \tag{2.7}$$

$$p_4 = sin\left(\psi/2\right) \tag{2.8}$$

The constraint equation for Euler parameters is easily found to be:

$$p_1^2 + p_2^2 + p_3^2 + p_4^2 = 1 \tag{2.9}$$

The Euler parameters have a very interesting set of properties summarized by Wittenburg (1977), Nikravesh et al. (1985), Nikravesh (1988), and Haug (1989). There are simple detailed expressions in these references to express the angular

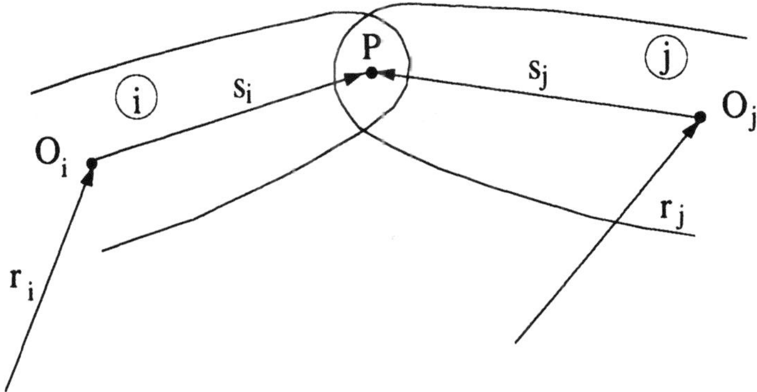

Figure 2.22. Spherical joint.

velocity and the angular acceleration of any element in accordance with the Euler parameters and their derivatives. Euler parameters are used in many computer programs for multibody simulation.

Some programs based on reference point coordinates use variables for the velocities (called *quasi-velocities*), which are different from the derivatives of the coordinates used to describe the position. Thus, Nikravesh et al. (1985) use Euler parameters to describe the position and components of the angular velocity vector ω to describe the velocities. The reason is that the indetermination encountered in position is not found in velocities and accelerations; thus the number of variables is minimized and the problem is simplified. Bear in mind that the angular velocity vector ω is not an integrable variable. There is no set of three parameters whose derivatives are the three components of the angular velocity vector. Therefore, in order to describe the position, one must rely on the Euler parameters.

In the case of three-dimensional multibody systems, the reference point coordinates have advantages and disadvantages similar to those encountered in planar systems. That the number of coordinates becomes larger is an added difficulty, and the description of spatial orientation and the formulation of constraint equations according to the terms of the Euler parameters is somewhat more complicated. The following references: Wittenburg (1977), Shabana (1989), and Huston et al. (1978), contain detailed descriptions of formulations based on the Euler parameters.

With reference point coordinates the constraint equations originate from the kinematic joints. The constraint equations corresponding to spherical (S), revolute (R), and prismatic (P) joints will be examined separately. Equations corresponding to other joints can be found in a similar way.

Spherical Joint. Two elements are linked by means of a spherical pair simply when they have one point in common. Let P be this point, and O_i and O_j be the reference points for the two elements, where two systems of coordinates rigidly attached to these elements are located (See Figure 2.22).

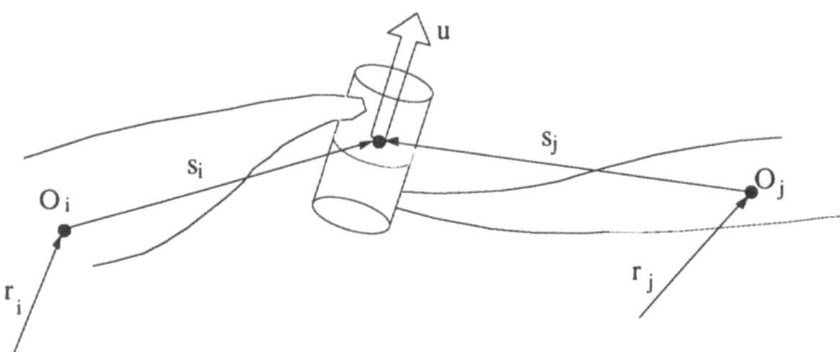

Figure 2.23. Revolute joint.

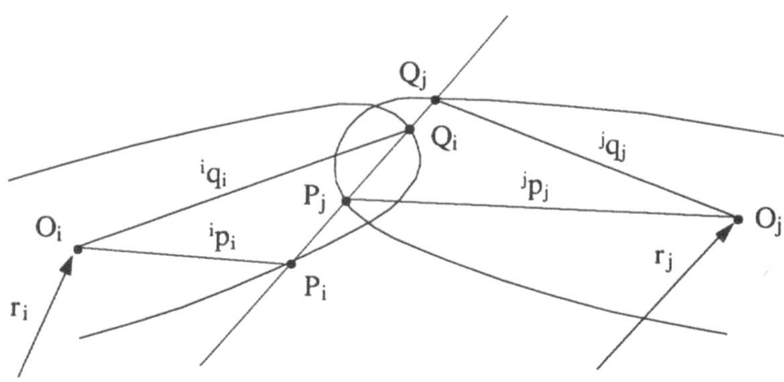

Figure 2.24. Prismatic joint.

The condition that P is a common point belonging to both elements can be written vectorially:

$$\mathbf{r}_i + \mathbf{s}_i - \mathbf{r}_j - \mathbf{s}_j = 0 \tag{2.10}$$

where all the vectors are expressed in the absolute coordinate system. Expressing the vectors $\mathbf{s}_i$ and $\mathbf{s}_j$ in their local frames and using the rotation matrices $\mathbf{A}_i$ and $\mathbf{A}_j$, equation (2.10) becomes

$$\mathbf{r}_i + \mathbf{A}_i{}^i\mathbf{s}_i - \mathbf{r}_j - \mathbf{A}_j{}^j\mathbf{s}_j = 0 \tag{2.11}$$

which is the constraint vector equation equivalent to the three algebraic equations that restrict the three degrees of freedom prevented by the spherical joint. In equation (2.11), the two local position vectors for point P ($^i\mathbf{s}_i$ and $^j\mathbf{s}_j$) are constant. The reference point coordinates are given by vectors $\mathbf{r}_i$ and $\mathbf{r}_j$; whereas the Euler angles, or Euler parameters, appear in matrices $\mathbf{A}_i$ and $\mathbf{A}_j$.

Revolute Joint. The revolute joint (See Figure 2.23) restricts five degrees of freedom; and therefore five constraint equations must be found.

The revolute pair can be defined in various ways. The first and perhaps the easiest way, is by means of two spherical pairs, that is, by making the two elements share two points positioned on the rotational axis. In this case, two vector equations formulated like the one in equation (2.11) can be obtained, thus resulting in six algebraic equations, out of which only five are independent. Whether or not it will be necessary to eliminate one or the other of the equations from one of the points will depend on the orientation of the joint's axis.

However, the most natural way to impose a revolute joint between two contiguous bodies is to establish the compatibility condition between the axis direction and one common point on this axis. According to Figure 2.23 we can write:

$$\mathbf{r}_i + \mathbf{A}_i \, {}^i\mathbf{s}_i - \mathbf{r}_j - \mathbf{A}_j \, {}^j\mathbf{s}_j = 0 \tag{2.12}$$

$$\mathbf{A}_i \, {}^i\mathbf{u} - \mathbf{A}_j \, {}^j\mathbf{u} = 0 \tag{2.13}$$

where ${}^i\mathbf{u}$ and ${}^j\mathbf{u}$ are the coordinates of a vector on the axis expressed in the local reference frames attached to bodies (i) and (j). Only two equations are independent in equation (2.13).

Prismatic Joint. The prismatic joint also generates five constraint equations, but in this case no point is shared. A possible modeling of a prismatic joint will be described next.

Let $\mathbf{r}_i$ and $\mathbf{r}_j$ be the reference point position vectors of the two elements shown in Figure 2.24, and P_i and Q_i be two points belonging to element (i), whose position vectors on the local axes of the element are ${}^i\mathbf{p}_i$ and ${}^i\mathbf{q}_i$, respectively. P_i and Q_i are positioned on the axis of the prismatic pair. P_j and Q_j are two points similar to the previous ones, belonging to element (j) and also positioned on the axis of the prismatic joint.

Four of the five constraint equations originate from imposing the condition that the four points P_i, Q_i, P_j and Q_j remain aligned. This condition can be imposed with the corresponding proportionality among the vector coordinates being aligned or with the following cross products of vectors:

$$(\mathbf{A}_i \, ({}^i\mathbf{q}_i - {}^i\mathbf{p}_i)) \wedge ((\mathbf{r}_j + \mathbf{A}_j \, {}^j\mathbf{p}_j) - (\mathbf{r}_i + \mathbf{A}_i \, {}^i\mathbf{p}_i)) = 0 \tag{2.14}$$

$$(\mathbf{A}_j \, ({}^j\mathbf{q}_j - {}^j\mathbf{p}_j)) \wedge ((\mathbf{r}_i + \mathbf{A}_i \, {}^i\mathbf{p}_i) - (\mathbf{r}_j + \mathbf{A}_j \, {}^j\mathbf{p}_j)) = 0 \tag{2.15}$$

where the symbol $\wedge$ stands for cross product.

Equation (2.14) guarantees that points P_i, Q_i and P_j are aligned; whereas equation (2.15) guarantees the same for points Q_j, P_j and Q_i.

Each one of the equations (2.14)-(2.15) give rise to two independent algebraic equations, which should be chosen among the three available equations in accordance with the direction of the axis of the prismatic pair (the component corresponding to the largest direction cosine for this axis shall be eliminated).

Equations (2.14) and (2.15) correspond exactly to the equations generated by a cylindrical joint. In order to find the additional 5th equation, that is characteristic of the prismatic joint, it is necessary to avoid the possibility that elements (i) and (j) have relative rotation with respect to the joint's axis. This is achieved by imposing the condition that two vectors, each fixed at a body and not parallel to the axis, will maintain a constant angle. Assuming that the angle does not have

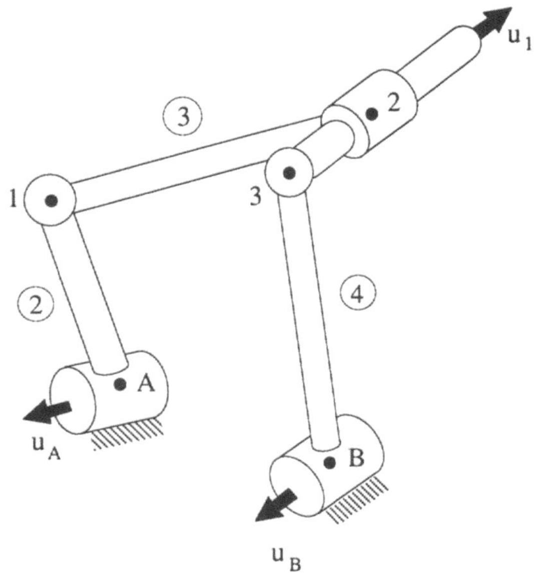

Figure 2.25. RSCR spatial mechanism.

a value close to 0°, this condition can be imposed by means of the scalar product of vectors as follows:

$$(\mathbf{A}_i \, {}^i\mathbf{p}_i) \cdot (\mathbf{A}_j \, {}^j\mathbf{p}_j) - L_{P_iO_i} L_{P_jO_j} \cos \alpha = 0 \qquad (2.16)$$

whereupon the joint is perfectly defined.

The reference point coordinates and the generation of the corresponding constraint equations have been extensively dealt with in Nikravesh (1988) and Haug (1989).

2.2.3 Natural Coordinates

In the case of three-dimensional multibody systems, the natural coordinates describe the position of each element by means of the Cartesian coordinates of the *basic points* distributed throughout the elements and by means of the Cartesian components of several *unit vectors* as seen in the example of Figure 2.25. Each element of the system should have a sufficient number of points and vectors linked to it; so that their motion completely defines that of the element.

Example 2.10

Figure 2.25 shows an RSCR spatial mechanism with four elements and one degree of freedom. There are three basic moving points (1, 2 and 3) and two fixed points (A and B). There is one moving unit vector $\mathbf{u}_1$ and two fixed vectors $\mathbf{u}_A$ and $\mathbf{u}_B$. Element 2 is made up of basic points A and 1, and the unit vector $\mathbf{u}_A$. Element 3 is

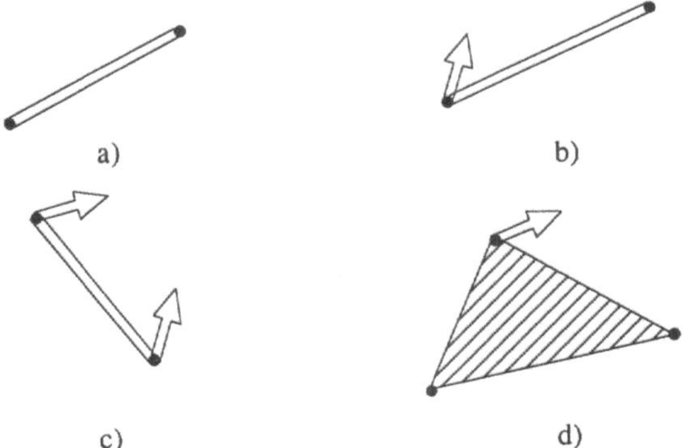

Figure 2.26. Modeling different spatial elements with natural coordinates.

made up of points 1 and 2 and vector $\mathbf{u}_1$. Element 4 is made up of points 3 and B and vectors $\mathbf{u}_1$ and $\mathbf{u}_B$. Each element has at least two points and one unit vector not aligned with the points; therefore their position and motion completely defines that of the element. The mechanism in Figure 2.25 has a total of 12 natural coordinates. This number is an average between the number of relative and reference point coordinates, since the same mechanism would have four, five or six relative coordinates (depending on how it is modeled) and 18 or 21 reference point coordinates, depending on whether Euler parameters or Euler angles were used.

Figure 2.26 shows several possible elements of a three-dimensional mechanism modeled with natural coordinates. There are many more possible combinations of unit vectors and points. One indispensable condition is that the motion of the element be defined by means of the motion of its points and vectors. This does not occur in the element in Figure 2.26a, as the coordinates for the element's two points are not capable of describing the angular position or the rotation around the line connecting these points. This rotation is determined with all the remaining elements in Figure 2.26, except for Figure 2.26b, which requires the unit vector to not be collinear with the direction determined by the two points.

The modeling of a three-dimensional mechanism with natural coordinates can be carried out following these general rules and recommendations:

1. The elements must contain a sufficient number of points and unit vectors so that their motion is completely defined.
2. A basic point shall be located on those joints in which there is a point common to the two linked elements. This happens at the spherical joint (S), at the revolute joint (R), at the universal joint (U), and at other kinematic joints.
3. A unit vector must be positioned at those joints having a rotational or translational axis and should have the direction of the corresponding axis. Sometimes the role performed by a unit vector can also be performed by a couple of basic points.

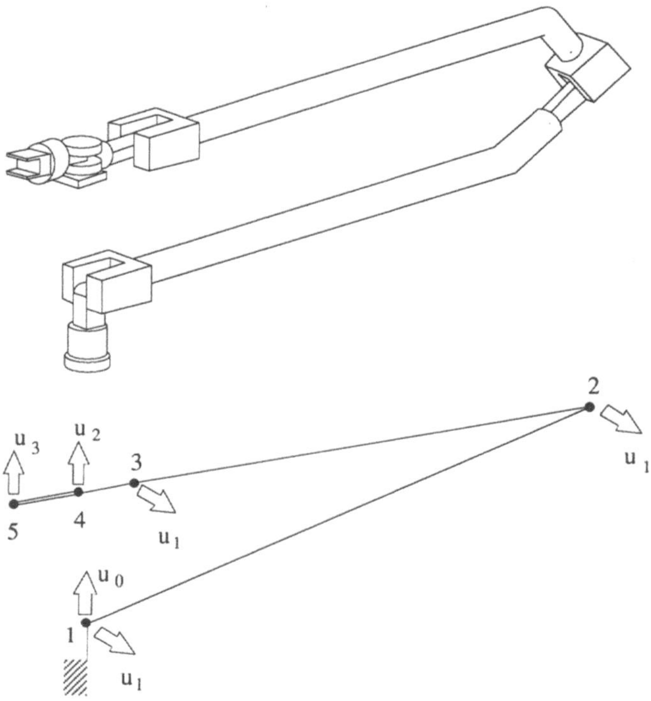

Figure 2.27. Modeling of a spatial robot arm with natural coordinates.

4. Some joints, such as the universal joint (U), have their own particular requirements concerning the introduction of points and unit vectors. These requirements will be studied later on when the constraint equations introduced by each joint are discussed.

5. All points of interest, whose positions are to be considered as a primary unknown variable of the problem, can likewise be defined as basic points.

6. Each unit vector is associated with a specific basic point, and the same single unit vector can be associated with several basic points. For example, on the robot's arm of Figure 2.27 there are three rotational joints whose axes have the same direction. It is not necessary to enter three different unit vectors which would substantially increase the number of unknown variables, but to enter only a single unit vector that is associated with three basic points. A minimum of 21 Cartesian variables are required for this robot.

In the case of three-dimensional multibody systems, the natural coordinates also provide a simple formulation and implementation. The complexity of the mathematical formulation increases *linearly* when moving from 2-D to 3-D applications, because it only suffices to add new points to the model and a new term to the equations coming from the scalar product of vectors. This may become more advantageous than the formulations based on *rotational* variables, in

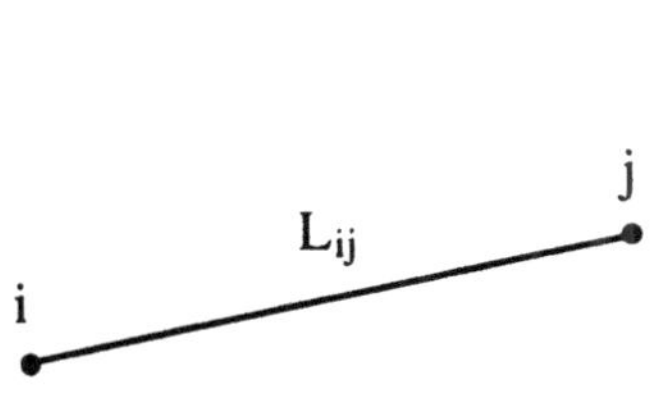

Figure 2.28. Element with two basic points.

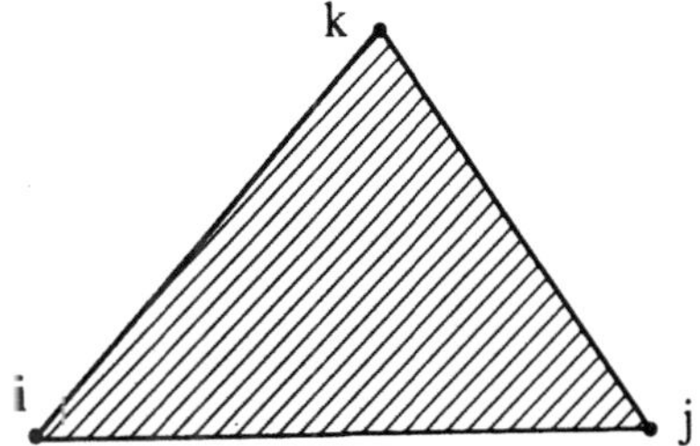

Figure 2.29. Element with three non-collinear basic points.

which the complexity increases at a faster rate by use of rotation matrices with transcendental functions, etc. As in the case of planar multibody systems, the need for preprocessing and postprocessing is minimal when using natural coordinates.

The fundamental topics of the formulation of the kinematic constraint equations will be addressed next. In the case of three-dimensional multibody systems, the constraint equations with natural coordinates also originate in two ways:

1. from the rigid body condition of the elements and
2. from some of the kinematic joints that exist among them.

In the sequel the constraint equations corresponding to both cases will be formulated separately.

2.2.3.1 Rigid Body Constraints

The natural coordinates corresponding to three-dimensional multibody systems have been introduced previously. These coordinates are made up of the Cartesian coordinates of certain points and by the Cartesian components of certain unit vectors. Each element and its motion are defined by a set of points and unit vectors rigidly attached to it. There are many different combinations of points and unit vectors that can be formed when defining an element. Some of the most commonly used combinations can be seen in Figures 2.28-2.33. It will be explained below how rigid body constraints can be found for the elements in these figures. Some cases of particular interest will be studied.

Element with two points. The element in Figure 2.28 has only two basic points and does not have any unit vector. This means that its rotation around the line connecting these basic points is not defined. At any position, the element will behave as if it had only five degrees of freedom. Taking into account that it has six natural coordinates (three Cartesian coordinates for each point), it must have one rigid body constraint equation. This equation is precisely the constant distance condition between points i and j that can be imposed using the scalar product of the relative position vector between both points:

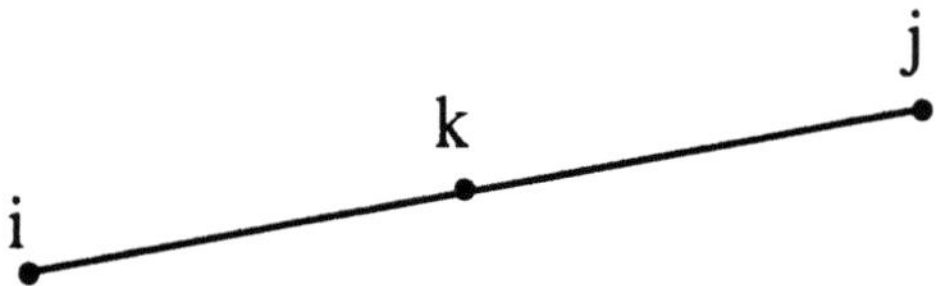

Figure 2.30. Element with three collinear basic points.

$$\mathbf{r}^{ij} \cdot \mathbf{r}^{ij} - L_{ij}^2 = 0 \tag{2.17}$$

where $\mathbf{r}^{ij}$ is the relative position vector. Equation (2.17) can be formulated as follows:

$$\left(x_j - x_i\right)^2 + \left(y_j - y_i\right)^2 + \left(z_j - z_i\right)^2 - L_{ij}^2 = 0 \tag{2.18}$$

that is a quadratic equation in the natural coordinates.

Element with three non-collinear points. The three-dimensional motion of the element in Figure 2.29 is fully represented by the motion of its three basic points. It has nine natural coordinates and six rigid body degrees of freedom. Therefore, it will be necessary to formulate three constraint equations which could correspond to the three following constant distance conditions:

$$\mathbf{r}^{ij} \cdot \mathbf{r}^{ij} - L_{ij}^2 = 0 \tag{2.19}$$

$$\mathbf{r}^{jk} \cdot \mathbf{r}^{jk} - L_{jk}^2 = 0 \tag{2.20}$$

$$\mathbf{r}^{ki} \cdot \mathbf{r}^{ki} - L_{ki}^2 = 0 \tag{2.21}$$

Element with three collinear points. When the three points are aligned (Figure 2.30), the three equations (2.19)-(2.21) are not independent, since point k can move infinitesimally without varying its distance to the other two points. Since the element has nine natural coordinates and five degrees of freedom, it will be necessary to find four constraint equations. One of them is the constant distance condition between points i and j:

$$\mathbf{r}^{ij} \cdot \mathbf{r}^{ij} - L_{ij}^2 = 0 \tag{2.22}$$

The other three equations originate from imposing the condition that vector $\mathbf{r}^{ij}$ is a specific constant α multiplied by vector $\mathbf{r}^{ik}$:

$$\mathbf{r}^{ij} - \alpha\,\mathbf{r}^{ik} = 0 \tag{2.23}$$

Equation (2.22) is quadratic, and the three algebraic equations that originate from the vector equation (2.23) are linear.

Element with two points and one unit vector. The element in Figure 2.31 contains two basic points and one non collinear unit vector. It has nine natural coordinates and six degrees of freedom which give rise to three constraint equations. These equations are the result of the constant distance condition between points i and j:

$$\mathbf{r}^{ij} \cdot \mathbf{r}^{ij} - L_{ij}^2 = 0 \qquad (2.24)$$

the constant angle condition between unit vector $\mathbf{u}^m$ and vector $\mathbf{r}^{ij}$:

$$\mathbf{r}^{ij} \cdot \mathbf{u}^m - L_{ij} \cos \phi = 0 \qquad (2.25)$$

and the unit module condition of vector $\mathbf{u}^m$:

$$\mathbf{u}^m \cdot \mathbf{u}^m - 1 = 0 \qquad (2.26)$$

Equations (2.25) and (2.26) can be formulated algebraically as follows:

$$(x_j - x_i)\, u_x^m + (y_j - y_i)\, u_y^m + (z_j - z_i)\, u_z^m - L_{ij} \cos \phi = 0 \qquad (2.27)$$

$$(u_x^m)^2 + (u_y^m)^2 + (u_z^m)^2 - 1 = 0 \qquad (2.28)$$

If the unit vector is aligned with points i and j (angle ϕ equal to zero), the element will have five degrees of freedom. In this case, the four constraint equations will be

$$\mathbf{r}^{ij} \cdot \mathbf{r}^{ij} - L_{ij}^2 = 0 \qquad (2.29)$$

$$\mathbf{r}^{ij} - \alpha\, \mathbf{u}^m = 0 \qquad (2.30)$$

where α is a constant. The three algebraic equations corresponding to equation (2.30) are linear. There is not much need in defining a unit vector in the direction of a known segment, because unit vectors are used for determining directions. In this case, the direction has already been determined. It is always possible that the unit vector may be introduced for other reasons such as the condition of compatibility with an adjacent body.

Element with two points and two unit vectors. The body of Figure 2.32 has two basic points and two non-coplanar unit vectors. Thus, it has 12

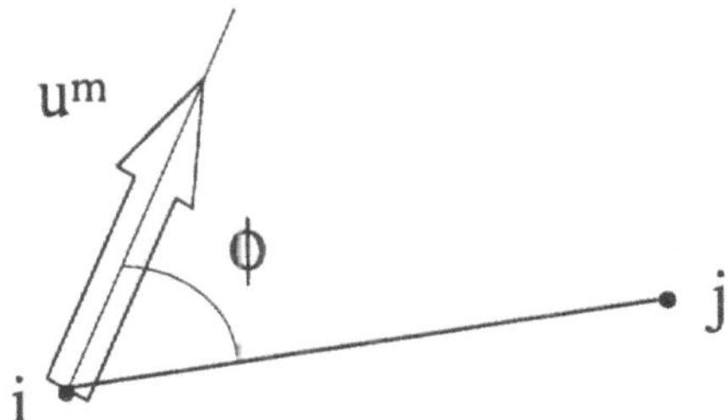

Figure 2.31. Element with two basic points and a unit vector.

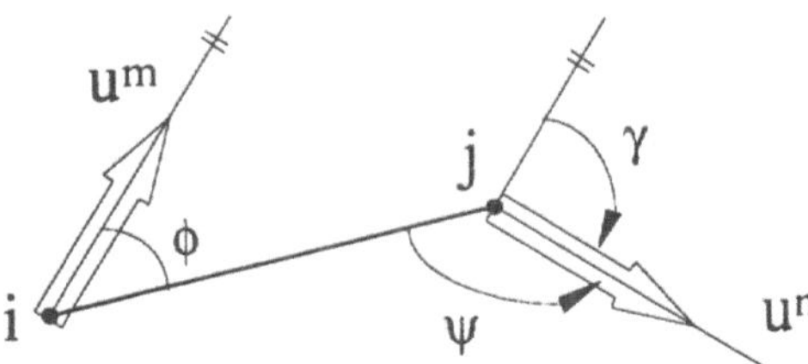

Figure 2.32. Element with two basic points and two unit vectors.

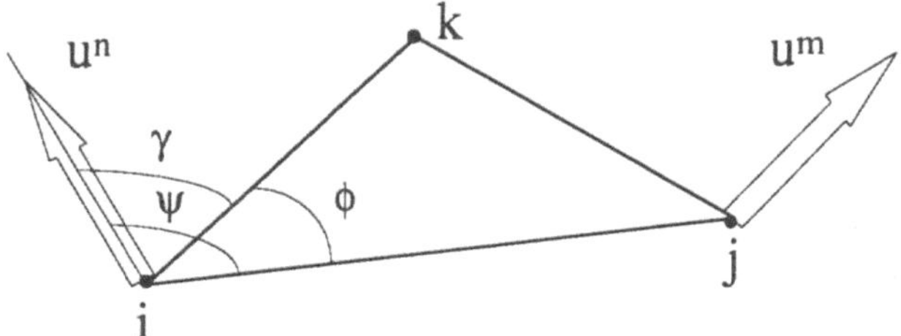

Figure 2.33. A more complex body representation.

natural coordinates and six degrees of freedom. It will be necessary to find six constraint equations. These six conditions are: one constant distance equation, three constant angle conditions (between the two vectors and the segment, and between the two vectors themselves), and two unit module conditions for the unit vectors. The corresponding equations become:

$$\mathbf{r}^{ij} \cdot \mathbf{r}^{ij} - L_{ij}^2 = 0 \tag{2.31}$$

$$\mathbf{r}^{ij} \cdot \mathbf{u}^m - L_{ij} \cos \phi = 0 \tag{2.32}$$

$$\mathbf{r}^{ij} \cdot \mathbf{u}^n - L_{ij} \cos \psi = 0 \tag{2.33}$$

$$\mathbf{u}^n \cdot \mathbf{u}^m - \cos \gamma = 0 \tag{2.34}$$

$$\mathbf{u}^m \cdot \mathbf{u}^m - 1 = 0 \tag{2.35}$$

$$\mathbf{u}^n \cdot \mathbf{u}^n - 1 = 0 \tag{2.36}$$

Several interesting cases can be discussed concerning this element. If one of the angles between segment (i-j) and unit vectors $\mathbf{u}^m$ and $\mathbf{u}^n$ is zero, it will be necessary to proceed as stated in the previous case. The same will occur if the angle γ between the two unit vectors is zero (or 180°); then the two unit vectors will be equal (or one is the opposite of the other). Strictly speaking, it would not be necessary to consider two different unit vectors. It could be assumed that it is the only vector associated to two different points. This would be the case of the element with two points and one vector studied previously. If it is wished that the two vectors be included, the following considerations will have to be made.

If the two unit vectors are coplanar, equations (2.31)-(2.36) are not linearly independent and do not guarantee a rigid body condition for the element. In order to find the constraint equations corresponding to this case, one should take into account that if $\mathbf{u}^n$ and $\mathbf{u}^m$ are coplanar, one of them (for example $\mathbf{u}^n$) can be expressed as a linear combination of $\mathbf{u}^m$ and segment (i-j). The constraint equations would then be:

$$\mathbf{r}^{ij} \cdot \mathbf{r}^{ij} - L_{ij}^2 = 0 \tag{2.37}$$

$$\mathbf{r}^{ij} \cdot \mathbf{u}^m - L_{ij} \, cos \, \phi = 0 \qquad (2.38)$$

$$\mathbf{u}^m \cdot \mathbf{u}^m - 1 = 0 \qquad (2.39)$$

$$\mathbf{u}^n - \alpha_1 \, \mathbf{r}^{ij} - \alpha_2 \, \mathbf{u}^m = 0 \qquad (2.40)$$

where α_1 and α_2 are the constant scalar coefficients of the linear combination. Equation (2.40) is the equivalent to three linear algebraic equations.

More complex elements. The example in Figure 2.33 is one of the many examples that may be created with a large number of points and unit vectors. For all the elements, the number of required constraint equations is always equal to the number of natural coordinates minus the number of rigid body degrees of freedom, which normally will be six. The constraint equations are always determined in the same way: 1. impose the constant distance conditions, 2. impose the necessary constant angle conditions so that the direction of the unit vectors is established, and 3. impose the unit module conditions.

For more complicated elements, the following steps can simplify the process of obtaining the constraint equations and improve the results:

1. Three vectors that can generate a base in the three-dimensional space are chosen. These vectors can be $\mathbf{r}^{ij}$ segments that link two basic points or unit vectors.
2. The constraint equations which guarantee that the three vectors chosen form a rigid body are formulated.
3. The remaining vectors of the body (segments and unit vectors) are expressed as a linear combination of the three vectors that form the base frame. The advantage is that all the equations obtained in this way are linear.

For example, in the body of Figure 2.33, segments (*i-k*) and (*i-j*) and the unit vector $\mathbf{u}^n$ can be used as base vectors. The equations that guarantee that these three vectors form a rigid body are the six indicated below:

$$\mathbf{r}^{ij} \cdot \mathbf{r}^{ij} - L_{ij}^2 = 0 \qquad (2.41)$$

$$\mathbf{r}^{ik} \cdot \mathbf{r}^{ik} - L_{ik}^2 = 0 \qquad (2.42)$$

$$\mathbf{r}^{ij} \cdot \mathbf{r}^{ik} - L_{ij} \, L_{ik} \, cos \, \phi = 0 \qquad (2.43)$$

$$\mathbf{r}^{ij} \cdot \mathbf{u}^n - L_{ij} \, cos \, \psi = 0 \qquad (2.44)$$

$$\mathbf{r}^{ik} \cdot \mathbf{u}^n - L_{ik} \, cos \, \gamma = 0 \qquad (2.45)$$

$$\mathbf{u}^n \cdot \mathbf{u}^n - 1 = 0 \qquad (2.46)$$

The remaining vectors (such as vector $\mathbf{u}^m$) can be expressed as a linear combination of the base "vectors"

$$\mathbf{u}^m - \alpha_1 \, \mathbf{r}^{ij} - \alpha_2 \, \mathbf{r}^{ik} - \alpha_3 \, \mathbf{u}^n = 0 \qquad (2.47)$$

where α_1, α_2, and α_3 are the coefficients of the linear combination.

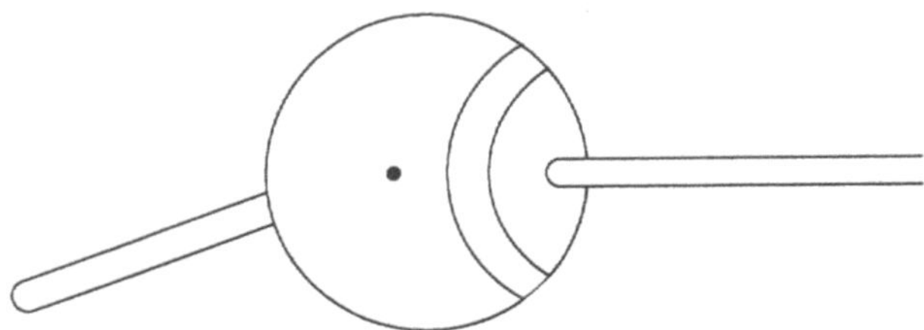

Figure 2.34. Spherical joint.

2.2.3.2 Joint Constraints

Once all the constraint equations which guarantee that each element moves as a rigid body have been entered, it is necessary to formulate the constraints that also guarantee that the bodies have relative motions in accordance with the kinematic joints that link them. It will be shown how, in the case of certain joints, it is not necessary to introduce any additional equations. In other cases this will have to be done. The spherical (S), revolute (R), cylindrical (C), prismatic (P), and the universal (U) joints will be considered below. Other types of joints will be discussed later on in other sections of the book.

Spherical joint (S). The spherical joint (Figure 2.34) is one of the joints that does not generate any constraint equation. The kinematic constraints corresponding to the spherical joint are automatically entered when two adjacent bodies share a basic point as in the case of planar systems with revolute joints. In fact, when two bodies share a point, the only possibility for relative motion is a rotation around this point. This rotation could be any one at all, just as it should be with the spherical joint.

The constraint equations for the spherical joint can also be defined when the basic point is not shared, such as when the joint is going to be broken in a specific moment of the simulation. It will suffice to match the coordinates of points i and j belonging to different bodies

$$x_i - x_j = 0 \tag{2.48}$$

$$y_i - y_j = 0 \tag{2.49}$$

$$z_i - z_j = 0 \tag{2.50}$$

Revolute joint (R). The revolute joint is considered automatically (with no need for constraint equations) when two adjacent elements share a basic point and a unit vector. Then the only possibility of relative motion is the rotation around this unit vector (Figure 2.35).

Another possible way of automatically introducing the revolute joint is by making two adjacent elements share two basic points (Figure 2.36). In this case

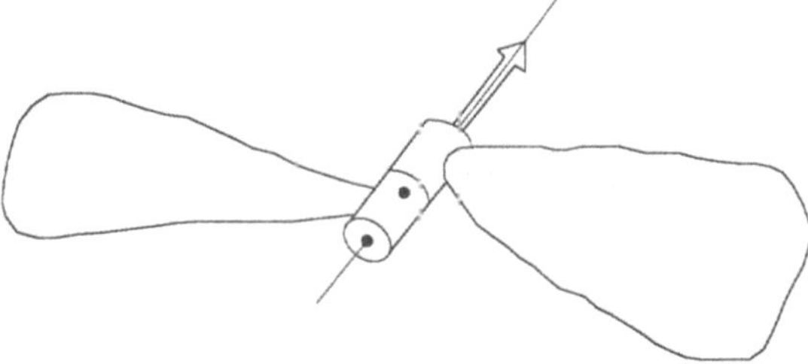

Figure 2.35. Revolute joint.

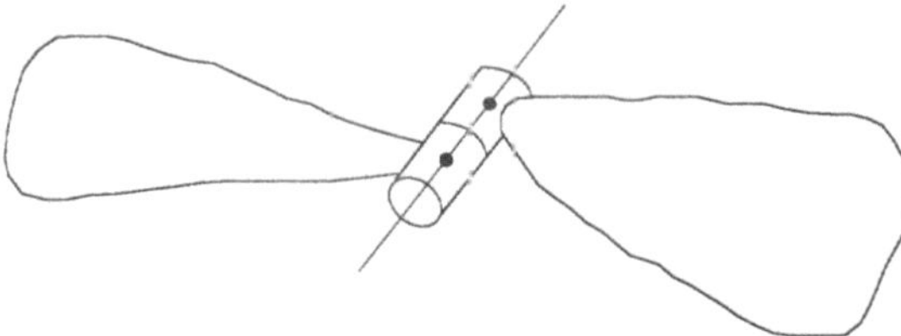

Figure 2.36. Revolute joint defined with two points.

the only possibility of relative motion is the rotation around the axis that goes through those two basic points.

A revolute joint can also be entered in the formulation (without sharing any variable) by matching the coordinates of two points and the components of two unit vectors, each one belonging to a different element.

Cylindrical joint (C). A cylindrical joint restricts four degrees of freedom, and should generate four constraint equations. In the cylindrical joint of Figure 2.37, two elements share a unit vector in the direction of the joint axis. This is equivalent to two constraint equations, the two independent equations that would originate from the cross product of two parallel vectors, each one belonging to a different element. The other two constraint equations originate from the condition that two basic points on the joint's axis, each one belonging to a different element, are aligned with the unit vector. Mathematically, this condition is expressed by the following cross product:

$$r^{ij} \wedge u = 0 \tag{2.51}$$

where only two of the three algebraic equations of (2.51) are independent.

Another way of introducing a cylindrical joint is by making four points (two in each element) permanently aligned on the joint's axis (Figure 2.38). In this case, the constraint equations originate from the following cross products of vectors:

$$r^{ij} \wedge r^{ik} = 0 \tag{2.52}$$

$$r^{kl} \wedge r^{kj} = 0 \tag{2.53}$$

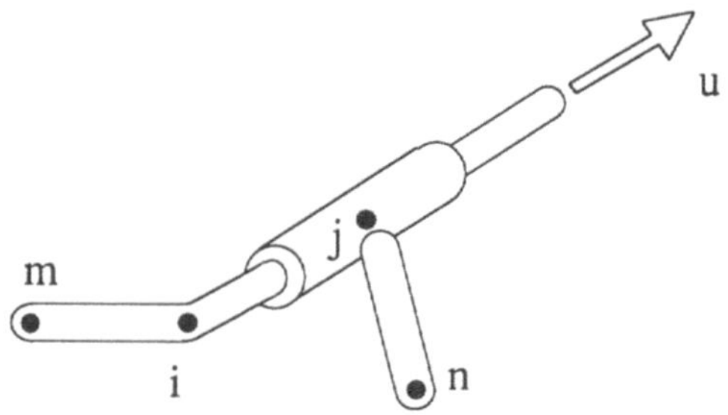

Figure 2.37. Cylindrical joint.

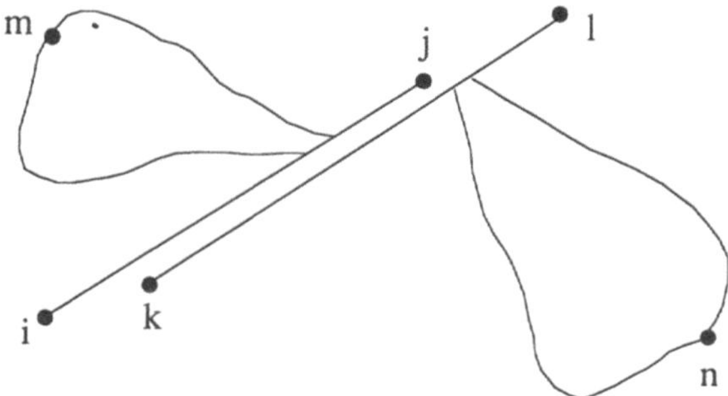

Figure 2.38. Alternative way to model a cylindrical joint.

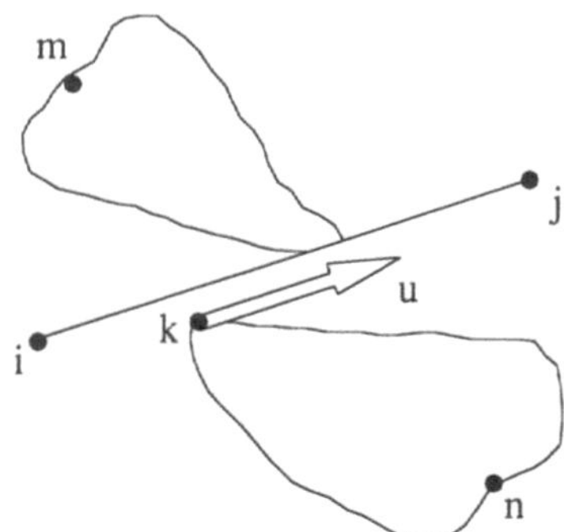

Figure 2.39. A third way to model a cylindrical joint.

The first one imposes the condition that the points i, j and k are aligned; and the second imposes the condition that points k, j, and l are aligned.

Figure 2.39 shows a third way of modeling a cylindrical joint. The four constraint equations are the result of imposing the conditions that point k and vector **u** remain aligned with points i and j. Mathematically, these conditions are expressed as follows:

$$\mathbf{r}^{ik} \wedge \mathbf{r}^{ij} = 0 \tag{2.54}$$

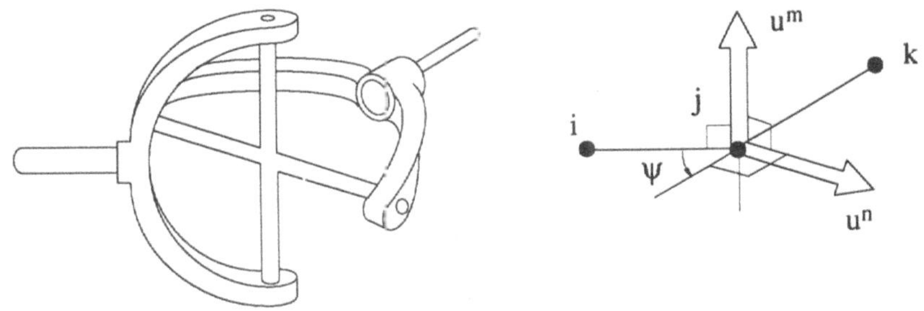

Figure 2.40. Universal joint.

$$\mathbf{r}^{ij} \wedge \mathbf{u} = 0 \tag{2.55}$$

Each one of equations (2.52)-(2.55) gives rise to two independent algebraic equations.

Prismatic joint (P). The prismatic joint P allows only one degree of freedom; and generates five constraint equations. These equations are the same ones that the cylindrical joint generates. In fact, all the degrees of freedom restricted by the cylindrical joint are also restricted by the prismatic joint. In addition one equation prevents relative rotation between the elements with respect to the joint axis.

The three configurations of Figures 2.37, 2.38, and 2.39 are all possible for the prismatic joint. The fifth equation (characteristic of the prismatic joint) can be obtained by means of a scalar product between two vectors (one of each element), which comply with the conditions of not being parallel to the joint's axis and not forming an angle close to 0° between them. If the angle is close to 0°, the scalar product should be substituted by the linear combination condition. For the joints in Figures 2.37, 2.38, and 2.39, the additional equations are respectively:

$$\mathbf{r}^{im} \cdot \mathbf{r}^{jn} - \alpha_1 = 0 \tag{2.56}$$

$$\mathbf{r}^{im} \cdot \mathbf{r}^{kn} - \alpha_2 = 0 \tag{2.57}$$

$$\mathbf{r}^{im} \cdot \mathbf{r}^{kn} - \alpha_3 = 0 \tag{2.58}$$

where m and n are appropriate points represented in Figure 2.38; and α_1, α_2, and α_3 are scalar constants.

Universal Joint (U). Figure 2.40 shows a drawing of a universal (Cardan) joint together with its modeling using natural coordinates. The universal joint restricts four degrees of freedom. If the angle is fixed between the two axes, then the universal joint will only allow for one degree of freedom (it restricts five).

In the model of the universal joint of Figure 2.40, vector $\mathbf{u}^m$ belongs to the same element as segment (i-j) and is orthogonal to it. Similarly, the unit vector $\mathbf{u}^n$ belongs to the segment (j-k) and is orthogonal to it. Therefore, point j is

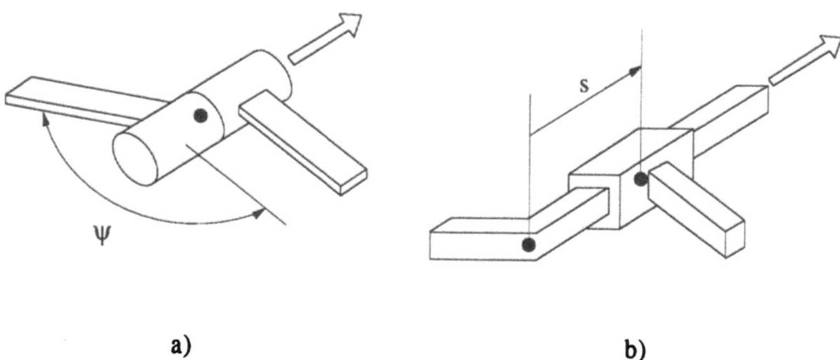

Figure 2.41. Introducing mixed coordinates in revolute and prismatic joints.

shared by both elements. This condition is equivalent to three constraint conditions even though no equation needs be formulated. Unit vectors $\mathbf{u}^m$ and $\mathbf{u}^n$ should be kept perpendicular to each other. This is the only equation that generates the universal joint to restrict the four degrees of freedom:

$$\mathbf{u}^m \cdot \mathbf{u}^n = 0 \tag{2.59}$$

If the angle formed by the two axes is to remain constant, the following equation must be considered:

$$\mathbf{r}^{ij} \cdot \mathbf{r}^{jk} - L_{ij}\, L_{jk}\, cos\, \psi = 0 \tag{2.60}$$

This equation should be substituted by a component of the cross product of vectors if the angle ψ is very small.

2.2.4 Mixed Coordinates

As in the planar case, the Cartesian coordinates of points and of unit vectors, which make up the set of natural coordinates, can also be supplemented with angles, distances, or any other type of variables related to the degrees of freedom that describe the relative motion of the kinematic joints. It is easier to simulate in this way the driving of a multibody system by means of motors or actuators located at the joints. There will be as many new constraint equations as there are new coordinates. Figures 2.41a and 2.41b show an R joint and a P joint respectively with their corresponding angle and distance defined as mixed or relative coordinates.

Consequently, mixed coordinates can be very useful for introducing variables as dependent coordinates which are directly related to the degrees of freedom permitted by the joints at which the actuators are connected. Only the prismatic and revolute joints will be considered here.

Prismatic joint. The distance s, between two basic points which belong to different elements in the joint's axis, becomes the new coordinate to be

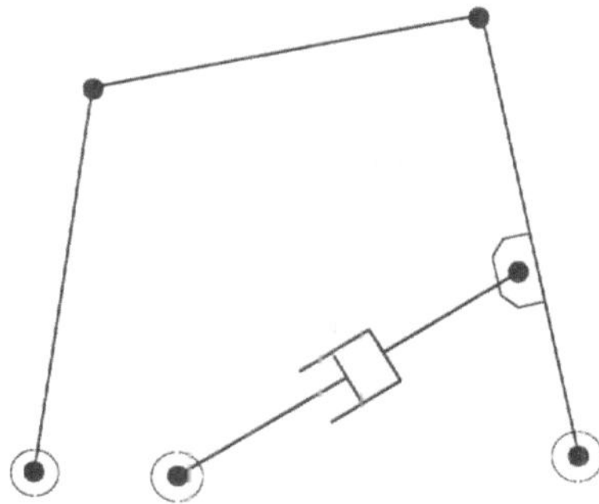

Figure 2.42. Mechanism with hydraulic actuators.

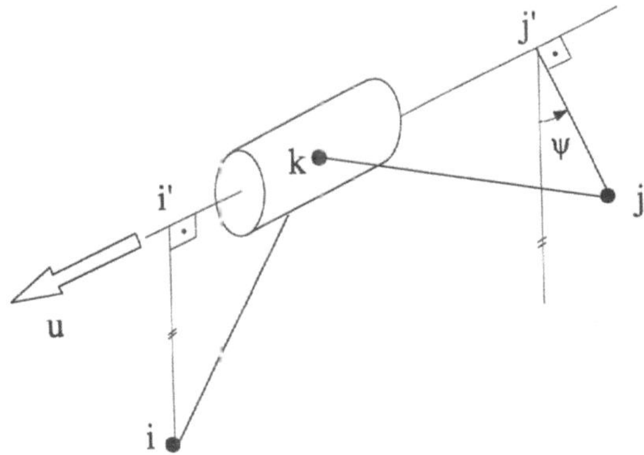

Figure 2.43. Detail of the angle description in a revolute joint.

introduced. Any modeling carried out for the prismatic joint (Figures 2.37, 2.38, and 2.39) contains the two basic points required to define the distance s. The additional constraint equation, assuming that i and j are located in the axis, is:

$$(x_i - x_j)^2 + (y_i - y_j)^2 + (z_i - z_j)^2 - s^2 = 0 \qquad (2.61)$$

This equation can also be used to define the conditions imposed by hydraulic actuators (or by linear motors), as may be seen in the mechanism of Figure 2.42. Here the distance s is related to the volume of fluid contained in the actuator, and its derivative depends on the corresponding flow. The variable s may or may not be known, depending on the conditions of the problem.

Revolute joint. Mixed coordinates in three dimensional revolute joints are more complicated than those of prismatic joints. Consider the revolute joint, which is shown in Figure 2.43 and defined by sharing a point k and a unit vector $\mathbf{u}$.

A problem with the angle definition in this joint occurs when points i, j, and k are not located on a plane perpendicular to the unit vector and to the joint axis, because then the angle between segments (i-k) and (k-j) does not exactly repre-

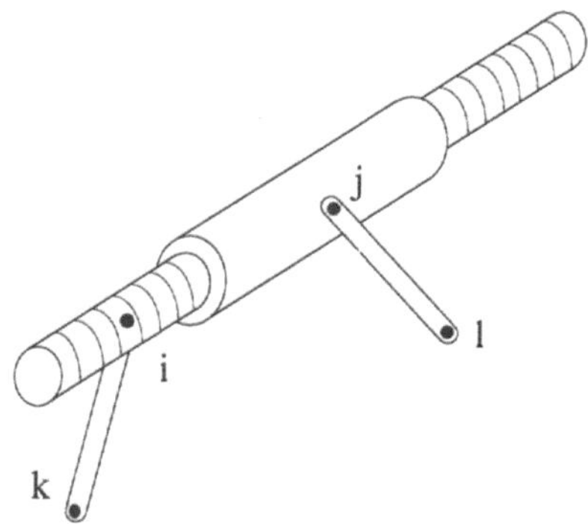

Figure 2.44. Helical joint defined with mixed coordinates.

sent the angle rotated by the joint. However, if i' and j' are the projections of points i and j on the joint axis, then the angle formed by the two bodies becomes equal to the angle ψ formed by segments $(i\text{-}i')$ and $(j\text{-}j')$.

Segment $(i'\text{-}i)$ is equal to segment $(k\text{-}i)$ minus the projection of $(k\text{-}i)$ on the joint axis. Segment $(j'\text{-}j)$ can be determined in a similar manner. When the angle ψ is not close to $0°$ or to $180°$, the corresponding additional constraint equation coming from the scalar product becomes:

$$(\mathbf{r}^{ki} - (\mathbf{r}^{ki}\cdot\mathbf{u})\,\mathbf{u}) \cdot (\mathbf{r}^{kj} - (\mathbf{r}^{kj}\cdot\mathbf{u})\,\mathbf{u}) - L_{i'i}\,L_{j'j}\,\cos\psi = 0 \qquad (2.62)$$

and by compacting this equation, one obtains

$$\mathbf{r}^{ki}\cdot\mathbf{r}^{kj} - (\mathbf{r}^{ki}\cdot\mathbf{u})\,(\mathbf{r}^{kj}\cdot\mathbf{u}) - L_{i'i}\,L_{j'j}\,\cos\psi = 0 \qquad (2.63)$$

Only the first term of this equation depends on the coordinates of the basic points. The second term is a constant that only needs to be calculated once. The third term is the one that causes the angle ψ to intervene. The two distances, $L_{i'i}$ and $L_{j'j}$, are also constant.

When the angle ψ is small or close to $180°$, the scalar product of equation (2.62) should be substituted by the cross product of vectors as follows:

$$(\mathbf{r}^{ki} - (\mathbf{r}^{ki}\cdot\mathbf{u})\,\mathbf{u}) \wedge (\mathbf{r}^{kj} - (\mathbf{r}^{kj}\cdot\mathbf{u})\,\mathbf{u}) - \mathbf{u}\,L_{i'i}\,L_{j'j}\,\sin\psi = 0 \qquad (2.64)$$

after expanding this equation, one arrives at

$$\mathbf{r}^{ki} \wedge \mathbf{r}^{kj} - (\mathbf{r}^{ki}\cdot\mathbf{u})\,\mathbf{u} \wedge \mathbf{r}^{kj} - (\mathbf{r}^{kj}\cdot\mathbf{u})\,\mathbf{r}^{ki} \wedge \mathbf{u} - \mathbf{u}\,L_{i'i}\,L_{j'j}\,\sin\psi = 0 \qquad (2.65)$$

which is also a quadratic equation in the natural coordinates. The coefficients of the second and third terms (scalar products between brackets) are constants. Out of the three algebraic equations corresponding to the vector equation (2.65), it is only necessary to consider the equation corresponding to the largest component of vector $\mathbf{u}$.

Mixed coordinates will be used next to formulate the constraint equations corresponding to other types of joints, such as gear or helical joints.

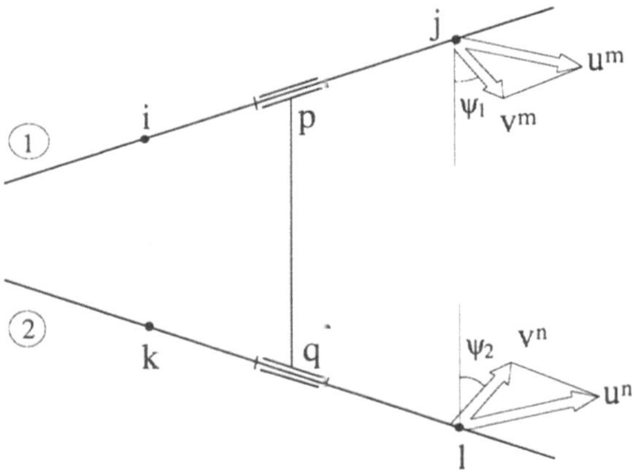

Figure 2.45. Gear joint defined with mixed coordinates.

Helical joint. In Figure 2.44 a helical joint is represented. Basically, a helical joint can be seen as a cylindrical joint in which the translational and rotational degrees of freedom are not independent but related by the linear equation:

$$s = s_0 + n\,\psi \tag{2.66}$$

where s_0 is a constant, giving the value of s when $\psi = 0$. In equation (2.66) the translational and rotational mixed coordinates s and ψ previously introduced have been used. Equation (2.66), together with the constraints of a cylindrical joint described in Section 2.2.3.2, are the constraint equations of the helical joint.

Gear joint. Figure 2.45 shows a possible modeling with mixed coordinates of a three-dimensional gear joint between axes 1 and 2 which cross but do not intersect in space. The axis 1 is defined by points i and j, and the axis 2 is defined by points k and l. The two unit vectors u^n and u^m are fixed to axes 1 and 2, respectively, and are used to complete the kinematic definition of the bodies related by the gear joint.

Two points p and q located on axes 1 and 2, respectively, are used to define the line that is the common normal to both axes. In a gear joint, the angles rotated by the axes are not independent but are related by a constant n defined by the quotient between the number of gear teeth. If ψ_1 and ψ_2 are the angles that measure both rotations and ψ_{20} is a constant initial value, the linear relation between angles ψ_1 and ψ_2 can be written as

$$\psi_2 = \psi_{20} - n\,\psi_1 \tag{2.67}$$

Angles ψ_1 and ψ_2 must be measured between the common normal line (p-q) and two straight lines that are normal to the axes 1 and 2 and that rotate with each of the gears. If vector u^m is normal to axis 1 and vector u^n is normal to axis 2, then angles ψ_1 and ψ_2 can be measured between these vectors and the normal line (p–q). If this is not the case, then two non-unit vectors v^m and v^n,

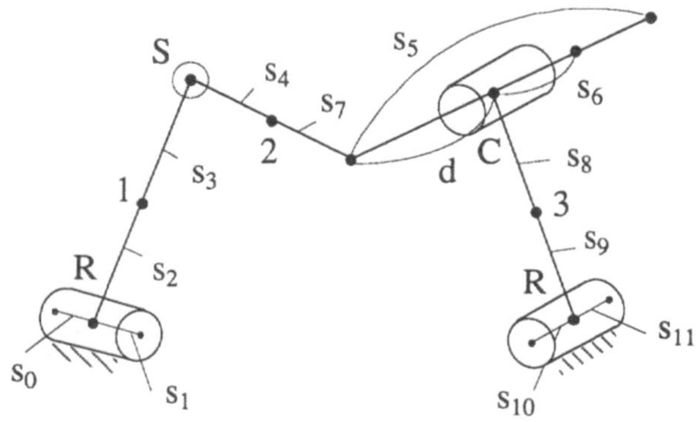

Figure 2.46. *RSCR* mechanism with reference point coordinates.

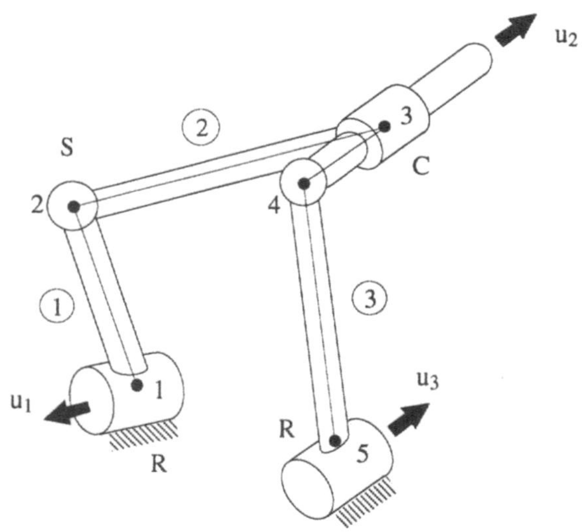

Figure 2.47. *RSCR* mechanism with natural coordinates.

belonging to bodies 1 and 2, respectively, can be obtained from $\mathbf{u}^m$ and $\mathbf{u}^n$ as follows:

$$\mathbf{v}^n = \mathbf{u}^n - \frac{\mathbf{u}^n \cdot \mathbf{r}^{ij}}{L_{ij}^2} \mathbf{r}^{ij} \tag{2.68}$$

$$\mathbf{v}^m = \mathbf{u}^m - \frac{\mathbf{u}^m \cdot \mathbf{r}^{kl}}{L_{kl}^2} \mathbf{r}^{kl} \tag{2.69}$$

The position of points p and q can be established from the positions of points $i, j, k,$ and l, in terms of two known constant coefficients α and β, as

$$r_p = r_i + \alpha \, (r_j - r_i) \tag{2.70}$$

$$r_q = r_k + \beta \, (r_l - r_k) \tag{2.71}$$

The constraint equations of the gear joint can now be written. This joint restricts five degrees of freedom and it is necessary to formulate five constraint equations. One of the constraints is given by equation (2.67), while the remaining four equations must force the axes to maintain their relative spatial position, i.e., to maintain their angle and their relative distance:

$$r^{pq} \cdot r^{pq} - C_1 = 0 \tag{2.72}$$

$$r^{pq} \cdot r^{ji} - C_2 = 0 \tag{2.73}$$

$$r^{pq} \cdot r^{lk} - C_3 = 0 \tag{2.74}$$

$$r^{ji} \cdot r^{lk} - C_4 = 0 \tag{2.75}$$

where C_1 through C_4 are constants. By substituting equations (2.70)-(2.71) into (2.72)-(2.75), four quadratic equations in the natural coordinates are obtained.

2.3 Comparison Between Reference Point and Natural Coordinates

In this section we present a comparative example between the reference point coordinates and the natural coordinates. In the previous sections we have shown how the natural coordinates represent a suitable choice of dependent coordinates and how to write the constraint equations. However, since the reference point coordinates are used in most of the multibody formulations, it is believed that a fully developed example can give a better idea to the reader of how both formulations compare.

Consider again an *RSCR* linkage similar to the one presented in Figure 2.25, this time modeled using reference point coordinates with Euler parameters. The mechanism is composed of four links and has a single degree of freedom. Since seven variables (three coordinates of the center of gravity and four Euler parameters) are used for each movable link, a total of 21 dependent variables are necessary. The constant geometrical data, composed of vectors s_i that appears in the constraint equations, have also been represented in Figure 2.46.

Following the notation of Nikravesh (1988), it can be shown that the *Jacobian matrix* (matrix of partial derivatives of the constraint equations) has the form shown in Table 2.1.

The Jacobian matrix is of size (20x21) with 128 non-zero elements. Matrices G_i, $\widetilde{S}_i$ and $\overline{S}_i'$ are

$$G_i = \begin{bmatrix} -e_1 & e_0 & -e_3 & e_2 \\ -e_2 & e_3 & e_0 & -e_1 \\ -e_3 & -e_2 & e_1 & e_0 \end{bmatrix} \tag{2.76}$$

Table 2.1. Jacobian matrix of an *RSCR* mechanism with reference point coordinates and Euler parameters.

		Φ_{r_1}	Φ_{p_1}	Φ_{r_2}	Φ_{p_2}	Φ_{r_3}	Φ_{p_3}
R	3	I	$2\,G_1\,\overline{S}_2'$	0	0	0	0
	3	0	$2\,\widetilde{S}_0\,G_1\,\overline{S}_1'$	0	0	0	0
S	3	I	$2\,G_1\,\overline{S}_3'$	$-I$	$-2\,G_2\,\overline{S}_4'$	0	0
C	2	0	0	0	$-2\,\widetilde{S}_6\,G_2\,\overline{S}_5'$	0	$2\,\widetilde{S}_5\,G_3\,\overline{S}_6'$
	2	0	0	$\widetilde{S}_5$	$-(2\,\widetilde{S}_5\,G_2\,\overline{S}_7' + 2\,\widetilde{d}\,G_2\,\overline{S}_5')$	$\widetilde{S}_5$	$2\,\widetilde{S}_5\,G_3\,\overline{S}_8'$
R	3	0	0	0	0	I	$2\,G_3\,\overline{S}_9'$
	2	0	0	0	0	0	$-2\,\widetilde{S}_{10}\,G_3\,\overline{S}_{11}'$

$$\widetilde{S}_i = \begin{bmatrix} 0 & -s_{iz} & s_{iy} \\ s_{iz} & 0 & -s_{ix} \\ -s_{iy} & s_{ix} & 0 \end{bmatrix} \tag{2.77}$$

$$\overline{S}_i' = \begin{bmatrix} 0 & -s'_{ix} & -s'_{iy} & -s'_{iz} \\ s'_{ix} & 0 & s'_{iz} & -s'_{iy} \\ s'_{iy} & -s'_{iz} & 0 & s'_{ix} \\ s'_{iz} & s'_{iy} & -s'_{ix} & 0 \end{bmatrix} \tag{2.78}$$

All the primed symbols are referred to the moving frame of the link to which it belongs. Therefore, all the vectors s'_i are constant, and their components in the absolute reference frame can be obtained through the rotation matrix as

$$s_i = A_i\, s_i' \tag{2.79}$$

Matrix A_i can be written in terms of the four Euler parameters as

$$A_i = (2\,e_{i0}^2 - 1)\,I_3 + 2\,(e_i\,e_i^T + e_{i0}\,\widetilde{e}_i) \tag{2.80}$$

with $e_i^T = \{e_{i1}\ e_{i2}\ e_{i3}\}$.

The number of floating point arithmetic operations (products, additions, and substractions) required to compute the Jacobian matrix of Table 2.1 is 988.

Let us now consider the same mechanism modeled with natural coordinates, as represented in Figure 2.47. In this case, 12 dependent variables are used which are the coordinates of points 2, 3, and 4, and the components of vector u_2. The Jacobian matrix, of size (12×13), is presented in Table 2.2.

Table 2.2. Jacobian matrix of an *RSCR* mechanism with natural coordinates.

	r_2	r_3	u_2	r_4
$r_{12} \cdot r_{12} = c_1$	x_{21} y_{21} z_{21}	0 0 0	0 0 0	0 0 0
$r_{12} \cdot u_1 = c_2$	u_{1x} u_{1y} u_{1z}	0 0 0	0 0 0	0 0 0
$r_{23} \cdot r_{23} = c_3$	x_{23} y_{23} z_{23}	x_{32} y_{32} z_{32}	0 0 0	0 0 0
$r_{32} \cdot u_2 = c_4$	$-u_{2x}$ $-u_{2y}$ $-u_{2z}$	u_{2x} u_{2y} u_{2z}	x_{32} y_{32} z_{32}	0 0 0
$u_2 \cdot u_2 = 1$	0 0 0	0 0 0	u_{2x} u_{2y} u_{2z}	0 0 0
$r_{43} \wedge u_2 = 0$	0 0 0 0 0 0 0 0 0	0 $-u_{2x}$ u_{2y} u_{2z} 0 $-u_{2x}$ $-u_{2y}$ u_{2x} 0	0 $-z_{43}$ y_{43} z_{43} 0 $-x_{43}$ $-y_{43}$ x_{43} 0	0 $-u_{2x}$ u_{2y} $-u_{2z}$ 0 u_{2x} u_{2y} $-u_{2x}$ 0
$r_{45} \cdot r_{45} = c_5$	0 0 0	0 0 0	0 0 0	x_{45} y_{45} z_{45}
$r_{45} \cdot u_2 = c_6$	0 0 0	0 0 0	x_{45} y_{45} z_{45}	u_{2x} u_{2y} u_{2z}
$r_{45} \cdot u_3 = c_7$	0 0 0	0 0 0	0 0 0	u_{3x} u_{3y} u_{3z}
$u_3 \cdot u_2 = 1$	0 0 0	0 0 0	u_{3x} u_{3y} u_{3z}	0 0 0

Table 2.3. *RSCR* mechanism: comparative results

RSCR	Euler parameters	Natural coordinates
Size	(20×21)	(11×12)
Non-zero elements	128	51
No. of fl.-point ops.	988	12

This Jacobian has 57 non-zero elements and requires 12 floating point operations to be calculated. All these results are summarized in Table 2.3.

2.4 Concluding Remarks

The *fully Cartesian* or *natural* coordinates described in this chapter have some interesting features that are convenient to summarize at this stage. Some of these features have been previously mentioned and others will be described for the first time:

1. Natural coordinates are composed of purely Cartesian variables and therefore are easy to define and to represent geometrically.
2. The rotation matrix of a rigid body whose motion is described with natural coordinates is a linear function of these coordinates (See Chapter 4). Note that with reference point coordinates the rotation matrix is a quadratic function of Euler parameters and a transcendental function (sine and cosine) of Euler or Bryant angles.

3. Natural coordinates can be defined at the joints and then shared by contiguous bodies, contributing to define the position of both bodies and significantly simplifying the definition of joint constraint equations. At the same time, the total number of variables is kept moderate.

4. With other kinds of coordinates, it is necessary to keep two sets of information: the variables that define the position and orientation of the reference frame attached to the moving body with respect to the inertial or fixed frame, and the local variables that define the body geometry (position and orientation of axis, etc.) with respect to the moving frame. With natural coordinates, a single set of variables define the geometry and the position of the body directly in the global reference frame. It is only necessary to keep some constant values—distances, angles, etc.—that are independent of the reference frame.

5. With natural coordinates the constraint equations that arise from the rigid body and joint conditions are *quadratic* (or linear); so their Jacobian matrix (of partial derivatives) is a *linear* (or constant) function of the natural coordinates.

6. Natural coordinates can be complemented easily with relative angles and distances defined at the joints to yield a mixed set of Cartesian and relative coordinates. Driving an angle or a distance, and defining forces and/or torques in joints become rather straightforward. Relative coordinates also simplify the task of defining the constraint equations for some particular joints, such as the helical and gear joints.

7. In the constraint equations arising from natural coordinates, the design variables—lengths, angles, etc.—appear explicitly, not hidden by coordinate transformations. Thus, parametric and variational design, kinematic synthesis, sensitivity analysis, and optimization may benefit from the use of these coordinates.

References

Argyris, J., "An Excursion into Large Rotations", *Computer Methods in Applied Mechanics and Engineering*, Vol. 32, pp. 85-155, (1982).

García de Jalón, J., Serna, M.A., and Avilés, R., "A Computer Method for Kinematic Analysis of Lower-Pair Mechanisms. Part I: Velocities and Accelerations and Part II: Position Problems", *Mechanism and Machine Theory*, Vol. 16, pp. 543-566, (1981).

García de Jalón, J., Unda, J., and Avello, A., "Natural Coordinates for the Computer Analysis of Multibody Systems", *Computer Methods in Applied Mechanics and Engineering*, Vol. 56, pp. 309-327, (1986).

García de Jalón, J., Unda, J., Avello, A., and Jiménez, J.M., "Dynamic Analysis of Three-Dimensional Mechanisms in Natural Coordinates", *ASME J. of Mechanisms, Transmissions and Automation in Design*, Vol. 109, pp. 460-465, (1987).

Hartenberg, R.S. and Denavit, I., *Kinematic Synthesis of Linkages*, McGraw-Hill, New York, (1964).

Haug, E.J., *Computer-Aided Kinematics and Dynamics of Mechanical Systems, Volume I: Basic Methods*, Allyn and Bacon, (1989).

Huston, R.L., Passerello, C.E., and Harlow, M.W., "Dynamics of Multirigid-Body Systems", *ASME Journal of Applied Mechanics*, Vol. 45, No. 4, pp. 889-894, (1978).

Jerkovsky, W., "The Structure of Multibody Dynamic Equations", *Journal of Guidance and Control*, Vol. 1, pp. 173-182, (1978).

Kim, S.S. and Vanderploeg, M.J., "QR Decomposition for State Space Representation of Constrained Mechanical Dynamic Systems", *ASME Journal on Mechanisms, Transmissions, and Automation in Design*, Vol. 108, pp. 176-182, (1986).

Nikravesh, P.E., Wehage, R.A., and Kwon, O.K., "Euler Parameters in Computational Dynamics and Kinematics. Parts I and II", *ASME Journal on Mechanisms, Transmissions, and Automation in Design*, Vol. 107, No. 3, pp. 358-369, (1985).

Nikravesh, P.E., *Computer-Aided Analysis of Mechanical Systems*, Prentice-Hall, (1988).

Orlandea, N., Chace, M.A., and Calahan, D.A., "A Sparsity Oriented Approach to the Dynamic Analysis and Design of Mechanical Systems, Parts I and II", *Journal of Engineering for Industry*, Vol. 99, pp. 773-784, (1977).

Paul, B. and Krajcinovic, K., "Computer Analysis of Machines With Planar Motion. Part 1 - Kinematics; Part 2 - Dynamics", *ASME Journal of Applied Mechanics*, Vol. 37, pp. 697-712, (1970).

Serna, M.A., Avilés, R., and García de Jalón, J., "Dynamic Analysis of Plane Mechanisms with Lower-Pairs in Basic Coordinates", *Mechanism and Machine Theory*, Vol. 17, pp. 397-403, (1982).

Shabana, A.A., *Dynamics of Multibody Systems*, John Wiley & Sons, New-York, (1989).

Sheth, P.N. and Uicker, J.J., "IMP (Integrated Mechanism Program): A Computer-Aided Design Analysis System for Mechanisms and Linkages", *ASME Journal of Engineering for Industry*, Vol. 94, pp. 454-464, (1972).

Smith, D.A, Chace, M.A., and Rubens, A.C., "The Automatic Generation of a Mathematical Model for Machinery Systems", *ASME Journal of Engineering for Industry*, Vol. 95, pp. 629-635, (1973).

Unda, J., García de Jalón, J., Losantos, F., and Enparantza, R., "A Comparative Study on Some Different Formulations of the Dynamic Equations of Constrained Mechanical Systems", *ASME Journal of Mechanisms, Transmissions, and Automation in Design*, Vol. 109, pp. 466-474, (1987).

Van der Werff, K., "A Finite Element Approach to Kinematics and Dynamics of Mechanisms", *5th World Congress on the Theory of Machines and Mechanisms*, Montreal (Canada), (1979).

Van der Werff, K. and Jonker, J.B., "Dynamics of Flexible Mechanisms", *Computer Aided Analysis and Optimization of Mechanical System Dynamics*, ed. by E.J. Haug, Springer-Verlag, pp. 381-400, (1984).

Wittenburg, J., *Dynamics of Systems of Rigid Bodies*, Teubner, Stuttgart, (1977).

Wittenburg, J. and Wolz, U., "MESA VERDE: A Symbolic Program for Nonlinear Articulated-Rigid-Body Dynamics", *ASME Conference on Mechanical Vibration and Noise*, Cincinnati (USA), (1985).

Problems

2/1   Write the constraint equations of the mechanism shown in the figure when modeled with: a) Relative coordinates, b) Reference point coordinates, c) Natural coordinates, d) Mixed coordinates, with relative coordinates in all the joints.

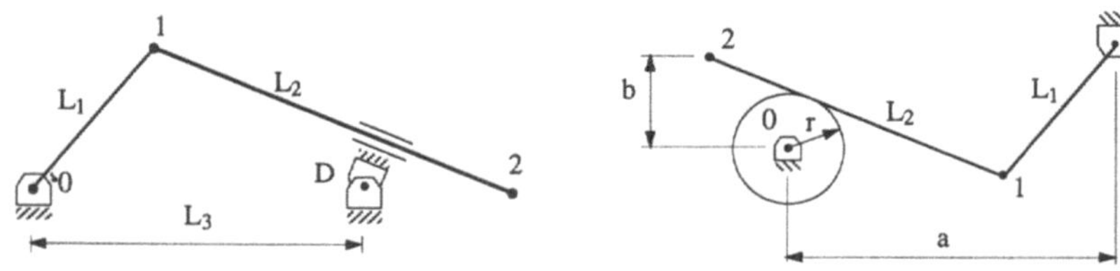

Figure P2/1.   Figure P2/2.

2/2   Assuming that there is rolling with no slipping between the disk and the rod, select an appropriate set of mixed coordinates (natural and relative) and write the constraint conditions.

2/3   The wheel on the figure rolls without slipping. Use mixed coordinates (natural and relative) and find the constraint equations. It is suggested that the contact between the wheel and ground be modeled by means of a rack and pinion type of kinematic joint (a particular case of the gear joint).

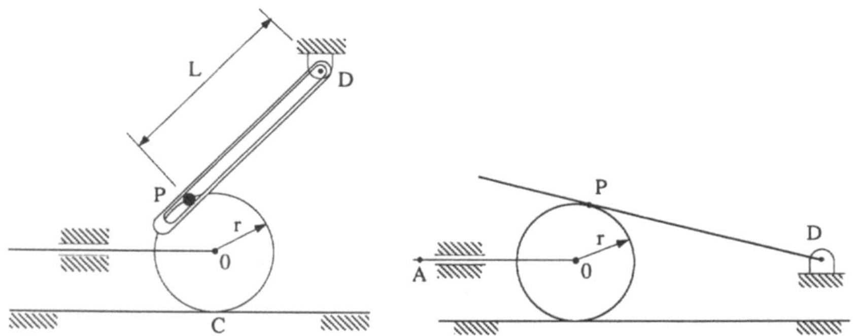

Figure P2/3.   Figure P2/4.

2/4   Write the constraint equations of the mechanism shown, knowing that the wheel rolls on the ground with no slipping and slides on the rod DP.

2/5   Given the mechanism shown in the figure, find the constraint equations that relate the input distance d with the output angle q and distance s.

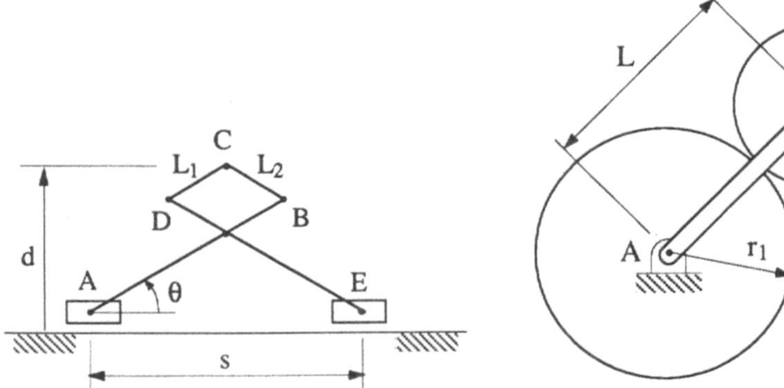

Figure P2/5.   Figure P2/6.

2/6   The centers of the two gears shown in the figure are connected by means of a rod with point A being fixed. Considering mixed coordinates, find the constraint equations that relate the angular positions (relative or absolute) of the three elements.

2/7   Consider the mechanism in the figure and find the constraint equations that relate the angles φ_1 and φ_2 with the parameter s that measures the relative position between elements 3 and 4.

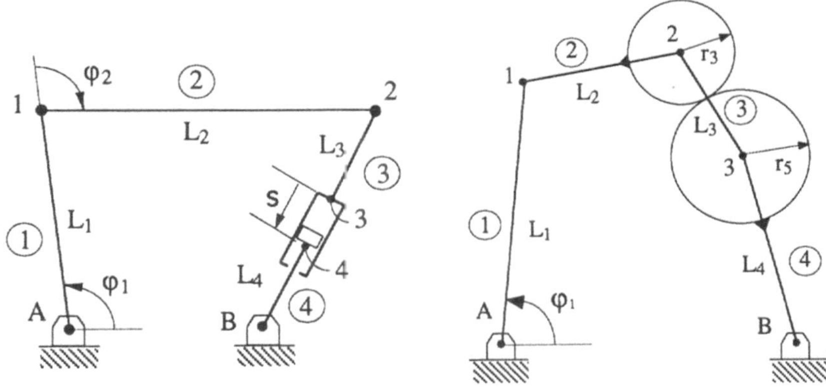

Figure P2/7.   Figure P2/8.

2/8   Consider the mechanism shown to be modeled with natural coordinates. Rods 2 and 4 are attached to the gears with radius r_3 and r_5 whose centers are connected by means of rod 3. Write the constraint equations that relate the position of the complete system in terms of the input angle φ_1.

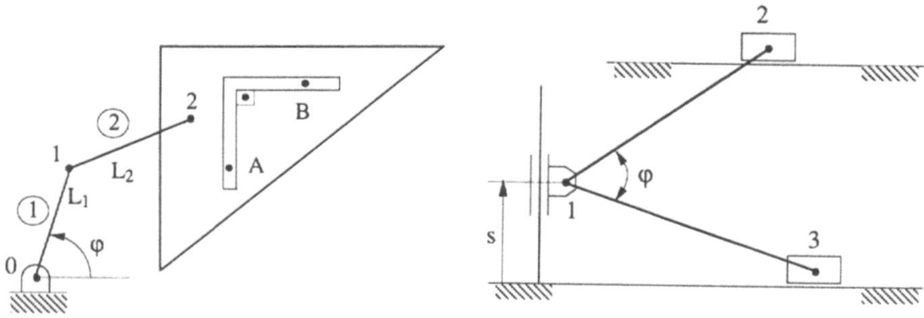

Figure P2/9.   Figure P2/10.

2/9   Two perpendicular slots have been cut in a plate that is placed so that the slots fit two fixed points A and B. Determine the constraint equations that allow one to obtain the position and orientation of the plate as a function of the input angle φ.

2/10   Determine in the mechanism shown the constraint equations that relate the distance s with the angle φ.

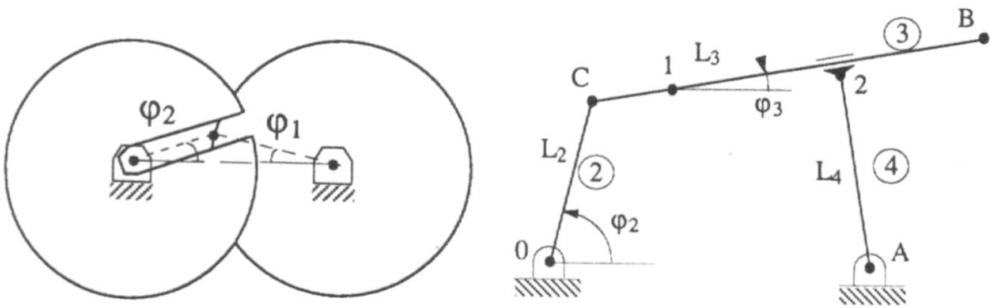

Figure P2/11.   Figure P2/12.

2/11   Consider the Geneva wheel of the figure and, using natural coordinates, find the equations that relate the input angle φ_1 with the output angle φ_2.

2/12   Element 2 of the mechanism in the figure is represented by the relative coordinate φ_2, element 3 by the reference point coordinates (x_1, y_1, ψ_3), and element 4 by the natural coordinates (x_2, y_2) and the fixed point A. Determine the five constraint equations that relate these six coordinates.

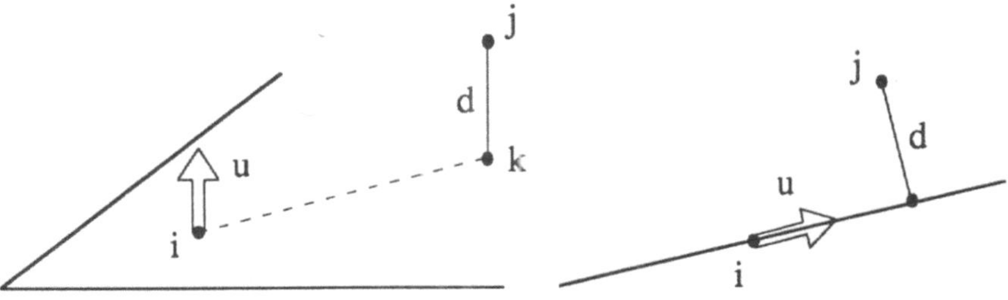

Figure P2/13.   Figure P2/14.

2/13 A plane may be defined by a point i and a unit perpendicular vector **u** as seen in the figure. Determine the constraint equations corresponding to the motion of a point j that moves parallel to the plane.

2/14 A straight line can be defined by a point i and a unit vector **u** as seen in the figure. Determine the constraint equations for a point j that moves at a constant distance d from the line. Discuss the case when $d=0$. How can you enforce that j moves on along the straight line?

2/15 Determine the constraint equations for a solid defined by two points and a unit vector that moves parallel to a plane, defined by a point i and a unit vector **u**, as seen in the figure.

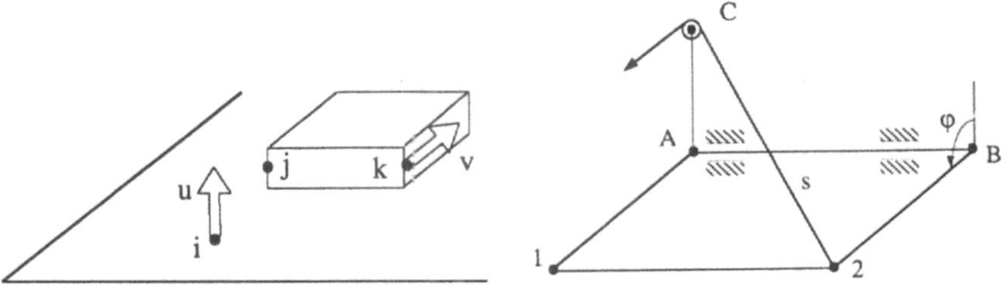

Figure P2/15.   Figure P2/16.

2/16 The figure shows the frame $A12B$ that can rotate about the fixed axis AB by the action of the string attached to point 2 that goes through a pulley located at C. Find the constraint equations that relate the angle φ with the length s of the cable between points 2 and C.

2/17 The ends of a slender rod of length $\sqrt{2}$ move on the sides of a cube with sides of unit length. Find the constraint equations that relate the distances s_1 and s_2.

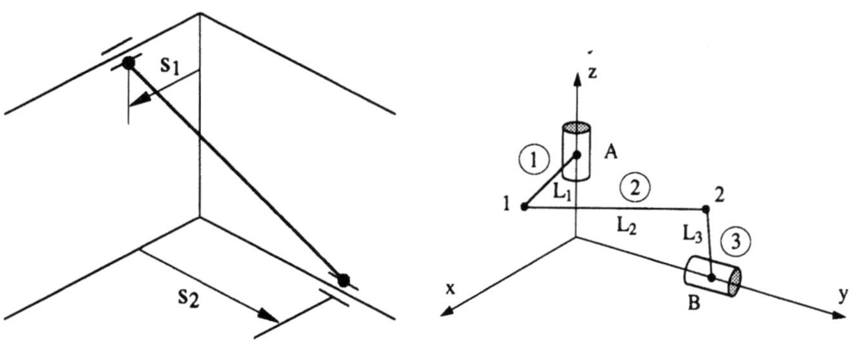

Figure P2/17.   Figure P2/18.

2/18   Use natural coordinates to model the given mechanism and find the constraint equations of the *RSSR* mechanism shown in the figure.

2/19   The mechanism shown has a composite kinematic joint R-C (Revolute-Cylindrical). Find the constraint equations that relate the angle φ with the distance s.

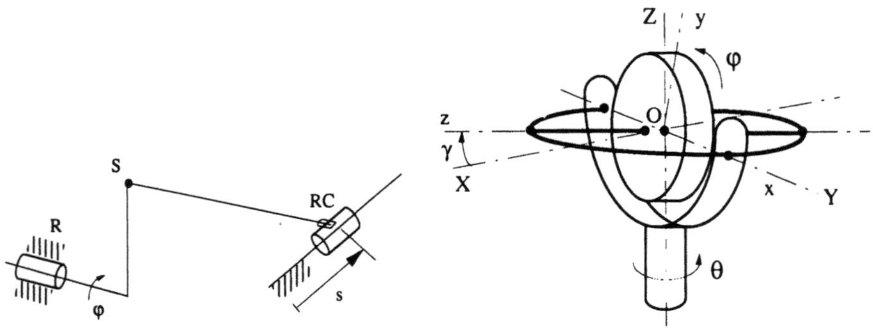

Figure P2/19.   Figure P2/20.

2/20   Find the constraint equations of the gyroscope shown, considering the angles of relative motion.

2/21   A six degree of freedom spatial manipulator is depicted in the Figure 2.27. Using natural coordinates and knowing that the axes defined by unit vector $\mathbf{u}_1$ are parallel, find the corresponding constraint equations.

3
Kinematic Analysis

The kinematic constraint equations corresponding to the natural coordinates were explained in detail in Chapter 2, both for planar and three-dimensional multibody systems. They were then compared to other types of coordinates. Attention was also given to the main sources of constraint equations with natural coordinates: rigid body constraints, joint constraints, and the optional definition of relative or joint coordinates.

In this chapter we will make use of those constraint equations to solve what is usually called kinematic problems, namely, the initial position or assembly problem, the finite displacement problem, and the velocity and acceleration analysis. The first two problems require an iterative solution of a system of nonlinear equations. Some special techniques to improve the convergence will be explained. Special attention will be addressed to the important case of *over constrained* multibody systems or, in general, to systems with non-independent constraint equations. The last section of this chapter is devoted to the case of *non-holonomic* joints.

3.1 Initial Position Problem

The initial position problem was defined in Section 1.3. It basically consists of determining the position of all the bodies in the system by knowing the positions of the fixed and the *input* bodies which can also be called *guided* or *driven* elements. Mathematically, the initial position problem is reduced to determining from the known coordinates corresponding to the input elements the vector of dependent coordinates that satisfies the nonlinear system of *constraint equations*. Note that the input can also be specified as the externally guided or driven linear or angular coordinates corresponding to several joints (as many joints as there are degrees of freedom) on which mixed coordinates have been defined. This basic notion is explained by means of the following examples:

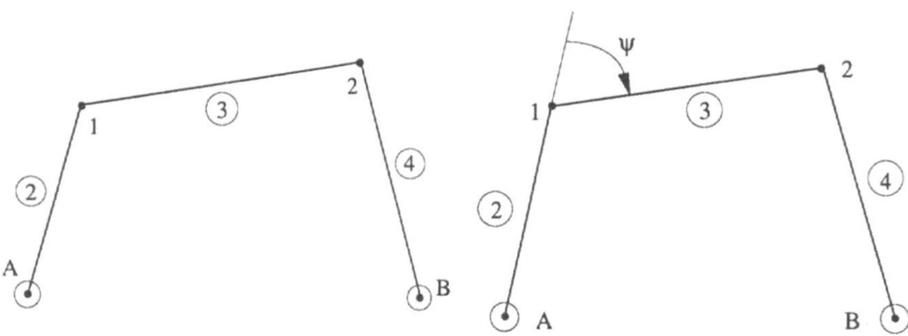

Figure 3.1. Four-bar mechanism modeled with natural coordinates.

Figure 3.2. Four-bar mechanism modeled with mixed coordinates.

Example 3.1

As a first example, the four-bar mechanism of Figure 3.1 will be considered. This system has four natural coordinates (x_1, y_1, x_2, y_2).

The constraint equations corresponding to this mechanism are following constant distance conditions:

$$(x_1 - x_A)^2 + (y_1 - y_A)^2 - L_2^2 = 0$$
$$(x_2 - x_1)^2 + (y_2 - y_1)^2 - L_3^2 = 0$$
$$(x_2 - x_B)^2 + (y_2 - y_B)^2 - L_4^2 = 0$$

These three equations are not sufficient to determine the four unknown variables of the problem. In fact, it is still necessary to enter the condition that the position of the input element (element 2) is known. If both coordinates of point 1 are known, then only two unknown variables are left. In this case, it is obvious that the first constraint equation which establishes the constant length condition of element 2 no longer makes any sense, because it does not contain any unknown variable. Consequently the problem reduces to the finding of x_2 and y_2, using the last two nonlinear quadratic constraint conditions.

Example 3.2

Let us consider the four-bar mechanism shown in Figure 3.2, which uses mixed coordinates; that is, the same coordinates as in example 3.1 plus the angle ψ between elements 2 and 3 at joint 1. Let's assume that, instead of the position of the input element, one knows the angle ψ. In this case the constraint equations will be as follows (assuming a suitable value for ψ to be able to use the scalar product):

$$(x_1 - x_A)^2 + (y_1 - y_A)^2 - L_2^2 = 0$$
$$(x_2 - x_1)^2 + (y_2 - y_1)^2 - L_3^2 = 0$$
$$(x_2 - x_B)^2 + (y_2 - y_B)^2 - L_4^2 = 0$$
$$(x_1 - x_A)(x_2 - x_1) + (y_1 - y_A)(y_2 - y_1) - L_2 L_3 \cos \psi = 0$$

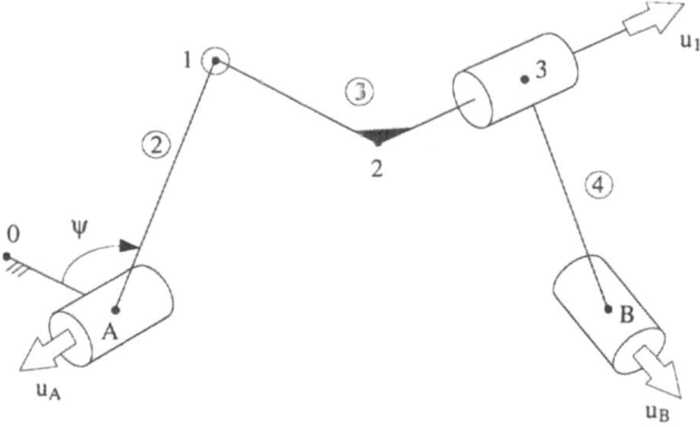

Figure 3.3. RSCR mechanism modeled with natural coordinates.

which is a system with four equations and four unknown variables, assuming that the externally driven angle ψ is known.

Example 3.3

Figure 3.3 depicts a three-dimensional four-bar mechanism *RSCR* (Revolute-Spherical-Cylindrical-Revolute) modeled with natural coordinates. This mechanism has three movable points and one movable unit vector; that is, twelve dependent Cartesian coordinates and one degree of freedom. Also the input angle ψ has been introduced as an additional externally driven coordinate. The constraint equations corresponding to this mechanism are the following:

1. $(x_1 - x_A)(x_0 - x_A) + (y_1 - y_A)(y_0 - y_A) + (z_1 - z_A)(z_0 - z_A) - k_1 \cos \psi = 0$

2. $(x_1 - x_A)^2 + (y_1 - y_A)^2 + (z_1 - z_A)^2 - k_2 = 0$

3. $(x_1 - x_A) u_{Ax} + (y_1 - y_A) u_{Ay} + (z_1 - z_A) u_{Az} - k_3 = 0$

4. $(x_2 - x_1)^2 + (y_2 - y_1)^2 + (z_2 - z_1)^2 - k_4 = 0$

5. $(x_2 - x_1) u_{1x} + (y_2 - y_1) u_{1y} + (z_2 - z_1) u_{1z} - k_5 = 0$

6. $u_{1x}^2 + u_{1y}^2 + u_{1z}^2 - 1 = 0$

7. $(x_3 - x_B)^2 + (y_3 - y_B)^2 + (z_3 - z_B)^2 - k_6 = 0$

8. $(x_3 - x_B) u_{1x} + (y_3 - y_B) u_{1y} + (z_3 - z_B) u_{1z} - k_7 = 0$

9. $(x_3 - x_B) u_{Bx} + (y_3 - y_B) u_{By} + (z_3 - z_B) u_{Bz} - k_8 = 0$

10. $u_{Bx} u_{1x} + u_{By} u_{1y} + u_{Bz} u_{1z} - k_9 = 0$

11. $(y_3 - y_2) u_{1z} - (z_3 - z_2) u_{1y} = 0$

12. $(z_3 - z_2) u_{1x} - (z_3 - x_2) u_{1z} = 0$

13. $(x_3 - x_2) u_{1y} - (y_3 - y_2) u_{1x} = 0$

This is the system of nonlinear equations that governs the position problem for the *RSCR* mechanism. The first equation corresponds to the input angle definition; equations 2 and 3 represent rigid body condition for element 2; equations 4 to 6 represent rigid body constraints for element 3; equations 6 to 10 represent the same for element 4, and equations 11 to 13 (only two of them are independent) contribute to define the cylindrical joint. Finally, k_i $(i=1,...,9)$ represents constant values.

The above examples clearly indicate that irrespective of the multibody systems being considered, the position problem is always based on solving the constraint equations, which make up the following set of equations:

$$\Phi(\mathbf{q}, t) = 0 \tag{3.1}$$

where $\mathbf{q}$ is the vector of the system dependent coordinates. It will be assumed that there are at least as many equations as there are unknown variables or coordinates. To solve systems of nonlinear equations such as (3.1), it is customary to resort to the well-known Newton-Raphson method which has quadratic convergence in the neighborhood of the solution (the error in each iteration is proportional to the square of the error in the previous iteration) and does not usually cause serious difficulties if one starts with a good initial approximation.

The Newton-Raphson method is based on a linearization of the system (3.1) and consists in replacing this system of equations with the first two terms of its expansion in a Taylor series around a certain approximation $\mathbf{q}_i$ to the desired solution. Once the substitution has been made, the system (3.1) becomes

$$\Phi(\mathbf{q}, t) \cong \Phi(\mathbf{q}_i) + \Phi_\mathbf{q}(\mathbf{q}_i)\,(\mathbf{q} - \mathbf{q}_i) = 0 \tag{3.2}$$

where the time variable has not been accounted for, so that in this problem has a constant value. Matrix $\Phi_\mathbf{q}$ is the Jacobian matrix for constraint equations; that is to say, the matrix of partial derivatives of these equations with respect to the dependent coordinates. This matrix takes the following form:

$$
\Phi_\mathbf{q} =
\begin{bmatrix}
\dfrac{\partial \phi_1}{\partial q_1} & \dfrac{\partial \phi_1}{\partial q_2} & \cdots\cdots & \dfrac{\partial \phi_1}{\partial q_n} \\[2mm]
\dfrac{\partial \phi_2}{\partial q_1} & \dfrac{\partial \phi_2}{\partial q_2} & \cdots\cdots & \dfrac{\partial \phi_2}{\partial q_n} \\[2mm]
\cdots\cdots & \cdots\cdots & \cdots\cdots & \cdots\cdots \\[2mm]
\dfrac{\partial \phi_m}{\partial q_1} & \dfrac{\partial \phi_m}{\partial q_2} & \cdots\cdots & \dfrac{\partial \phi_m}{\partial q_n}
\end{bmatrix}
\tag{3.3}
$$

In equation (3.3), m is the number of constraint equations and n the number of dependent coordinates. If the constraint equations are independent, $f = n-m$ is the number of degrees of freedom of the multibody system.

Equation (3.2) represents a system of linear equations constituting an approximation to the nonlinear system (3.1). The vector $\mathbf{q}$, obtained from the solution of equation (3.2), will be an approximation of the solution in (3.1). By calling this approximate solution $\mathbf{q}_{i+1}$, a recursive formula is obtained as follows:

$$\Phi(\mathbf{q}_i) + \Phi_\mathbf{q}(\mathbf{q}_i)\,(\mathbf{q}_{i+1} - \mathbf{q}_i) = 0 \tag{3.4}$$

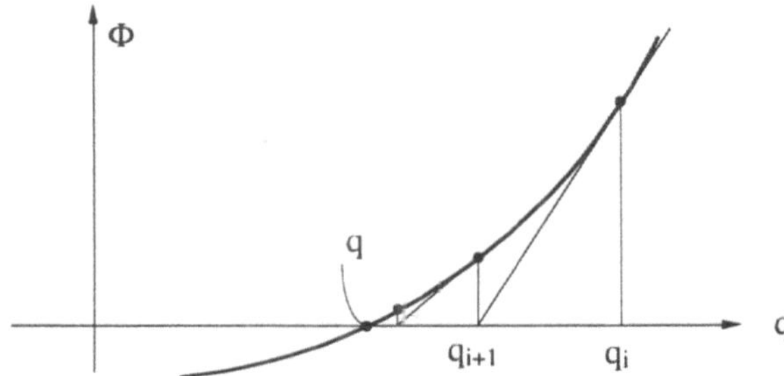

Figure 3.4. Iteration process of the Newton-Raphson method.

which can be used repeatedly until the error in the system of equations (3.1) is insignificant, or until the difference between the results of two successive iterations is smaller than a pre-specified tolerance. Figure 3.4 shows the geometric representation of the Newton-Raphson method for the case of a nonlinear equation with one unknown. The function $\Phi(q)$ is linearized at point q_i, i.e. substituted by its tangent linear space, which are the first two terms of the Taylor expansion formula. The point where the tangent space intersects the horizontal axis is the approximate solution q_{i+1}. The function $\Phi(q)$ is again replaced at point q_{i+1} by the new tangent space and a new approximate solution q_{i+2} is found. One arrives ultimately within the desired accuracy to the exact solution q.

Note that the Newton-Raphson iteration will not always converge to a solution. It has been pointed out that if the initial approximation is not close enough to a solution, the algorithm may diverge. There is still another source of difficulties. If the values of the input variables do not correspond to a possible physical solution, the mathematical algorithm will fail irrespective of how the initial approximation has been chosen.

The Jacobian matrix of the constraint equations, defined by means of equation (3.3), plays an extremely important role in all kinematic and dynamic analysis problems. In the equation (3.4), the Jacobian matrix determines the linear equation system used to find the successive approximations for solving the initial position problem. Evaluating and triangularizing this matrix easily and quickly are characteristics of all good multibody system analysis methods. The natural coordinates permit the performance of these operations in the best possible way.

In Section 1.2, it was stated that the initial position problem had multiple solutions, and this is indeed the case. Depending on the vector q_o where the iteration begins, some solution will be attained.

Example 3.4

To complete this section on the initial position problem, the equations (3.4) corresponding to the four-bar mechanism studied in Examples 3.1 and 3.2 will be com-

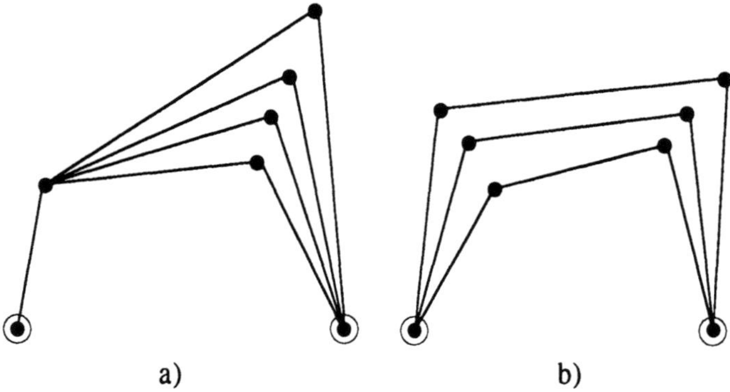

Figure 3.5. Iteration process of the Newton-Raphson method in a four-bar mechanism.

pletely developed. The constraint equations of this case were presented in Example 3.1, and consequently the equation (3.4) takes the following form:

$$
2 \begin{bmatrix} (x_1 - x_A)(y_1 - y_A) & 0 & 0 \\ (x_1 - x_2)(y_1 - y_2) & (x_2 - x_1)(y_2 - y_1) \\ 0 & 0 & (x_2 - x_B)(y_2 - y_B) \end{bmatrix}_i \left(\begin{Bmatrix} x_1 \\ y_1 \\ x_2 \\ y_2 \end{Bmatrix}_{i+1} - \begin{Bmatrix} x_1 \\ y_1 \\ x_2 \\ y_2 \end{Bmatrix}_i \right) =
$$

$$
= - \begin{Bmatrix} (x_1 - x_A)^2 + (y_1 - y_A)^2 - L_2^2 \\ (x_2 - x_1)^2 + (y_2 - y_1)^2 - L_3^2 \\ (x_2 - x_B)^2 + (y_2 - y_B)^2 - L_4^2 \end{Bmatrix}_i
$$

In this system of equations, at least one of the four unknown coordinates must be known ahead of time in order to be able to solve the problem. If, for example, x_1 is known, then:

$$
(x_1)_{i+1} - (x_1)_i = 0
$$

and the first column of the Jacobian matrix is multiplied by zero, meaning that it can be eliminated.

In the case of the four-bar mechanism of Figure 3.2, modeled with mixed coordinates and whose constraint equations are presented in Example 3.2, equation (3.4) becomes:

$$
\begin{bmatrix} 2(x_1 - x_A) & 2(y_1 - y_A) & 0 & 0 & 0 \\ 2(x_1 - x_2) & 2(y_1 - y_2) & 2(x_2 - x_1) & 2(y_2 - y_1) & 0 \\ 0 & 0 & 2(x_2 - x_B) & 2(y_2 - y_B) & 0 \\ (x_2 - x_1 + x_A - x_1) & (y_2 - y_1 + y_A - y_1) & (x_1 - x_A) & (y_1 - y_A) & (L_2 L_3 \sin\psi) \end{bmatrix}_i \cdot
$$

$$
\cdot \left(\begin{Bmatrix} x_1 \\ y_1 \\ x_2 \\ y_2 \\ \psi \end{Bmatrix}_{i+1} - \begin{Bmatrix} x_1 \\ y_1 \\ x_2 \\ y_2 \\ \psi \end{Bmatrix}_i \right) = \begin{Bmatrix} (x_1 - x_A)^2 + (y_1 - y_A)^2 - L_2^2 \\ (x_2 - x_1)^2 + (y_2 - y_1)^2 - L_3^2 \\ (x_2 - x_B)^2 + (y_2 - y_B)^2 - L_4^2 \\ (x_1 - x_A)(x_2 - x_1) + (y_1 - y_A)(y_2 - y_1) - L_2 L_3 \cos\psi \end{Bmatrix}_i
$$

Usually, the angle ψ is known; therefore, the last unknown variable $(\psi_{i+1} - \psi_i)$ has a zero value. Thus the fifth column of the Jacobian matrix can be eliminated.

One characteristic common to the Jacobian matrices shown in this example (and in all the Jacobian matrices calculated with natural coordinates) is that they are linear functions of the dependent variables. For example, Figures 3.5a and 3.5b include drawings of the initial approximation, the first iterations, and the final solution of the initial position problem in the two four-bar mechanisms of Figures 3.1 and 3.2 computed according to the above expressions.

The Newton-Raphson method, explained in this section, converges rather quickly (quadratic convergence) when it is close to the desired solution. At times, and during the first iterations, it can give very abrupt jumps as a result of not having started from a sufficiently good initial approximation. Figure 3.6 shows what could happen in this case. The approximate solution q_{i+1} is further away from the true solution q than the previous approximation q_i. It is even possible that the value of function $\Phi(q)$, a function that should be equal to zero, could increase when moving from q_i to q_{i+1}.

This problem is not easy to solve without resorting to much more complicated numerical methods. In general, one should do everything possible to start from good initial approximations. If this cannot be achieved, then one should try to apply a reduction to the coordinates modification given by equation (3.4) and to apply it to the previous approximate solution q_i. As this often works, a correction factor of 1/2 or 1/3 is recommended. Finally, one should always make sure that the module of $\Phi(q)$ decreases when going from point q_i to q_{i+1}.

Some authors have solved the position problem at times by calculating different solutions numerically by means of the so-called *continuation methods* (Tsai and Morgan (1985)). Continuation methods start out from a position where the multibody system complies with all the constraint equations, although the input elements might not be at the desired position and the fixed joints might not be at

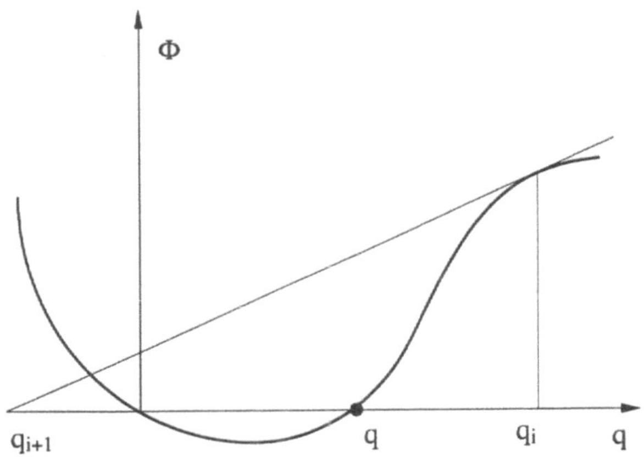

Figure 3.6. Possible divergence in the Newton-Raphson iteration

their final position. With relaxed conditions regarding the input elements and the fixed element, it is not difficult to find a position on the multibody system that satisfies the constraint equations. Then, by means of small finite displacements whose convergence is guaranteed by the Newton-Raphson method, an attempt is made to move the input elements and the fixed joints to their correct position. At times, the bifurcation points (points at which two or more possible movements can occur) provide a way of finding different solutions to the position problem.

3.2 Velocity and Acceleration Analysis

3.2.1 Velocity Analysis

The equations that permit solving the velocity problem originate after one differentiates with respect to time the constraint equations. If these equations are represented symbolically as

$$\Phi(\mathbf{q},\, t) = 0 \tag{3.5}$$

by differentiating with respect to time, the following equation is obtained:

$$\Phi_\mathbf{q}(\mathbf{q},\, t)\, \dot{\mathbf{q}} = -\, \Phi_t \equiv \mathbf{b} \tag{3.6}$$

where $\Phi_\mathbf{q}$ is the Jacobian matrix defined by means of equation (3.3). Vector $\dot{\mathbf{q}}$ is the vector of dependent velocities (derivative with respect to the time of the vector of dependent coordinates or position variables). Vector $(-\Phi_t = \mathbf{b})$ is the partial derivative of the constraint equations with respect to time. If all the constraints are scleronomous, meaning that there are no rheonomous or time-dependent constraints, this derivative will be zero) If the position of the multibody system is known, equation (3.6) allows us to determine the velocities of the multibody system by starting from the velocity of the input elements. Just as in the position problem, the matrix that controls the velocity problem is the Jacobian matrix of the constraint equations. The essential difference between both problems is that where the position problem is nonlinear, the equations governing the velocity problem are linear. This means that the equations do not have to be iterated, and there is only one solution to a properly posed problem. The following example illustrates these concepts:

Example 3.5

As an example of this, the velocity equations of the four-bar mechanism of Figure 3.1 will be determined below by using: a) relative coordinates, b) reference point coordinates, c) natural coordinates, and d) mixed coordinates.

a) Using *relative* coordinates, the constraint equations are given by (See Section 2.1.1),

$$L_1 \cos \Psi_1 + L_2 \cos (\Psi_1 + \Psi_2) + L_3 \cos (\Psi_1 + \Psi_2 + \Psi_3) - OD = 0$$

$$L_1 \sin \Psi_1 + L_2 \sin (\Psi_1 + \Psi_2) + L_3 \sin (\Psi_1 + \Psi_2 + \Psi_3) = 0$$

Differentiating these equations with respect to time, we obtain:

$$-L_1 \sin \psi_1 \; \dot{\psi}_1 - L_2 \sin (\psi_1 + \psi_2) (\dot{\psi}_1 + \dot{\psi}_2) -$$
$$- L_3 \sin (\psi_1 + \psi_2 + \psi_3) (\dot{\psi}_1 + \dot{\psi}_2 + \dot{\psi}_3) = 0$$

$$L_1 \cos \psi_1 \; \dot{\psi}_1 + L_2 \cos (\psi_1 + \psi_2) (\dot{\psi}_1 + \dot{\psi}_2) +$$
$$+ L_3 \cos (\psi_1 + \psi_2 + \psi_3) (\dot{\psi}_1 + \dot{\psi}_2 + \dot{\psi}_3) = 0$$

and by rearranging these equations, we arrive at:

$$\begin{bmatrix} -L_1 s_1 - L_2 s_{12} - L_3 s_{123} & -L_2 s_{12} - L_3 s_{123} & -L_3 s_{123} \\ L_1 c_1 + L_2 c_{12} + L_3 c_{123} & L_2 c_{12} + L_3 c_{123} & L_3 c_{123} \end{bmatrix} \begin{Bmatrix} \dot{\psi}_1 \\ \dot{\psi}_2 \\ \dot{\psi}_3 \end{Bmatrix} = \begin{Bmatrix} 0 \\ 0 \end{Bmatrix}$$

where $s_1 = \sin \psi_1$, $s_{12} = \sin (\psi_1 + \psi_2)$, and so forth.

If one of the three velocities in the previous equation is known such as the one corresponding to the input coordinate, the corresponding column of the Jacobian matrix can be moved to the right-hand side of the equation. This results in a system of two linear equations with two unknown velocities that can be solved with no difficulties.

b) Using *reference point* coordinates, the constraint equations are represented by (See Section 2.1.2):

$$(x_1 - x_0) - L_1/2 \; \cos\Psi_1 = 0$$
$$(y_1 - y_0) - L_1/2 \; \sin\Psi_1 = 0$$
$$(x_2 - x_1) - L_1/2 \; \cos\Psi_1 - L_2/2 \; \cos\Psi_2 = 0$$
$$(y_2 - y_1) - L_1/2 \; \sin\Psi_1 - L_2/2 \; \sin\Psi_2 = 0$$
$$(x_3 - x_2) - L_2/2 \; \cos\Psi_2 - L_3/2 \; \cos\Psi_3 = 0$$
$$(y_3 - y_2) - L_2/2 \; \sin\Psi_2 - L_3/2 \; \sin\Psi_3 = 0$$
$$(x_3 - x_D) - L_3/2 \; \cos\Psi_3 = 0$$
$$(y_3 - y_D) - L_3/2 \; \sin\Psi_3 = 0$$

and the time derivatives are:

$$\dot{x}_1 + L_1/2 \; \dot{\Psi}_1 \sin\Psi_1 = 0$$
$$\dot{y}_1 - L_1/2 \; \dot{\Psi}_1 \cos\Psi_1 = 0$$
$$\dot{x}_2 - \dot{x}_1 + L_1/2 \; \dot{\Psi}_1 \sin\Psi_1 + L_2/2 \; \dot{\Psi}_2 \sin\Psi_2 = 0$$
$$\dot{y}_2 - \dot{y}_1 - L_1/2 \; \dot{\Psi}_1 \cos\Psi_1 - L_2/2 \; \dot{\Psi}_2 \cos\Psi_2 = 0$$
$$\dot{x}_3 - \dot{x}_2 + L_2/2 \; \dot{\Psi}_2 \sin\Psi_2 + L_3/2 \; \dot{\Psi}_3 \sin\Psi_3 = 0$$
$$\dot{y}_3 - \dot{y}_2 - L_2/2 \; \dot{\Psi}_2 \cos\Psi_2 - L_3/2 \; \dot{\Psi}_3 \cos\Psi_3 = 0$$
$$\dot{x}_3 + L_3/2 \; \dot{\psi}_3 \sin\Psi_3 = 0$$
$$\dot{y}_3 - L_3/2 \; \dot{\Psi}_3 \cos\Psi_3 = 0$$

These equations can be expressed in matrix form as follows:

$$\Phi_q \, \dot{q} = 0$$

where the matrix Φ_q is

$$
\begin{bmatrix}
1 & 0 & s_1\,L_1/2 & 0 & 0 & 0 & 0 & 0 & 0 \\
0 & 1 & -c_1\,L_1/2 & 0 & 0 & 0 & 0 & 0 & 0 \\
-1 & 0 & s_1\,L_1/2 & 1 & 0 & s_2\,L_2/2 & 0 & 0 & 0 \\
0 & -1 & -c_1\,L_1/2 & 0 & 1 & -c_2\,L_2/2 & 0 & 0 & 0 \\
0 & 0 & 0 & -1 & 0 & s_2\,L_2/2 & 1 & 0 & s_3\,L_3/2 \\
0 & 0 & 0 & 0 & -1 & -c_2\,L_2/2 & 0 & 1 & -c_3\,L_3/2 \\
0 & 0 & 0 & 0 & 0 & 0 & 1 & 0 & s_3\,L_3/2 \\
0 & 0 & 0 & 0 & 0 & 0 & 0 & 1 & -c_3\,L_3/2
\end{bmatrix}
$$

which is a system of eight equations with nine unknown velocities. If the angular velocity $\dot{\psi}_1$ is known for element 2, the third column of the Jacobian matrix will be moved to the right side member. The result will be a system of eight linear equations with eight unknown velocities.

c) With *natural* coordinates the constraint equations (Section 2.1.3) are represented by:

$$(x_1 - x_A)^2 + (y_1 - y_A)^2 - L_2{}^2 = 0$$

$$(x_2 - x_1)^2 + (y_2 - y_1)^2 - L_3{}^2 = 0$$

$$(x_3 - x_B)^2 + (y_3 - y_B)^2 - L_4{}^2 = 0$$

whose time derivatives are:

$$(x_1 - x_A)\,\dot{x}_1 + (y_1 - y_A)\,\dot{y}_1 = 0$$

$$(x_2 - x_1)\,(\dot{x}_2 - \dot{x}_1) + (y_2 - y_1)\,(\dot{y}_2 - \dot{y}_1) = 0$$

$$(x_2 - x_B)\,\dot{x}_2 + (y_2 - y_B)\,\dot{y}_2 = 0$$

and in matrix form yields:

$$
\begin{bmatrix}
(x_1{-}x_A) & (y_1{-}y_A) & 0 & 0 \\
(x_1{-}x_2) & (y_1{-}y_2) & (x_2{-}x_1) & (y_2{-}y_1) \\
0 & 0 & (x_2{-}x_B) & (y_2{-}y_B)
\end{bmatrix}
\begin{Bmatrix}
\dot{x}_1 \\ \dot{y}_1 \\ \dot{x}_2 \\ \dot{y}_2
\end{Bmatrix}
=
\begin{Bmatrix}
0 \\ 0 \\ 0
\end{Bmatrix}
$$

By knowing one of the four natural velocities and by moving the corresponding column to the right-hand side of this equation, one can find the remaining velocities with the resulting set of three linear equations and three unknown variables.

d) Using *mixed* coordinates, the constraint equations (Section 2.1.4) are:

$$(x_1 - x_A)^2 + (y_1 - y_A)^2 - L_2{}^2 = 0$$

$$(x_2 - x_1)^2 + (y_2 - y_1)^2 - L_3{}^2 = 0$$

$$(x_3 - x_B)^2 + (y_3 - y_B)^2 - L_4{}^2 = 0$$

$$(x_1 - x_A)\,(x_2 - x_1) + (y_1 - y_A)\,(y_2 - y_1) - L_2\,L_3\,\cos\Psi = 0$$

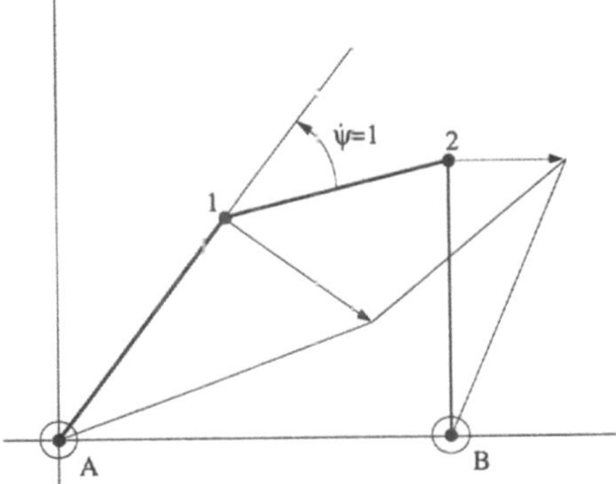

Figure 3.7. Results of a velocity analysis in a four-bar mechanism.

differentiating with respect to time:

$$\left(x_1-x_A\right)\dot{x}_1 + \left(y_1-y_A\right)\dot{y}_1 = 0$$

$$\left(x_2-x_1\right)\left(\dot{x}_2-\dot{x}_1\right) + \left(y_2-y_1\right)\left(\dot{y}_2-\dot{y}_1\right) = 0$$

$$\left(x_2-x_B\right)\dot{x}_2 + \left(y_2-y_B\right)\dot{y}_2 = 0$$

$$\left(x_2-x_1\right)\dot{x}_1 + \left(x_1-x_A\right)\left(\dot{x}_2-\dot{x}_1\right) + \left(y_2-y_1\right)\dot{y}_1 +$$
$$+ \left(y_1-y_A\right)\left(\dot{y}_2-\dot{y}_1\right) + L_2 L_3 \, sin\psi \, \dot{\psi} = 0$$

which can be expressed in matrix form as

$$\begin{bmatrix} \left(x_1-x_A\right) & \left(y_1-y_A\right) & 0 & 0 & 0 \\ \left(x_1-x_2\right) & \left(y_1-y_2\right) & \left(x_2-x_1\right) & \left(y_2-y_1\right) & 0 \\ 0 & 0 & \left(x_2-x_B\right) & \left(y_2-y_B\right) & 0 \\ \left(x_2-2x_1+x_A\right) & \left(y_2-2y_1+y_A\right) & \left(x_1-x_A\right) & \left(y_1-y_A\right) & L_2L_3sin\psi \end{bmatrix} \begin{Bmatrix} \dot{x}_1 \\ \dot{y}_1 \\ \dot{x}_2 \\ \dot{y}_2 \\ \dot{\psi} \end{Bmatrix} = \begin{Bmatrix} 0 \\ 0 \\ 0 \\ 0 \end{Bmatrix}$$

If $\dot{\psi}$ is known, the fifth column will be moved to the right-hand side and will leave four equations with four unknowns. Figure 3.7 shows the result of a velocity analysis in accordance with an input velocity of $\dot{\psi}=1$.

3.2.2 Acceleration Analysis

The finding of the dependent acceleration vector $\ddot{q}$ becomes apparent by simply differentiating with respect to time the velocity equation (3.6). This yields the following result:

$$\Phi_q\left(q, t\right)\ddot{q} = -\dot{\Phi}_t - \dot{\Phi}_q\,\dot{q} \equiv c \qquad (3.7)$$

If the position vector q and the velocity vector $\dot{q}$ are known, by solving the system of linear equations (3.7), one can find the dependent acceleration vector $\ddot{q}$. Note that the leading matrix of the systems of linear equations (3.6) and (3.7) is exactly the same. As a consequence, if it has been formed and triangularized

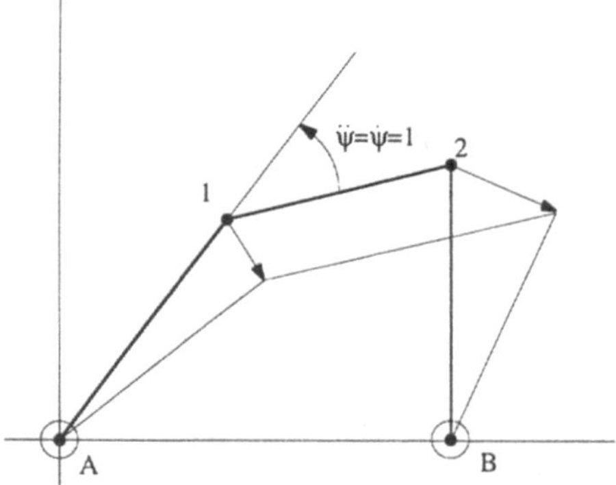

Figure 3.8. Results of an acceleration analysis in a four-bar mechanism.

to solve the velocity problem, the acceleration analysis can be carried out by simply forming the right-hand side and by performing a forward reduction and backward substitution. When there are no rheonomous or time-dependent constraints, the velocity problem is homogeneous; whereas the acceleration problem is always non-homogeneous as long as the velocities are not equal to zero.

Equation (3.7) can be differentiated once again to obtain the *jerk* or *over acceleration* equation:

$$\Phi_q \frac{d}{dt}\left(\ddot{q}\right) = -\dddot{\Phi}_t - 2\,\dot{\Phi}_q\,\ddot{q} - \ddot{\Phi}_q\,\dot{q} \tag{3.8}$$

Once again a system of linear equations has been obtained whose leading matrix is the Jacobian matrix of the constraint equations.

Example 3.6

Included below are the acceleration equations for the four-bar mechanism of Example 3.5d modeled with mixed coordinates. These equations are obtained by differentiating the corresponding velocity equations:

$$\begin{bmatrix} (x_1-x_A) & (y_1-y_A) & 0 & 0 & 0 \\ (x_1-x_2) & (y_1-y_2) & (x_2-x_1) & (y_2-y_1) & 0 \\ 0 & 0 & (x_2-x_B) & (y_2-y_B) & 0 \\ (x_2-2x_1+x_A) & (y_2-2y_1+y_A) & (x_1-x_A) & (y_1-y_A) & L_2L_3\,sin\psi \end{bmatrix} \begin{Bmatrix} \ddot{x}_1 \\ \ddot{y}_1 \\ \ddot{x}_2 \\ \ddot{y}_2 \\ \ddot{\psi} \end{Bmatrix} =$$

$$= -\begin{bmatrix} \dot{x}_1 & \dot{y}_1 & 0 & 0 & 0 \\ (\dot{x}_1-\dot{x}_2) & (\dot{y}_1-\dot{y}_2) & (\dot{x}_2-\dot{x}_1) & (\dot{y}_2-\dot{y}_1) & 0 \\ 0 & 0 & \dot{x}_2 & \dot{y}_2 & 0 \\ (\dot{x}_2-2\dot{x}_1) & (\dot{y}_2-2\dot{y}_1) & \dot{x}_1 & \dot{y}_1 & L_2L_3\,cos\psi\,\dot{\psi} \end{bmatrix} \begin{Bmatrix} \dot{x}_1 \\ \dot{y}_1 \\ \dot{x}_2 \\ \dot{y}_2 \\ \dot{\psi} \end{Bmatrix}$$

If all the velocities $\dot{q}$ and input accelerations $\ddot{\psi}$ are known, the remaining accelerations $\ddot{q}$ can be calculated by means of the four equations with four unknowns resulting from moving the fifth column, multiplied by $\ddot{\psi}$, to the RHS of the acceleration equation. Figure 3.8 graphically shows the result of an acceleration analysis that corresponds to the expression developed in this example.

3.3 Finite Displacement Analysis

The finite displacement analysis is closely related to the initial position problem, and is controlled by the same system of nonlinear equations (the kinematic constraint equations). The velocity and acceleration analyses are used at times in finite displacement analysis to improve the initial approximation with which the iterative process begin, which explains the reason for including it here and not immediately after the initial position problem.

3.3.1 Newton-Raphson Iteration

As explained in Section 1.2, once one knows a position of the multibody system where all the constraint equations are satisfied, the finite displacement problem consists of finding the new position that the system takes when a finite displacement is applied to each one of the input elements or externally driven relative coordinates. Finite displacement is understood to be any movement other than infinitesimal.

The main problem dealt with in this section is of the same nature and consequently controlled by the same equations of the position problem. Therefore, the Newton-Raphson method can be used for solving it. The difference between both problems lies in the fact that *the finite displacement problem usually relies on a good initial approximation* which is obtained from a previous exact position where all the elements satisfy the constraints. It is possible to improve upon the approximation by means of a velocity and acceleration analysis, as will be described in the next section.

These advantages do away with many of the convergence problems encountered in the initial position problem. In addition, the problem of multiple solutions becomes marginal. If the displacement of the input elements is small enough, then of all the possible solutions for the constraint equations, the correct one will be the closest to the starting position. This is precisely the one obtained by the Newton-Raphson iterations. However, there still remains the possibility of driving or trying to drive the multibody system to unfeasible positions, that is, positions that cannot be reached without violating some constraints equations. Trying to move the end-effector of a robot out of its workspace is an example of a finite displacement problem where the Newton-Raphson method will necessarily fail to find a correct solution.

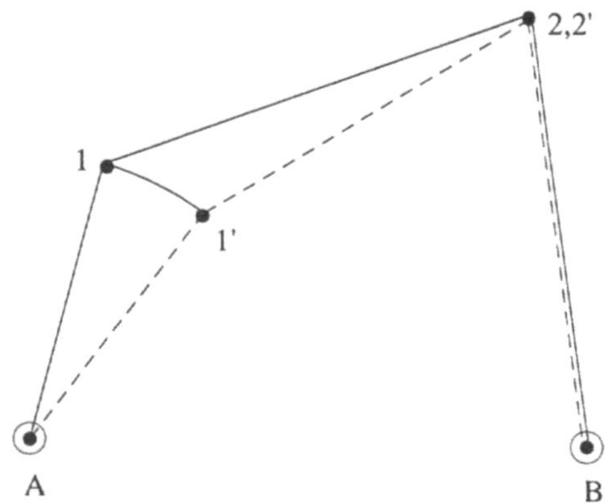

Figure 3.9. Improving the initial approximation.

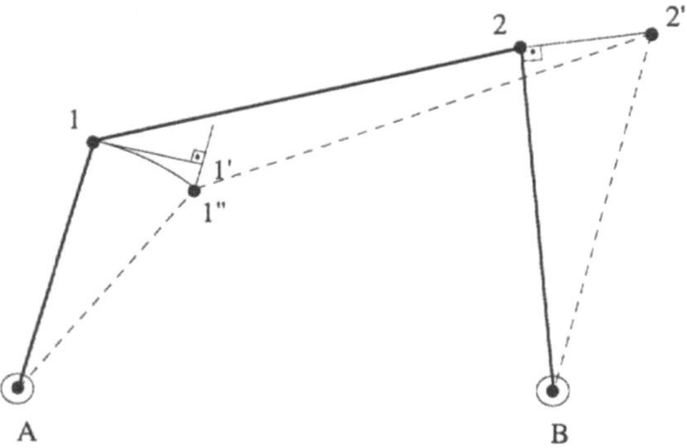

Figure 3.10. Better estimate for an initial approximation.

3.3.2 Improved Initial Approximation

In order to determine and improve the initial or starting approximation, the example of the four-bar mechanism will be used once again. This will clearly describe the method without any loss of generality.

Figure 3.9 shows a four-bar mechanism, in which the input element has been rotated a finite angle. One possible way of generating an initial approximation is by not varying the remaining natural coordinates as in the starting position shown in Figure 3.9. This approximation leads to a severe violation of the constraint equations.

The initial approximation shown in Figure 3.9 can be improved upon by means of *velocity analysis*, as indicated in Figure 3.10. The velocity analysis is carried out by imposing a velocity at the input element so that the endpoint 1' of the velocity vector of 1 is the closest point to 1" over the perpendicular to A-1 (1'-1" is parallel to A-1).

Since 1" is known, it is not difficult to determine the velocity of the input element such that the end of the velocity vector at point 1 is 1'. Using this velocity as input, a velocity analysis is performed, and the ends of the velocity vectors are determined for all the basic points of the mechanism (in Figure 3.10, 2' is the end of the velocity vector of point 2).

The initial approximation used to start the iterations for the Newton-Raphson method is indicated by the dotted lines in Figure 3.10 It is an improvement over the one in Figure 3.9. Note that the initial approximation is (A-1"-2'-B) and not (A-1'-2'-B). The exact position 1" of point 1 is known because it belongs to the input element and this exact position should be used.

It is not essential that point 1' be the closest one to 1" on the tangent to the trajectory of 1. Another simpler possibility for calculating point 1' and the velocity of the input element, is to assume that point 1 changes to position 1" in an arbitrary period of time such as 1 second. Next, calculate the angular velocity of the input element by dividing angle 1-1" (in radians) by the said amount of time where the quotient is the said angular velocity. The position at the initial approximation of any point P can be calculated by means of the following expression:

$$q = q_0 + \dot{q}\, \Delta t \tag{3.9}$$

Equation (3.9) is an approximate integral of velocities starting from the previous position. An approximate integration which also causes the accelerations to intervene can be obtained in a similar manner:

$$q = q_0 + \dot{q}\, \Delta t + \frac{1}{2} \ddot{q}\, \Delta t^2 \tag{3.10}$$

This formula suggests that the initial approximation can be constructed starting from a velocity analysis and an acceleration analysis. To calculate the velocity and acceleration of the input element one may proceed as follows:

1. Apply one of the previously studied methods and determine the velocity of the input elements.
2. Knowing the initial and final position of the input elements and their velocity, determine the acceleration to be applied to them applying equation (3.10) to the input elements

Determination of the initial approximation by means of velocity and acceleration analysis allows the iterations to begin with a *better approximation* to the final solution. The cost of an acceleration analysis is small if a velocity analysis has already been performed. The matrix for both systems of equations is the same, and one only needs to form it and triangularize it once. Based on the experience gained through numerical experiments performed by the authors, the initial approximation constructed with velocities and accelerations does not always give better results than the one determined from velocities only.

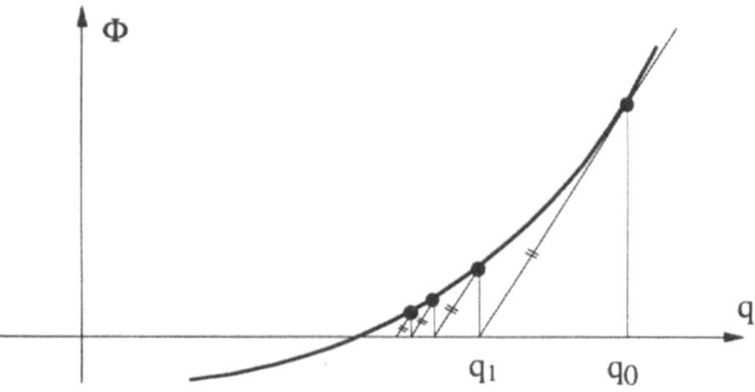

Figure 3.11. Iteration process of the Modified Newton-Raphson method.

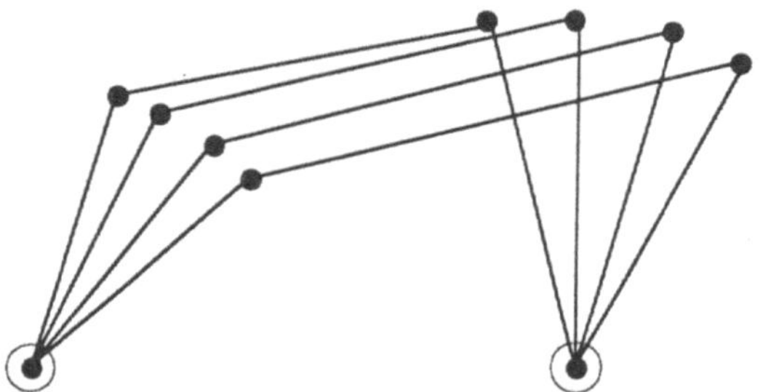

Figure 3.12. Kinematic simulation of a four-bar mechanism.

3.3.3 Modified Newton-Raphson Iteration

The Newton-Raphson method of solving systems of nonlinear equations proceeds as indicated in equation (3.4) and in Figure 3.4. It has already been mentioned that this iterative scheme has second order convergence in the neighborhood of the solution. The most important computational burden in the solution of equation (3.4) is the factorization of the Jacobian matrix.

The idea behind the modified Newton-Raphson method consists of applying the same iterative scheme but with a constant Jacobian matrix (See Figure 3.11),

$$\Phi(q_i) + \Phi_q(q_o) \, (q_{i+1} - q_i) = 0 \tag{3.11}$$

The main advantage of the modified Newton-Raphson method is the reduced computational cost of each iteration. More iterations may be necessary to satisfy the convergence criterion, but in general the total CPU time can be reduced. However, if the motion increments (finite displacements of the input elements) are not small, the modified Newton-Raphson method is bound to have more convergence difficulties than the standard Newton-Raphson method. Sometimes, a mixed strategy such as a new Jacobian factorization every few iterations may give the best results.

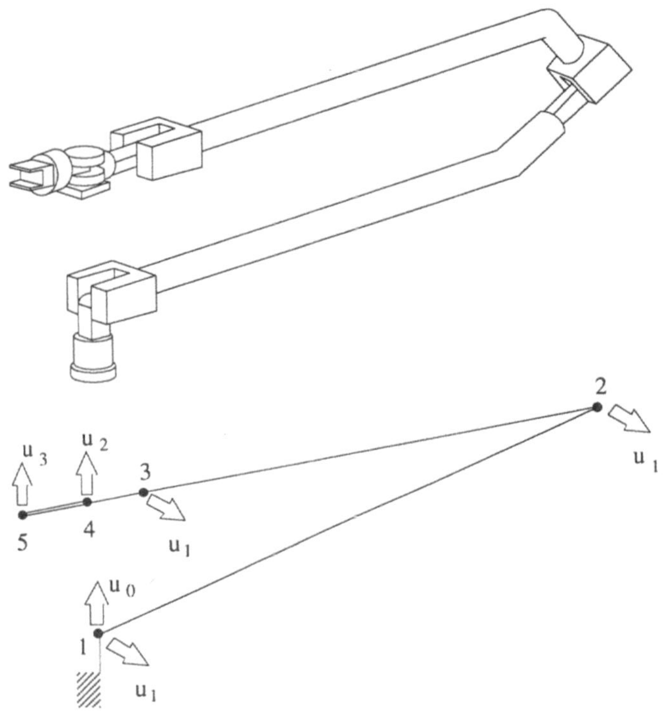

Figure 3.13. Spatial robotic manipulator modeled with natural coordinates.

Table 3.1. CPU time in seconds for 1000 finite displacement analyses of the robot in Figure 3.13.

(SNR) Standard Newton Raphson	(MNR) Modified Newton Raphson	MNR with improved initial approx.
53.1	21.7	15.5

3.3.4 Kinematic Simulation

Kinematic simulation is merely a repetition of the finite displacement problem, with the object of generating a sequence of positions that represent its movement in a specific time period or range of the input variables. This sequence of positions can later be depicted in animated form on the computer screen, as long as a system with sufficiently fast graphics is available.

The only problem with kinematic simulation is to find easy and general means of defining the movement of the input elements for the time interval in which one wants to simulate the motion of the multibody system. The increment between two consecutive positions depends on the speed with which one wants to visualize the motion afterwards.

Kinematic simulation may simply involve the animated display of the system's motion, or it may also deal with the study of trajectories, possible collisions or geometrical interferences between solid models of the mechanism elements, and even the necessary driving forces and reactions that occur in a specific movement. This last problem is truly dynamic (an inverse dynamic problem) even though it may seem to be just a kinematic problem, since no dynamic differential equations need to be integrated.

All data previously stated for the finite displacement problem and the kinematic simulation is basically valid only when the displacements of the input elements are small. If the input displacements are very large, then it is desirable to split them into a series of smaller ones and to solve them sequentially. As an example Figure 3.12 shows the kinematic simulation of a four-bar mechanism where there are various consecutive positions of the system.

Example 3.7

Figure 3.13 shows a spatial 6R robot modeled with natural coordinates. It has four movable points and three movable vectors, with a total number of 21 dependent coordinates and six degrees of freedom. The kinematic simulation consists of imposing an end-effector translation on an elliptic path contained in a plane perpendicular to the robot initial position plane. One thousand finite increments in the end-effector position have been imposed. The corresponding CPU times on an HP 9000/834 computer (14 Mips and 1.8 DP Linpack Mflops) are shown in Table 3.1 for different conditions: standard Newton-Raphson method, modified Newton-Raphson method, and modified Newton-Raphson method with initial approximation obtained from a velocity analysis. In this case the improvements that result from using modified Newton-Raphson method and the velocity approximation are quite important. These figures could be modified for other computers according to the DP Linpack megaflops ratio.

3.4 Redundant Constraints

It has been seen in the previous sections of this chapter that the nonlinear kinematic constraint equations that govern the initial position or finite displacement problems can be formulated as,

$$\Phi(\mathbf{q}, t) = 0 \tag{3.12}$$

In order to solve the previously mentioned kinematic problems departing from equation (3.12), such as initial position and finite displacement problems using Newton-Raphson iterations, or velocity and acceleration analysis, it is necessary to solve linear systems of equations in the form:

$$\Phi_{\mathbf{q}}(\mathbf{q}, t)\, \mathbf{x} = \mathbf{d} \tag{3.13}$$

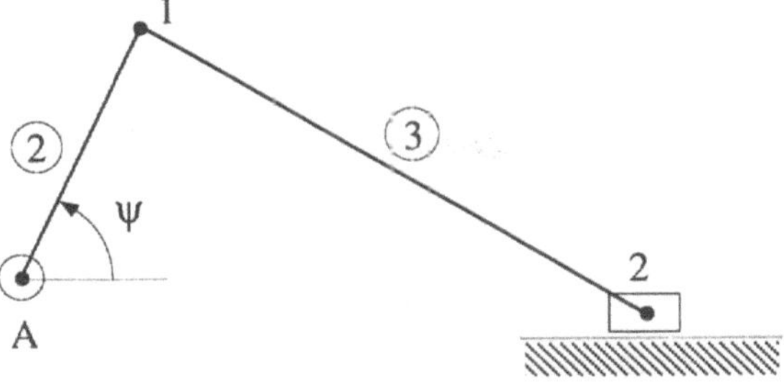

Figure 3.14. Slider-crank mechanism modeled with mixed coordinates.

The vector **x** may represent incremental displacements, velocities, or accelerations, depending on the type of problem being analyzed. The corresponding RHS term is **d**.

In practice, the multibody software developer and the engineer analyst very often face the problem of an excess of equations (more equations than the strictly necessary number) in (3.12), reflected by the fact that some of these equations are not independent from the remaining ones. This lack of equation independence in system (3.12) may lead to the following troubles in the solution of the linear system of equations (3.13):

a) a rank deficiency in the Jacobian matrix Φ_q, if an inadequate subset of equations is chosen, and

b) an over-constrained system of linear equations (more equations than unknowns) which will not have a solution that satisfy all the equations.

This section is addressed to consider the ways on which redundant equations appear in system (3.12), the consequences that this fact has on the system of linear equations (3.13) and the practical solutions or numerical strategies that can be followed to eliminate the resulting difficulties. Some very simple examples will be used to explain these points.

Multibody systems with n dependent coordinates and f degrees of freedom will be considered in the sequel. If the analyst is able to find $m=n-f$ independent constraint equations, no redundant constraints appear in the formulation and the standard formulations of previous sections in this chapter have full validity.

However, if $m>n-f$ consistent constraint equations are found, it is clear that there are $m-(n-f)$ redundant constraint equations. The following examples illustrate two possible origins for this situation.

Example 3.8

Figure 3.14 shows a planar slider-crank mechanism driven by the angle ψ. The constraint equations corresponding to the rigid body condition and the prismatic joint are

$$\left(x_1-x_A\right)^2 + \left(y_1-y_A\right)^2 - L_2^2 = 0 \qquad \text{(i)}$$

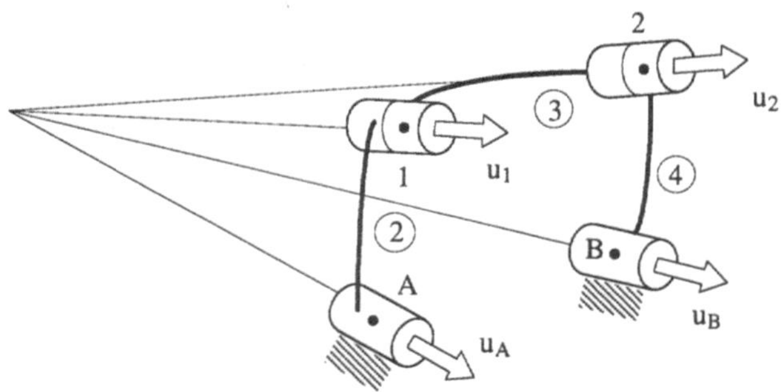

Figure 3.15. Spherical mechanism with natural coordinates.

$$\left(x_2{-}x_1\right)^2 + \left(y_2{-}y_1\right)^2 - L_3^2 = 0 \tag{ii}$$

$$y_2 = 0 \tag{iii}$$

It remains to formulate the constraint equation corresponding to the driven angle ψ. If angle ψ is near $90°$ or $(-90°)$ the cosine equation that can be used is

$$\left(x_1 - x_A\right) - L_2 \cos\psi = 0 \tag{iv}$$

However, if angle ψ is near $0°$ or $180°$, this equation is not valid (See Section 2.1.4) and the sine function that shall be used instead is

$$\left(y_1 - y_A\right) - L_2 \sin\psi = 0 \tag{v}$$

After these facts we arrive at the following situation. There are four dependent coordinates (four unknown ones (x_1, y_1, x_2, y_2) and one externally driven ψ) and four equations ((i), (ii), (iii), and (iv) or (v) depending on ψ value). If the user takes care of switching between equations (iv) and (v) according to the value of ψ, then the relation $m=n{-}f$ is always met and there is no problem. However, the user can decide to include always both equations (iv) and (v) with $m>n{-}f$, leaving to the equation solver the responsibility of disregarding the less appropriate equation in each position. Note that equations (i)-(v) constitute a system of nonlinear redundant but compatible equations: If equations (iv) and (v) are squared and added together, equation (i) is obtained.

Example 3.9

Consider the spherical four-bar mechanisms of Figure 3.15. It is well known that spherical mechanisms (all the revolute joint axes pass on a common point) are exceptions to the Grübler criterion, because they have more degrees of freedom than foreseen by the Grübler formula. In particular, the mechanism of Figure 3.15 has one degree of freedom, but the Grübler formula predicts (-2). This is due to the particular orientation of joint axes. Note that for arbitrary joint orientations the Grübler result has full sense.

Consider this mechanism in terms of natural coordinates. There are two movable points and two movable unit vectors; hence there are $n=12$ dependent coordinates. The following constraint equations shall be formulated:

– Unit module conditions:

1.- $$u_{1x}^2 + u_{1y}^2 + u_{1z}^2 - 1 = 0$$

2.- $$u_{2x}^2 + u_{2y}^2 + u_{2z}^2 - 1 = 0$$

– Constant distance conditions:

3.- $$(x_1-x_A)^2 + (y_1-y_A)^2 + (z_1-z_A)^2 - L_2^2 = 0$$

4.- $$(x_2-x_1)^2 + (y_2-y_1)^2 + (z_2-z_1)^2 - L_3^2 = 0$$

5.- $$(x_2-x_B)^2 + (y_2-y_B)^2 + (z_2-z_B)^2 - L_4^2 = 0$$

– Constant angle conditions:

6.- $$(x_1-x_A)\,u_{Ax} + (y_1-y_A)\,u_{Ay} + (z_1-z_A)\,u_{Az} - k_1 = 0$$

7.- $$(x_2-x_1)\,u_{1x} + (y_2-y_1)\,u_{1y} + (z_2-z_1)\,u_{1z} - k_2 = 0$$

8.- $$(x_2-x_B)\,u_{Bx} + (y_2-y_B)\,u_{By} + (z_2-z_B)\,u_{Bz} - k_3 = 0$$

– Linear combination conditions:

9., 10., 11.- $$\mathbf{u}_1 - k_4\,\mathbf{u}_A - k_5\,(\mathbf{r}_1 - \mathbf{r}_A) = 0$$

12., 13., 14.- $$\mathbf{u}_2 - k_6\,\mathbf{u}_1 - k_7\,(\mathbf{r}_2 - \mathbf{r}_1) = 0$$

15., 16., 17.- $$\mathbf{u}_2 - k_8\,\mathbf{u}_B - k_9\,(\mathbf{r}_2 - \mathbf{r}_B) = 0$$

There are 17 equations. Taking into account that vectors $\mathbf{u}_1$ and $\mathbf{u}_2$ have unit length, only two of each three linear combination conditions are necessary. This gives a total number of 14 constraint equations on 12 dependent coordinates.

There is another way to arrive at the same result. Each element with two points and two unit vectors generates six rigid body constraint equations, including two unit module conditions. This gives a total number of 18 equations. If it is taken into account that vectors $\mathbf{u}_A$ and $\mathbf{u}_B$ are constant, that is no unit module conditions for them are necessary and the unit module condition for vectors $\mathbf{u}_1$ and $\mathbf{u}_2$ has been considered twice because they belong to two different elements, one arrives again to a total number of constraint equations $m=14$.

Then, this mechanism has 12 dependent coordinates, one degree of freedom, and 14 constraint equations, which gives an excess of three constraint equations, in accordance with the wrong prediction of Grübler criterion (-2 instead of 1). Of these 14 constraint equations, only 11 are independent.

The two previous examples demonstrate without any lack of generality the two ways from which redundant constraint equations arise:

a) Due to convenience of implementation, as in Example 3.8.
b) In over constrained multibody systems that are exceptions to the Grübler criterion, as in Example 3.9.

Once a system has been characterized using systems of redundant constraint equations, the search of solutions can follow two different avenues:

1. Systems of equations (3.12) and (3.13) can be preprocessed with the aim of determining and eliminating the dependent equations, to keep only $m=n-f$ independent constraint equations, and then to use the standard formulations of

Sections 3.1-3.3. The main disadvantage of this method is the need to repeat the dependent equation elimination process each time the multibody system changes its configuration or, in the case of Example 3.7, after large changes in the position of the multibody system. Thus this procedure is not suitable for real-time applications (there is no time to repeat the dependent equations elimination process) or even for interactive simulation.

2. The second possibility is to solve system (3.12) directly, with a procedure capable of directly tackling redundant constraints on a strictly standard form. This way will be explained next.

Let us assume that system (3.12) has m nonlinear equations, of which only $(n-f)<m$ are independent. As a consequence, one may be tempted to think that the redundant equations in (3.12) just produce an excess of compatible equations in the linear system (3.13). If this were true no particular difficulties would appear during the solution, because there are a lot of ways and numerical routines to solve linear systems of equations with an excess of compatible equations. However, the problem is a little more complicated than assumed previously.

The redundant but compatible nonlinear equations in system (3.12) can induce an excess of non-compatible linear equations in system (3.13). This does not happen in velocity or acceleration analysis, because in these cases the Jacobian matrix is evaluated in the exact position $\mathbf{q}$, a position in which all constraint equations (3.12) are satisfied.

However, in the initial position and finite displacement problems, the following Newton-Raphson iteration formula is used:

$$\left(\Phi_{\mathbf{q}}\right)_i \left(\mathbf{q}_{i+1} - \mathbf{q}_i\right) = -\left(\Phi\right)_i \tag{3.14}$$

In this expression the Jacobian matrix $\Phi_{\mathbf{q}}$ is evaluated at an intermediate approximate position $\mathbf{q}_i$ at which the constraint equations (3.12) are not fulfilled. This makes the linear system (3.14) over-constrained and non-compatible which does not have an exact solution that satisfies every equation. There are again two ways to circumvent this difficulty:

a) Sometimes this problem can be solved using Gaussian elimination with column pivoting and row scaling. Then, as long as $\mathbf{q}_i$ is approaching the true solution at which the constraint equations are fulfilled, the algorithm tends to disregard automatically the dependent equations. However, this procedure can not be considered in general sufficiently robust and reliable.

b) A reliable algorithm to solve the redundant system of linear equations (3.14) is the least-square formulation (Strang (1980)). Let us consider the normal equations corresponding to system (3.14):

$$\left(\Phi_{\mathbf{q}}^{T} \Phi_{\mathbf{q}}\right)_i \left(\mathbf{q}_{i+1} - \mathbf{q}_i\right) = -\left(\Phi_{\mathbf{q}}^{T}\right)_i \left(\Phi\right)_i \tag{3.15}$$

This algorithm converges on a very reliable way to the exact solution of all constraint equations. It has been gathered from numerous simulations that it allows large displacements in the input coordinates with fast and reliable convergence.

It can be argued that the solution of system (3.15) is less efficient than the solution of equation (3.14), mainly because the product $(\Phi_q^T \Phi_q)_i$ needs to be performed prior to the solution. However, practical experience has shown that even for non-redundant systems, equation (3.15) can be more efficient than its counterpart (3.14). In large multibody systems, matrix Φ_q tends to be very sparse, and then the product $(\Phi_q^T \Phi_q)_i$ can be carried out very efficiently. System (3.15), although perhaps less sparse than system (3.14), has the advantage of being symmetric with the possibility of saving storage and using simpler pivoting strategies.

3.5 Subspace of Allowable Motions

In kinematic problems, the motion of the input elements or driven degrees of freedom is already known. From this knowledge the motion of the remaining bodies (or elements) is calculated. In the case of the direct or forward dynamic problem, the motion of the input elements is not known, or at least, it is not known for all of them, and the motion is obtained as a solution to the dynamic differential equations. Before entering the study of the dynamic problems treated in Chapters 4 and 5, we will study in this section the *possible* or *allowable* motions that the multibody system may have in accordance with the constraint equations. The study of these possible motions and the methods of expressing them is a purely kinematic problem that has important implications in the formulation of the differential equations of motion. These allowable motions will be studied next, and it will be distinguished, in order to introduce the subject progressively, between scleronomous and rheonomous constraints.

We will see in this section that the actual velocity vector $\dot{q}$ of a constrained multibody system is always a vector that belongs to a very particular vector space called the *space of allowable motions*. The term *motions* should actually be *velocities*. The study of this vector space and the ability to find a basis for it constitute very important points for both kinematics and dynamics multibody formulations. Many authors have been explicitly or implicitly referring to it. See for instance: Kamman and Huston (1984), Kim and Vanderploeg (1986), Many et al. (1985), Agrawal (1984), Kane and Levinson (1985), Ider and Amirouche (1988), Huston (1990), and others. However, we find that the concept of the *space of allowable motions* allows for a simpler and more general way to explain, on a unified background, many different ideas and formulations that have been introduced in the last years. This concept is also the key towards the understanding of the improved real time dynamic formulations that will be studied in Chapters 5 and 8.

3.5.1 Scleronomous Systems

Consider a system with m constraint equations that do not depend explicitly on the time variable t, n dependent coordinates, and $f=n-m$ degrees of freedom. The constraint equations only depend on the dependent coordinates vector $\mathbf{q}$, and can be formulated as

$$\Phi(\mathbf{q}) = 0 \tag{3.16}$$

The velocity and acceleration equations are obtained by differentiating (3.16) with respect to time:

$$\Phi_{\mathbf{q}}(\mathbf{q})\,\dot{\mathbf{q}} = 0 \tag{3.17}$$

$$\Phi_{\mathbf{q}}(\mathbf{q})\,\ddot{\mathbf{q}} = -\dot{\Phi}_{\mathbf{q}}\,\dot{\mathbf{q}} \tag{3.18}$$

Equation (3.17) indicates that the velocity vector $\dot{\mathbf{q}}$ of a multibody system, at a specific position, belongs to the *nullspace* of the Jacobian matrix $\Phi_{\mathbf{q}}$ of the constraint equations. The theory of linear systems of equations (Strang (1980)) establishes that if the matrix $\Phi_{\mathbf{q}}$ has m independent rows and n columns ($m+f=n$), (it is of rank m, because the rank is equal to the number of rows), then the nullspace of $\Phi_{\mathbf{q}}$ is the *subspace of the possible or allowable motions* (velocities), in the sense that any possible velocity vector (compatible with the constraint equations) must belong to this subspace. The dimension of the space of allowable motions is the number of degrees of freedom $f=n-m$ of the multibody system.

Example 3.10

Consider again the four-bar mechanism with four dependent coordinates and one degree of freedom of Figure 3.1; thus, it has three constraint equations corresponding to the three constant distance conditions. The Jacobian matrix of the constraint equations for this mechanism (as already shown in Example 3.1) is

$$\Phi_{\mathbf{q}} = \begin{bmatrix} x_1-x_A & y_1-y_A & 0 & 0 \\ x_1-x_2 & y_1-y_2 & x_2-x_1 & y_2-y_1 \\ 0 & 0 & x_2-x_B & y_2-y_B \end{bmatrix}$$

By symbolically performing a Gaussian elimination of this matrix, it can easily be demonstrated that its nullspace is defined by the following column vector:

$$\mathbf{r} = \begin{Bmatrix} (y_1-y_A)[(y_2-y_1)(x_2-x_B)-(x_2-x_1)(y_2-y_B)] \\ -(x_1-x_A)[(y_2-y_1)(x_2-x_B)-(x_2-x_1)(y_2-y_B)] \\ (y_2-y_B)[(y_1-y_2)(x_1-x_A)-(x_1-x_2)(y_1-y_A)] \\ (x_2-x_B)[(y_1-y_2)(x_1-x_A)-(x_1-x_2)(y_1-y_A)] \end{Bmatrix}$$

Since in this case the nullspace has dimension 1, the vector $\mathbf{r}$ completely defines this subspace. In other words, any possible velocity vector $\dot{\mathbf{q}}$ shall be equal to the vector $\mathbf{r}$ multiplied by a specific factor.

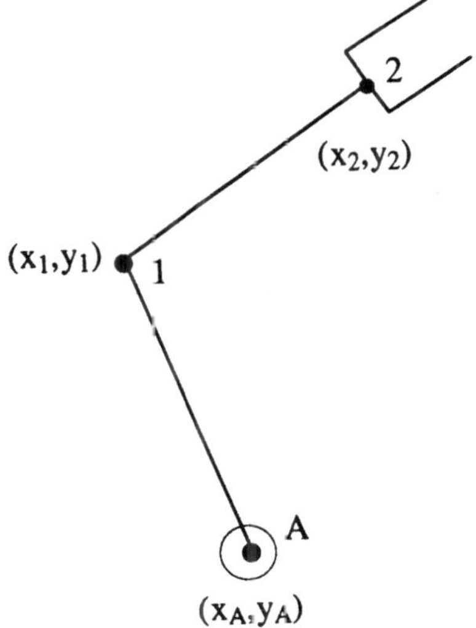

Figure 3.16. Planar robotic manipulator modeled with natural coordinates.

Example 3.11

Figure 3.16 shows a planar robot arm with four dependent natural coordinates and two degrees of freedom. There must be two independent constraint equations, which in this case are the corresponding constant distance equations. Their Jacobian matrix is similar to that of Example 3.10, except for the last row which is in this case eliminated,

$$\Phi_q = \begin{bmatrix} x_1-x_A & y_1-y_A & 0 & 0 \\ x_1-x_2 & y_1-y_2 & x_2-x_1 & y_2-y_1 \end{bmatrix}$$

The velocity equation will be

$$\begin{bmatrix} x_1-x_A & y_1-y_A & 0 & 0 \\ x_1-x_2 & y_1-y_2 & x_2-x_1 & y_2-y_1 \end{bmatrix} \begin{Bmatrix} \dot{x}_1 \\ \dot{y}_1 \\ \dot{x}_2 \\ \dot{y}_2 \end{Bmatrix} = \begin{Bmatrix} 0 \\ 0 \end{Bmatrix}$$

In order to find a basis of the nullspace of this matrix, one may find two linearly independent vectors that belong to the said subspace.

This can be done as follows: by making, $\dot{x}_2 = 1$, $\dot{y}_2 = 0$ the following vector is obtained:

$$\mathbf{r}^1 = \begin{Bmatrix} \dfrac{-(x_2-x_1)(x_1-x_A)}{(y_1-y_2)(x_1-x_A)-(x_1-x_2)(y_1-y_A)} \\[2ex] \dfrac{(x_2-x_1)(y_1-y_A)}{(y_1-y_2)(x_1-x_A)-(x_1-x_2)(y_1-y_A)} \\[2ex] 1 \\[2ex] 0 \end{Bmatrix}$$

Similarly, making $\dot{x}_2 = 0$, $\dot{y}_2 = 1$, the following vector is obtained:

$$\mathbf{r}^2 = \left(\begin{array}{c} \dfrac{(y_2-y_1)(y_1-y_A)}{(y_1-y_2)(x_1-x_A)-(x_1-x_2)(y_1-y_A)} \\[2ex] \dfrac{-(y_2-y_1)(x_1-x_A)}{(y_1-y_2)(x_1-x_A)-(x_1-x_2)(y_1-y_A)} \\[2ex] 0 \\[2ex] 1 \end{array} \right)$$

It is evident that vectors $\mathbf{r}^1$ and $\mathbf{r}^2$ are independent in that one of them can never be obtained by multiplying the other by a constant. Therefore, a basis of the subspace of allowable motions can be formed. Any possible velocity vector of the mechanism in Figure 3.16 can be expressed as a linear combination of $\mathbf{r}^1$ and $\mathbf{r}^2$ as follows:

$$\dot{\mathbf{q}} = \mathbf{r}^1\,\dot{z}_1 + \mathbf{r}^2\,\dot{z}_2$$

where $\dot{z}_1$ and $\dot{z}_2$ are the coefficients of the linear combination, namely, *the independent velocities* of the mechanism.

An attempt will be made further on to generalize all that stated in the previous examples. The vector $\dot{\mathbf{q}}$ characterizes the velocity of the system with n dependent coordinates. To represent the velocity of the multibody system with a lower number of variables, one should construct a basis for the nullspace, so that the velocity of the system can be represented by means of a new vector $\dot{z}$, whose components are those of the vector $\dot{\mathbf{q}}$ on the chosen nullspace basis. Vector $\dot{z}$ will have only $f=n-m$ components which will be independent.

Let $\mathbf{r}^i$ $(i=1, 2, ..., f)$ be a set of f linearly independent vectors that constitute a basis of the nullspace of $\Phi_\mathbf{q}$. Any dependent velocity vector $\dot{\mathbf{q}}$ can be expressed as a linear combination of this basis as follows:

$$\dot{\mathbf{q}} = \mathbf{r}^1\,\dot{z}_1 + \mathbf{r}^2\,\dot{z}_2 + ... + \mathbf{r}^f\,\dot{z}_f \tag{3.19}$$

Introducing an $(n\math{\times}f)$ matrix $\mathbf{R}$, whose columns are the vectors $\mathbf{r}^i$, this expression can be written as

$$\dot{\mathbf{q}} = \mathbf{R}\,\dot{z} \tag{3.20}$$

Matrix $\mathbf{R}$ thus defined plays a very important role in some of the most efficient formulations for dynamic analysis. Since vectors $\mathbf{r}^i$ are the components of a basis of the nullspace of the Jacobian matrix, it can be verified that

$$\Phi_\mathbf{q}(\mathbf{q})\,\mathbf{r}^i = 0 \qquad (i = 1, 2, ..., f) \tag{3.21}$$

and consequently

$$\Phi_\mathbf{q}(\mathbf{q})\,\mathbf{R} = 0 \tag{3.22}$$

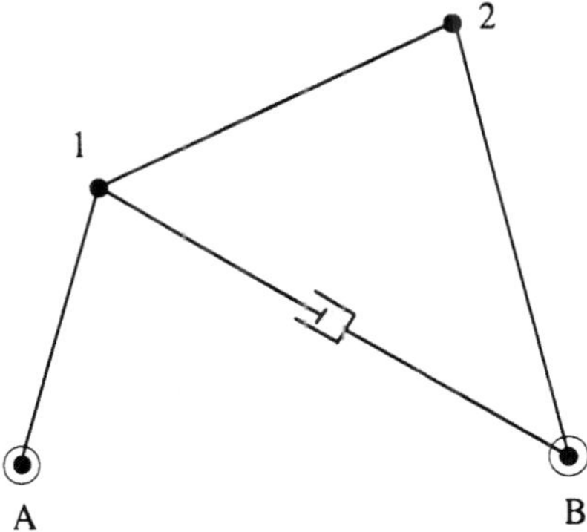

Figure 3.17. Hydraulically-driven four-bar mechanism.

The previous expression is a reminder that the matrix $\mathbf{R}$ depends on the position vector $\mathbf{q}$, and therefore there is a different matrix $\mathbf{R}$ for each of the positions of the multibody system.

The results obtained for the velocities can also be extended to the accelerations. We must search for a way of expressing the dependent accelerations $\ddot{\mathbf{q}}$ in terms of f independent accelerations $\ddot{\mathbf{z}}$. Differentiating (3.20) with respect to time we arrive at the following expression:

$$\ddot{\mathbf{q}} = \mathbf{R}\,\ddot{\mathbf{z}} + \dot{\mathbf{R}}\,\dot{\mathbf{z}} \tag{3.23}$$

Note that now both matrices $\mathbf{R}$ and $\dot{\mathbf{R}}$ are needed for the acceleration transformation. The calculation method for matrix $\dot{\mathbf{R}}$ depends on the method adopted to form $\mathbf{R}$ and will be seen later on in this chapter.

Even though in the simple examples presented in this section the matrix $\mathbf{R}$ has been calculated symbolically. In practice, this matrix needs to be calculated numerically. However, both the concepts and applications concerning the matrix $\mathbf{R}$ remain of general validity.

3.5.2 Rheonomous Systems

Rheonomous systems are characterized by the fact that some of the constraint equations depend on the time variable. This general case will be used to generalize the concepts and ideas introduced previously. For rheonomous systems the analytical expression for the constraint equations become:

$$\Phi(\mathbf{q}, t) = 0 \tag{3.24}$$

The velocity and acceleration equations are obtained by differentiating this equation with respect to time once and twice, respectively:

$$\Phi_{\mathbf{q}}(\mathbf{q}, t)\,\dot{\mathbf{q}} = -\Phi_t \equiv \mathbf{b} \tag{3.25}$$

$$\Phi_{\mathbf{q}}(\mathbf{q}, t)\,\ddot{\mathbf{q}} = -\dot{\Phi}_t - \dot{\Phi}_{\mathbf{q}}\,\dot{\mathbf{q}} \equiv \mathbf{c} \tag{3.26}$$

where the dot indicates total derivative; and the sub index t, the partial derivative with respect to time. Equations (3.25) and (3.26) serve as definitions for the right-hand side vectors $\mathbf{b}$ and $\mathbf{c}$, which will be extensively used in the dynamic formulations of Chapter 5.

If all the degrees of freedom of the multibody system are controlled kinematically, that is, if the motion of all the input elements is known as a function of time, equations (3.25) and (3.26) constitute two systems of m equations with m unknowns controlled by rank m matrices. The solution of these equation systems is perfectly determined, and there should be no problem in finding this solution.

Example 3.12

Such is the case with all the mechanism's degrees of freedom controlled kinematically in the hydraulically driven four-bar mechanism shown in Figure 3.17, whose constraint equations are:

$$\left(x_1 - x_A\right)^2 + \left(y_1 - y_A\right)^2 - L_2^2 = 0$$

$$\left(x_2 - x_1\right)^2 + \left(y_2 - y_1\right)^2 - L_3^2 = 0$$

$$\left(x_2 - x_B\right)^2 + \left(y_2 - y_B\right)^2 - L_4^2 = 0$$

$$\left(x_1 - x_B\right)^2 + \left(y_1 - y_B\right)^2 - f(t)^2 = 0$$

It may be seen that the last constraint is time dependent, thus rheonomous.

It is noted that the general case, where some of the input elements can be controlled kinematically (their motions prescribed by means of rheonomous constraint equations), and others have their motion kinematically undetermined, constitutes a dynamic problem determined by the differential equations of motion. From here on, it will be assumed that the equation (3.25) has a total number of m independent constraint equations, corresponding to the constraints of rigid body, joints, and degrees of freedom kinematically controlled by means of rheonomous equations. If there are n dependent coordinates, there will be $(n-m)$ free or kinematically undetermined degrees of freedom.

We will introduce now a large family of methods in which the independent velocities $\dot{z}$ can be defined as the projection of the dependent velocities $\dot{q}$ on the rows of a constant (not time or position dependent) matrix $\mathbf{B}$

$$\dot{z} = \mathbf{B}\,\dot{q} \tag{3.27}$$

Equation (3.26) can be augmented by equation (3.27) to yield

$$\begin{bmatrix} \Phi_q \\ \mathbf{B} \end{bmatrix} \dot{q} = \begin{Bmatrix} \mathbf{b} \\ \dot{z} \end{Bmatrix} \tag{3.28}$$

Let us assume at this point that matrix $\mathbf{B}$, in addition to being constant, also fulfills the condition of having $f = n - m$ rows that are linearly independent from one another and also linearly independent of the m rows of Φ_q.

With these assumptions, the matrix in equation (3.28) can be inverted, and finding the vector $\dot{q}$ involves the solution of the following equation:

$$\dot{\mathbf{q}} = \left[\begin{array}{c} \mathbf{\Phi_q} \\ \mathbf{B} \end{array} \right]^{-1} \left\{ \begin{array}{c} \mathbf{b} \\ \dot{\mathbf{z}} \end{array} \right\} \equiv \mathbf{S}\mathbf{b} + \mathbf{R}\dot{\mathbf{z}} \tag{3.29}$$

where $\mathbf{S}$ is a matrix constituted by the m first columns of the inverse matrix of equation (3.29), and $\mathbf{R}$ is the matrix constituted by the $f=n-m$ last columns of the said inverse matrix. It can be verified that

$$\left[\begin{array}{c} \mathbf{\Phi_q} \\ \mathbf{B} \end{array} \right] \left[\begin{array}{c} \mathbf{\Phi_q} \\ \mathbf{B} \end{array} \right]^{-1} = \left[\begin{array}{c} \mathbf{\Phi_q} \\ \mathbf{B} \end{array} \right] [\, \mathbf{S}\ \mathbf{R}\,] = \left[\begin{array}{cc} \mathbf{\Phi_q S} & \mathbf{\Phi_q R} \\ \mathbf{B S} & \mathbf{B R} \end{array} \right] = \left[\begin{array}{cc} \mathbf{I} & \mathbf{0} \\ \mathbf{0} & \mathbf{I} \end{array} \right] \tag{3.30}$$

which demonstrates that the columns of matrix $\mathbf{R}$ pertain to and generate the nullspace of $\mathbf{\Phi_q}$.

Regarding the linear equation system (3.25) which is undetermined as long as a value is not given to the input velocities, equation (3.29) indicates that the general solution of the system is obtained as the sum of a *particular solution* of the complete equation (term $\mathbf{S}\mathbf{b}$) in addition to the *general solution* of the homogeneous equation (term $\mathbf{R}\dot{\mathbf{z}}$).

The result of equation (3.29) may be compared with the terminology commonly used in Kane's method (Kane and Levinson (1985)). The columns of matrix $\mathbf{R}$ are the *partial velocities* with respect to the generalized coordinates $\mathbf{z}$, and the term $\mathbf{S}\mathbf{b}$ constitutes the *partial velocities with respect to time*. However, the approach presented herein includes a more general algebraic method to compute these partial velocities for all kind of multibody systems: open or closed chains, overconstrained, singular positions, and so forth.

The acceleration equation can be obtained in a similar manner. Augmenting equation (3.26) with the derivative with respect to time of equation (3.27), we obtain:

$$\left[\begin{array}{c} \mathbf{\Phi_q} \\ \mathbf{B} \end{array} \right] \{ \ddot{\mathbf{q}} \} = \left\{ \begin{array}{c} \mathbf{c} \\ \ddot{\mathbf{z}} \end{array} \right\} \tag{3.31}$$

and the inversion of this matrix:

$$\ddot{\mathbf{q}} = \left[\begin{array}{c} \mathbf{\Phi_q} \\ \mathbf{B} \end{array} \right]^{-1} \left\{ \begin{array}{c} \mathbf{c} \\ \ddot{\mathbf{z}} \end{array} \right\} = \mathbf{S}\,\mathbf{c} + \mathbf{R}\,\ddot{\mathbf{z}} \tag{3.32}$$

This expression, analogous to expression (3.29), indicates that matrix $\mathbf{R}$ can be calculated by triangularizing the leading matrix of systems (3.28) or (3.31), and performing f successive forward and backward substitutions with the f last columns of unit matrix $\mathbf{I}$ as the RHS terms.

Some of the dynamic formulations that will be seen in Chapter 5 require the calculation of the term $(\mathbf{S}\mathbf{c})$ in expression (3.32). This term generalizes the role of the term $(\mathbf{R}\dot{\mathbf{z}})$ in equation (3.23). In order to determine $(\mathbf{S}\mathbf{c})$, one possibility is to calculate the matrix $\mathbf{S}$ by the same method used to calculate $\mathbf{R}$ and then multiply by the known vector $\mathbf{c}$. This way is valid but not very efficient. It is not necessary to calculate matrix $\mathbf{S}$, but just to calculate the product $(\mathbf{S}\mathbf{c})$. From expression (3.32), it is concluded that the product $(\mathbf{S}\mathbf{c})$ is $\ddot{\mathbf{q}}$ when $\ddot{\mathbf{z}}$ is zero. By making $\ddot{\mathbf{z}}$ equal to zero in expression (3.31) and finding $\ddot{\mathbf{q}}$, one can arrive at the desired

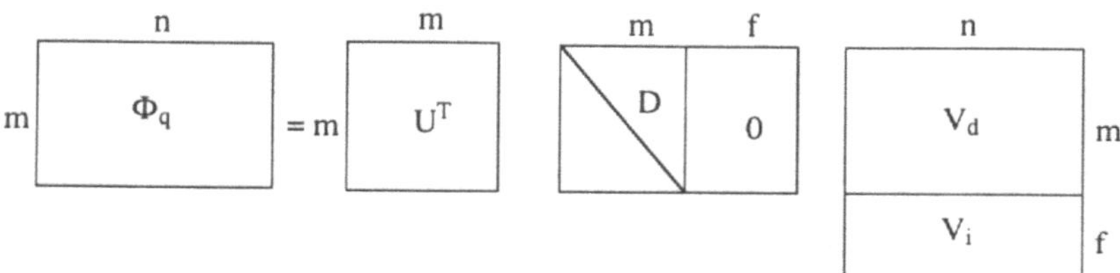

Figure 3.18. Singular value decomposition of the matrix Φ_q.

term. Since the leading matrix of system (3.31) has been previously triangularized when finding matrix **R**, the calculation of the term being considered requires very little additional effort.

During the preceding development one has been able to see that the inclusion of rheonomous links in the analysis can be carried out very simply and efficiently. The methods currently used to determine a basis of the subspace of allowable motions (matrix **R**) are divided into two large groups, the *projection* methods and the *orthogonalization* methods, which will be studied in the following sections.

3.5.3 Calculation of Matrix R: Projection Methods

Projection methods are based on defining the independent velocities $\dot{z}$ as the projection of the dependent velocities $\dot{q}$ on the rows of a known constant matrix **B**:

$$\dot{z} = B\,\dot{q} \tag{3.33}$$

Following the mathematical formulation of Section 3.5.2 for the general case of rheonomous systems, the following expressions have been obtained:

$$\begin{bmatrix} \Phi_q \\ B \end{bmatrix} \dot{q} = \begin{Bmatrix} b \\ \dot{z} \end{Bmatrix} \tag{3.34}$$

$$\dot{q} = \begin{bmatrix} \Phi_q \\ B \end{bmatrix}^{-1} \begin{Bmatrix} b \\ \dot{z} \end{Bmatrix} = Sb + R\dot{z} \tag{3.35}$$

$$\begin{bmatrix} \Phi_q \\ B \end{bmatrix} \ddot{q} = \begin{Bmatrix} c \\ \ddot{z} \end{Bmatrix} \tag{3.36}$$

$$\ddot{q} = \begin{bmatrix} \Phi_q \\ B \end{bmatrix}^{-1} \begin{Bmatrix} c \\ \ddot{z} \end{Bmatrix} = Sc + R\ddot{z} \tag{3.37}$$

It is clear that these expressions completely define the transformation between dependent and independent variables.

This only leaves matrix **B** to be determined. Once this matrix is calculated, it remains constant during a large range of motion of the multibody system. The condition with which matrix $\mathbf{B}$ must comply in order for the inverse matrix in expressions (3.35) and (3.37) to exist, is that its $n{-}m$ rows must be independent from one another and independent from the m rows of matrix Φ_q. We can identify and describe in this context three methods that have been proposed in the literature to construct the matrix $\mathbf{B}$. These will be reviewed below.

1. *Method based on the Singular Value decomposition.* Singular Value decomposition (SV) is a generalization of the eigenvalue and eigenvector concept applicable to rectangular matrices. The SV decomposes a rectangular matrix such as Φ_q, as indicated in the sketch of Figure 3.18, or shortly:

$$\Phi_q = \mathbf{U}^T \mathbf{D} \mathbf{V} \tag{3.38}$$

where matrix $\mathbf{U}$ is orthogonal (its inverse is equal to its transpose and its rows are mutually orthogonal) of size $(m{\times}m)$. Matrix $\mathbf{D}$ is composed of a diagonal matrix of size $(m{\times}m)$ that contains the singular values and a zero matrix given by $f{=}n{-}m$ last columns. Matrix $\mathbf{V}$ is orthogonal of size $(n{\times}n)$ and can be decomposed into two sub-matrices $\mathbf{V}_d$ and $\mathbf{V}_i$ of sizes $(m{\times}n)$ and $(f{\times}n)$ respectively, according to the partition in $\mathbf{D}$. The most important property of the SV decomposition that pertains to the problem at hand is that the rows of the matrix $\mathbf{V}_i$ constitute an *orthogonal basis* of the nullspace of matrix Φ_q. In other words, it is verified that

$$\Phi_q \mathbf{V}_i^T = 0 \tag{3.39}$$

In view of this expression, Singh and Likins (1985) proposed constructing the matrix $\mathbf{R}$ directly from the SV decomposition. The problem is that the SVD is essentially an iterative process in some ways similar to the calculation of all the eigenvalues and eigenvectors of a matrix. This process consumes a great deal of computer time, and it is absolutely impractical to carry out at each position q of the system. Other authors (Mani, Haug, and Atkinson (1985)) have proposed using the SV decomposition to calculate the matrix $\mathbf{B}$. This operation only needs to be performed once or at most a few times throughout the entire range of the motion of the multibody system. Remember that many matrices $\mathbf{R}$, corresponding to different positions q of the multibody system, can be calculated from only one matrix $\mathbf{B}$. Matrix $\mathbf{B}$ continues to be valid as long as its rows are independent from those of $\Phi_q(q)$.

Equation (3.39) indicates that the rows of matrix $\mathbf{V}_i$ are orthogonal to the rows of Φ_q at the position q, for which the SV decomposition has been performed. This means that matrix $\mathbf{V}_i$ complies with the conditions required for matrix $\mathbf{B}$, so long as no large changes are produced in the positions q and, thus in matrix Φ_q, that the linear independence condition between the rows of the said matrix and those of matrix $\mathbf{B}$ is lost. The following example helps to clarify this point.

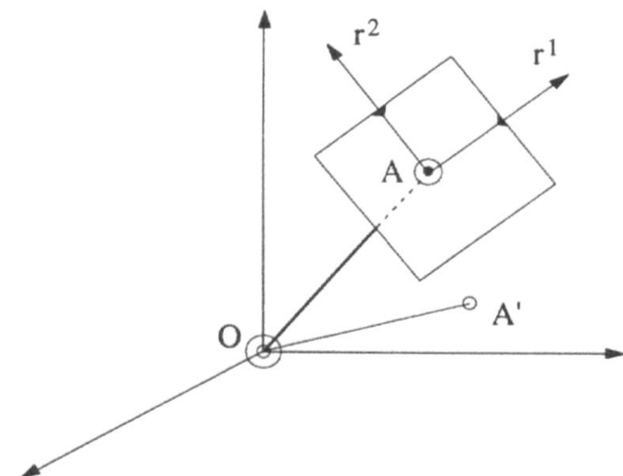

Figure 3.19. Graphical representation of the columns of the matrix **B**.

Example 3.13

Demonstrate that after performing the singular value decomposition of Φ_q, under the condition that $\mathbf{R} = \mathbf{V}_i^T$, the following relation $\mathbf{B} = \mathbf{V}_i$ is satisfied.

After the singular value decomposition Φ_q is orthogonal to $\mathbf{V}_i^T$, we can take $\mathbf{R} = \mathbf{V}_i^T$, and also

$$\dot{\mathbf{q}} = \mathbf{R} \ \dot{\mathbf{z}}$$

and

$$\dot{\mathbf{z}} = \mathbf{B} \ \dot{\mathbf{q}} = \mathbf{B} \mathbf{R} \ \dot{\mathbf{z}}$$

Therefore $\mathbf{B} \mathbf{R} = \mathbf{I}$. Since $\mathbf{R} = \mathbf{V}_i^T$ and $\mathbf{V}_i$ is an orthogonal matrix, the following relationship: $\mathbf{B} = \mathbf{V}_i$ immediately follows.

This method of calculating the matrix **R** has a very simple geometric interpretation that can be seen in Figure 3.19. Let's assume that bar OA is fixed at O by means of a spherical pair. If the rotation around the axis OA is not considered, this mechanism has two degrees of freedom and three natural coordinates, the Cartesian coordinates of point A. Thus, it will have one constraint equation which will be the constant distance condition between points O and A. In this case $m=1$, $n=3$ and $f=2$.

The subspace of the possible movements has a dimension 2, and is formed by the plane perpendicular to OA through A, since in fact all the possible velocity vectors of point A are contained in the said plane. Vectors $\mathbf{r}^1$ and $\mathbf{r}^2$ constitute an orthogonal basis of the subspace. It has been possible to calculate them by means of the SV decomposition of matrix Φ_q at the position OA.

Now it can be assumed that the mechanism moves and changes to a new position OA'. Matrix **B** continues being defined by vectors $\mathbf{r}^1$ and $\mathbf{r}^2$, calculated at the position OA. The independent velocities $\dot{z}_1$ and $\dot{z}_2$ are the projections of the velocity of A' on the axes $\mathbf{r}^1$ and $\mathbf{r}^2$, respectively. Note that the velocity of A' is no longer contained in the tangent plane at A but can be easily determined

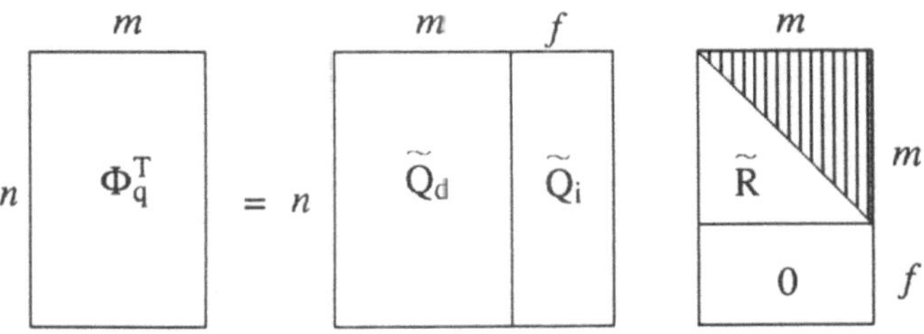

Figure 3.20. QR Decomposition of the matrix Φ_q.

from its projections on $\mathbf{r}^1$ and $\mathbf{r}^2$, and by the condition that it is perpendicular to OA' (constraint condition for velocities).

2. *Method based on the QR decomposition.* This method of constructing matrix **B** is similar to the previous one, but it uses the QR instead of the SV decomposition. The main advantage is that QR decomposition is a direct, not iterative, process which requires considerably fewer arithmetical operations as indicated by Kim and Vanderploeg (1986).

The QR method decomposes the matrix Φ_q^T as indicated in the sketch of Figure 3.20, or briefly,

$$\Phi_q^T = \widetilde{Q}\,\widetilde{R} \tag{3.40}$$

where $\widetilde{Q}$ is an orthogonal $(n{\times}n)$ matrix, and $\widetilde{R}$ is a rectangular $(n{\times}m)$ matrix formed by an upper triangular matrix $(m{\times}m)$ and a zero matrix of order $(f{\times}m)$. Note that a tilde has been used to distinguish the result of the QR decomposition from the matrix **Q** that symbolizes the forcing vector in dynamic analysis (Chapter 4) and the matrix **R** (basis of the nullspace of the Jacobian matrix). The application of this decomposition to the problem at hand is straightforward when considering that the f last columns of $\widetilde{Q}$ which define the sub-matrix $\widetilde{Q}_i$ constitute an orthogonal basis of the nullspace of the matrix Φ_q. This matrix can be written as

$$\mathbf{B} = \widetilde{Q}_i^T \tag{3.41}$$

and likewise verified that

$$\Phi_q(\mathbf{q})\,\widetilde{Q}_i = 0 \tag{3.42}$$

This matrix **B** is used in exactly the same way as that calculated by means of the SV decomposition. The QR decomposition is carried out at a determined position **q** of the multibody system. The matrix **B** is formed, and the matrix **R** is calculated for the successive positions of the multibody system using equation (3.35), without recalculating the matrix **B**. This recalculation will have to be done when its rows become a linear combination of those of Φ_q.

The geometric interpretation of the method based on QR decomposition is similar to that of the SV decomposition. Returning to Figure 3.19, matrix **B** formed by vectors $\mathbf{r}^1$ and $\mathbf{r}^2$ gives inadequate results when the bar moves to a position perpendicular to OA and therefore, parallel to the plane defined by $\mathbf{r}^1$ and $\mathbf{r}^2$. At this position the derivative of the constant distance condition in matrix Φ_q is a linear combination of $\mathbf{r}^1$ and $\mathbf{r}^2$ and the matrix of equations (3.34) and (3.36) cannot be inverted.

Both the QR and SV decomposition can be carried out by means of standard Fortran or C subroutines, contained in the readily available IMSL, Harwell, NAG, and other mathematical libraries.

3. *Method based on Gaussian triangularization.* This method, described by Serna et al. (1982), is based on the triangularization of matrix Φ_q by means of the Gauss method with total pivoting. This triangularization implies decomposition of the Jacobian matrix in sub-matrices, as shown below:

$$\Phi_q \equiv \begin{bmatrix} \Phi_q^d & \Phi_q^i \end{bmatrix} \tag{3.43}$$

where matrix Φ_q^d is a square matrix ($m{\times}m$) that contains the columns of Φ_q in which the pivots have appeared. Matrix Φ_q^i contains the columns in which the pivots have not appeared and has the size ($m{\times}f$). In the theory of linear equation systems, the variables associated with columns Φ_q^i are called *independent variables*, and those associated with columns Φ_q^d are called *dependent variables*. The reason for this nomenclature is that to solve a system of m equations with n unknowns, with a matrix such as that in equation (3.43), it is necessary to assign a value to the independent variables and then, with matrix Φ_q^d reduced to triangular form, calculate the dependent variables with the corresponding forward and backward substitutions.

Once matrix Φ_q is triangularized as shown in equation (3.43), matrix $\mathbf{B}$ is a Boolean matrix constructed as follows:

$$\mathbf{B} \equiv \begin{bmatrix} \mathbf{0} & | & \mathbf{I} \end{bmatrix} \tag{3.44}$$

whereupon the matrix from which the inverse of matrix $\mathbf{R}$ is calculated is

$$\begin{bmatrix} \Phi_q \\ \mathbf{B} \end{bmatrix} = \begin{bmatrix} \Phi_q^d & | & \Phi_q^i \\ \mathbf{0} & | & \mathbf{I} \end{bmatrix} \tag{3.45}$$

The rows of matrix $\mathbf{B}$ defined in this way are formed by ones and zeros. Since matrix Φ_q^d is triangularizable, it is guaranteed that the rows of matrix $\mathbf{B}$ are independent from those of Φ_q. Note that the triangularization of matrix (3.45) is simpler than with the SV or QR decomposition. In the part corresponding to matrix $\mathbf{B}$, no additional work is necessary, since the zeros have already been obtained. With this method, matrix $\mathbf{R}$ is calculated more easily and with fewer arithmetical operations.

Other repercussions from choosing matrix $\mathbf{B}$, in accordance with equation (3.44), will be analyzed below. Particularizing equation (3.27) for this case,

$$\dot{z} = \begin{bmatrix} \mathbf{0} & | & \mathbf{I} \end{bmatrix} \dot{q} \tag{3.46}$$

This expression indicates that the independent velocities $\dot{z}$ are chosen as a *subset* or *extraction* of the dependent velocities $\dot{q}$. In other words, f elements of $\dot{q}$ have been chosen to form vector $\dot{z}$.

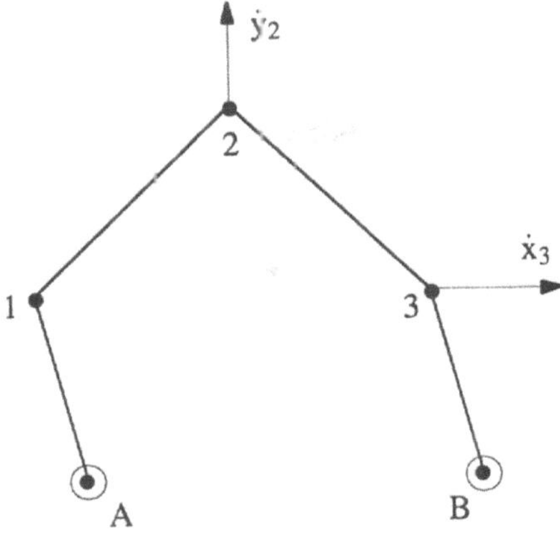

Figure 3.21. Five-bar mechanism.

Example 3.14

Let as consider a planar mechanism with five bars and two degrees of freedom, as shown in Figure 3.21. The independent coordinates are a subset of the dependent coordinates. Let's assume that the independent velocities are

$$\dot{z}_1 \equiv \dot{y}_2$$

$$\dot{z}_2 \equiv \dot{x}_3$$

According to expression (3.35), matrix $\mathbf{R}$ can be formed thusly. The first column of matrix $\mathbf{R}$ (vector $\mathbf{r}^1$) is the velocity vector of the mechanism, when the independent velocities have the following value:

$$\dot{y}_2 = 1 \qquad \dot{x}_3 = 0$$

The second column of matrix $\mathbf{R}$ (vector $\mathbf{r}^2$) is the velocity vector of the mechanism, when the independent velocities are:

$$\dot{y}_2 = 0 \qquad \dot{x}_3 = 1$$

Thus in this case and with this type of matrix $\mathbf{B}$, matrix $\mathbf{R}$ is especially easy to construct. It will suffice to alternatively give value 1 to each of the independent velocities, keeping all the others at the value 0. Therefore, the partition of matrix $\mathbf{R}$, which corresponds to the independent velocities, becomes the unit matrix $\mathbf{I}$.

This geometrical interpretation and significance of the independent velocities contrasts with that carried out for the SVD, starting from Figure 3.19.

3.5.4. Orthogonalization Methods

One can use the orthogonalization methods to try to obtain a matrix $\mathbf{R}$, whose columns are in some way orthogonal to one another at all times. The method that will be explained here was developed by Liang and Lance (1985) and is based on the previously considered matrix:

$$P \equiv \begin{bmatrix} \Phi_q \\ B \end{bmatrix} \tag{3.47}$$

where the matrix B can be constructed by means of any of the methods shown in previous sections, but preferably by means of the third method. This produces a Boolean matrix B in accordance with the partition of Φ_q, and determined by the Gaussian triangularization method with total pivoting.

The first step in this method consists of orthogonalizing the n rows of matrix P by means of the Gram-Schmidt orthogonalization method. This yields an (nxn) matrix as follows:

$$V \equiv \begin{bmatrix} V^d \\ V^i \end{bmatrix} \tag{3.48}$$

where the rows of V^d are the rows of Φ_q with each of them orthogonalized in relation to the previous ones, and where the rows of V^i are the rows of matrix B with each one of them orthogonalized in relation to the rows of V^i and to the previous rows of V^d.

According to the standard Gram-Schmidt orthogonalization process, all the orthonormal vectors v^i are calculated by means of the general expressions:

$$v^i = \alpha_i \left(p^i - \sum_1^{i-1} (v^j \cdot p^i) v^j \right) \tag{3.49}$$

and

$$\alpha_i = 1 \left/ \left| p_i - \sum_1^{i-1} \left(v^j \cdot p^i \right) v^j \right| \right. \tag{3.50}$$

The f last vectors obtained in this way correspond to the rows of matrix B, orthogonalized with respect to those of Φ_q and in relation to the previous rows of B. These f rows form matrix V^i (See equation (3.48)). This matrix is orthogonal to Φ_q. In addition, the rows of V^i are independent and mutually orthogonal. By the first condition, the rows of V^i pertain to the Φ_q nullspace, and by the second condition, they constitute a basis. Thus, matrix V^{iT} can be taken as matrix R.

The final step is the calculation of the term $(R\dot{z})$ or (Sc). Liang and Lance (1985) calculate matrix $\dot{R}$ explicitly, differentiating equations (3.49) and (3.50) with respect to time:

$$\dot{v}_i = \dot{\alpha}_i \left(p^i - \sum_1^{i-1} \left(v^j \cdot p^i \right) v^j \right) +$$
$$+ \alpha_i \left(\dot{p}^i - \sum_1^{i-1} \left(\dot{v}^j \cdot p^i + v^j \cdot \dot{p}^i \right) v^j - \sum_1^{i-1} \left(v^j \cdot p^i \right) \dot{v}^j \right) \tag{3.51}$$

and

$$\dot{\alpha}_i = - \left(p^i \cdot p^i - \sum_1^{i-1} \left(v^j \cdot p^i \right)^2 \right)^{-\frac{3}{2}} \left(\dot{p}^i \cdot p^i - \sum_1^{i-1} \left(v^j \cdot p^i \right) \left(\dot{v}^j \cdot p^i + v^j \cdot \dot{p}^i \right) \right) \tag{3.52}$$

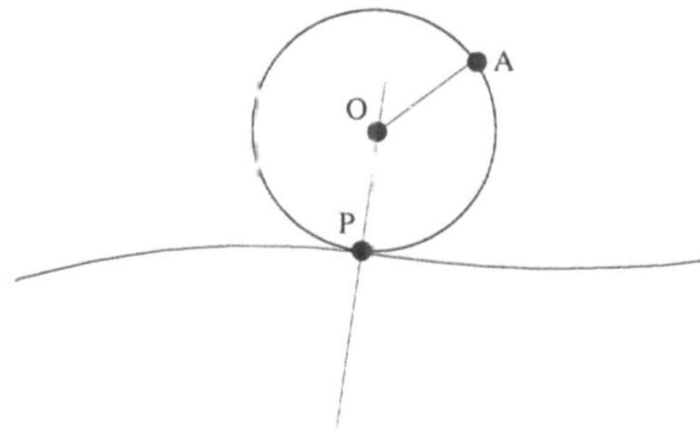

Figure 3.22. Disc modeled with two basic points, in contact with a surface.

In these expressions, the derivatives of the $\mathbf{P}$ rows are obtained from the derivatives of the rows of $\Phi_{\mathbf{q}}$. The derivatives or the $\mathbf{B}$ rows are zero.

The calculation of matrix $\mathbf{R}$ by means of equations (3.49) and (3.50) and of matrix $\dot{\mathbf{R}}$ by means of equations (3.51) and (3.52) requires an enormous computational effort, which is far greater than that required for the previously explained projection methods. Another important feature of this method is that the matrix $\mathbf{R}$ obtained depends on the order in which rows $\Phi_{\mathbf{q}}$ and $\mathbf{B}$ are considered, since the Gram-Schmidt orthogonalization depends on this order. The effect of this order is not yet known.

3.6 Multibody Systems with Non-Holonomic Joints

Non-holonomic pairs or joints are those whose constraint equation (joint constraint equations) do not depend on the dependent coordinates only but also on the dependent velocities by means of non-integrable equations. The rolling of a disc or wheel on a surface is a typical example of the non-holonomic constraint. It is the only one that will be discussed here. A distinction will be made between the planar case and the three-dimensional one.

3.6.1 Wheel Element in the Planar Case: First Method

Figure 3.22 shows a disc in contact with a surface. In principle, there can be two types of movement between the surface of the wheel and that of the track: *rolling* and rolling plus sliding (simply referred to as *sliding*).

The fact that the movement is one type or the other depends on the dynamic conditions of the problem (coefficient of friction and contact force), which will not be studied here. In this section, only those constraint equations corresponding to the rolling and sliding conditions will be studied.

There are at least two ways of establishing the constraint equations of the system shown in Figure 3.22. One way is to directly establish the non-holonomic

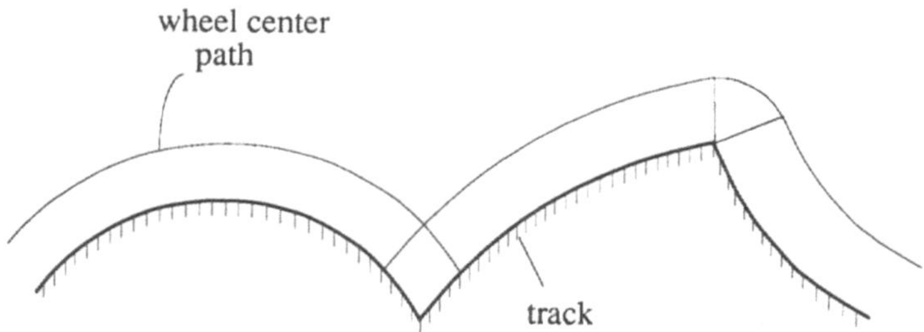

Figure 3.23. Wheel center path.

constraint equations in terms of dependent velocities. The second way is to substitute the non-holonomic joint with one or more equivalent holonomic joints for that position of the multibody system. Both ways will be seen further on. In either case, it will be assumed that the wheel is modeled by means of two basic points, one at the center and the other at the end of a radius. The movement of both points completely defines the movement of the wheel.

The constraint equations of non-holonomic joints are directly generated in terms of the velocities. Therefore, *there are no constraint equations for the position problem*. To solve the position problem with wheels and tracks, it is necessary to dispense with the wheels and find the position of their centers. It is known that they are located on a curve, as shown in Figure 3.23, which is the geometric place of the possible positions of the wheel center. Once the position of the center of the wheel is known, the wheel can be placed at the desired angular position.

Rolling. In the case that there is a rolling motion between the wheel and the track, the joint kinematic constraint equation establishes that the velocity of the wheel point in contact with the track shall be zero. If the track moves, the condition is that the velocity of the two points in contact is the same. From here on, it will be assumed that the track is stationary and all that stated previously will be applicable to the relative movement between the rolling wheel and the track.

Depending on the velocities of the basic points O and A, the condition that the velocity of point P be zero is equivalent to the conditions that the velocities of O and A be perpendicular to PO and PA respectively. When the coordinates of P, O, and A are known, these conditions are easily established by means of the scalar product of vectors:

$$(x_O - x_P)\,\dot{x}_O + (y_O - y_P)\,\dot{y}_O = 0 \tag{3.53}$$

$$(x_A - x_P)\,\dot{x}_A + (y_A - y_P)\,\dot{y}_A = 0 \tag{3.54}$$

Point P can be determined by means of the normal line traced from point O to the track.

Sliding. In the case that there is a sliding motion, the constraint condition establishes that point P does not have any velocity in the direction normal to the sur-

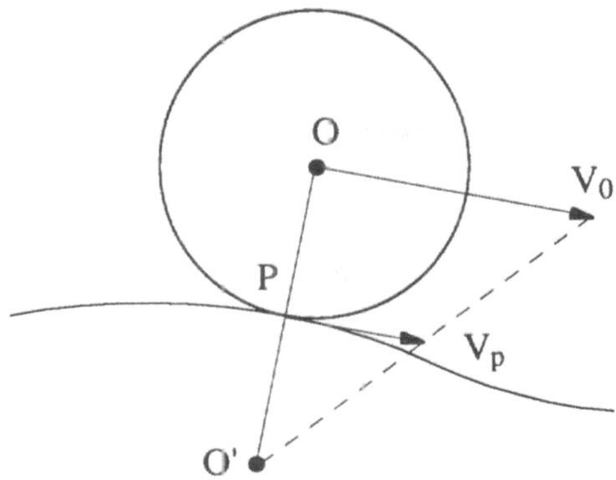

Figure 3.24. Description of the center of curvature of the wheel.

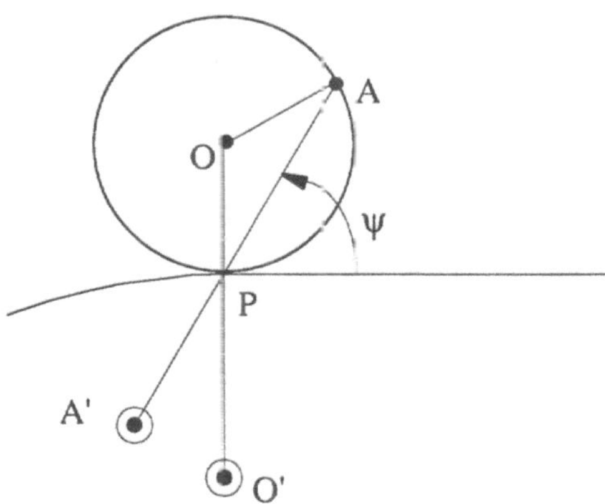

Figure 3.25. Substituting a non-holonomic constraint by a holonomic one.

faces in contact. This condition is equivalent to equation (3.53), which establishes that the velocity of point O is perpendicular to PO. Note that in this case the equation (3.54) is no longer valid. It should be noted that when there is a rolling movement, two degrees of freedom are restricted. If there is sliding, only one degree of freedom is restricted. This is in accordance with the number of equations that should be satisfied.

To derive equations (3.53) and (3.54) and to find the equations corresponding to the accelerations, one should bear in mind that in order for the equations to be valid at all times, point P must not belong to either the wheel or to the rolling track, but should be the mathematical point that always coincides with the pole (center of velocities). If equations (3.53) and (3.54) are differentiated with respect to the time, one obtains

$$(x_O{-}x_P)\,\ddot{x}_O + (\dot{x}_O{-}\dot{x}_P)\,\dot{x}_O + (y_O{-}y_P)\,\ddot{y}_O + (\dot{y}_O{-}\dot{y}_P)\,\dot{y}_O = 0 \qquad (3.55)$$

$$\left(x_A - x_P\right)\ddot{x}_A + \left(\dot{x}_A - \dot{x}_P\right)\dot{x}_A + \left(y_A - y_P\right)\ddot{y}_A + \left(\dot{y}_A - \dot{y}_P\right)\dot{y}_A = 0 \qquad (3.56)$$

In these equations, (x_P, y_P) are the components of the pole velocity which must be calculated, since equations (3.55) and (3.56) should not have unknowns other than the natural accelerations.

To calculate the pole velocity $\dot{r}_P$, one discovers the only solution is to consider the geometry of the trajectory. If O' is the center of curvature of the wheel center O trajectory (See Figure 3.24), points P and O are always aligned with O'. The ends of the corresponding velocity vectors are aligned also. From this it may be deduced that:

$$\dot{r}_P = \left(O'P\right)/\left(O'O\right) \cdot \dot{r}_O \qquad (3.57)$$

By decomposing this expression into its Cartesian coordinates and substituting them in equations (3.55) and (3.56), the constraint equations corresponding to the accelerations can be obtained.

3.6.2 Wheel Element in the Planar Case: Second Method

Another way of entering the rolling and sliding conditions is by substituting the non-holonomic joint for one or more equivalent holonomic joints.

In Figure 3.25, the said substitution is being carried out and is based on knowledge of the centers of curvature of the trajectories of the basic points O and A. The center of curvature of the trajectory of O is the center of curvature of the track at the point of contact P. The center of curvature of the trajectory of A can be calculated using the Euler-Savary formula:

$$\left(\frac{1}{PA} + \frac{1}{PA'}\right) sin\,\Psi = \frac{1}{PO} + \frac{1}{PO'} \qquad (3.58)$$

Rolling. In the case of a rolling motion, the non-holonomic joint is replaced by the articulated quadrilateral $A'AOO'$, since the distance between the point and the center of curvature of its trajectory is constant and has first and second constant derivatives. Therefore, this equivalence is instantaneously valid for velocities and accelerations. Note that at another time, the equivalent four-bar mechanism will be different. For the position of Figure 3.25, the constraint equations are

$$\left(x_A - x_{A'}\right)^2 + \left(y_A - y_{A'}\right)^2 - L_{AA'}^2 = 0 \qquad (3.59)$$

$$\left(x_O - x_{O'}\right)^2 + \left(y_O - y_{O'}\right)^2 - L_{OO'}^2 = 0 \qquad (3.60)$$

Sliding. In the case of a sliding movement, only the constant distance condition between points O and O' should be imposed. The equations corresponding to velocities and accelerations are obtained by differentiating equations (3.59) and (3.60) and considering A' and O' as fixed points.

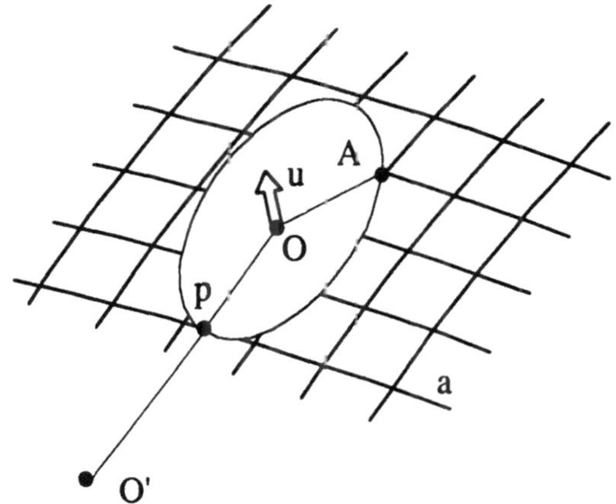

Figure 3.26. Modeling of a three-dimensional wheel with natural coordinates.

3.6.3 Wheel Element in the Three-Dimensional Case

Consider the three-dimensional wheel element shown in Figure 3.26, which is formed by two basic points O and A and a unit vector **u** perpendicular to the plane of the wheel and associated to point O.

Rolling. The constraint equations corresponding to this non-holonomic joint are defined by the condition that the velocity of material point P (pertaining to the wheel) is zero:

$$\dot{\mathbf{r}}_P = 0 \tag{3.61}$$

The vector equation (3.61) represents three algebraic equations. Therefore three degrees of freedom are restricted by this joint when there is a rolling movement, thus only permitting the three rotations about point P. It is necessary now to consider equation (3.61), in accordance with the natural velocities and coordinates. The velocities of points O and A can be expressed in terms of the angular velocity vector ω:

$$\dot{\mathbf{r}}_O = \omega \wedge \left(\mathbf{r}_O - \mathbf{r}_P\right) \tag{3.62}$$

$$\dot{\mathbf{r}}_A = \omega \wedge \left(\mathbf{r}_A - \mathbf{r}_P\right) \tag{3.63}$$

which can be expanded to yield

$$\dot{x}_O = \omega_y \left(z_O - z_P\right) - \omega_z \left(y_O - y_P\right) \tag{3.64}$$

$$\dot{y}_O = \omega_z \left(x_O - x_P\right) - \omega_x \left(z_O - z_P\right) \tag{3.65}$$

$$\dot{z}_O = \omega_x \left(y_O - y_P\right) - \omega_y \left(x_O - x_P\right) \tag{3.66}$$

$$\dot{x}_A = \omega_y \left(z_A - z_P\right) - \omega_z \left(y_A - y_P\right) \tag{3.67}$$

$$\dot{y}_A = \omega_z \left(x_A - x_P\right) - \omega_x \left(z_A - z_P\right) \tag{3.68}$$

$$\dot{z}_A = \omega_x \left(y_A - y_P\right) - \omega_y \left(x_A - x_P\right) \tag{3.69}$$

Three conveniently selected equations among the previous ones, which are two corresponding to one point and one to the other, permit determining the vector ω. For example, by selecting equations (3.64), (3.65), and (3.69), the following matrix equation can be written:

$$\begin{bmatrix} 0 & \left(z_O - z_P\right) & -\left(y_O - y_P\right) \\ -\left(z_O - z_P\right) & 0 & \left(x_O - x_P\right) \\ \left(y_A - y_P\right) & -\left(x_A - x_P\right) & 0 \end{bmatrix} \begin{Bmatrix} \omega_x \\ \omega_y \\ \omega_z \end{Bmatrix} = \begin{Bmatrix} \dot{x}_O \\ \dot{y}_O \\ \dot{z}_A \end{Bmatrix} \tag{3.70}$$

If the matrix of this system is not singular, one can find the angular velocity vector ω and substitute its value in the remaining equations (3.66), (3.67), and (3.68), which are those not used to determine the angular velocity vector. Consequently

$$\begin{Bmatrix} \dot{z}_O \\ \dot{x}_A \\ \dot{y}_A \end{Bmatrix} =$$

$$= \begin{bmatrix} \left(y_O - y_P\right) & -\left(x_O - x_P\right) & 0 \\ 0 & \left(z_A - z_P\right) & -\left(y_A - y_P\right) \\ -\left(z_A - z_P\right) & 0 & -\left(x_A - x_P\right) \end{bmatrix} \begin{bmatrix} 0 & \left(z_O - z_P\right) & -\left(y_O - y_P\right) \\ -\left(z_O - z_P\right) & 0 & -\left(x_O - x_P\right) \\ \left(y_A - y_P\right) & -\left(x_A - x_P\right) & 0 \end{bmatrix}^{-1} \begin{Bmatrix} \dot{x}_O \\ \dot{y}_O \\ \dot{z}_A \end{Bmatrix} \tag{3.71}$$

This equation can be considered as the constraint equation in velocities for the rolling joint. In practice, these equations must be numerically evaluated.

In the case of accelerations, it is necessary to differentiate equations (3.64)-(3.69), considering that P is the mathematical point of contact:

$$\ddot{x}_O = \dot{\omega}_y \left(z_O - z_P\right) - \dot{\omega}_z \left(y_O - y_P\right) + \omega_y \left(\dot{z}_O - \dot{z}_P\right) - \omega_z \left(\dot{y}_O - \dot{y}_P\right) \tag{3.72}$$

$$\ddot{y}_O = \dot{\omega}_z \left(x_O - x_P\right) - \dot{\omega}_x \left(z_O - z_P\right) + \omega_z \left(\dot{x}_O - \dot{x}_P\right) - \omega_x \left(\dot{z}_O - \dot{z}_P\right) \tag{3.73}$$

$$\ddot{z}_O = \dot{\omega}_x \left(y_O - y_P\right) - \dot{\omega}_y \left(x_O - x_P\right) + \omega_x \left(\dot{y}_O - \dot{y}_P\right) - \omega_y \left(\dot{x}_O - \dot{x}_P\right) \tag{3.74}$$

$$\ddot{x}_A = \dot{\omega}_y \left(z_A - z_P\right) - \dot{\omega}_z \left(y_A - y_P\right) + \omega_y \left(\dot{z}_A - \dot{z}_P\right) - \omega_z \left(\dot{y}_A - \dot{y}_P\right) \tag{3.75}$$

$$\ddot{y}_A = \dot{\omega}_z \left(x_A - x_P\right) - \dot{\omega}_x \left(z_A - z_P\right) + \omega_z \left(\dot{x}_A - \dot{x}_P\right) - \omega_x \left(\dot{z}_A - \dot{z}_P\right) \tag{3.76}$$

$$\ddot{z}_A = \dot{\omega}_x \left(y_A - y_P\right) - \dot{\omega}_y \left(x_A - x_P\right) + \omega_x \left(\dot{y}_A - \dot{y}_P\right) - \omega_y \left(\dot{x}_A - \dot{x}_P\right) \tag{3.77}$$

From equations (3.72), (3.73), and (3.77), the angular acceleration vector $\dot{\omega}$ can be found. It will be a function of the coordinates, the velocities, and vector ω. By substituting in the three remaining equations, one obtains three ratios between the natural accelerations of points A and O. Vector ω, which can be replaced by the equation (3.70), intervenes in these ratios. In the resulting equation, the only unknown term is the pole velocity $\dot{r}_p$. Point O´ is the center of curvature of the intersection of the disc plane with the contact surface (line (a) in Figure 3.26). Point P is a mathematical point that instantaneously moves along the line of intersection. Of the three components of vector ω at point P, only one perpendicular to the disc plane produces a velocity at this point. This velocity will be:

$$\dot{\mathbf{r}}_P = \left(\dot{\mathbf{r}}_O - \left(\dot{\mathbf{r}}_O \cdot \mathbf{u}\right)\mathbf{u}\right)\frac{PO'}{PO} \tag{3.78}$$

All the ratios are defined. The only thing left to do is to formulate the constraint equations for the three-dimensional wheel. These equations are more complicated than in the planar case, and they should be numerically formulated.

Sliding. In the case that there is sliding, only the component of the velocity of material point P in a direction normal to the surfaces in contact should be cancelled. In this case, the joint permits five degrees of freedom. One way of entering this constraint equation is by expressing the equations in a local system of coordinates located at the point of contact and only cancelling the normal component. Another possibility is to project the previous equations by means of their scalar product for a unit vector in the direction normal to the surfaces at the point of contact. As this value does not depend on the natural coordinates, it does not increase the degree of the polynomial equations obtained.

For the three-dimensional wheel, the problem is more complicated than in the planar case and should be solved by considering the center of the wheel, the unit vector normal to the wheel on which it is located, and the contact surface.

Likewise, in the case of the three-dimensional wheel, it is possible to substitute the rolling or sliding joint for one or more equivalent holonomic joints, remembering the differential geometry of the surfaces in contact and the three-dimensional generalization of the Euler-Savary formula. This approach is more complicated than in the planar case, and will not be developed here.

References

Agrawal, O.P., "Dynamic Analysis of Multibody Systems Using Tangent Coordinates", Pergamon Press, *The Theory of Machines and Mechanisms* (Proc. 7th World Congress), pp. 533-536, (1987).

Agrawal, O.P. and Saigal, S., "Dynamic Analysis of Multibody Systems Using Tangent Coordinates", *Computers & Structures*, Vol. 31, pp. 349-355, (1989).

García de Jalón, J., Avello, A., and Jiménez, J.M., "Basis for the Nullspace of the Jacobian Matrix in Constrained Multibody Systems", Pergamon Press, *The Theory of Machines and Mechanisms* (Proc. 7th World Congress), pp. 503-504, (1987).

Huston, R.L., *Multibody Dynamics.*, Butterworth-Heinemann., (1990).

Ider, S.K. and Amirouche, F.M.L., "Coordinate Reduction in the Dynamics of Constrained Multibody Systems– A New Approach", *ASME Journal of Applied Mechanics*, Vol. 55, pp. 899-904, (1988).

Kamman, J.W. and Huston, R.L., "Dynamics of Constrained Multibody Systems", *ASME Journal of Applied Mechanics*, Vol. 51, pp. 899-903, (1984).

Kane, T.R. and Levinson, D.A., *Dynamics: Theory and Applications*, McGraw-Hill Book Company, (1985).

Kim, S.S. and Vanderploeg, M.J., "QR Decomposition for State Space Representation of Constrained Mechanical Dynamic Systems", *ASME Journal on Mechanisms, Transmissions, and Automation in Design*, Vol. 108, pp. 176-182, (1986).

Liang, C.G. and Lance, G.M., "A Differentiable Null Space Method for Constrained Dynamic Analysis", ASME paper No. 85-DET-86, (1985).

Mani, N.K., Haug, E.J., and Atkinson, K.E., "Application of Singular Value Decomposition for Analysis of Mechanical System Dynamics", *ASME Journal on Mechanisms, Transmissions, and Automation in Design*, Vol. 107, pp. 82-87, (1985).

Serna, M.A., Avilés, R., and García de Jalón, J., "Dynamic Analysis of Plane Mechanisms with Lower-Pairs in Basic Coordinates", *Mechanism and Machine Theory*, Vol. 17, pp. 397-403, (1982).

Singh, R.P. and Likins, P.W., "Singular Value Decomposition for Constrained Dynamic Systems", *ASME Journal of Applied Mechanics*, Vol. 52, pp. 943-948, (1985).

Strang, G., *Linear Algebra and its Applications*, Academic Press, New York, (1976).

Tsai, L.-W. and Morgan, A.P., "Solving the Kinematics of the Most General Six-and Five-Degree-of-Freedom Manipulators by Continuation Methods", *ASME Journal of Mechanisms, Transmissions, and Automation in Design*, Vol. 107, pp. 189-200, (1985).

Problems

3/1   Starting from the constraint equations resulting from Problem 2/1, write the velocity equations of the mechanism shown in the Figure P3/1 when modeled with:
a) Relative coordinates. b) Reference point coordinates. c) Natural coordinates. d) Mixed coordinates, with relative coordinates in all the pairs.

3/2   Write the velocity and acceleration equations for the mechanism shown in the Figure.

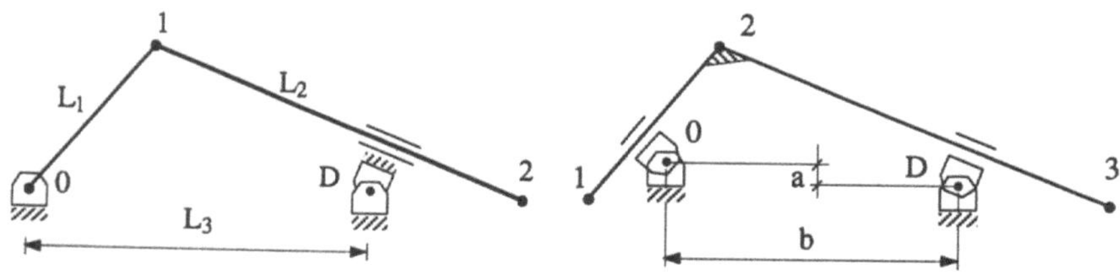

Figure P3/1.   Figure P3/2.

3/3   Assuming that there is rolling with no slipping between the disk and the rod and using the set of mixed coordinates chosen in Problem 2/2 (natural and relative), write the Newton-Raphson iteration equations for the finite displacement problem. Write the driving constraint equations for the cases when the rotation of the disk and the relative rotation in joint 1 are the externally driven variables.

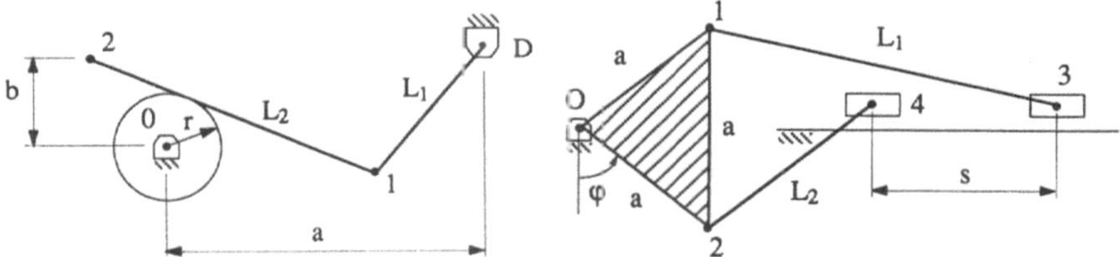

Figure P3/3.   **Figure P3/4.**

3/4   Write the velocity equations for the mechanism of the figure.

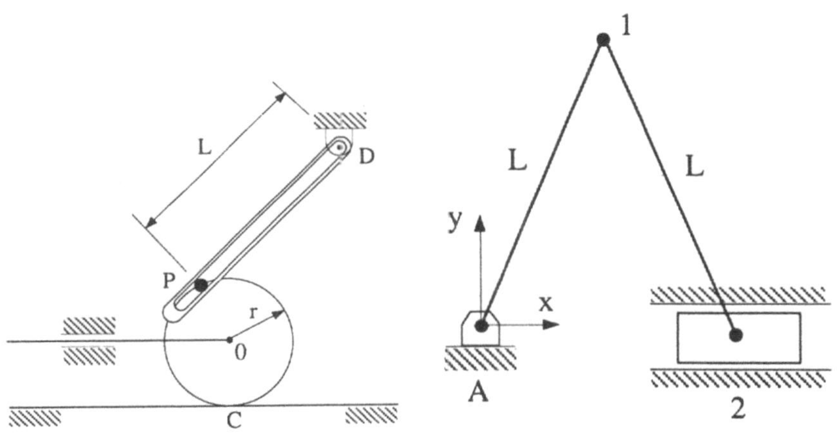

Figure P3/5.   Figure P3/6

3/5   The wheel on the figure rolls without slipping. Use mixed coordinates (natural and relative) and find the velocity equations. It is suggested that the contact between the wheel and the ground be modeled by means of a rack and pinion type of kinematic joint.

3/6   The figure shows a slider-crank mechanism with the two elements of equal length. Write the velocity equations of this mechanism when the driven velocity is:
a) x-velocity of point 1. b) x-velocity of point 2.
What will happen to the rank of the Jacobian matrix when both bars are in the vertical position?

3/7   For the slider-crank mechanism of Problem 3/6, find analytically the matrix $\mathbf{R}$ (null-space of the Jacobian matrix without driving constraints) both for the general position and for the case when both bars are placed vertically (singular position).

3/8   Find the velocity equations of the quick return mechanism of the figure when using mixed coordinates, so that the input angle φ is directly related to the output distance s.

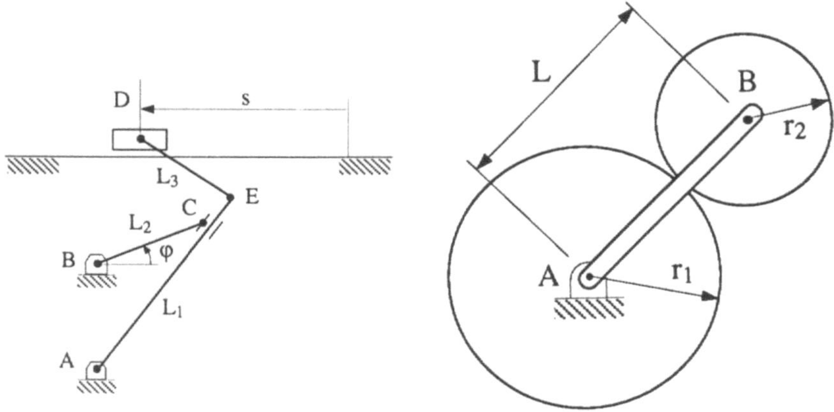

Figure P3/8.   Figure P3/9.

3/9   The centers of the two gears shown in the figure are connected by means of a rod with point A being fixed. Considering mixed coordinates and using the constraint equations of Problem 2/6, find the equations that relate the angular velocities (relative or absolute) of the three elements.

3/10 Consider the mechanism in the figure and find the constraint equations that relate the velocities and accelerations of the angles φ_1 and φ_2 with the parameter s, that measures the relative position between elements 3 and 4 and its derivatives with respect to time.

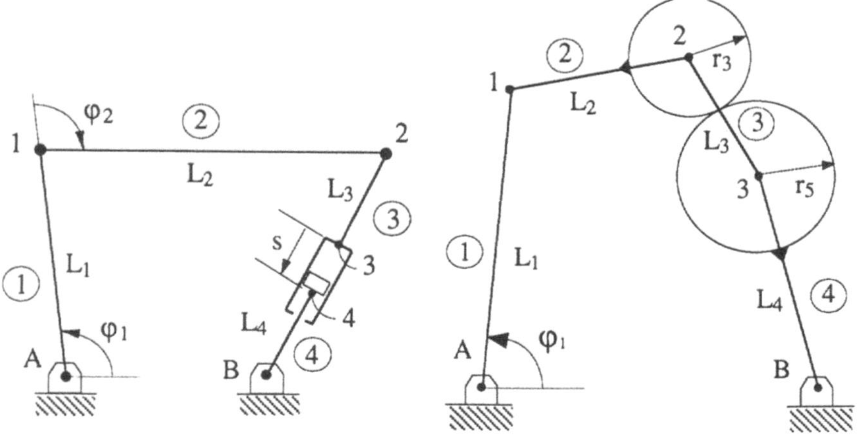

Figure P3/10.   Figure P3/11.

3/11 Consider the mechanism shown (See Problem 2/8 for the constraint equations) to be modeled with natural coordinates. Rods 2 and 4 are attached to the gears with radius

r_3 and r_5 whose centers are connected by means of rod 3. Write the velocity equations for the complete system in terms of the input angle φ_1.

3/12 Determine in the mechanism shown the constraint equations that relate the velocities and accelerations of coordinate s with the angle φ.

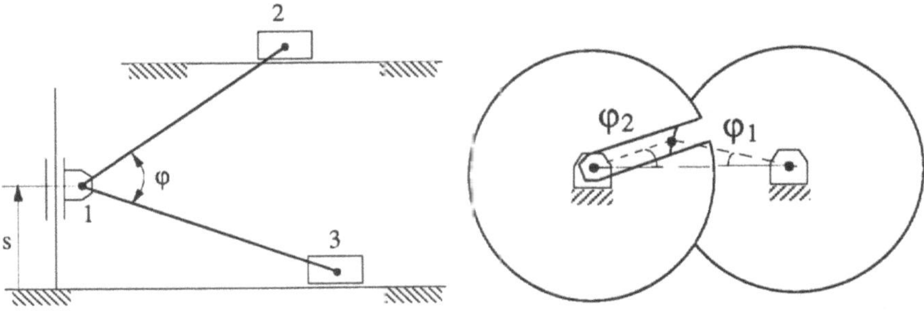

Figure P3/12.   Figure P3/13.

3/13 Consider the Geneva wheel of the figure. Using natural coordinates, find the equations that relate the velocities of input angle φ_1 with the output angle φ_2.

3/14 Find for the clam-shell bucket of the figure the equations that relate the speed of the control cable v with the angle θ.

3/15 The figure shows the frame $A12B$ that can rotate about the fixed axis AB by the action of the string attached to point 2 that goes through a pulley located at C (See Problem 2/16). Find the equations that allow one to relate the angular velocity $\dot{\psi}$ with the cable speed $\dot{s}$ at the pulley.

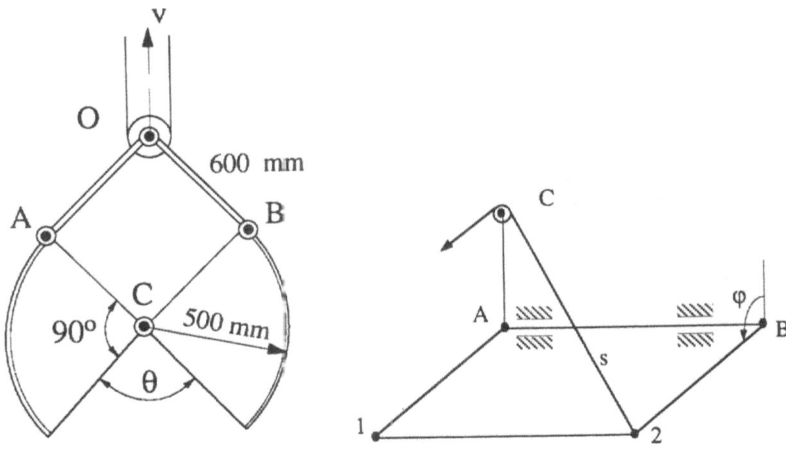

Figure P3/14.   Figure P3/15.

3/16 The ends of a slender rod of length $\sqrt{2}$ move on the sides of a cube with sides of unit length. Find the equation that relates the speed of both ends of the rod.

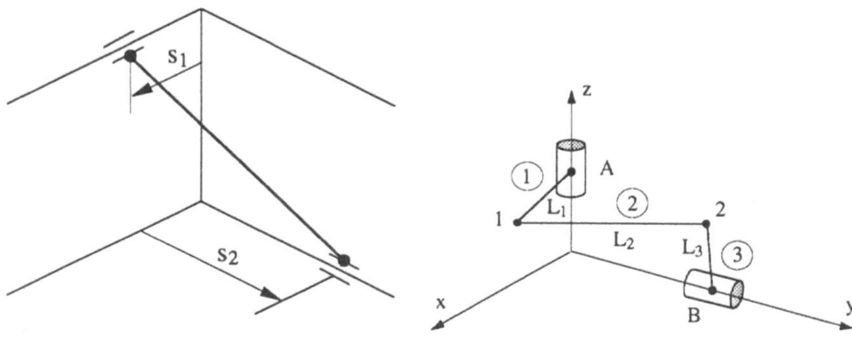

Figure P3/16.   Figure P3/17.

3/17 Use natural coordinates and the result of Problem 2/18 to find the Jacobian matrix of the *RSSR* mechanism shown in the figure. How many arithmetic operations are necessary to evaluate this Jacobian matrix?

3/18 The mechanism shown has a revolute joint, a spherical joint, and a composite joint RC. Find the constraint equations that relate the time derivatives of angle φ and distance s.

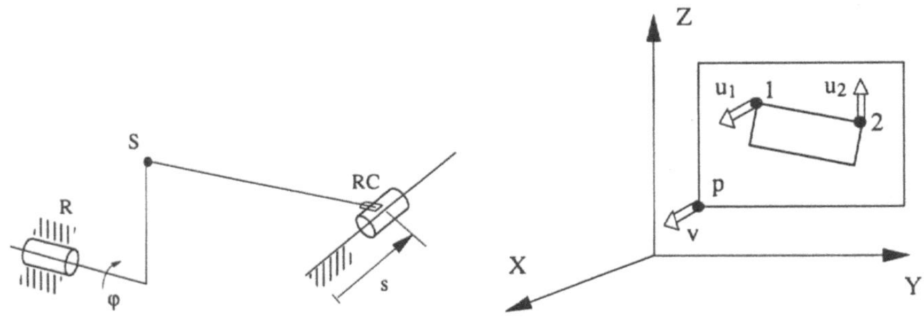

Figure P3/18.   Figure P3/19.

3/19 For the 3-D planar joint shown in the figure, find the constraint equations for positions, velocities, and accelerations. The plane is defined by point P and its normal unit vector **v**, and the body is defined by points 1 and 2, and unit vectors $\mathbf{u}_1$ and $\mathbf{u}_2$. Discuss what difficulties may arise with other possible body positions and/or configurations.

3/20 Find the velocity constraint equations of the gyroscope shown, considering the angles of relative motion (See Problem 2/20).

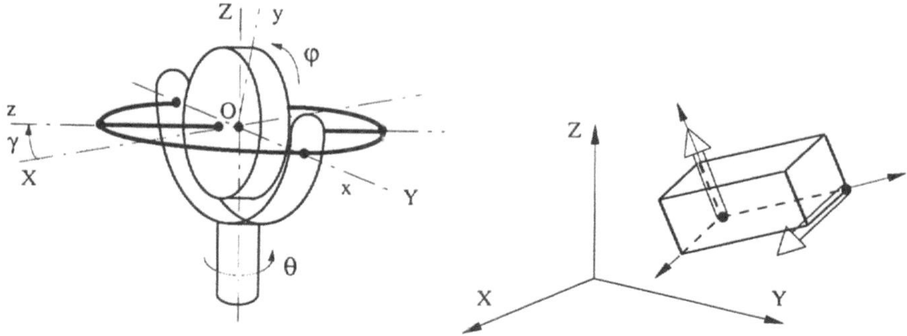

Figure P3/20.   Figure P3/21.

3/21 Find the six kinematic constraint equations (and their derivatives) that allow one to guide kinematically the three translational displacements and the three roll, pitch and yaw rotations (in the local reference frame) of the rigid body of the figure.

3/22 A six degree of freedom spatial manipulator is depicted in Figure 3.13 (See also Problem 2/21). Use natural coordinates to find the finite displacement Newton-Raphson iteration equations.

4
Dynamic Analysis. Mass Matrices and External Forces

The formulation of the inertia and external forces appearing at any of the elements of a multibody system, in terms of the dependent coordinates that describe their position, velocity, and acceleration, is of fundamental importance for the solution of the dynamic analysis.

The external and inertial forces of a body subjected to an acceleration field may be expressed in a diversity of ways. The most general way is finding the resultant force and torque about a specific point of the element. However, there are many ways of doing this. All of the ways are based on representing the inertial and external forces by means of equivalent force systems (the same resultant force and torque about any point). This will depend on the type of representation (coordinates) used for the multibody system. We deal in this chapter with the representation of the inertia and external forces generated in the elements of planar and three-dimensional multibody systems that are characterized by *natural coordinates*. The formulation of the simpler planar element is the start, which will serve as an introduction to the more complicated three-dimensional development.

The reader needs a minimum background in analytical dynamics for the understanding of this and subsequent chapters. For this reason, a background on this topic is provided in the first section, that can certainly be skipped by those with a sufficient knowledge.

4.1 Background on Analytical Dynamics

This section is intended to provide the reader with a basic background on some fundamental principles of analytical dynamics that are important for the understanding of the rest of this chapter and subsequent ones. The interested reader who wants to attain a deeper knowledge on this topic is referred to other works dedicated to the study of classical mechanics, such as Goldstein (1980), and Bastero and Casellas (1976).

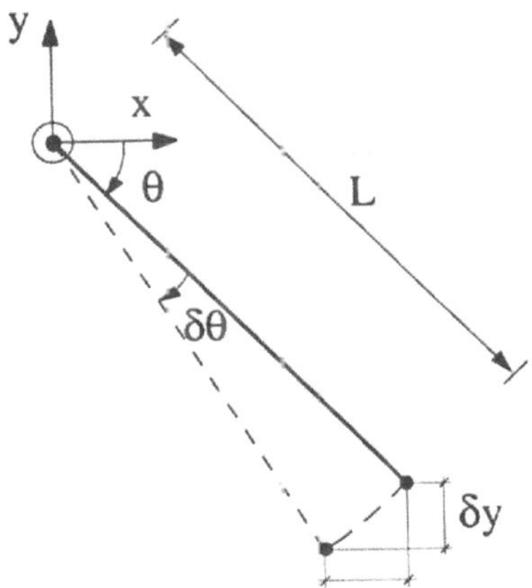

Figure 4.1. A virtual displacement on a single pendulum.

4.1.1 Principle of Virtual Displacements

The principle of virtual displacements is a powerful principle which is suitable
for the dynamic analysis of connected rigid and flexible multibody systems. Prior
to the definition of the principle, we need to introduce some concepts. A *virtual
displacement* is defined as an infinitesimal imaginary change of configuration of
a system at a stationary time that is consistent with its boundary and constraint
conditions. If the configuration is represented by the position vector $\mathbf{q}$, the vector
of generalized virtual displacement is customarily denoted by $\delta\mathbf{q}$.

Figure 4.1 shows a single pendulum of length L with a mass m at its tip. If
the system is characterized by the independent coordinate θ, a virtual displacement $\delta\theta$ consistent with the boundary condition is simply an imaginary infinitesimal rotation at the hinge. If the system is characterized by the dependent
Cartesian coordinates x and y of its tip, the virtual displacements δx and δy are
not independent but interrelated through the constraint condition:

$$\phi \equiv x^2 + y^2 - L^2 = 0$$

The relationship between δx and δy can be obtained by imposing the condition that a virtual variation of the constraint be zero:

$$\delta\phi = \phi_q\,\delta\mathbf{q} = \begin{bmatrix} 2x & 2y \end{bmatrix} \begin{Bmatrix} \delta x \\ \delta y \end{Bmatrix} = 0 \tag{4.1}$$

Virtual quantities operate the same way as *variations*. Without entering into
the details of calculus of variations (Reddy (1984)), a virtual quantity or variation
δ, acts like a differential operator but with respect to the dependent variables
only, since the time variable is considered fixed. In other words, the multibody

system is considered at a stationary position. As an example, if ϕ is a function of $\mathbf{q}$, $\dot{\mathbf{q}}$, and t, its variation is

$$\delta\phi = \phi_\mathbf{q}\,\delta\mathbf{q} + \phi_{\dot{\mathbf{q}}}\,\delta\dot{\mathbf{q}} \tag{4.2}$$

whereas its differential is

$$d\phi = \phi_\mathbf{q}\,d\mathbf{q} + \phi_{\dot{\mathbf{q}}}\,d\dot{\mathbf{q}} + \phi_t\,dt \tag{4.3}$$

The laws of variation of sums, products, ratios, and so forth are the same as those of differentiation. In addition, the variational operator can be interchanged with the differential and integral operators.

Virtual work δW is defined as the work done by all the forces acting on a system, including the inertial forces, that undergoes a virtual displacement and can be expressed as

$$\delta W = \sum_1^n Q_i\,\delta q_i \tag{4.4}$$

Each of the generalized forces Q_i represents the virtual work done when $\delta q_i=1$ and $\delta q_j=0$ for $j\neq i$. When the system is characterized by n independent coordinates, the *principle of virtual displacements* can be defined as

$$\delta W = \sum_1^n Q_i\,\delta q_i = 0 \tag{4.5}$$

meaning that the virtual work of all the forces acting on the system, including the inertial forces, must be zero. In multibody dynamics the forces Q_i do not include the reaction forces, because these do not produce any virtual work. The reason being that reaction forces are couples of internal forces that act along the vector connecting their position coordinates but with opposite signs, thus canceling the corresponding virtual work.

4.1.2 Hamilton's Principle

Consider a system characterized by a set of n independent coordinates q_i. Let $L=T-V$ be the system *Lagrangian*, where T and V are the kinetic and potential energy, respectively, and W_{nc} is the work done by the non-conservative forces. Hamilton's principle (Hamilton (1834)) establishes that the motion of the system from time t_1 to time t_2, at which the motion is specified, is such that the integral action

$$A = \int_{t_1}^{t_2} L\,dt + \int_{t_1}^{t_2} W_{nc}\,dt \tag{4.6}$$

has a stationary value for the correct path of the motion. This means that the variation of the action A has to vanish:

$$\delta A = \int_{t_1}^{t_2} \delta L \, dt \; + \int_{t_1}^{t_2} \delta W_{nc} \, dt = 0 \tag{4.7}$$

where the property that the variation is interchangeable with the integral operator has been used.

In most cases, the representation of multibody systems is done by means of dependent coordinates that are interrelated through the constraint conditions. Let us assume that the system is characterized by a vector $\mathbf{q}$ of n dependent coordinates that satisfy m constraint conditions $\phi_k(\mathbf{q}, t)=0$, which we will assume are of the holonomic type. Hamilton's principle can still be generalized for these cases by means of the Lagrange multipliers technique. Accordingly, the action A is augmented with an additional term:

$$A = \int_{t_1}^{t_2} L \, dt + \int_{t_1}^{t_2} W_{nc} \, dt - \int_{t_1}^{t_2} \sum_{k=1}^{m} (\phi_k \, \lambda_k) \, dt \tag{4.8}$$

where λ_k are the Lagrange multipliers affected by a minus sign for convenience of the formulation. The stationary condition $\delta A = 0$ now leads to

$$\delta A = \int_{t_1}^{t_2} \delta L \, dt + \int_{t_1}^{t_2} \delta W_{nc} \, dt - \int_{t_1}^{t_2} \sum_{k=1}^{m} \sum_{i=1}^{n} \left(\delta q_i \frac{\partial \phi_k}{\partial q_i} \lambda_k \right) dt = 0 \tag{4.9}$$

The summation in the last term of (4.9) can also be expressed in matrix form as $(\delta \mathbf{q}^{\mathrm{T}} \Phi_{\mathbf{q}}^{\mathrm{T}} \lambda)$, where λ is the vector of the Lagrange multipliers and $\Phi_{\mathbf{q}}$ is the Jacobian matrix of the constraints.

4.1.3 Lagrange's Equations

The Lagrange's equations can be directly obtained from Hamilton's principle. Knowing that $T=T(\mathbf{q}, \dot{\mathbf{q}})$ and $V=V(\mathbf{q})$ their variations become

$$\delta T = \sum_{i=1}^{n} \frac{\partial T}{\partial q_i} \delta q_i + \sum_{i=1}^{n} \frac{\partial T}{\partial \dot{q}_i} \delta \dot{q}_i \equiv \delta \mathbf{q}^{\mathrm{T}} \frac{\partial T}{\partial \mathbf{q}} + \delta \dot{\mathbf{q}}^{\mathrm{T}} \frac{\partial T}{\partial \dot{\mathbf{q}}} \tag{4.10a}$$

$$\delta V = \sum_{i=1}^{n} \frac{\partial V}{\partial q_i} \delta q_i \equiv \delta \mathbf{q}^{\mathrm{T}} \frac{\partial V}{\partial \mathbf{q}} \tag{4.10b}$$

$$\delta W_{nc} = \delta \mathbf{q}^{\mathrm{T}} \mathbf{Q}_{ex} \tag{4.10c}$$

where $\mathbf{Q}_{ex}$ represents the external forces and those not coming from a potential. Note that both index and matrix notation have been used simultaneously. Continuing with matrix notation, the application of Hamilton's principle defined by equation (4.9) leads to

$$\int_{t_1}^{t_2}\left[\delta\mathbf{q}^{\mathrm{T}}\left(\frac{\partial T}{\partial\mathbf{q}}-\frac{\partial V}{\partial\mathbf{q}}+\mathbf{Q}_{ex}-\Phi_{\mathbf{q}}^{\mathrm{T}}\,\lambda\right)+\delta\dot{\mathbf{q}}^{\mathrm{T}}\frac{\partial T}{\partial\dot{\mathbf{q}}}\right]dt=0 \qquad (4.11)$$

The last term can be integrated by parts to yield

$$\int_{t_1}^{t_2}\delta\dot{\mathbf{q}}^{\mathrm{T}}\frac{\partial T}{\partial\dot{\mathbf{q}}}\,dt=\left[\delta\mathbf{q}^{\mathrm{T}}\frac{\partial T}{\partial\dot{\mathbf{q}}}\right]_{t_1}^{t_2}-\int_{t_1}^{t_2}\delta\mathbf{q}^{\mathrm{T}}\frac{d}{dt}\left(\frac{\partial T}{\partial\dot{\mathbf{q}}}\right)dt \qquad (4.12)$$

The first term on the RHS vanishes because the motion is specified at the two ends t_1 and t_2. Thus the variations will be zero: $\delta\mathbf{q}(t_1)=\delta\mathbf{q}(t_2)=0$. The substitution of (4.12) into (4.11) yields

$$\int_{t_1}^{t_2}\left[\delta\mathbf{q}^{T}\left(\frac{d}{dt}\left(\frac{\partial L}{\partial\dot{\mathbf{q}}}\right)-\frac{\partial L}{\partial\mathbf{q}}+\Phi_{\mathbf{q}}^{\mathrm{T}}\,\lambda-\mathbf{Q}_{ex}\right)\right]dt=0 \qquad (4.13)$$

Although the coordinates $\mathbf{q}$ are not independent, the expression between parenthesis can always be made zero through the selection of the m Lagrange multipliers λ. According to the fundamental lemma of the calculus of variations (Reddy (1984)), equation (4.13) leads to

$$\frac{d}{dt}\left(\frac{\partial L}{\partial\dot{\mathbf{q}}}\right)-\frac{\partial L}{\partial\mathbf{q}}+\Phi_{\mathbf{q}}^{\mathrm{T}}\,\lambda=\mathbf{Q}_{ex} \qquad (4.14)$$

which along with the m constraint equations $\Phi(\mathbf{q})=0$ constitutes a set of $(m+n)$ differential algebraic equations of motion. In the next example we apply these equations to a mechanical system.

Example   4.1

Use the Lagrange's equations to write the equations of motion of a mechanical system with kinetic energy $T = 1/2\ \dot{\mathbf{q}}^{\mathrm{T}}\mathbf{M}\ \dot{\mathbf{q}}$, potential energy $V = V(\mathbf{q})$, external forces $\mathbf{Q}_{ex}$, and whose constraint conditions are $\Phi(\mathbf{q})=0$.

The partial derivatives of the kinetic and potential energies are

$$\frac{\partial L}{\partial\dot{\mathbf{q}}}=\mathbf{M}(\mathbf{q})\ \dot{\mathbf{q}}\ \ ;\ \ \frac{\partial L}{\partial\mathbf{q}}=\frac{\partial T}{\partial\mathbf{q}}-\frac{\partial V}{\partial\mathbf{q}}$$

then

$$\frac{d}{dt}\left(\frac{\partial L}{\partial\dot{\mathbf{q}}}\right)=\dot{\mathbf{M}}(\mathbf{q})\ \dot{\mathbf{q}}+\mathbf{M}(\mathbf{q})\ \ddot{\mathbf{q}}$$

and the application of equation (4.14) leads to

$$\mathbf{M}(\mathbf{q})\ \ddot{\mathbf{q}}+\Phi_{\mathbf{q}}^{\mathrm{T}}\,\lambda=\mathbf{Q}_{ex}-\dot{\mathbf{M}}(\mathbf{q})\ \dot{\mathbf{q}}+T_{\mathbf{q}}-V_{\mathbf{q}}$$

It is worth pointing out that the terms $\dot{\mathbf{M}}(\mathbf{q})\ \dot{\mathbf{q}}$ and $T_{\mathbf{q}}$ are quadratic in the velocities with coefficients that may depend on $\mathbf{q}$ (See Example 4.2). The terms that in-

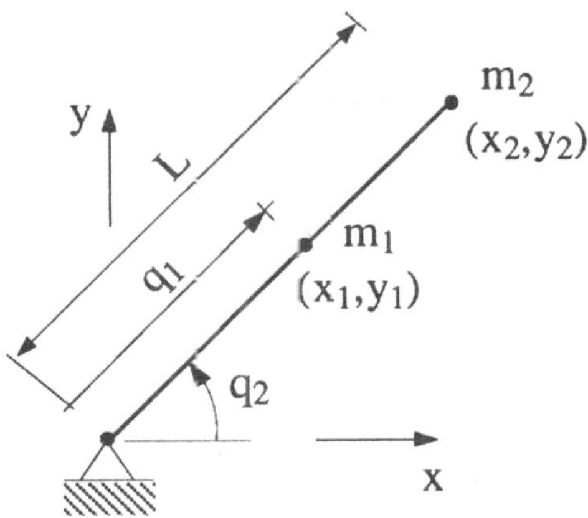

Figure 4.2. Mass sliding along a single pendulum.

volve $\dot{q}_i^2$ are called *centrifugal*, and those that involve $(\dot{q}_i\,\dot{q}_j)$ are called *Coriolis* terms. The term V_q involves q but not its derivatives.

Example 4.2

Figure 4.2 depicts a mass m_1 that slides along a massless rod which also has a mass m_2 attached at its tip. Find the equations of motion of this system subject to gravity using the two independent coordinates q_1 and q_2.

The kinetic energy of this system is

$$T = \frac{1}{2}\,m_1\left(\dot{x}_1^2 + \dot{y}_1^2\right) + \frac{1}{2}\,m_2\left(\dot{x}_2^2 + \dot{y}_2^2\right)$$

Knowing that $x_1 = q_1\,cos\,q_2$, $y_1 = q_1\,sin\,q_2$, $x_2 = L\,cos\,q_2$, and $y_2 = L\,sin\,q_2$ the kinetic energy in terms of q_1 and q_2 becomes

$$T = \frac{1}{2}\left\{\dot{q}_1\;\dot{q}_2\right\}\begin{bmatrix} m_1 & 0 \\ 0 & m_2\,L^2 + m_1\,q_1^2 \end{bmatrix}\begin{Bmatrix} \dot{q}_1 \\ \dot{q}_2 \end{Bmatrix}$$

Similarly, the potential energy is

$$V = m_1\,g\,q_1\,sin\,q_2 + m_2\,g\,L\,sin\,q_2$$

and the application of equation (4.14) leads to

$$\begin{bmatrix} m_1 & 0 \\ 0 & m_2 L^2 + m_1 q_1^2 \end{bmatrix}\begin{Bmatrix} \ddot{q}_1 \\ \ddot{q}_2 \end{Bmatrix} = \begin{Bmatrix} m_1 q_1 \dot{q}_2^2 \\ -2m_1 q_1 \dot{q}_1 \dot{q}_2 \end{Bmatrix} - \begin{Bmatrix} m_1\,g\,sin\,q_2 \\ (m_1 q_1 + m_2 L)\,g\,cos\,q_2 \end{Bmatrix}$$

Since q_1 and q_2 are independent coordinates, note that there are no Lagrange multipliers involved in the equations. Note also that the mass matrix depends on the coordinate q_2, and this leads to velocity-dependent nonlinear terms on the RHS of the equation. In addition, the gravity forces contain nonlinear terms that involve transcendental functions with q_2 as argument.

Example 4.3

Repeat Example 4.2 using the Cartesian (natural) coordinates of the two masses m_1 and m_2.

The kinetic energy of the system in Cartesian coordinates takes a simpler expression:

$$T = \frac{1}{2} \{\dot{x}_1 \dot{y}_1 \dot{x}_2 \dot{y}_2\} \begin{bmatrix} m_1 & 0 & 0 & 0 \\ 0 & m_1 & 0 & 0 \\ 0 & 0 & m_2 & 0 \\ 0 & 0 & 0 & m_2 \end{bmatrix} \begin{Bmatrix} \dot{x}_1 \\ \dot{y}_1 \\ \dot{x}_2 \\ \dot{y}_2 \end{Bmatrix}$$

Similarly the potential energy is

$$V = m_1\, g\, y_1 + m_2\, g\, y_2$$

The constraint conditions to be satisfied by the Cartesian coordinates are

$$x_2^2 + y_2^2 - L^2 = 0$$
$$x_1 y_2 - x_2 y_1 = 0$$

Since the mass matrix is now constant, the equations of motion take this simpler form:

$$\begin{bmatrix} m_1 & 0 & 0 & 0 \\ 0 & m_1 & 0 & 0 \\ 0 & 0 & m_2 & 0 \\ 0 & 0 & 0 & m_2 \end{bmatrix} \begin{Bmatrix} \ddot{x}_1 \\ \ddot{y}_1 \\ \ddot{x}_2 \\ \ddot{y}_2 \end{Bmatrix} + \begin{bmatrix} 0 & y_2 \\ 0 & -x_2 \\ 2x_2 & -y_1 \\ 2y_2 & x_1 \end{bmatrix} \begin{Bmatrix} \lambda_1 \\ \lambda_2 \end{Bmatrix} = \begin{Bmatrix} 0 \\ -m_1 g \\ 0 \\ -m_2 g \end{Bmatrix}$$

Although the number of equations in Example 4.3 has increased compared to the results of Example 4.2, first the mass matrix and gravity forces are constant. Secondly, the degree of nonlinearity has decreased, since there are neither velocity-dependent terms nor transcendental functions in the RHS. In addition, the Jacobian matrix of the constraints is linear in $\mathbf{q}$.

This simple example already illustrates a general fact: the formulation of the equations of motion in independent coordinates leads to a minimum set of highly nonlinear and coupled ordinary differential equations. On the other hand, the formulation using natural coordinates, at the expense of increasing the number of equations, results in a simpler and less coupled set of equations with milder nonlinearities.

4.1.4 Virtual Power

Virtual power also constitutes a powerful principle that will be extensively used in the formulations of this chapter. A virtual velocity vector is defined as a set of imaginary velocities at a stationary time that is consistent with the homogeneous form of the velocity constraint conditions, that is, having no RHS term including the partial derivatives with respect to time. Following the example of Figure 4.1, let $\dot{\mathbf{q}}^*$ be a set of virtual velocities which must satisfy the velocity constraint conditions:

$$\dot{\phi} = \phi_{\mathbf{q}}^{T}\, \dot{\mathbf{q}}^{*} = \begin{bmatrix} 2x & 2y \end{bmatrix} \begin{Bmatrix} \dot{x}^{*} \\ \dot{y}^{*} \end{Bmatrix} = 0 \qquad (4.15)$$

Contrary to the virtual displacements, the virtual velocities need not be infinitesimal since equation (4.15) involves $\dot{\mathbf{q}}$ and not $\delta\mathbf{q}$. A virtual velocity (finite) is a virtual displacement (infinitesimal) divided by the infinitesimal scalar δt. The principle of virtual power is heavily used for the algorithms and formulations presented in this book; hence it is worthwhile to describe it in some detail. Virtual power can be applied with dependent or with independent coordinates. Both ways will be presented next.

Dependent coordinates. If $\dot{\mathbf{q}}^{*}$ constitutes a set of n dependent virtual velocities, the principle of virtual power can be formulated as:

$$W^{*} = \sum_{i=1}^{n} F_i\, \dot{q_i}^{*} \equiv \dot{\mathbf{q}}^{*T} \mathbf{F} = 0 \qquad (4.16)$$

where $\mathbf{F}$ is the vector of all the forces that produce virtual power, including the inertial ones:

$$\mathbf{F} = \mathbf{M}\,\ddot{\mathbf{q}} - \mathbf{Q} \qquad (4.17)$$

Vector $\mathbf{Q}$ includes the external forces and the velocity-dependent inertia forces (centrifugal and Coriolis), but it does not include internal constraint forces, since they do not produce virtual power. Therefore, equation (4.17) leads to a set of equilibrium equations ($\mathbf{M}\,\ddot{\mathbf{q}} - \mathbf{Q} = 0$) in which the internal constraint forces are missing. These forces should appear in the equilibrium equations. In order to find the equilibrium equations from (4.16) and (4.17), we need to add a set of forces in the direction of the constraint violations ($\Phi_{\mathbf{q}}^{T}\,\lambda$), where the columns of $\Phi_{\mathbf{q}}^{T}$ (rows of $\Phi_{\mathbf{q}}$) give the direction of constraint forces and λ is the vector of their unknown magnitudes. As the virtual velocity vector $\dot{\mathbf{q}}^{*}$ belongs to the nullspace of $\Phi_{\mathbf{q}}$, the product ($\dot{\mathbf{q}}^{*T}\Phi_{\mathbf{q}}^{T}\,\lambda$) is zero and can be added to equation (4.16) yielding:

$$W^{*} = \dot{\mathbf{q}}^{*T}\,(\mathbf{M}\,\ddot{\mathbf{q}} - \mathbf{Q} + \Phi_{\mathbf{q}}^{T}\,\lambda) = 0 \qquad (4.18)$$

Only $n-m$ elements of the virtual velocity vector $\dot{\mathbf{q}}^{*}$ can be arbitrarily selected. It is always possible to find the m components λ (Lagrange multipliers) in such a way that the parenthesis of (4.18) becomes zero. Consequently, the complete set of force equilibrium equations is:

$$\mathbf{M}\,\ddot{\mathbf{q}} + \Phi_{\mathbf{q}}^{T}\,\lambda = \mathbf{Q} \qquad (4.19)$$

Equation (4.19) is analogous to the equations of motion of Example 4.1, that were obtained from the Lagrange's equations (4.14).

Independent coordinates. The virtual power principle can be applied also with independent virtual velocities. From equations (4.16) and (4.17), it could not be concluded that $(M \ddot{q} - Q = 0)$, because the multiplying virtual velocities $\dot{q}^*$ were not independent; hence they could not be chosen arbitrarily. However, it is possible to use now the transformations between dependent and independent velocities and accelerations introduced in Chapter 3 as carried out in the following example.

Example 4.4

Starting from equation (4.16), obtain the equations of motion with independent coordinates, using the velocity and acceleration transformations defined in Section 3.5.

The virtual velocity vector $\dot{q}^*$ must satisfy the homogeneous version of the velocity constraint equations:

$$\Phi_q \dot{q}^* = 0 \tag{i}$$

and according to equation (3.29) there is an $(n \times f)$ matrix R such that

$$\dot{q}^* = R \dot{z}^* \tag{ii}$$

On the other hand, according to equation (3.32), the following relationship between dependent and independent accelerations can be established:

$$\ddot{q} = R \ddot{z} + Sc \tag{iii}$$

where R is the same matrix that appears in the previous expression and (Sc) is a term that depends on the actual velocities and can be computed easily. Introducing equation (iii) and the transpose of equation (ii) in equation (4.16), we get

$$\dot{z}^{*T} R^T (M R \ddot{z} + M Sc - Q) = 0 \tag{iv}$$

Since the virtual velocities $\dot{z}^*$ can be chosen arbitrarily, it is possible to conclude that the term that multiplies them in expression (iv) must be zero. Consequently, we arrive at the following set of equations:

$$R^T M R \ddot{z} = R^T (Q - M Sc) \tag{v}$$

This is an important result that will be developed with more detail in Chapter 5. In Chapter 8, it will be used as the basis of very efficient dynamic formulations. The application of the virtual power method with independent coordinates is also referred to in the bibliography as Kane's method or Kane's equations (Kane and Levinson (1985), and Huston (1990)). Other authors (Schiehlen (1984)) refer to it as Jourdains' principle.

4.1.5 Canonical Equations

The Lagrange's equations lead to a set of n second order differential equations in the coordinates q. Hamilton introduced a transformation that leads to a set of $2n$ first order differential equations and they are called the *canonical* or *Hamilton's equations*. The study of these equations is important because it gives one a further insight into the multibody problems. In addition, they provide an alternative to the acceleration-based formulations at the time of their numerical implementation. Also, the canonical equations constitute the foundation for the study of quantum and relativistic mechanics.

The *canonical momenta* is defined as

$$\mathbf{p} = \frac{\partial L}{\partial \dot{\mathbf{q}}} \tag{4.20}$$

where L is the Lagrangian and $\mathbf{q}$ a set of dependent coordinates that characterize the system. According to this new variable, the Lagrange's equations (4.14) take the following form:

$$\dot{\mathbf{p}} = \mathbf{Q}_{ex} + \frac{\partial L}{\partial \mathbf{q}} - \Phi_{\mathbf{q}}^{T} \lambda \tag{4.21}$$

The Lagrangian is a function of $\mathbf{q}$, $\dot{\mathbf{q}}$, and t, and, consequently, its differential becomes

$$dL = d\dot{\mathbf{q}}^{T} \frac{\partial L}{\partial \dot{\mathbf{q}}} + d\mathbf{q}^{T} \frac{\partial L}{\partial \mathbf{q}} + \frac{\partial L}{\partial t} dt \tag{4.22}$$

Using equations (4.20) and (4.21):

$$dL = d\dot{\mathbf{q}}^{T} \mathbf{p} + d\mathbf{q}^{T} \left(\dot{\mathbf{p}} - \mathbf{Q}_{ex} + \Phi_{\mathbf{q}}^{T} \lambda \right) + \frac{\partial L}{\partial t} dt \tag{4.23}$$

Knowing that $d\dot{\mathbf{q}}^{T} \mathbf{p} = d(\mathbf{p}^{T} \dot{\mathbf{q}}) - d\mathbf{p}^{T} \dot{\mathbf{q}}$, equation (4.23) can be transformed into

$$d\left(\mathbf{p}^{T} \dot{\mathbf{q}} - L \right) = d\mathbf{p}^{T} \dot{\mathbf{q}} + d\mathbf{q}^{T} \left(\mathbf{Q}_{ex} - \Phi_{\mathbf{q}}^{T} \lambda - \dot{\mathbf{p}} \right) - \frac{\partial L}{\partial t} dt \tag{4.24}$$

The expression $H = \mathbf{p}^{T} \dot{\mathbf{q}} - L$ is called the *Hamiltonian function*. The RHS of (4.24) tells us that it is an explicit function of $\mathbf{p}$, $\mathbf{q}$, and t. Consequently,

$$dH = d\mathbf{p}^{T} \frac{\partial H}{\partial \mathbf{p}} + d\mathbf{q}^{T} \frac{\partial H}{\partial \mathbf{q}} + dt \frac{\partial H}{\partial t} \tag{4.25}$$

Finally, identifying the terms on the RHS of equations (4.24) and (4.25), we arrive at the canonical equations:

$$\frac{\partial H}{\partial \mathbf{p}} = \dot{\mathbf{q}} \tag{4.26}$$

$$\frac{\partial H}{\partial \mathbf{q}} = \mathbf{Q}_{ex} - \Phi_{\mathbf{q}}^{T} \lambda - \dot{\mathbf{p}} \tag{4.27}$$

In the case of mechanical systems, the Lagrangian L is defined in terms of $\mathbf{q}$, $\dot{\mathbf{q}}$, and t. Rather than following a lengthy process to form the Hamiltonian as an explicit function of $\mathbf{q}$, $\mathbf{p}$, and t, and then differentiating as in (4.26) and (4.27), the canonical equations can be directly obtained from (4.20) and (4.27). Since the system kinetic energy is a quadratic function of the generalized velocities, equations (4.20) and (4.27) directly lead to the following set of equations in matrix form:

$$\mathbf{p} = \mathbf{M}\,\dot{\mathbf{q}} \tag{4.28}$$

$$\dot{\mathbf{p}} = L_{\mathbf{q}} + \mathbf{Q}_{ex} - \Phi_{\mathbf{q}}^{\mathrm{T}}\,\lambda \tag{4.29}$$

where $\mathbf{M}$ is the mass matrix, $L_{\mathbf{q}} = T_{\mathbf{q}} - V_{\mathbf{q}}$ is the partial derivative of the Lagrangian with respect to the coordinates, $\Phi_{\mathbf{q}}$ is the Jacobian matrix of the constraints, and $\mathbf{Q}_{ex}$ the vector of applied external forces. The combination of equations (4.28)-(4.29) and the constraints conditions constitutes a system of $2n+m$ differential and algebraic equations (DAEs). Although we arrive at n more equations than with the acceleration-based formulation (4.14), $\dot{\mathbf{p}}$ can be obtained explicitly by (4.29). When comparing equations (4.29) and (4.14), one may see that the term $(\mathbf{M}\,\dot{\mathbf{q}})$ that appears in the Lagrange's equations is not present in their canonical counterparts.

Example 4.5

Repeat Example 4.2 using the canonical equations.

The application of equations (4.28) and (4.29) along with the expressions obtained in Example 4.2 directly leads to the following equations:

$$\begin{Bmatrix} \dot{p}_1 \\ \dot{p}_2 \end{Bmatrix} = -\begin{Bmatrix} m_1\,g\,\sin q_2 \\ (m_1 q_1 + m_2 L)\,g\,\cos q_2 \end{Bmatrix} + \begin{Bmatrix} m_1 q_1\,\dot{q}_2^2 \\ 0 \end{Bmatrix}$$

$$\begin{Bmatrix} p_1 \\ p_2 \end{Bmatrix} = \begin{bmatrix} m_1 & 0 \\ 0 & m_2 L^2 + m_1 q_1^2 \end{bmatrix} \begin{Bmatrix} \dot{q}_1 \\ \dot{q}_2 \end{Bmatrix}$$

Note how the term related to $(\dot{\mathbf{M}}\,\dot{\mathbf{q}})$ that appears in Example 4.2 is not now present.

4.2 Inertial Forces. Mass Matrix

We study in this section the formation of the inertial forces that arise in each of the elements of a multibody system undergoing a given motion. The aim of this study is the formation of the mass matrix of the most common elements that appear in the analysis of multibody systems. The form of these mass matrices will undoubtedly depend on the type of coordinates chosen for the representation of the multibody system, and we will only use the setting provided by the natural coordinates.

The inertial forces will be represented by means of equivalent forces that are congruent with the natural coordinates of the element. Of all the analytical methods exposed in Section 4.1, we will use the virtual power method, because it leads to a direct formulation of the inertial forces and avoids the differentiation process inherent in the Lagrange's equations. The Lagrange's equations can lead to rather involved computations in cases when the kinetic energy is position dependent. As mentioned in Section 4.1.4, if $\dot{\mathbf{q}}^*$ are the virtual velocities, the virtual power of the inertial forces can be written as the following dot product:

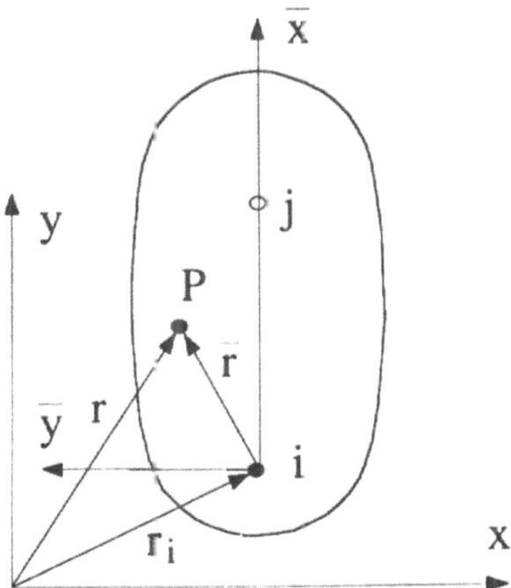

Figure 4.3. Inertial and local system of coordinates in a planar element with natural coordinates.

$$W^* = \dot{q}^{*T} Q_I \tag{4.30}$$

where Q_I are the inertial forces with respect to the natural coordinates, and W^* is the scalar virtual power.

The expressions of the inertial forces for the elements of the planar and three-dimensional multibody systems will be developed below. The formulation of the planar element is simpler and will serve as introduction for the three-dimensional cases. In both the planar and spatial cases, the aim is to establish the inertial forces as a product of a matrix (mass matrix) times the acceleration vector, so that the virtual power expression becomes

$$W^* = - \dot{q}^{*T} M \ddot{q} \tag{4.31}$$

where the matrix M is a square, symmetric, and positive definite matrix which depends on the inertial characteristics of the element: mass, position of the center of gravity, and inertial tensor, and in some cases, on the position q of the element as well.

4.2.1 Mass Matrix of Planar Bodies

Consider the planar element shown in Figure 4.3, whose motion is completely defined by that of the basic points i and j. Consider also an inertial coordinates system (x, y) and a moving one $(\bar{x}, \bar{y})$, with its origin fixed at the basic point i and axis $\bar{x}$ going through point j.

The location of a generic point P of the element is defined by a position vector r in the inertial system and $\bar{r}$ in the moving system, so that

$$\mathbf{r} = \mathbf{r}_i + \mathbf{A}\,\bar{\mathbf{r}} \qquad (4.32)$$

where $\mathbf{A}$ is the rotation matrix. Since the element is rigid, the local position vector $\bar{\mathbf{r}}$ remains constant with the motion of the element. If the position of an element is characterized by the Cartesian (natural) coordinates of the points i and j, that of the point P will be defined as follows:

$$\mathbf{r} = \mathbf{r}_i + \mathbf{A}\,\bar{\mathbf{r}} = \mathbf{r}_i + c_1\,(\mathbf{r}_j - \mathbf{r}_i) + c_2\,\mathbf{n} \qquad (4.33)$$

where c_1 and c_2 are the components of the vector $\bar{\mathbf{r}}$ in the basis formed by the orthogonal vectors $(\mathbf{r}_j - \mathbf{r}_i)$ and $\mathbf{n}$. The vector $\mathbf{n}$ follows the direction of $\bar{y}$ and has the same module as $(\mathbf{r}_j - \mathbf{r}_i)$. Accordingly, the components of the vector $\mathbf{r}$ become

$$x = x_i + c_1\,(x_j - x_i) - c_2\,(y_j - y_i) \qquad (4.34a)$$

$$y = y_i + c_1\,(y_j - y_i) + c_2\,(x_j - x_i) \qquad (4.34b)$$

Equations (4.34) can be expressed in matrix form as follows:

$$\mathbf{r} = \begin{Bmatrix} x \\ y \end{Bmatrix} = \begin{bmatrix} (1-c_1) & c_2 & c_1 & -c_2 \\ -c_2 & (1-c_1) & c_2 & c_1 \end{bmatrix} \begin{Bmatrix} x_i \\ y_i \\ x_j \\ y_j \end{Bmatrix} \equiv \mathbf{C}\,\mathbf{q} \qquad (4.35)$$

where $\mathbf{q}^T = \{x_i\ y_i\ x_j\ y_j\}$ is the vector containing the natural coordinates of the element. Note again that the matrix $\mathbf{C}$ is constant for a given point P. It does not change with the system's motion or with time. Consequently,

$$\dot{\mathbf{r}} = \mathbf{C}\,\dot{\mathbf{q}} \qquad (4.36)$$

$$\ddot{\mathbf{r}} = \mathbf{C}\,\ddot{\mathbf{q}} \qquad (4.37)$$

The coefficients c_1 and c_2 that define the matrix $\mathbf{C}$ can be simply expressed as a function of the coordinates of the points i and j in the local reference frame in the following way:

$$\bar{\mathbf{r}} = c_1\,(\bar{\mathbf{r}}_j - \bar{\mathbf{r}}_i) + c_2\,\bar{\mathbf{n}} \qquad (4.38)$$

Note, however, that $\bar{\mathbf{r}}_i{=}0$ in the local reference frame, therefore (4.38) becomes

$$\bar{\mathbf{r}} = \begin{bmatrix} \bar{\mathbf{r}}_j & | & \bar{\mathbf{n}} \end{bmatrix} \begin{Bmatrix} c_1 \\ c_2 \end{Bmatrix} \equiv \bar{\mathbf{X}}\,\mathbf{c} \qquad (4.39)$$

where the vector $\mathbf{c}$ contains the coefficients c_1 and c_2, and the matrix $\bar{\mathbf{X}}$ has as columns the components of the vectors $\bar{\mathbf{r}}_j$ and $\bar{\mathbf{n}}$. Therefore, the following form is taken:

$$\bar{\mathbf{X}} = \begin{bmatrix} \bar{x}_j & -\bar{y}_j \\ \bar{y}_j & \bar{x}_j \end{bmatrix} = \begin{bmatrix} L_{ij} & 0 \\ 0 & L_{ij} \end{bmatrix} \qquad (4.40)$$

where L_{ij} is the distance between points i and j. The matrix $\bar{\mathbf{X}}$ can always be inverted unless points i and j are coincident. Equation (4.39) can be used to solve for $\mathbf{c}$:

$$\mathbf{c} = \bar{\mathbf{X}}^{-1}\,\bar{\mathbf{r}} \qquad (4.41)$$

We now can formulate the virtual power of the inertial forces generated within the element. These can be obtained as the integral, extended to all the elements, of the virtual power of the inertial force of a differential mass located at the generic point P:

$$W^* = -\rho \int_\Omega \dot{\mathbf{r}}^{*T}\, \ddot{\mathbf{r}}\, d\Omega \qquad (4.42)$$

where ρ is the mass density. Making use of equations (4.36) and (4.37), equation (4.42) becomes

$$W^* = -\rho \int_\Omega \dot{\mathbf{q}}^{*T}\, \mathbf{C}^T \mathbf{C}\, \ddot{\mathbf{q}}\, d\Omega \qquad (4.43)$$

Since the vectors $\dot{\mathbf{q}}^{*T}$ and $\ddot{\mathbf{q}}$ are independent of Ω, they can be moved out of the integral to yield

$$W^* = -\dot{\mathbf{q}}^{*T}\left(\rho \int_\Omega \mathbf{C}^T \mathbf{C}\, d\Omega\right)\ddot{\mathbf{q}} \qquad (4.44)$$

and comparing (4.31) with (4.44) the mass matrix can be established as

$$\mathbf{M} = \rho \int_\Omega \mathbf{C}^T \mathbf{C}\, d\Omega \qquad (4.45)$$

Performing the product $\mathbf{C}^T\mathbf{C}$ in equation (4.44), we obtain

$$\mathbf{M} = \int_\Omega \rho \begin{bmatrix} (1-c_1)^2 + c_2^2 & 0 & (1-c_1)c_1 - c_2^2 & -c_2 \\ 0 & (1-c_1)^2 + c_2^2 & c_2 & (1-c_1)c_1 - c_2^2 \\ (1-c_1)c_1 - c_2^2 & c_2 & c_1^2 + c_2^2 & 0 \\ -c_2 & (1-c_1)c_1 - c_2^2 & 0 & c_1^2 + c_2^2 \end{bmatrix} d\Omega \qquad (4.46)$$

Note that the equation (4.46) involves the following integrals:

$$\int_\Omega \rho\, d\Omega = m \qquad (4.47a)$$

$$\int_\Omega \rho\, \mathbf{c}\, d\Omega = \overline{\overline{\mathbf{X}}}^{-1} \int_\Omega \rho\, \bar{\mathbf{r}}\, d\Omega = m\, \overline{\mathbf{X}}^{-1}\, \bar{\mathbf{r}}_G \qquad (4.47b)$$

$$\int_\Omega \rho\, \mathbf{c}\, \mathbf{c}^T d\Omega = \overline{\mathbf{X}}^{-1}\left(\int_\Omega \rho\, \bar{\mathbf{r}}\, \bar{\mathbf{r}}^T\, d\Omega\right)\overline{\mathbf{X}}^{-T} = \frac{1}{L^2}\begin{bmatrix} I_{\bar{x}} & I_{\bar{x}\bar{y}} \\ I_{\bar{x}\bar{y}} & I_{\bar{y}} \end{bmatrix} \qquad (4.47c)$$

where m is the total mass of the element, $\bar{\mathbf{r}}_G$ represents the local coordinates of the center of gravity, and $I_{\bar{x}}$, $I_{\bar{y}}$ and $I_{\bar{x}\bar{y}}$ are the moments and products of inertia with respect to $\bar{x}$ and $\bar{y}$ respectively. The integral in (4.47b) is the static moment or first order area moment which is equal to the mass times the coordinates of the center of gravity. Similarly, the integral in (4.47c) yields the moments of inertia. Substituting the results of (4.47) into (4.46), we obtain the final expression for the mass matrix $\mathbf{M}$:

$$
\mathbf{M} = \begin{bmatrix}
m - \dfrac{2m\bar{x}_G}{L_{ij}} + \dfrac{I_i}{L_{ij}^2} & 0 & \dfrac{m\,\bar{x}_G}{L_{ij}} - \dfrac{I_i}{L_{ij}^2} & -\dfrac{m\,\bar{y}_G}{L_{ij}} \\[2.5ex]
0 & m - \dfrac{2m\bar{x}_G}{L_{ij}} + \dfrac{I_i}{L_{ij}^2} & \dfrac{m\,\bar{y}_G}{L_{ij}} & \dfrac{m\,\bar{x}_G}{L_{ij}} - \dfrac{I_i}{L_{ij}^2} \\[2.5ex]
\dfrac{m\,\bar{x}_G}{L_{ij}} - \dfrac{I_i}{L_{ij}^2} & \dfrac{m\,\bar{y}_G}{L_{ij}} & \dfrac{I_i}{L_{ij}^2} & 0 \\[2.5ex]
-\dfrac{m\,\bar{y}_G}{L_{ij}} & \dfrac{m\,\bar{x}_G}{L_{ij}} - \dfrac{I_i}{L_{ij}^2} & 0 & \dfrac{I_i}{L_{ij}^2}
\end{bmatrix} \qquad (4.48)
$$

where I_i is the polar moment of inertia with respect to the basic point i. The mass matrix defined by equation (4.48) is completely general for any planar element, since any planar element will have at least two basic points with which the mass matrix can be formulated. The mass matrix thus obtained is always constant and this constitutes an important fact.

Example 4.6

Derive the mass matrix of a single bar element of total mass m and length L_{ij} that has the center of gravity located at the middle point.

Choosing the points i and j as the end points, one finds that the different parameters that appear in (4.48) take the following values:

$$
\bar{x}_G = \frac{L_{ij}}{2} \; ; \quad \bar{y}_G = 0 \; ; \quad I_i = m\frac{L_{ij}^2}{3}
$$

and the direct application of (4.48) yields

$$
\mathbf{M} = \begin{bmatrix}
\dfrac{m}{3} & 0 & \dfrac{m}{6} & 0 \\[2ex]
0 & \dfrac{m}{3} & 0 & \dfrac{m}{6} \\[2ex]
\dfrac{m}{6} & 0 & \dfrac{m}{3} & 0 \\[2ex]
0 & \dfrac{m}{6} & 0 & \dfrac{m}{3}
\end{bmatrix}
$$

4.2.2 Mass Matrix of Three-Dimensional Bodies

The determination of the mass matrix of the three-dimensional bodies (or elements) follows a method similar to that used in the planar case. Three-dimensional bodies not only require more complicated algebraic manipulations than the planar ones, but in some cases the mass matrix is not constant and, therefore, requires a special study. The analysis will start with the most general element defined by two points and two non-coplanar unit vectors, which happens to have a constant mass matrix. The mass matrix of the other elements including those

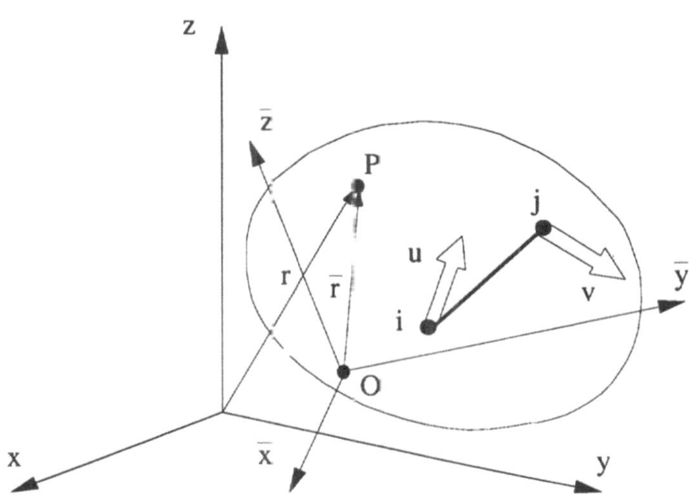

Figure 4.4. Inertial and local system of coordinates in a spatial element with natural coordinates.

that are not constant, may be derived from this main one by coordinate transformations in which the virtual power method plays a key role.

Element with two points and two non-coplanar unit vectors. Consider the element of Figure 4.4 defined by two basic points i and j and two unit vectors $\mathbf{u}$ and $\mathbf{v}$. Similar to the planar case, consider an inertial reference frame (x,y,z) and a moving (or local) one $(\bar{x},\bar{y},\bar{z})$ rigidly attached to the element that has its origin located at a point O. Again, P represents a generic point of the element, and its location is defined by the position vector $\mathbf{r}$ in the inertial frame and $\bar{\mathbf{r}}$ in the local frame. Since the element is rigid, the local relative position vector $(\bar{\mathbf{r}} - \bar{\mathbf{r}}_i)$ remains constant with the motion of the element. If the position of the three dimensional body is characterized by the Cartesian coordinates of the points i and j and the unit vectors $\mathbf{u}$ and $\mathbf{v}$, the position of the point P relative to point i will be defined as follows:

$$\mathbf{r} - \mathbf{r}_i = c_1 (\mathbf{r}_j - \mathbf{r}_i) + c_2 \mathbf{u} + c_3 \mathbf{v} \tag{4.49}$$

where c_1, c_2, and c_3 are the components of the vector $(\bar{\mathbf{r}} - \bar{\mathbf{r}}_i)$ in the basis formed by vectors $(\mathbf{r}_j - \mathbf{r}_i)$, $\mathbf{u}$, and $\mathbf{v}$. Equation (4.49) may be represented in matrix form as

$$\mathbf{r} = \begin{Bmatrix} x \\ y \\ z \end{Bmatrix} = \begin{bmatrix} (1-c_1)\,\mathbf{I}_3 & c_1\,\mathbf{I}_3 & c_2\,\mathbf{I}_3 & c_3\,\mathbf{I}_3 \end{bmatrix} \begin{Bmatrix} \mathbf{r}_i \\ \mathbf{r}_j \\ \mathbf{u} \\ \mathbf{v} \end{Bmatrix} = \mathbf{C}\,\mathbf{q}^e \tag{4.50}$$

where $\mathbf{I}_3$ is the identity matrix of order (3x3) and $\mathbf{q}^e$ is the vector of the natural coordinates. As in the planar case, the matrix $\mathbf{C}$ is independent of the system's motion and therefore remains constant with time. Again, the following relations are obtained:

$$\dot{\mathbf{r}} = \mathbf{C}\,\dot{\mathbf{q}}^e \tag{4.51}$$

$$\ddot{\mathbf{r}} = \mathbf{C}\,\ddot{\mathbf{q}}^e \tag{4.52}$$

The coefficients c_1, c_2, and c_3 that define $\mathbf{C}$ can be expressed as a function of the coordinates of the points i and j and unit vectors $\mathbf{u}$ and $\mathbf{v}$ in the local reference frame in the following way:

$$\bar{\mathbf{r}} - \bar{\mathbf{r}}_i = c_1\,(\bar{\mathbf{r}}_j - \bar{\mathbf{r}}_i) + c_2\,\bar{\mathbf{u}} + c_3\,\bar{\mathbf{v}} \tag{4.53}$$

Equation (4.53) expressed in matrix form becomes

$$\bar{\mathbf{r}} - \bar{\mathbf{r}}_i = \left[\bar{\mathbf{r}}_j - \bar{\mathbf{r}}_i \mid \bar{\mathbf{u}} \mid \bar{\mathbf{v}}\right] \begin{Bmatrix} c_1 \\ c_2 \\ c_3 \end{Bmatrix} = \bar{\mathbf{X}}\,\mathbf{c} \tag{4.54}$$

where the vector $\mathbf{c}$ contains the coefficients c_1, c_2, and c_3, and the matrix $\bar{\mathbf{X}}$ has as columns the components of the vectors $(\bar{\mathbf{r}}_j - \bar{\mathbf{r}}_i)$, $\bar{\mathbf{u}}$, and $\bar{\mathbf{v}}$. As in the planar case, the matrix $\bar{\mathbf{X}}$ can always be inverted (provided $(\bar{\mathbf{r}}_j - \bar{\mathbf{r}}_i)$, $\bar{\mathbf{u}}$, and $\bar{\mathbf{v}}$ are non-coplanar). Equation (4.54) can be used to solve for $\mathbf{c}$:

$$\mathbf{c} = \bar{\mathbf{X}}^{-1}\,(\bar{\mathbf{r}} - \bar{\mathbf{r}}_i) \tag{4.55}$$

We have gathered all the information necessary to formulate the virtual power of the inertial forces generated within the spatial element. The integral form of the virtual power becomes

$$W^* = -\rho \int_\Omega \dot{\mathbf{r}}^{*T}\,\ddot{\mathbf{r}}\,d\Omega \tag{4.56}$$

where ρ is the mass density. Making use of equations (4.51) and (4.52), equation (4.56) becomes

$$W^* = -\rho \int_V \dot{\mathbf{q}}^{*T}\,\mathbf{C}^T\mathbf{C}\,\ddot{\mathbf{q}}\,dV = -\dot{\mathbf{q}}^{*T}\left(\rho \int_V \mathbf{C}^T\mathbf{C}\,dV\right)\ddot{\mathbf{q}} \tag{4.57}$$

where again the vectors $\dot{\mathbf{q}}^{*T}$ and $\ddot{\mathbf{q}}$, which are independent of V, have been moved out of the integral. Comparing (4.31) with (4.57) we obtain the expression for the mass matrix:

$$\mathbf{M} = \rho \int_V \mathbf{C}^T\mathbf{C}\,dV \tag{4.58}$$

The substitution of $\mathbf{C}$ into equation (4.57) yields

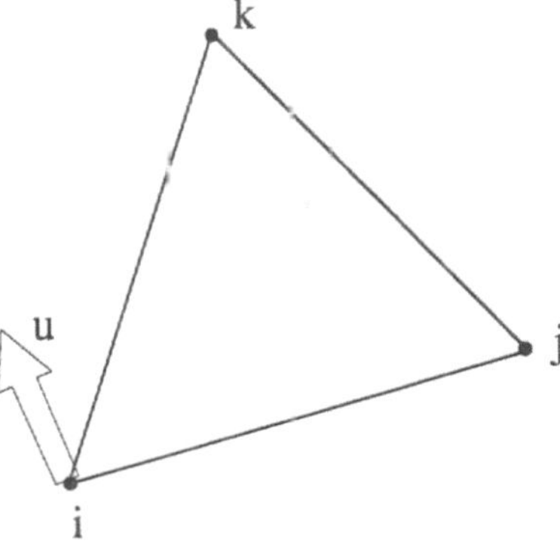

Figure 4.5. Element with three basic points and a unit vector.

$$\mathbf{M} = \int_{V} \rho \begin{bmatrix} (1-c_1)^2\,\mathbf{I}_3 & (1-c_1)\,c_1\,\mathbf{I}_3 & (1-c_1)\,c_2\,\mathbf{I}_3 & (1-c_1)\,c_3\,\mathbf{I}_3 \\ (1-c_1)\,c_1\,\mathbf{I}_3 & c_1^2\,\mathbf{I}_3 & c_1 c_2\,\mathbf{I}_3 & c_1 c_3\,\mathbf{I}_3 \\ (1-c_1)\,c_2\,\mathbf{I}_3 & c_1 c_2\,\mathbf{I}_3 & c_2^2\,\mathbf{I}_3 & c_2 c_3\,\mathbf{I}_3 \\ (1-c_1)\,c_3\,\mathbf{I}_3 & c_1 c_3\,\mathbf{I}_3 & c_2 c_3\,\mathbf{I}_3 & c_3^2\,\mathbf{I}_3 \end{bmatrix} dV \qquad (4.59)$$

The integration over the element of the product $\mathbf{C}^T\mathbf{C}$ involves the following integrals:

$$\int_{V} \rho\,dV = m \qquad (4.60a)$$

$$\int_{V} \rho\,\mathbf{c}\,dV = \overline{\mathbf{X}}^{-1} \int_{V} \rho\,(\bar{\mathbf{r}} - \bar{\mathbf{r}}_i)\,dV = m\,\overline{\mathbf{X}}^{-1}\,(\bar{\mathbf{r}}_G - \bar{\mathbf{r}}_i) \equiv m\,\mathbf{a} \qquad (4.60b)$$

$$\int_{V} \rho\,\mathbf{c}\,\mathbf{c}^T dV = \overline{\mathbf{X}}^{-1} \left(\int_{V} \rho\,(\bar{\mathbf{r}} - \bar{\mathbf{r}}_i)(\bar{\mathbf{r}} - \bar{\mathbf{r}}_i)^T dV \right) \overline{\mathbf{X}}^{-T} = \overline{\mathbf{X}}^{-1}\,\mathbf{J}_i\,\overline{\mathbf{X}}^{-T} \equiv \mathbf{Z} \qquad (4.60c)$$

where m is the total mass of the element, and $\bar{\mathbf{r}}_G$ represents the local coordinates of the element's center of gravity. The matrix $\mathbf{Z}$ can be formed from the matrix $\mathbf{J}_i$ which contains all the information about the moments and products of inertia of the element as in the planar case. Substituting the results of (4.60) into (4.49), the following expression for the mass matrix is obtained:

$$\mathbf{M} = \begin{bmatrix} (m-2ma_1+z_{11})\,\mathbf{I}_3 & (ma_1-z_{11})\,\mathbf{I}_3 & (ma_2-z_{12})\,\mathbf{I}_3 & (ma_3-z_{13})\,\mathbf{I}_3 \\ (ma_1-z_{11})\,\mathbf{I}_3 & z_{11}\,\mathbf{I}_3 & z_{12}\,\mathbf{I}_3 & z_{13}\,\mathbf{I}_3 \\ (ma_2-z_{21})\,\mathbf{I}_3 & z_{21}\,\mathbf{I}_3 & z_{22}\,\mathbf{I}_3 & z_{23}\,\mathbf{I}_3 \\ (ma_3-z_{31})\,\mathbf{I}_3 & z_{31}\,\mathbf{I}_3 & z_{32}\,\mathbf{I}_3 & z_{33}\,\mathbf{I}_3 \end{bmatrix} \qquad (4.61)$$

The mass matrix defined by equation (4.61) is a *constant* and symmetric matrix formed by diagonal sub-matrices of size (3×3). This matrix depends on a minimum set of ten different values, since the inertial characteristics of a three-

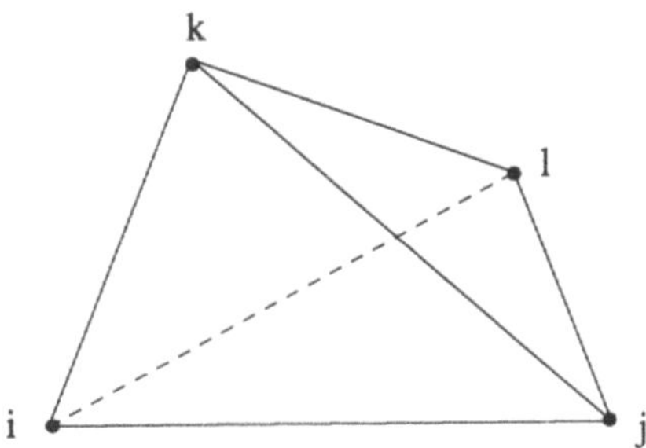

Figure 4.6. Element with four basic points.

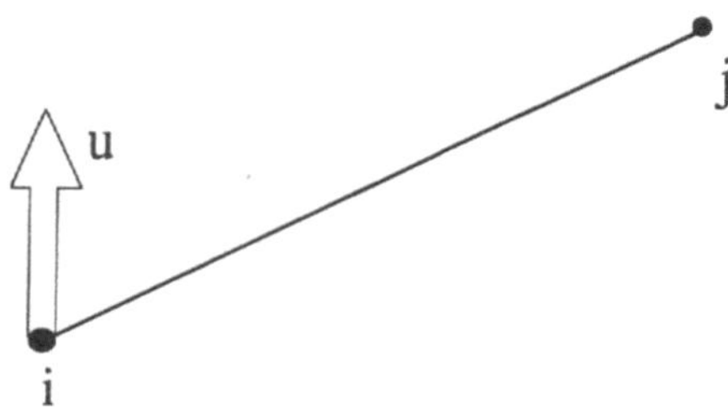

Figure 4.7. Element with two basic points and a unit vector.

dimensional solid depend on ten parameters: the mass, the coordinates of the center of gravity, and the six different elements of the inertial tensor.

Other Elements with Constant Mass Matrix. Consider the element of Figure 4.5, determined by three non-aligned basic points and by one unit vector not contained in the plane determined by the three points. This element may actually be made similar to the one considered in the previous section (two points and two unit vectors), by simply defining the missing unit vector $\mathbf{v}$ at point j as the vector that goes from j to k. Thus

$$\mathbf{v} = \left(\mathbf{r}_k - \mathbf{r}_j\right) / L_{jk} \tag{4.62}$$

Now we can use (4.61) to obtain the virtual velocity and acceleration vectors:

$$\begin{Bmatrix} \ddot{\mathbf{r}}_i \\ \ddot{\mathbf{r}}_j \\ \ddot{\mathbf{u}} \\ \ddot{\mathbf{v}} \end{Bmatrix} = \begin{bmatrix} \mathbf{I}_3 & 0 & 0 & 0 \\ 0 & \mathbf{I}_3 & 0 & 0 \\ 0 & 0 & 0 & \mathbf{I}_3 \\ 0 & -\dfrac{1}{L_{jk}}\mathbf{I}_3 & \dfrac{1}{L_{jk}}\mathbf{I}_3 & 0 \end{bmatrix} \begin{Bmatrix} \ddot{\mathbf{r}}_i \\ \ddot{\mathbf{r}}_j \\ \ddot{\mathbf{r}}_k \\ \ddot{\mathbf{u}} \end{Bmatrix} \equiv \mathbf{V} \begin{Bmatrix} \ddot{\mathbf{r}}_i \\ \ddot{\mathbf{r}}_j \\ \ddot{\mathbf{r}}_k \\ \ddot{\mathbf{u}} \end{Bmatrix} \tag{4.63}$$

where the transformation matrix $\mathbf{V}$ is a (12×12) constant matrix. Equation (4.63) may be expressed as

$$\ddot{\mathbf{q}} = \mathbf{V}\,\ddot{\mathbf{q}}_{new} \tag{4.64}$$

where $\mathbf{q}$ represents the components of the two points and two unit vectors, and $\mathbf{q}_{new}$ are those of the new element represented by three points and one unit vector. Similarly, the relation between the virtual velocities will be

$$\dot{\mathbf{q}}^* = \mathbf{V}\,\dot{\mathbf{q}}^*_{new} \tag{4.65}$$

The expression for the virtual power given by equation (4.31) can be used to obtain the mass matrix of the element at hand. This can be done by simply substituting (4.64) and (4.65) into (4.31) to yield

$$W^* = -\dot{\mathbf{q}}^{*T}\mathbf{M}\,\ddot{\mathbf{q}} = -\dot{\mathbf{q}}^{*T}_{new}\,\mathbf{V}^T\mathbf{M}\,\mathbf{V}\,\dot{\mathbf{q}}^*_{new} \tag{4.66}$$

and, therefore, the new mass matrix becomes

$$\mathbf{M}_{new} = \mathbf{V}^T\,\mathbf{M}\,\mathbf{V} \tag{4.67}$$

Since $\mathbf{V}$ is constant the new mass matrix will also be constant.

A similar situation arises with the type of element depicted in Figure 4.6, which can also be made equivalent to an element with two points and two non-coplanar unit vectors. In fact, the two unit vectors $\mathbf{u}$ and $\mathbf{v}$ located at points i and j may be defined as

$$\mathbf{u} = \left(\mathbf{r}_k - \mathbf{r}_i\right)/L_{ik} \tag{4.68a}$$

$$\mathbf{v} = \left(\mathbf{r}_l - \mathbf{r}_j\right)/L_{jl} \tag{4.68b}$$

Equations (4.68) allow for the definition of a new transformation matrix $\mathbf{V}$ as:

$$\begin{Bmatrix}\mathbf{r}_i\\\mathbf{r}_j\\\mathbf{u}\\\mathbf{v}\end{Bmatrix} = \begin{bmatrix} \mathbf{I}_3 & 0 & 0 & 0 \\ 0 & \mathbf{I}_3 & 0 & 0 \\ -\dfrac{1}{L_{jk}}\mathbf{I}_3 & 0 & \dfrac{1}{L_{jk}}\mathbf{I}_3 & 0 \\ 0 & -\dfrac{1}{L_{jk}}\mathbf{I}_3 & 0 & \dfrac{1}{L_{jk}}\mathbf{I}_3 \end{bmatrix}\begin{Bmatrix}\mathbf{r}_i\\\mathbf{r}_j\\\mathbf{r}_k\\\mathbf{r}_l\end{Bmatrix} = \mathbf{V}\,\mathbf{q} \tag{4.69}$$

The matrix $\mathbf{V}$ is again constant and, therefore, it also relates virtual velocities and accelerations as in (4.64) and (4.65). The new mass matrix is thus obtained by the same equation (4.66) with $\mathbf{V}$ defined by (4.69).

Element with two points and one non-aligned unit vector. Figure 4.7 shows an element with two basic points and one unit vector not aligned with them. The coordinate transformation between the standard element consisting of two points and two non-coplanar unit vectors, and the new one is not as simple to establish as with the elements considered in the previous section. In the element of Figure 4.7, the vector $\mathbf{v}$ cannot be constructed non-coplanar with $\mathbf{u}$ by just a linear combination of the position vectors of points i, j, and $\mathbf{u}$.

A geometric transformation is needed in this case, and the method of constructing the vector $\mathbf{v}$ is to start from the following cross product of vectors:

$$\mathbf{v} = c\left(\mathbf{r}_i - \mathbf{r}_j\right) \wedge \mathbf{u} \tag{4.70}$$

where c is a constant that makes the vector $\mathbf{v}$ have a unit module. The differentiation of equation (4.70) with respect to time yields

$$\dot{\mathbf{v}} = c\left(\mathbf{r}_i - \mathbf{r}_j\right) \wedge \dot{\mathbf{u}} + c\left(\dot{\mathbf{r}}_i - \dot{\mathbf{r}}_j\right) \wedge \mathbf{u} \tag{4.71}$$

A transformation matrix $\mathbf{V}$ for velocities can now be constructed from (4.71) as follows:

$$\begin{Bmatrix} \dot{\mathbf{r}}_i \\ \dot{\mathbf{r}}_j \\ \dot{\mathbf{u}} \\ \dot{\mathbf{v}} \end{Bmatrix} = \begin{bmatrix} \mathbf{I}_3 & 0 & 0 \\ 0 & \mathbf{I}_3 & 0 \\ 0 & 0 & \mathbf{I}_3 \\ -c\,\tilde{\mathbf{u}} & c\,\tilde{\mathbf{u}} & c\,\tilde{\mathbf{r}}_{ij} \end{bmatrix} \begin{Bmatrix} \dot{\mathbf{r}}_i \\ \dot{\mathbf{r}}_j \\ \dot{\mathbf{u}} \end{Bmatrix} = \mathbf{V}\,\dot{\mathbf{q}}_{new} \tag{4.72}$$

where $\tilde{\mathbf{u}}$ and $\tilde{\mathbf{r}}_{ij}$ are skew-symmetric matrices of order (3x3) that correspond to the cross product of vectors obtained with $\mathbf{u}$ and $(\mathbf{r}_i-\mathbf{r}_j)$, respectively. These matrices are written as follows:

$$\tilde{\mathbf{u}} = \begin{bmatrix} 0 & -u_z & u_y \\ u_z & 0 & -u_x \\ -u_y & u_x & 0 \end{bmatrix} \tag{4.73}$$

$$\tilde{\mathbf{r}}_{ij} = \begin{bmatrix} 0 & -(z_i-z_j) & (y_i-y_j) \\ (z_i-z_j) & 0 & -(x_i-x_j) \\ -(y_i-y_j) & (x_i-x_j) & 0 \end{bmatrix} \tag{4.74}$$

Equation (4.72) can be differentiated with respect to time so as to obtain the corresponding equation for the accelerations. However, the matrix $\mathbf{V}$ defined by equation (4.72) is no longer constant and, therefore, the differentiation process is different from those appearing in the previous section:

$$\begin{Bmatrix} \ddot{\mathbf{r}}_i \\ \ddot{\mathbf{r}}_j \\ \ddot{\mathbf{u}} \\ \ddot{\mathbf{v}} \end{Bmatrix} = \begin{bmatrix} \mathbf{I}_3 & 0 & 0 \\ 0 & \mathbf{I}_3 & 0 \\ 0 & 0 & \mathbf{I}_3 \\ -c\tilde{\mathbf{u}} & c\tilde{\mathbf{u}} & c\tilde{\mathbf{r}}_{ij} \end{bmatrix} \begin{Bmatrix} \ddot{\mathbf{r}}_j \\ \ddot{\mathbf{r}}_i \\ \ddot{\mathbf{u}} \end{Bmatrix} + \begin{bmatrix} \mathbf{I}_3 & 0 & 0 \\ 0 & \mathbf{I}_3 & 0 \\ 0 & 0 & \mathbf{I}_3 \\ -c\dot{\tilde{\mathbf{u}}} & c\dot{\tilde{\mathbf{u}}} & c\dot{\tilde{\mathbf{r}}}_{ij} \end{bmatrix} \begin{Bmatrix} \dot{\mathbf{r}}_i \\ \dot{\mathbf{r}}_j \\ \dot{\mathbf{u}} \end{Bmatrix} \tag{4.75}$$

or

$$\ddot{\mathbf{q}} = \mathbf{V}\,\ddot{\mathbf{q}}_{new} + \dot{\mathbf{V}}\,\dot{\mathbf{q}}_{new} \tag{4.76}$$

The relation between the virtual velocities is

$$\dot{\mathbf{q}}^* = \mathbf{V}\,\dot{\mathbf{q}}_{new}^* \tag{4.77}$$

By substituting expressions (4.76) and (4.77) into the virtual power expression (4.31), we obtain

$$W^* = -\dot{\mathbf{q}}^{*T}\mathbf{M}\,\ddot{\mathbf{q}} = -\dot{\mathbf{q}}_{new}^{*T}\,\mathbf{V}^T\mathbf{M}\,(\mathbf{V}\,\ddot{\mathbf{q}}_{new} + \dot{\mathbf{V}}\,\dot{\mathbf{q}}_{new}) \tag{4.78}$$

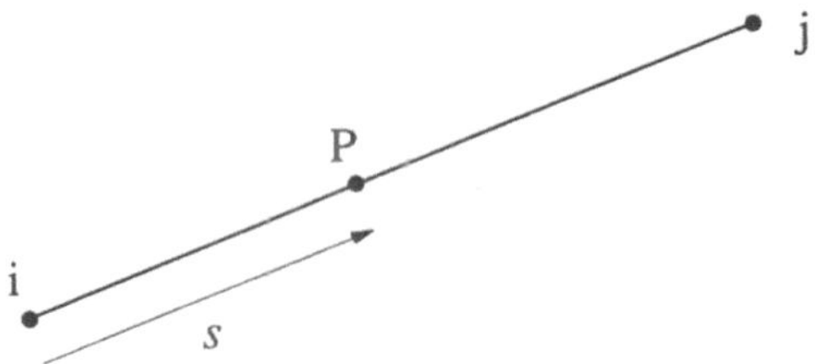

Figure 4.8. Element with two basic points.

where $\mathbf{M}$ is the constant mass matrix corresponding to the element with two points and two unit vectors. This equation contains the new mass matrix $\mathbf{M}_{new}=\mathbf{V}^{T}\mathbf{M}\mathbf{V}$ and velocity-dependent inertial terms defined by $\mathbf{V}^{T}\mathbf{M}\dot{\mathbf{V}}\dot{\mathbf{q}}_{new}$ to be added to the external forces in the right-hand side of the equations of motion. Note the simplicity by which these forces are obtained using the virtual power method in which only the computation of $\dot{\mathbf{V}}$ is required, as compared to the Lagrange's equations. The latter would have required the differentiation of the new mass matrix with respect to both time and q.

Element with two basic points. The use of an element with two basic points only makes sense when it is used: first, to maintain a constant distance between two points; and second, when its dimensions, other than the length, are negligible. As in the planar case, the motion of an element of this type is fully characterized by the motion of the two basic points (See Figure 4.8). Any rotation around the element axis is disregarded, an assumption that is valid only if the moment of inertia about that axis is negligible.

The position vector of a generic point P can be defined in terms of the variable s (Figure 4.8) as:

$$\mathbf{r} = \mathbf{r}_i \frac{L-s}{L} + \mathbf{r}_j \frac{s}{L} \tag{4.79}$$

which may be expressed in matrix form as:

$$\mathbf{r} = \begin{bmatrix} \dfrac{L-s}{L} & \dfrac{s}{L} \end{bmatrix} \begin{Bmatrix} \mathbf{r}_i \\ \mathbf{r}_j \end{Bmatrix} = \mathbf{N}(s)\,\mathbf{q} \tag{4.80}$$

Since the matrix $\mathbf{N}$ is independent of the motion, the velocity and acceleration vectors will be given by

$$\dot{\mathbf{r}} = \mathbf{N}(s)\,\dot{\mathbf{q}} \tag{4.81}$$

$$\ddot{\mathbf{r}} = \mathbf{N}(s)\,\ddot{\mathbf{q}} \tag{4.82}$$

Consequently, the virtual power of the inertial forces is in this case:

$$W^{*} = \int_{0}^{L} \dot{\mathbf{r}}^{*T}\,\ddot{\mathbf{r}}\,\rho\,A\,ds \tag{4.83}$$

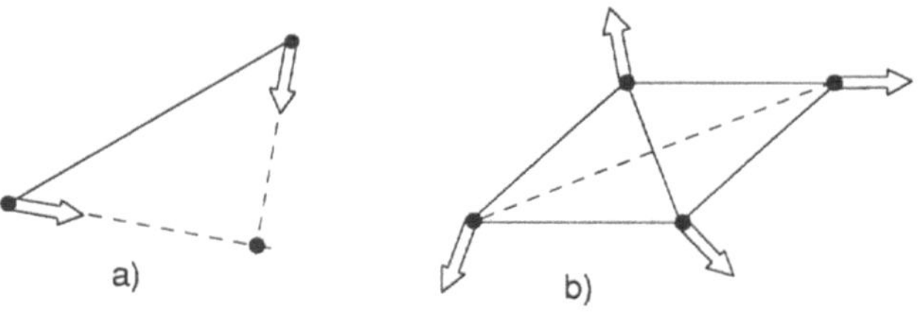

Figure 4.9. a) Element with two basic points and two coplanar unit vectors,
b) Element with four basic points and four coplanar unit vectors.

where A is the cross-sectional area of the bar. By substituting the results of expressions (4.81) and (4.82) in equation (4.83), the following expression is obtained:

$$W^* = -\int_0^L \dot{\mathbf{q}}^{*T} \mathbf{N}^T \rho\, A\, \mathbf{N}\, \ddot{\mathbf{q}}\, ds = -\dot{\mathbf{q}}^{*T}\left(\int_0^L \mathbf{N}^T \rho\, A\, \mathbf{N}\, ds\right)\ddot{\mathbf{q}} \qquad (4.84)$$

Therefore the mass matrix is

$$\mathbf{M} = \int_0^L \mathbf{N}^T \mathbf{N}\, \rho\, A\, ds =$$

$$= \begin{bmatrix} \mathbf{I}_3 \int_0^L \left(\dfrac{L-s}{L}\right)^2 \rho\, A\, ds & \mathbf{I}_3 \int_0^L \dfrac{L-s}{L}\,\dfrac{s}{L}\,\rho\, A\, ds \\[2em] \mathbf{I}_3 \int_0^L \dfrac{L-s}{L}\,\dfrac{s}{L}\,\rho\, A\, ds & \mathbf{I}_3 \int_0^L \dfrac{s^2}{L^2}\,\rho\, A\, ds \end{bmatrix} \qquad (4.85)$$

where $\mathbf{I}_3$ is the unit (3x3) matrix. Assuming a constant density and cross sectional area for the element, the following final expression for the mass matrix is obtained:

$$\mathbf{M} = \begin{bmatrix} \dfrac{m}{3}\mathbf{I}_3 & \dfrac{m}{6}\mathbf{I}_3 \\[1.2em] \dfrac{m}{6}\mathbf{I}_3 & \dfrac{m}{3}\mathbf{I}_3 \end{bmatrix} \qquad (4.86)$$

Mass matrix of other three-dimensional elements. All the three-dimensional elements that may appear in practice, belong either to one of the groups studied previously or contain a set of points and vectors whose mass matrix is known. For example, the body in Figure 4.9b contains four basic points and four unit vectors. Fortunately, one does not have to worry about calculating mass matrices of elements as complicated as this one. It will be sufficient to take any two points with which two non-coplanar unit vectors are associated and attribute the corresponding mass matrix to the subset of points and unit vectors given by equation (4.61). The body's system of local coordinates will be located according to the selected points and unit vectors. The inertial properties of the element, such as center of gravity and inertial tensor, must be referred to those coordinates.

The element shown in Figure 4.9a, contains two basic points and two unit vectors that are coplanar. For this reason, the mass matrix of equation (4.61), which assumes that the unit vectors are non-coplanar, is not applicable. Therefore, one of the two unit vectors must be selected and used in the virtual power equation (4.78) corresponding to an element with two points and only one unit vector.

Irrespective of the geometry and number of basic points and unit vectors of any element of a multibody system, a subset of points and vectors can always be found whose mass matrix corresponds to one of those calculated in the previous sections.

4.2.3 Kinetic Energy of an Element

The formation of the mass matrices of different elements has been studied in the previous sections through the application of the principle of virtual power. Expressions for the kinetic energy of a body may be convenient, either because it is of direct interest to evaluate the energy, or because one wishes to formulate the equations of motion through another type of formulation, such as the method of Lagrange.

The way of formulating the kinetic energy is rather similar to that used for the virtual power of the inertial forces. The kinetic energy is defined by:

$$T = \frac{1}{2} \int_v \dot{\mathbf{r}}^T \dot{\mathbf{r}}\, \rho\ dV = \frac{1}{2} \rho \int_v \dot{\mathbf{q}}^T \mathbf{C}^T \mathbf{C}\ \dot{\mathbf{q}}\ dV \qquad (4.87)$$

Since the natural velocities are independent of V, they can be moved out of the integral to yield

$$T = \frac{1}{2} \dot{\mathbf{q}}^T \mathbf{M}\ \dot{\mathbf{q}} \qquad (4.88)$$

where $\dot{\mathbf{q}}$ are the natural velocities of the element. The matrix $\mathbf{M}$ given by equation (4.61) corresponds to the element that has two non-coplanar points and two unit vectors. In the other cases, the mass matrix may be formed by the coordinate transformation matrix $\mathbf{V}^T \mathbf{M} \mathbf{V}$. The use of the Lagrange's equations for the formulation of the equation of motions with non-constant mass matrices will lead to the differentiation process explained in Example 4.1, which may become involved in those cases where the mass matrix is coordinate-dependent. The kinetic energy is then given by

$$T = \frac{1}{2} \dot{\mathbf{q}}^T \mathbf{V}^T(\mathbf{q})\ \mathbf{M}\ \mathbf{V}(\mathbf{q})\ \dot{\mathbf{q}} \qquad (4.89)$$

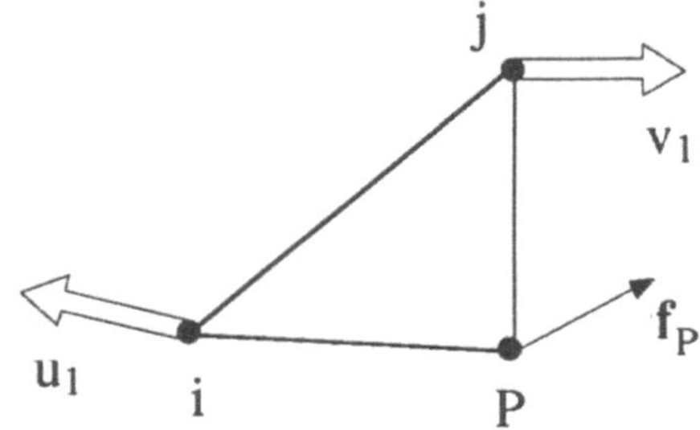

Figure 4.10. Concentrated force $\mathbf{f}_P$ at point P.

4.3 External Forces

We deal in this section with the formulation of external forces, such as concentrated loads and moments that are applied to the elements of the multibody system, and those generated from specific types of elements, like translational and rotational springs. Although the forces generated by springs are not really external, it is customary in multibody dynamics to include them as applied external forces, as opposed to structural mechanics and the finite element formulation in which those forces are considered as internal forces acting by means of the stiffness matrix.

4.3.1 Concentrated Forces and Torques

This section deals with the analysis of concentrated forces and torques. The problem is the representation of those forces and torques in terms of the natural coordinate system used to represent the motion of an element.

Concentrated Forces. When a concentrated force is applied at a point P of an element that is not a basic point (See Figure 4.10), a transformation needs to be introduced that will transform the forces at point P to the natural coordinates (basic points and unit vectors) of the element. A typical case is the gravity force, applied at the center of gravity of an element, which needs to be referred to the natural coordinate system used to represent the motion of the given element. In order to set up the required transformation, we can directly use equation (4.50) applied to point P, so that

$$\mathbf{r}_P = \mathbf{C}_P \, \mathbf{q}^e \tag{4.90}$$

where $\mathbf{q}^e$ represents the natural coordinates of the element. The coefficients that form the matrix $\mathbf{C}_P$ can be obtained using equation (4.55), such that

$$\mathbf{c} = \overline{\mathbf{X}}_P^{-1} \, (\bar{\mathbf{r}}_P - \bar{\mathbf{r}}_i) \tag{4.91}$$

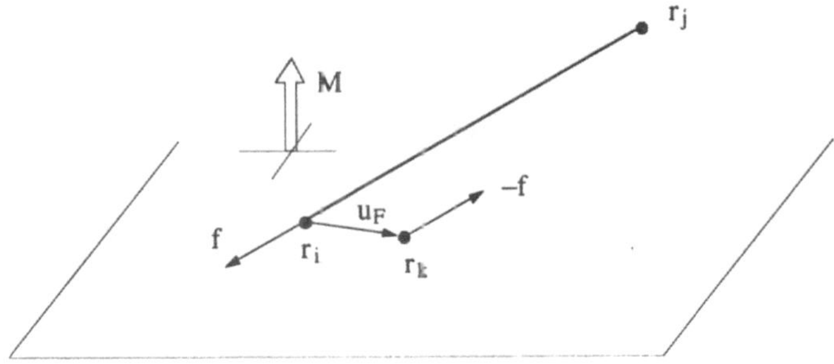

Figure 4.11. Concentrated torque **M** acting on element (*i-j*).

The vector $\bar{r}_P$ contains the coordinates of P in the local frame. The matrix C_P in equation (4.90) acts as a coordinate transformation matrix that may also be used to transform the forces f_p into equivalent forces, Q^e_{ex}, expressed in terms of the natural coordinates of the element. We use for this purpose the principle of virtual work and impose the condition that both sets of forces perform an equal amount of virtual work. Accordingly,

$$\delta W = \delta r_P^T \, f_P = \delta q^{eT} \, Q^e_{ex} \tag{4.92}$$

Since $\delta r_P^T = \delta q^{eT} C_P^T$, equation (4.92) becomes

$$\delta W = \delta q^{eT} \, C_P^T \, f_P \tag{4.93}$$

Comparing equations (4.92) and (4.93), we obtain the following equation for the force transformation:

$$Q^e_{ex} = C_P^T \, f_P \tag{4.94}$$

The *potential* of this concentrated external force f_p, acting at point P, is defined by the expression:

$$V = -\int_{q_0}^{q} dr_P^T \, f(q) = -\int_{q_0}^{q} dq^T C_P^T \, f(q) = -\int_{q_0}^{q} dq^T Q \tag{4.95}$$

that is also valid for the case in which the force depends on the position. The differential term dq has been intentionally placed in the left side of the integral, because this order leads to simpler and more congruent expressions for the derivatives of V that are calculated in Chapter 6.

Concentrated Torques. The case of a concentrated torque can be dealt with in a similar way as that for the concentrated force except for the addition of a preceding and important transformation. In basic statics, any torque **M** may be replaced by an equivalent pair of forces, **f** and **−f**, of equal magnitude and opposite directions, acting on a plane perpendicular to the direction of **M**, and separated by a vector **d**. The result is: **M=d∧f**.

Figure 4.11 shows a torque $\mathbf{M}$ acting on an element with basic points i and j, which is substituted by a pair of forces $\mathbf{f}$ and $-\mathbf{f}$ applied at the beginning and end of a unit vector $\mathbf{u}_f$ with origin at $\mathbf{r}_i$. This unit vector is defined by:

$$\mathbf{u}_f = \frac{(\mathbf{r}_j - \mathbf{r}_i) \wedge \mathbf{M}}{\left|(\mathbf{r}_j - \mathbf{r}_i) \wedge \mathbf{M}\right|} \tag{4.96}$$

and

$$\mathbf{f} = \mathbf{u}_f \wedge \mathbf{M} \tag{4.97}$$

Note that vector $\mathbf{u}_f$ does not belong to the natural coordinates vector $\mathbf{q}^e$. The forces $\mathbf{f}$ and $-\mathbf{f}$ can now be treated as concentrated forces following the results of the previous section. Accordingly, the virtual work expression becomes

$$\delta W = \delta \mathbf{q}^{e\mathrm{T}} \, (\mathbf{C}_i^{\mathrm{T}} \, \mathbf{f} - \mathbf{C}_{i+\mathbf{u}_f}^{\mathrm{T}} \, \mathbf{f}) \tag{4.98}$$

Therefore the equivalent generalized force with respect to the natural coordinates finally results in

$$\mathbf{Q}_{ex}^e = (\mathbf{C}_i^{\mathrm{T}} - \mathbf{C}_{i+\mathbf{u}_f}^{\mathrm{T}}) \, \mathbf{f} \tag{4.99}$$

where $\mathbf{C}_i$ is very simply obtained in this case because i is a basic point. It will be shown that the force $\mathbf{f}$ given by equation (4.97) is the force variable conjugated with the displacement variable $\mathbf{u}_f$. The *potential* of the torque $\mathbf{M}$ can be calculated as the sum of the potentials of forces $\mathbf{f}$ and $-\mathbf{f}$. This potential is

$$V = -\int_{\mathbf{q}_0}^{\mathbf{q}} \left(d\mathbf{r}_i^{\mathrm{T}} \, \mathbf{f} - \left(d\mathbf{r}_i^{\mathrm{T}} + d\mathbf{u}_f^{\mathrm{T}}\right) \mathbf{f}\right) = +\int_{\mathbf{q}_0}^{\mathbf{q}} d\mathbf{u}_f^{\mathrm{T}} \, \mathbf{f} = -\int_{\mathbf{q}_0}^{\mathbf{q}} d\mathbf{u}_f^{\mathrm{T}} \, (\mathbf{M} \wedge \mathbf{u}_f) \tag{4.100}$$

The forces $\mathbf{f}$ and $-\mathbf{f}$ can be treated as concentrated forces following the results of the previous section. Accordingly, the expression for the potential becomes

$$V = -\int_{\mathbf{q}_0}^{\mathbf{q}} d\mathbf{r}_i^{\mathrm{T}} \, \mathbf{f} - \left(d\mathbf{r}_i^{\mathrm{T}} + d\mathbf{u}_f^{\mathrm{T}}\right) \mathbf{f} = \int_{\mathbf{q}_0}^{\mathbf{q}} d\mathbf{q}^{e\mathrm{T}} (\mathbf{C}_i^{\mathrm{T}} - \mathbf{C}_{i+\mathbf{u}_f}^{\mathrm{T}}) \, \mathbf{f} \tag{4.101}$$

4.3.2 Forces Exerted by Springs

Springs are elements capable of storing elastic potential energy and of exerting forces that are a function of their positions. In addition, springs play an important role in all but the kinematic problems. We study in this section the forces exerted by both translational and rotational springs as well as the potential energy stored in them. As in previous sections, the study will be conducted within the context of planar and three-dimensional natural coordinates.

Translational Springs Between Basic Points. Consider the translational spring shown in Figure 4.12 which connects the basic points i and j. Let L_{ij} and L_0 be the deformed and undeformed lengths of the spring, respectively. When the

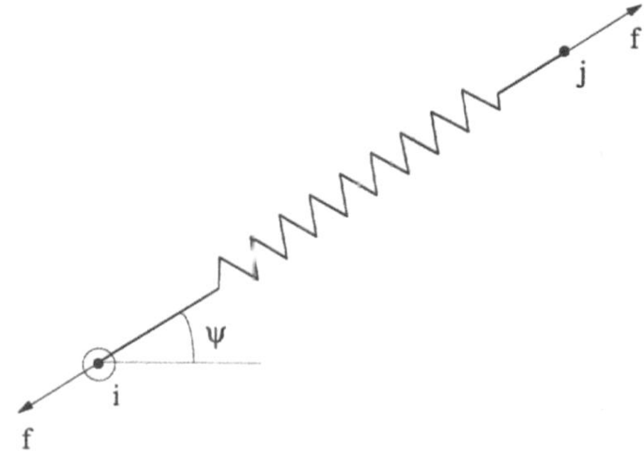

Figure 4.12. Translational spring with mixed coordinates.

spring is stretched (or compressed), it exerts a force between the basic points in
the directions of the spring that is a function of the elongation. The value of the
force is given by

$$f = k(L_{ij})(L_{ij} - L_0) \tag{4.102}$$

where $k(L_{ij})$ is the stiffness of the spring which will have a constant value k, if
the spring is linear.

In the planar case, according to Figure 4.12, the force vector that acts on the
basic points i and j can be expressed as follows:

$$\mathbf{Q} = \begin{Bmatrix} Q_{ix} \\ Q_{iy} \\ Q_{jx} \\ Q_{jy} \end{Bmatrix} = f \begin{Bmatrix} -\cos\psi \\ -\sin\psi \\ \cos\psi \\ \sin\psi \end{Bmatrix} = \frac{f}{L_{ij}} \begin{Bmatrix} x_i - x_j \\ y_i - y_j \\ x_j - x_i \\ y_j - y_i \end{Bmatrix} \tag{4.103}$$

and in the three-dimensional case:

$$\mathbf{Q}^{T} = \frac{f}{L_{ij}} \left\{ (x_i{-}x_j) \ (y_i{-}y_j) \ (z_i{-}z_j) \ (x_j{-}x_i) \ (y_j{-}y_i) \ (z_j{-}z_i) \right\} \tag{4.104}$$

Both f and L_{ij} are functions of the natural coordinates. If the spring is linear,
equation (4.104) becomes

$$\mathbf{Q}^{T} = k\left(1 - \frac{L_0}{L_{ij}}\right) \left\{ (x_i{-}x_j) \ (y_i{-}y_j) \ (z_i{-}z_j) \ (x_j{-}x_i) \ (y_j{-}y_i) \ (z_j{-}z_i) \right\} \tag{4.105}$$

It is important at times to evaluate the *potential energy* stored in spring elements (See Chapter 6). The potential energy stored by a translational spring is
equal to the integral of the force times the differential extension of the spring,
between the non-deformed state and the final deformed configuration. Considering
the coordinates of points i and j:

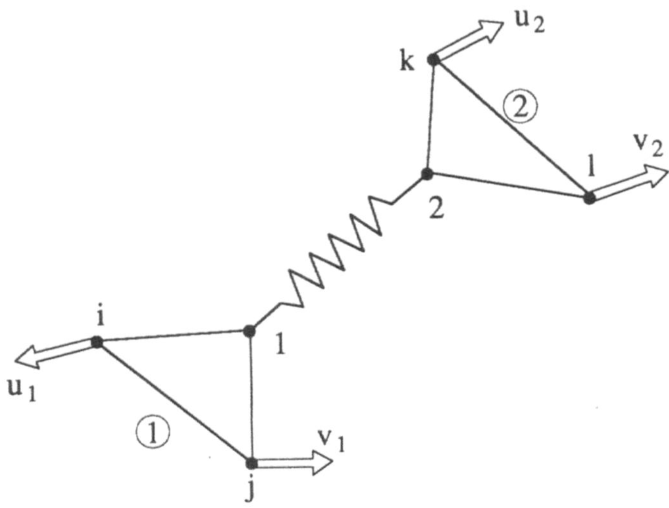

Figure 4.13. Translational spring between any two points.

$$V = \int_{o}^{L-L_o} d\mathbf{q}^T \mathbf{Q} = \int_{o}^{L-L_o} f\, dl \tag{4.106}$$

When the spring is linear, the integration of equation (4.106) yields

$$V = \frac{1}{2}\, k\, (L_{ij} - L_o)^2 = \frac{1}{2}\, k\left[\sqrt{(x_j-x_i)^2 + (y_j-y_i)^2 + (z_j-z_i)^2} - L_o\right]^2 \tag{4.107}$$

Translational Springs With Relative Coordinates. Equations (4.103) and (4.104) define the spring forces in terms of the natural coordinates. However, this formulation can be greatly simplified if the relative distance s between the points i and j is introduced as a new dependent (mixed) coordinate through the following constraint condition:

$$(x_i - x_j)^2 + (y_i - y_j)^2 + (z_i - z_j)^2 - s^2 = 0 \tag{4.108}$$

The force f can be directly entered into the formulation as the conjugate variable of distance s with a value:

$$f = k(s)\,(s - s_0) \tag{4.109}$$

where $s_o = L_o$, and s is equal to the deformed length. Proceeding with this mixed type of coordinate representation, the formulation of the forcing terms becomes much simpler at the expense of increasing the number of dependent variables by one variable.

If the dependent coordinate s is used, the potential energy of the spring is simply given by

$$V = \int_o^{s-so} f\, ds = \int_c^{s-so} k(s)\,(s-s_o)\, ds \tag{4.110}$$

Translational Springs Between Any Two Points. When the origin and end points of the spring are not basic points but any two points corresponding to two different elements, equations (4.104) and (4.107) cannot be used.

As shown in Figure 4.13, it is necessary to construct the position vectors of the origin and end of the spring, starting from the local coordinates of these points at the local coordinates frames attached to the elements to which they belong. In the case of Figure 4.13, we can write:

$$\mathbf{r}_1 = \mathbf{r}_i + \mathbf{A}_1\,(\bar{\mathbf{r}}_1 - \bar{\mathbf{r}}_i) \tag{4.111}$$

$$\mathbf{r}_2 = \mathbf{r}_k + \mathbf{A}_2\,(\bar{\mathbf{r}}_2 - \bar{\mathbf{r}}_k) \tag{4.112}$$

where $\mathbf{A}_1$ and $\mathbf{A}_2$ are rotation matrices that depend on the natural coordinates of the points and vectors of each element. In computing these rotation matrices, we can express the local and global coordinates of the basic points and vectors that belong to a rigid body as the columns of a (3x3) matrix $\mathbf{X}$ as follows:

$$\mathbf{X} \equiv \left[\mathbf{r}_i{-}\mathbf{r}_j \,|\, \mathbf{u} \,|\, \mathbf{v}\right] = \mathbf{A}\,\overline{\mathbf{X}} = \mathbf{A}\left[\bar{\mathbf{r}}_i{-}\bar{\mathbf{r}}_j \,|\, \bar{\mathbf{u}} \,|\, \bar{\mathbf{v}}\right] \tag{4.113}$$

Hence, the rotation matrix $\mathbf{A}$ can be found as

$$\mathbf{A} = \mathbf{X}\,\overline{\mathbf{X}}^{-1} = \left[\mathbf{r}_i{-}\mathbf{r}_j \,|\, \mathbf{u} \,|\, \mathbf{v}\right]\overline{\mathbf{X}}^{-1} \tag{4.114}$$

Consequently, the rotation matrices corresponding to the rigid bodies to which points 1 and 2 belong will be defined by

$$\mathbf{A}_1 = \mathbf{X}_1\,\overline{\mathbf{X}}_1^{-1} = \left[\mathbf{r}_i{-}\mathbf{r}_j \,|\, \mathbf{u}_1 \,|\, \mathbf{v}_1\right]\overline{\mathbf{X}}_1^{-1} \tag{4.115}$$

$$\mathbf{A}_2 = \mathbf{X}_2\,\overline{\mathbf{X}}_2^{-1} = \left[\mathbf{r}_k{-}\mathbf{r}_l \,|\, \mathbf{u}_2 \,|\, \mathbf{v}_2\right]\overline{\mathbf{X}}_2^{-1} \tag{4.116}$$

Using the result of equation (4.90), we can write

$$\mathbf{r}_1 = \mathbf{C}_1\,\mathbf{q}_1^e \tag{4.117}$$

$$\mathbf{r}_2 = \mathbf{C}_2\,\mathbf{q}_2^e \tag{4.118}$$

Matrices $\mathbf{C}_1$ and $\mathbf{C}_2$, defined by equation (4.50), are constant matrices that permit finding the global coordinates of points $\mathbf{r}_1$ and $\mathbf{r}_2$ in terms of the natural coordinates of the elements to which they belong. Using matrices $\mathbf{C}_1$ and $\mathbf{C}_2$, we can obtain two expressions similar to (4.94):

$$\mathbf{Q}_1^e = \mathbf{C}_1^T\,\mathbf{f}_1 \tag{4.119}$$

$$\mathbf{Q}_2^e = \mathbf{C}_2^T\,\mathbf{f}_2 \tag{4.120}$$

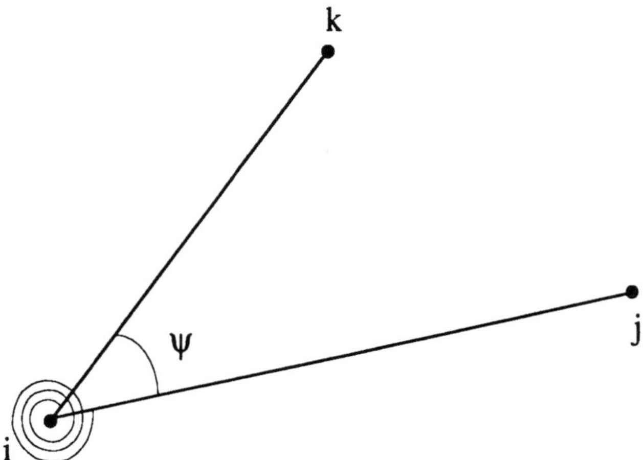

Figure 4.14. Rotational spring with mixed coordinates.

where $\mathbf{f}_2 = -\mathbf{f}_1$. The generalized forces $\mathbf{Q}_1^e$ and $\mathbf{Q}_2^e$ are conjugated with the virtual displacements $\delta\mathbf{r}_1$ and $\delta\mathbf{r}_2$, that can be defined as functions of $\delta\mathbf{q}$ by equations (4.117) and (4.118). It is possible to write an expression for the potential energy analogous to equation (4.106):

$$V = \int_{q_0}^{q} k(L)\left(1 - \frac{L_o}{L}\right) \cdot \left\{ \begin{array}{c} \mathbf{C}_1\,\mathbf{q}_1 - \mathbf{C}_2\,\mathbf{q}_2 \\ -\mathbf{C}_1\,\mathbf{q}_1 + \mathbf{C}_2\,\mathbf{q}_2 \end{array} \right\}^T \left\{ \begin{array}{c} \mathbf{C}_1\,\delta\mathbf{q}_1 \\ \mathbf{C}_2\,\delta\mathbf{q}_2 \end{array} \right\} \qquad (4.121)$$

and expanding the product of vectors,

$$V = \int_{q_0}^{q} k(L)\left(1 - \frac{L_o}{L}\right) \cdot \left\{ \mathbf{q}_1^T \;\; \mathbf{q}_2^T \right\} \begin{bmatrix} \mathbf{C}_1^T & -\mathbf{C}_1^T \\ -\mathbf{C}_2^T & \mathbf{C}_2^T \end{bmatrix} \begin{bmatrix} \mathbf{C}_1 & 0 \\ 0 & \mathbf{C}_2 \end{bmatrix} \left\{ \begin{array}{c} \delta\mathbf{q}_1 \\ \delta\mathbf{q}_2 \end{array} \right\} =$$

$$= \int_{q_0}^{q} k(L)\left(1 - \frac{L_o}{L}\right) \cdot \left\{ \delta\mathbf{q}_1^T \;\; \delta\mathbf{q}_2^T \right\} \begin{bmatrix} \mathbf{C}_1^T\mathbf{C}_1 & -\mathbf{C}_1^T\mathbf{C}_2 \\ -\mathbf{C}_2^T\mathbf{C}_1 & \mathbf{C}_2^T\mathbf{C}_2 \end{bmatrix} \left\{ \begin{array}{c} \mathbf{q}_1 \\ \mathbf{q}_2 \end{array} \right\} \qquad (4.122)$$

The length L that is the distance between points $\mathbf{r}_1$ and $\mathbf{r}_2$ can be computed by using the formulae (4.117) and (4.118) to solve for the coordinates of $\mathbf{r}_1$ and $\mathbf{r}_2$; and then to find the distance directly.

Rotational Springs. A rotational spring exerts, between the elements to which it is connected, a torque about the common articulation of both elements; that is a function of the relative angle twisted between them (Figure 4.14). Angles of more than 360º are possible; therefore, it is necessary to take into account not only the angle between both elements as calculated with the scalar or cross products of vectors, but also the number of complete turns that the rotational spring may have gone through.

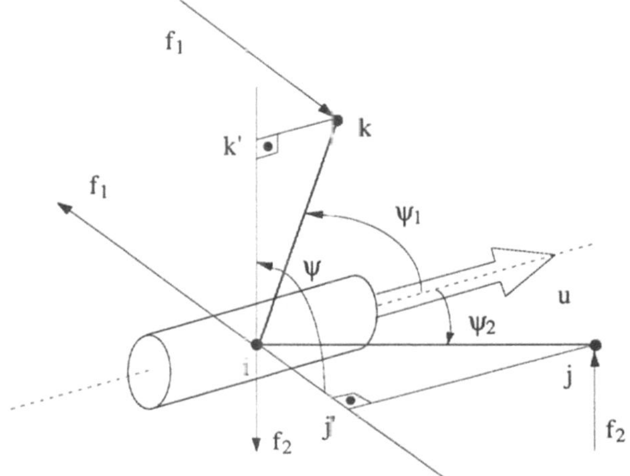

Figure 4.15. Special case of rotational springs.

Consider the planar rotational spring shown in Figure 4.14 in which ψ_0 is the angle corresponding to the non-deformed position of the spring. If the relative angular position of the elements is considered as a new mixed variable $(\psi+2n\pi)$, the torque exerted on both elements will be directly given by

$$M = k(f)\left(\psi + 2n\pi - \psi_0\right) \tag{4.123}$$

Note that the new dependent coordinate ψ may be introduced by either one or a combination of the following constraint conditions:

$$\mathbf{r}_{ij} \cdot \mathbf{r}_{jk} = L_{ij}\, L_{jk}\, cos\ \psi \tag{4.124}$$

$$\left|\mathbf{r}_{ij} \wedge \mathbf{r}_{jk}\right| = L_{ij}\, L_{jk}\, sin\ \psi \tag{4.125}$$

In addition, the potential energy is given by

$$V = \int_{\psi_0}^{\psi+2n\pi-\psi_0} M(\psi)\,d\psi = \int_{\psi_0}^{\psi+2n\pi-\psi_0} k(\psi)\left(\psi-\psi_0\right)d\psi \tag{4.126}$$

Since ψ is one of the dependent coordinates of the system, no additional transformation is required.

In the three-dimensional case, the situation is a little more complicated, particularly in the case in which the points i, j, and k are not in a plane perpendicular to the axis of the revolute joint. Here, the angle ψ in the joint is not the angle formed by the segments $(i-j)$ and $(i-k)$, but the angle determined by two straight lines normal to the axis of the pair. This can be seen in Figure 4.15. It is assumed in this figure that the axis of the pair is determined by the unit vector **u**. A similar and simpler formulation for the forces is obtained when the pair is determined by two basic points.

If the angle is introduced as a dependent coordinate, the expression for the potential energy is immediate and responds to the same equation as that of the planar case (equation (4.126)). When the spring is linear,

$$V = \frac{1}{2} k \left(\psi + 2n\pi - \psi_o \right)^2 \tag{4.127}$$

where k is the spring constant.

4.3.3 Forces Induced by Known Acceleration Fields

The simplest case of forces induced by known acceleration fields are gravitational forces. The gravitational force acting on an element is simply the product of its mass m times the gravitational acceleration $\mathbf{g}$, acting on the center of gravity $\mathbf{r}_G$:

$$\mathbf{f} = - m \, \mathbf{g} \tag{4.128}$$

and the potential of this force can be expressed as

$$V = - m \, \mathbf{r}_G^T \, \mathbf{g} \tag{4.129}$$

A matrix $\mathbf{C}_G$ similar to that of expression (4.50) can be constructed for the center of gravity to express its coordinates in terms of natural coordinates $\mathbf{q}^e$ of the element. Consequently the potential becomes

$$V = - m \, \mathbf{q}^{eT} \, \mathbf{C}_G^T \, \mathbf{g} \tag{4.130}$$

Another important case is the one in which the external forces originate from a *known accelerations field*. This situation arises when the fixed element is moving in a prescribed mode or when the entire multibody system is subjected to a rotation.

Let $\mathbf{v}_0$ and Ω be the velocity of the origin and the angular velocity vector, respectively, of the reference coordinate frame whose motion is known. The acceleration of this system is defined by $\dot{\mathbf{v}}_0$, Ω, and $\dot{\Omega}$ which are known. Using the *principles of relative motion* (Greenwood (1988)), the motion with respect to a moving reference frame can be studied as if it were an absolute motion, introducing as known external forces all the inertial forces corresponding to the motion of the frame. Assuming a standard element with two basic points i and j and two non-coplanar unit vectors $\mathbf{u}$ and $\mathbf{v}$, these accelerations are:

$$\ddot{\mathbf{r}}_i = \dot{\Omega} \wedge \mathbf{r}_i + \Omega \wedge \left(\Omega \wedge \mathbf{r}_i \right) + \dot{\mathbf{v}}_0 \tag{4.131}$$

$$\ddot{\mathbf{r}}_j = \dot{\Omega} \wedge \mathbf{r}_j + \Omega \wedge \left(\Omega \wedge \mathbf{r}_j \right) + \dot{\mathbf{v}}_0 \tag{4.132}$$

$$\ddot{\mathbf{u}} = \dot{\Omega} \wedge \mathbf{u} + \Omega \wedge \left(\Omega \wedge \mathbf{u} \right) \tag{4.133}$$

$$\ddot{\mathbf{v}} = \dot{\Omega} \wedge \mathbf{v} + \Omega \wedge \left(\Omega \wedge \mathbf{v} \right) \tag{4.134}$$

As a consequence of these accelerations, the vector of inertial forces $\mathbf{Q}_{in}^e$ acting on the element is,

$$Q_{in}^e = - M^e \begin{Bmatrix} \ddot{r}_i \\ \ddot{r}_j \\ \ddot{u} \\ \ddot{v} \end{Bmatrix} = - M^e \ddot{q}^e \tag{4.135}$$

where M^e is the mass matrix developed in Section 4.2.2. The potential of these forces, which is position dependent, will be

$$V = - \int_{q_0}^{q} dq^{eT} Q_{in}^e = - \int_{q_0}^{q} dq^{eT} M^e \ddot{q}^e \tag{4.136}$$

One will find these expressions useful for carrying out the dynamic analysis of a multibody system evolving in a known field of centrifugal forces.

References

Bastero, J.M. and Casellas, J., *Curso de Mecánica*, Eunsa, (1976).

Goldstein, H., *Classical Mechanics*, 2nd Edition, Addison-Wesley, London, (1980).

Greenwood, T. J., *Principles of Dynamics*, 2nd Edition, Prentice Hall, (1988).

Hamilton, W.R., "On the General Method in Dynamics", *Philosophical Transactions of the Royal Society of London*, pp. 247-308, (1834).

Huston, R.L., *Multibody Dynamics*, Butterworth-Heinemann, (1990).

Kane, T.R. and Levinson, D.A., *Dynamics: Theory and Applications*, McGraw-Hill, (1985).

Reddy J.N., *Energy and Variational Methods in Applied Mechanics*, Wiley Interscience, (1984).

Schiehlen, W.O., "Dynamics of Complex Multibody Systems", *SM Archives*, Vol. 9, pp. 159-195, (1984).

Problems

4/1   Using equation (4.48), find the inertial matrix (with respect to points i and j) of the 2-D element shown in the figure, in the following cases:
a) The element has a concentrated unit mass located at i.
b) The element has a concentrated unit mass located at j.
c) The element has a concentrated unit mass located at point (0,1)
d) The element has a concentrated unit mass located at point (0,-1)
e) The element consists of a disk with its center at i, having unit radius, and uniformly distributed unit mass.

4/2   Using the results of Problem 4/1 and admitting the possibility of inertial matrices corresponding to negative masses, to eliminate mass from a real rigid body, find the concentrated masses at points (0,0), (1,0), (0,1), and (0,-1), so that the resulting mass matrix is:

$$\begin{bmatrix} 0 & 0 & 1 & 0 \\ 0 & 0 & 0 & 1 \\ 1 & 0 & 0 & 0 \\ 0 & 1 & 0 & 0 \end{bmatrix}$$

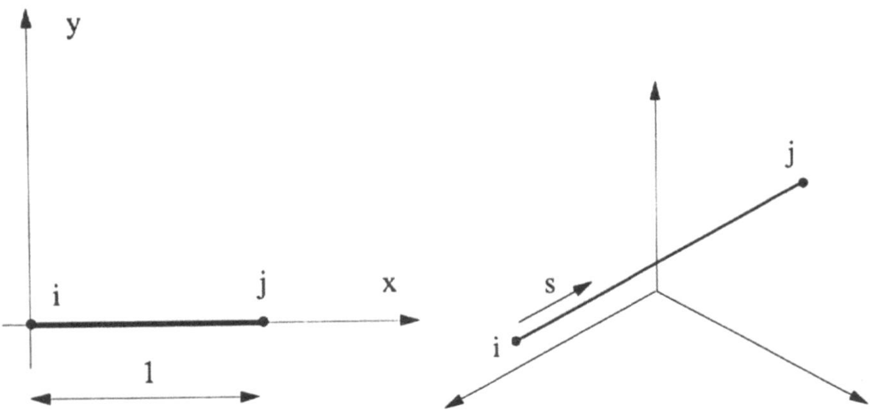

Figure P4/1.   Figure P4/3.

4/3   Find by integration of

$$\mathbf{M}\,\ddot{\mathbf{q}} \equiv \frac{m}{L} \int_0^L \dot{\mathbf{r}}(s)\, ds$$

the inertial matrix $\mathbf{M}$ of an homogeneous 3-D bar and show that $\mathbf{M}$ is independent of the position of the bar.

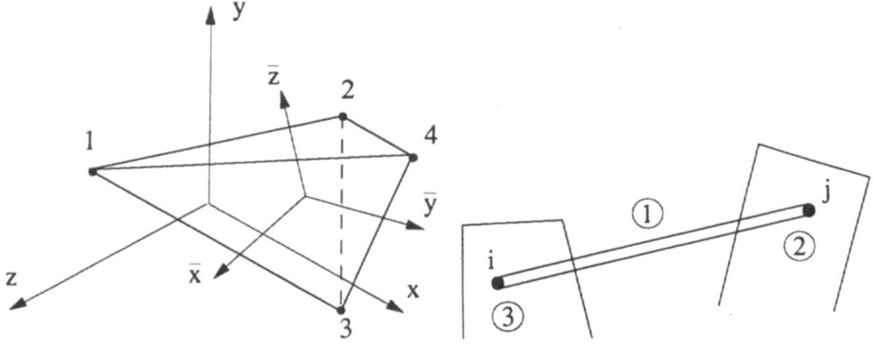

Figure P4/4.   Figure P4/5.

4/4   The inertial properties of a 3-D rigid body depend on ten parameters including the mass m, the position of center of gravity $\bar{\mathbf{r}}_G$, and the inertial tensor $\bar{\mathbf{I}}_G$, defined on a moving frame attached to the body. Show by intuitive reasoning (without algebraic calculations) that it is possible to find a (12x12) constant inertial matrix in the global frame. Use as acceleration variables the Cartesian accelerations of points 1, 2, 3, and 4.

4/5   A bar 1 connects two rigid bodies 2 and 3 through two spherical joints i and j. The motion of bodies 2 and 3 is known, so the motion of the bar is known except for the rotation around the direction i-j. There is a friction torque of constant magnitude T and direction opposed to the relative angular velocity applied at the joints i and j. Find the directions of torques T_i and T_j that guarantee the complete equilibrium of the bar.

5
Dynamic Analysis. Equations of Motion

This chapter deals with the *direct dynamic problem* which consists of determining the motion of a multibody system that results from the application of the external forces and/or the kinematically controlled or driven degrees of freedom. The direct dynamic analysis is also commonly referred to as the *dynamic simulation*. Its importance is steadily increasing in fields such as: automobile industry, aerospace, robotics, machinery, biomechanics, and others. The possibility of kinematically controlling some degrees of freedom in a dynamic problem has many practical applications. For example, in the analysis of vehicle suspensions, if the wheel is rigid, its center follows the trajectory determined by the rolling surface. The dynamic problem will determine the resulting motion of all the vehicle's remaining elements.

It is very important to emphasize the difference between the kinematically and dynamically controlled degrees of freedom. In the previously stated kinematic simulation, all the degrees of freedom were controlled kinematically; that is, the motion of as many input elements as degrees of freedom is known. There are as many additional kinematic or driving constraint equations such as known angles, and known distances, as degrees of freedom. In order for the problem to be truly dynamic, it is necessary that the number of unknown dependent variables be greater than the total number of independent geometric and driving constraint equations. As a result, the motion of the multibody system cannot be unequivocally defined by the geometric and driving constraint equations and by the known motion of points and vectors only. In order to determine the motion of the entire system, it is necessary to establish the dynamic equilibrium condition that leads to a system of second order differential equations generally called the *equations of motion*.

The direct dynamic problem pursues the determination of the system's motion during a period of time originated by known external forces and/or the kinematically driven degrees of freedom. The position of the multibody system is characterized by its dependent coordinates. It is not sufficient to know the values of a minimum set of independent coordinates, because the position is not unequivocally known by simply knowing the independent coordinates. However, at the time of formulating the equations of motion, it is possible to do it with both dependent or independent coordinates. There is not a consensus among the experts as to which method is the best for all cases. A method can be advantageous over another under certain conditions and vice versa. Continuing research is being carried out to find the best possible formulation in terms of efficiency and accuracy.

The dynamic formulation with independent coordinates calls for solving the position problem at each stage of the integration or adopting alternative methods that will be later explained below. In practice this is not a serious problem. As the positions of the system corresponding to two consecutive steps of the integration are very close, the position problem converges rapidly and the problem of multiple solutions does not present a serious practical difficulty.

We discuss in this chapter several methods concerning formulating and solving the direct dynamic problem with both dependent (Section 5.1) and independent coordinates (Section 5.2). In many instances we include algorithms which explain and facilitate their computer implementation. Some recent formulations that are based on velocity transformations and the canonical equations are discussed in Sections 5.3 and 5.4, respectively.

5.1 Formulations in Dependent Coordinates

Several methods of formulating the equations of motion with dependent coordinates will be developed below. In all cases, the desired end result can be obtained by either the Lagrange's equations or the method of virtual power. Hereafter, the vector $\mathbf{q}$ represents a set of n unknown dependent coordinates, m is the total number of independent constraint equations (geometric and kinematic), and $f=n-m$ is the number of dynamic degrees of freedom. The constraint conditions are written in the following general form:

$$\Phi(\mathbf{q}, t) = 0 \tag{5.1}$$

Let $T(\mathbf{q}, \dot{\mathbf{q}})$ be the kinetic energy of the system, $V(\mathbf{q})$ the potential energy and $\mathbf{Q}_{ex}(\mathbf{q})$ the vector of generalized external forces acting along the dependent coordinates $\mathbf{q}$ of a constrained mechanical system. The Lagrange's equations of such systems have been derived in Chapter 4 in the form:

$$\frac{d}{dt}\left(\frac{\partial L}{\partial \dot{\mathbf{q}}}\right) - \frac{\partial L}{\partial \mathbf{q}} + \Phi_{\mathbf{q}}^{T} \lambda = \mathbf{Q}_{ex} \tag{5.2}$$

where $L = T-V$ is the *Lagrangian* function. The third term on the LHS of equation (5.2) is introduced, because the coordinates $\mathbf{q}$ are not independent but interrelated by means of the constraint equations. The matrix $\Phi_{\mathbf{q}}$ is the Jacobian matrix of the nonlinear constraint equations (5.1). The vector λ in (5.2) represents the Lagrange multipliers. With this formulation the number of unknowns has increased to $n+m$, since not only $\mathbf{q}$ but also λ needs to be calculated.

As shown in Chapter 4, the kinetic energy of a multibody system can be written as follows:

$$T = \frac{1}{2} \dot{\mathbf{q}}^T \mathbf{M}(\mathbf{q}) \dot{\mathbf{q}} \tag{5.3}$$

where the mass matrix is constant as long as all the bodies have at least two points and two non-coplanar unit vectors or an equivalent structure. Otherwise, the mass matrix is dependent on the positions $\mathbf{q}$. For the general case in which the kinetic energy depends on $\mathbf{q}$, equation (5.2) becomes (See Example 4.1)

$$\mathbf{M}\ddot{\mathbf{q}} + \Phi_{\mathbf{q}}^T \lambda = \mathbf{Q}_{ex} + L_{\mathbf{q}} - \dot{\mathbf{M}}\dot{\mathbf{q}} \tag{5.4}$$

where $\mathbf{Q}_{ex}$ is the vector of external forces and L the Lagrangian. It may be seen that equation (5.2) involves the time differentiation of the mass matrix which leads in certain cases to rather involved computations.

Another way of formulating the equations of motion is by means of the *method of virtual power*. It was demonstrated in Chapter 4 that the virtual power of the forces acting on a multibody system can be written as

$$\dot{\mathbf{q}}^{*T} \left(\mathbf{M}\ddot{\mathbf{q}} - \mathbf{Q} \right) = 0 \tag{5.5}$$

where $\dot{\mathbf{q}}^*$ are the virtual velocities, which must satisfy the first derivative of the constraint equations at a stationary time. Therefore,

$$\Phi_{\mathbf{q}}(\mathbf{q}, t)\, \dot{\mathbf{q}}^* = 0 \tag{5.6}$$

It cannot be inferred from equation (5.5), that the part of the expression between parenthesis is zero. In addition to the inertial and external forces coming from a potential, the constraint forces, such as forces at the pairs, also act on the multibody system. Although they do not appear in a virtual power expression, they should appear in the equilibrium equations. In order to eliminate the virtual dependent velocities from equation (5.5), one should add to (5.5) the equation (5.6) transposed and multiplied by a vector of m unknown coefficients λ. This would yield:

$$\dot{\mathbf{q}}^{*T} \left(\mathbf{M}\ddot{\mathbf{q}} - \mathbf{Q} + \Phi_{\mathbf{q}}^T \lambda \right) = 0 \tag{5.7}$$

As mentioned in Section 4.1, it is always possible to find m values of the vector λ. This vector establishes the magnitude of the constraint forces; so that

$$\mathbf{M}\ddot{\mathbf{q}} + \Phi_{\mathbf{q}}^T \lambda = \mathbf{Q} \tag{5.8}$$

where now the vector $\mathbf{Q}$ contains the external forces plus all the velocity-dependent inertial terms obtained as explained in Chapter 4. Hence, equations (5.4) and (5.8) are equivalent. The first term corresponds to the inertial forces; the last, to the external forces, velocity-dependent inertial forces, and those obtained from a potential. The intermediate term corresponds to the forces associated with the imposed constraints, that is, the forces necessary for the constraint equations to be satisfied.

5.1.1 Method of the Lagrange Multipliers

Equation (5.8) represents n equations and $(n+m)$ unknowns: the n elements of vector $\ddot{q}$ and the m elements of vector λ. In order to have a sufficient number of equations, it is necessary to supply m more equations. The obvious choice is to use the algebraic constraint equations (5.1) which along with (5.8) constitute a set of differential algebraic equations, DAEs of index three (See Chapter 7). In order to avoid DAEs, one can use the acceleration kinematic equations which are obtained by differentiating the constraint equations (5.1) twice with respect to time:

$$\Phi_q \, \ddot{q} = -\dot{\Phi}_t - \dot{\Phi}_q \, \dot{q} \equiv c \tag{5.9}$$

This expression is used to define the vector c. By writing equations (5.8) and (5.9) jointly, one obtains:

$$\begin{bmatrix} M & \Phi_q^T \\ \Phi_q & 0 \end{bmatrix} \begin{Bmatrix} \ddot{q} \\ \lambda \end{Bmatrix} = \begin{Bmatrix} Q \\ c \end{Bmatrix} \tag{5.10}$$

which is a system of $(n+m)$ equations with $(n+m)$ unknowns, whose matrix is symmetrical and, in general, non-positive definite, and also very sparse in many practical cases.

The system of equations (5.10) can be used for the simultaneous solution of the accelerations and Lagrange multipliers. Alternatively, equation (5.8) can be solved first to obtain an expression for the accelerations:

$$\ddot{q} = M^{-1} Q - M^{-1} \Phi_q^T \lambda \tag{5.11}$$

which can only be used if the mass matrix is non-singular, as it will be in most of the cases. By substituting equation (5.11) in equation (5.9), one obtains:

$$\Phi_q M^{-1} \Phi_q^T \lambda = \Phi_q M^{-1} Q - c \tag{5.12}$$

from which the Lagrange multiplier vector λ can be found. In order to calculate the accelerations, it will suffice to substitute λ in equation (5.11).

In the majority of practical cases, the direct solution of equations (5.10) is preferable over the use of (5.11) and (5.12). The main advantage of the dynamic formulation in dependent coordinates using Lagrange multipliers, besides the conceptual simplicity of the method, is permitting the calculation of forces associated with the constraints (which depend on the Lagrange multipliers) with a minimum additional effort. The solution of (5.10) yields λ directly without the need for a special call to an inverse dynamic module (See Chapter 6).

Numerical Implementation. It will be assumed that a numerical integration subroutine for first order differential equations, such as those described in Chapter 7, is available. The operation of these subroutines can be summarized as follows: given the vector of derivatives $\dot{y}_t$ of the dependent variables at time t, the numerical integration subroutine (n.i.s.) returns the value of the vector **y** at time $(t+\Delta t)$. Schematically,

$$\dot{\mathbf{y}}_t \xrightarrow{\text{n.i.s.}} \mathbf{y}_{t+\Delta t}$$

Therefore, following the Lagrange multiplier method, the numerical integration of the equations of motion may proceed as follows:

Algorithm 5-1

1. Start at a time t in which the position **q** and velocity $\dot{\mathbf{q}}$ are known.
2. Use equations (5.10) to solve the accelerations at time t. We call this process a *function evaluation*.
3. The vector $\dot{\mathbf{y}}_t^T \equiv \{\ddot{\mathbf{q}}^T, \dot{\mathbf{q}}^T\}_t$ is given as input to the numerical integration subroutine (valid for first order differential equations), and the vector $\mathbf{y}_{t+\Delta t}^T \equiv \{\dot{\mathbf{q}}^T, \mathbf{q}^T\}_{t+\Delta t}$ is obtained:

$$\dot{\mathbf{y}}_t^T \equiv \{\ddot{\mathbf{q}}^T, \dot{\mathbf{q}}^T\}_t \xrightarrow{\text{n.i.s.}} \mathbf{y}_{t+\Delta t}^T \equiv \{\dot{\mathbf{q}}^T, \mathbf{q}^T\}_{t+\Delta t}$$

4. Upon convergence of the n.i.s., update the time variable and go to step 2.

This numerical integration algorithm has the advantage of being much simpler than those shown below corresponding to other methods. However, it may not be the most efficient. In addition, it will be explained below that as the numerical integration proceeds using this algorithm, the constraint conditions are progressively violated leading to unacceptable results in all but very short simulations.

5.1.2 *Method Based on the Projection Matrix* **R**

A second possibility of formulating the motion differential equations with dependent coordinates is based on the matrix **R** introduced in Chapter 3. Remember that the $f=n-m$ columns of the matrix **R** represent a basis of the nullspace of the Jacobian Φ_q; that is, a basis of the subspace of possible motions. The matrix **R** verifies the following relationship for holonomic systems:

$$\Phi_q \, \mathbf{R} = 0 \tag{5.13}$$

It also directly relates the dependent and independent velocities for the case in which there are no rheonomous constraints:

$$\dot{\mathbf{q}} = \mathbf{R} \, \dot{\mathbf{z}} \tag{5.14}$$

If in the virtual power equation (5.5) the dependent virtual velocities $\dot{\mathbf{q}}^*$ are substituted in terms of the independent virtual velocities $\dot{\mathbf{z}}^*$, the use of equation (5.14) leads to

$$\dot{\mathbf{z}}^{*T} \mathbf{R}^T (\mathbf{M} \, \ddot{\mathbf{q}} - \mathbf{Q}) = 0 \tag{5.15}$$

Since the previous expression should be verified for any arbitrary vector of independent virtual velocities, the following must also be satisfied:

$$\mathbf{R}^T \mathbf{M} \ddot{\mathbf{q}} = \mathbf{R}^T \mathbf{Q} \qquad (5.16)$$

Equation (5.16) contains $(n-m)$ equations with n unknowns. In order to have as many equations as unknowns, it is necessary to complete this system with the kinematic acceleration equations (5.9), resulting in

$$\begin{bmatrix} \boldsymbol{\Phi}_q \\ \mathbf{R}^T \mathbf{M} \end{bmatrix} \ddot{\mathbf{q}} = \left\{ \begin{array}{c} \mathbf{c} \\ \mathbf{R}^T \mathbf{Q} \end{array} \right\} \qquad (5.17)$$

which is a system of n equations with n unknowns which can be easily solved for the dependent accelerations $\ddot{\mathbf{q}}$. The upper part of equation (5.17), corresponding to matrix $\boldsymbol{\Phi}_q$, has been previously factored in order to calculate the matrix $\mathbf{R}$. Because of this, the system of equations can be solved with very little additional effort. The method based on equation (5.17) is sometimes more efficient than the one based on equations (5.10) (Unda et al. (1987)). The dynamic formulation whose end result is equation (5.17) was originally introduced by Kamman and Huston (1984), although they did not use a general matrix $\mathbf{R}$ but a set of eigenvectors associated with the zero eigenvalues of the matrix $(\boldsymbol{\Phi}_q^T \boldsymbol{\Phi}_q)$.

Matrix $\mathbf{R}$ can be calculated by means of any of the methods explained in Chapter 3. The simplest is the projection method (Section 5.2.3) based on the selection of the independent coordinates as a subset of the dependent ones.

Remarks:

*   The same result of equation (5.16) can be arrived at by eliminating the vector λ in equation (5.8). By multiplying equation (5.8) by the matrix $\mathbf{R}^T$, we can write

$$\mathbf{R}^T \mathbf{M} \ddot{\mathbf{q}} + \mathbf{R}^T \boldsymbol{\Phi}_q^T \lambda = \mathbf{R}^T \mathbf{Q} \qquad (5.18)$$

    but by virtue of equation (5.13), the term containing the Lagrange multipliers can be cancelled, thus equation (5.16) is obtained.

*   Equation (5.17) allows one to clearly distinguish the equations corresponding to the kinematics (the m first ones) from the equations corresponding to the dynamics (the $n-m$ last ones). Besides, system (5.17) does not explicitly contain any independent coordinates; rather they are implicitly considered via the matrix $\mathbf{R}$. Each matrix $\mathbf{R}$ implies a choice of independent coordinates in accordance with equation (5.14). The chosen set of independent coordinates must be changed anytime there is a need to guarantee the existence and perfect conditioning of the Jacobian factorization necessary to compute the matrix $\mathbf{R}$.

Numerical Implementation. Similar to the case of the Lagrange multiplier method, the numerical integration process of the equations of motion using the matrix $\mathbf{R}$ may proceed as follows:

Algorithm 5-2

1. Start at a time t in which the position q and velocity $\dot{q}$, are known.
2. Form the matrix Φ_q and triangularize it with column partial pivoting. From this triangularization, decide whether the current set of coordinates continue to be valid (independent) to form the matrix R, or if it is necessary to change them. In the latter case, make a new choice of independent columns by means of a triangularization with total pivoting.
3. Form the matrix R and the product $R^T M$.
4. Solve equation (5.17) for the dependent accelerations.
5. Obtain the vectors q and $\dot{q}$ at time $(t+\Delta t)$ by numerical integration:

$$\dot{y}_t^T \equiv \{\ddot{q}^T, \dot{q}^T\}_t \quad \xrightarrow{\text{n.i.s.}} \quad y_{t+\Delta t}^T \equiv \{\dot{q}^T, q^T\}_{t+\Delta t}$$

6. Upon convergence of the n.i.s., update the time variable and go to step 2.

Similar to Algorithm 5-1, this method also requires constraint stabilization for long simulations. This point is treated next.

5.1.3 Stabilization of the Constraint Equations

Once the position, velocity, and external forces are known, equations (5.10) and (5.17) permit calculating the dependent accelerations of the system at a specific time. Both equations use the kinematic acceleration equations (5.9) which are obtained by differentiating the constraint equations (5.2) twice with respect to time. This means that the following differential equation is also being integrated with respect to time:

$$\ddot{\Phi}(q,t) \equiv \Phi_q \ddot{q} + \dot{\Phi}_q \dot{q} + \dot{\Phi}_t = 0 \tag{5.19}$$

This system of differential equations has the following general solution:

$$\Phi(q,t) = a_1 t + a_2 \tag{5.20}$$

where a_1 and a_2 are constant vectors that depend on the initial conditions; that is, on the value of the constraint equations and on its first derivative with respect to time at $t=0$. If the position and initial velocity satisfy the constraint equations, both vectors a_1 and a_2 are null. Theoretically, equation (5.20) guarantees that the constraint equations will be satisfied at any time. The fact of the matter is very different.

Equation (5.19) is *unstable*, since for any vector a_1 different from zero the general solution given by equation (5.20) is not bounded and tends to increase indefinitely with time. Even though the initial conditions guarantee that $a_1 = 0$, during the course of the numerical integration, the round-off errors that appear during the integration do not satisfy the constraint equations. The effects of these errors increase with time, in accordance with expression (5.20). Therefore, the constant distances cease to be constant and the points of the same element progressively move closer to or further away from them. A similar situation happens with the other constraint equations.

This difficulty takes place during the integration of the differential equation (5.19). For example, let's take a look at a constraint equation of constant distance between two points i and j:

$$(\mathbf{r}_i - \mathbf{r}_j)^T (\mathbf{r}_i - \mathbf{r}_j) - L_{ij}^2 = 0 \tag{5.21}$$

By differentiating this equation twice with respect to time, one obtains

$$(\mathbf{r}_i - \mathbf{r}_j)^T (\ddot{\mathbf{r}}_i - \ddot{\mathbf{r}}_j) + (\dot{\mathbf{r}}_i - \dot{\mathbf{r}}_j)^T (\dot{\mathbf{r}}_i - \dot{\mathbf{r}}_j) = 0 \tag{5.22}$$

When numerically integrating equation (5.22), one loses the information about the distance L_{ij} that should be maintained between both of the points. The constant L_{ij} disappears during the differentiating process. Thus, expression (5.22) does not have the information corresponding to the distance that must be maintained, and errors accumulate on the distance between points i and j. The same situation happens with all the other constraint equations, which at the time of differentiation lose the information carried by the constant terms.

The instability during the numeric integration of kinematic acceleration equations ensures that Algorithms 5-1 and 5-2 may not be used directly to obtain the solution of the dynamic simulation problem. Two different methods have been proposed for overcoming this difficulty and are stated below.

5.1.3.1 Integration of a Mixed System of Differential and Algebraic Equations

The purpose of this method is to jointly solve the system of nonlinear algebraic equations (5.1) and differential equations (5.10) or (5.17). Thus, the constraint equations, and not only their second derivatives, are satisfied at any given time. Refer to Chapter 7 for a description on the general solution of differential algebraic equations or DAEs. There are numerical integration methods for mixed systems of differential equations that permit adding algebraic equations. However, they are not the most numerically efficient and are not free from stability problems in the simulation of mechanical systems (See Chapter 7 and Steigerwald (1990)).

5.1.3.2 Baumgarte Stabilization

The aim of the Baumgarte stabilization method (Baumgarte (1972)) is to replace the differential constraint equations (5.19) by the following system:

$$\ddot{\Phi} + 2\alpha \dot{\Phi} + \beta^2 \Phi = 0 \tag{5.23}$$

where α and β are appropriately chosen constants. The general solution to this differential equation is

$$\Phi = a_1 e^{s_1 t} + a_2 e^{s_2 t} \tag{5.24}$$

where $\mathbf{a}_1$ and $\mathbf{a}_2$ are constant vectors that depend on the initial conditions, and s_1 and s_2 are the roots of the characteristic equation, defined by the expression:

$$s_1, s_2 = -\alpha \pm \sqrt{\alpha^2 - \beta^2} \tag{5.25}$$

If α and β are positive constants, the two roots s_1 and s_2 have a real negative part which guarantees the stability of the general solution (5.24) in contrast with that in (5.20). The initial position and velocity conditions of the multibody system should guarantee that the vectors $\mathbf{a}_1$ and $\mathbf{a}_2$ are zero. If the numerical rounding-off errors alter this condition, the real negative part of the exponential terms damps out the possible errors occurring during the integration process. The constants α and β are usually equal to one another with values between 1 and 20, and it appears that the behavior of the method does not significantly depend on these values. Chang and Nikravesh (1985), and Bae and Yang (1990) proposed different methods for optimizing this choice.

By using equation (5.23) instead of equation (5.9), the differential equations of motion (5.10) and (5.17) are respectively transformed into:

$$\begin{bmatrix} \mathbf{M} & \boldsymbol{\Phi}_{\mathbf{q}}^{\mathrm{T}} \\ \boldsymbol{\Phi}_{\mathbf{q}} & \mathbf{0} \end{bmatrix} \begin{Bmatrix} \ddot{\mathbf{q}} \\ \lambda \end{Bmatrix} = \begin{Bmatrix} \mathbf{Q} \\ \mathbf{g} \end{Bmatrix} \tag{5.26}$$

and

$$\begin{bmatrix} \boldsymbol{\Phi}_{\mathbf{q}} \\ \mathbf{R}^{\mathrm{T}}\mathbf{M} \end{bmatrix} \ddot{\mathbf{q}} = \begin{Bmatrix} \mathbf{g} \\ \mathbf{R}^{\mathrm{T}}\mathbf{Q} \end{Bmatrix} \tag{5.27}$$

where

$$\mathbf{g} = -\dot{\boldsymbol{\Phi}}_t - \dot{\boldsymbol{\Phi}}_{\mathbf{q}}\,\dot{\mathbf{q}} - 2\alpha\left(\boldsymbol{\Phi}_{\mathbf{q}}\,\dot{\mathbf{q}} + \boldsymbol{\Phi}_t\right) - \beta^2\,\boldsymbol{\Phi} \tag{5.28}$$

Nikravesh (1984) has studied comparatively the numerical integration of the equations of motion with dependent coordinates without stabilization, and with Baumgarte stabilization, using the integration of mixed systems of differential algebraic equations. His conclusions indicate that the Baumgarte stabilization is twice as efficient as the integration of the mixed systems, even though not all the problems examined were satisfactorily solved with the said stabilization method. On the other hand, the direct integration of equations (5.10) or (5.17) produced unacceptable results. The Baumgarte stabilization is general, simple, and numerically efficient. Its computational cost is a small fraction of the total required. However, it does not solve all possible instabilities, such as near kinematic singular configurations (Haug (1989)). This aspect works in favor of other methods with dependent and independent coordinates that will be studied below.

5.1.4 Penalty Formulations

As shown in Section 5.1.1, the Lagrange multipliers technique allows for the solution of the dynamic problem at the expense of solving for an augmented set of

$(n+m)$ unknowns: **q** plus λ. In this section we will present an alternative penalty formulation proposed by Bayo et al. (1988) that eliminates the Lagrange multipliers from the equations of motion and leads to a set of n ordinary differential equations with $\ddot{q}$ as the only unknowns. In essence, this method directly incorporates the constraint equations as a dynamical system, penalized by a large factor, into the equations of motions. The larger the penalty factor the better the constraints will be achieved at the cost of introducing some numerical ill-conditioning. We will show next how this penalty formulation can be applied to holonomic and non-holonomic constraint conditions and how to avoid the numerical problems that may arise in the use of penalty factors. Theoretical studies of its convergence and stability have been carried out by Kurdila and Narcowich (1992). This penalty method has also been successfully extended to real time dynamics within the context of fully Cartesian coordinates in Bayo et al. (1991). This will be explained in Chapter 8.

Holonomic Systems. A holonomic system is characterized by constraint equations of the form given in (5.2) which represent a set of nonlinear algebraic equations in the coordinates and the time variable. The penalty formulation is derived by adding three terms to the *Lagrangian*: These terms include a fictitious potential:

$$V^* = \sum_k \frac{1}{2} \alpha_k \, \omega_k^2 \, \Phi_k^2 \equiv \frac{1}{2} \Phi^{\mathrm{T}} \alpha \, \Omega^2 \, \Phi \tag{5.29}$$

a set of Rayleigh's dissipative forces:

$$G_k = -2\alpha_k \, \omega_k \, \mu_k \, \frac{\partial \phi_k}{dt} \equiv -2 \, \alpha \, \Omega \, \mu \, \dot{\Phi} \tag{5.30}$$

and a fictitious kinetic energy term:

$$T^* = \sum_k \frac{1}{2} \alpha_k \left(\frac{d\Phi_k}{dt}\right)^2 \equiv \frac{1}{2} \dot{\Phi}^{\mathrm{T}} \alpha \, \dot{\Phi} \tag{5.31}$$

The α_k are very large real values (penalty numbers), and ω_k and μ_k represent the natural frequency and the damping ratio of the penalty system (mass, dashpot, and spring) corresponding to the constraint $\Phi_k=0$. Matrices α, Ω and μ are $(m \times m)$ diagonal matrices that contain the values of the penalty numbers, the natural frequencies, and the damping ratios of the penalty systems assigned to each constraint condition. If the same values are used for each constraint, these matrices become identity matrices multiplied by the respective penalty numbers. Note that in equations (5.29) through (5.31), we have used both index as well as matrix notation, hoping that this will lead to a better understanding of the physical significance of the different terms. In the following discussion, we will only use the matrix form in order to be consistent with the notation used so far in this book.

The differentiation of the new penalty terms that form the Lagrangian term $L^*=T^*-V^*$ leads to

$$\frac{\partial L^*}{\partial \mathbf{q}} = \dot{\Phi}_{\mathbf{q}}^{\mathrm{T}} \, \alpha \, \dot{\Phi} - \Phi_{\mathbf{q}}^{\mathrm{T}} \, \alpha \, \Omega^2 \, \Phi \tag{5.32}$$

$$\frac{\partial L^*}{\partial \dot{\mathbf{q}}} = \dot{\Phi}_{\dot{\mathbf{q}}}^{\mathrm{T}} \, \alpha \, \dot{\Phi} \tag{5.33}$$

$$\frac{d}{dt}\left(\frac{\partial L^*}{\partial \dot{\mathbf{q}}}\right) = \ddot{\Phi}_{\dot{\mathbf{q}}}^{\mathrm{T}} \, \alpha \, \dot{\Phi} + \dot{\Phi}_{\dot{\mathbf{q}}}^{\mathrm{T}} \, \alpha \, \ddot{\Phi} = \dot{\Phi}_{\mathbf{q}}^{\mathrm{T}} \, \alpha \, \dot{\Phi} + \Phi_{\mathbf{q}}^{\mathrm{T}} \, \alpha \, \ddot{\Phi} \tag{5.34}$$

where the easily verifiable relation $\ddot{\Phi}_{\dot{\mathbf{q}}}^{\mathrm{T}} = \dot{\Phi}_{\mathbf{q}}^{\mathrm{T}}$ is used.

The work done by the fictitious Rayleigh forces is

$$\delta W_R = -2\left(\delta\Phi\right)^{\mathrm{T}} \alpha \, \Omega \, \mu \, \dot{\Phi} = -2 \, \delta\mathbf{q}^{\mathrm{T}} \Phi_{\mathbf{q}}^{\mathrm{T}} \, \alpha \, \Omega \, \mu \, \dot{\Phi} \tag{5.35}$$

and, therefore, the final expression obtained by the application of the Lagrange's equations is

$$\mathbf{M}\,\ddot{\mathbf{q}} + \Phi_{\mathbf{q}}^{\mathrm{T}} \, \alpha \left(\ddot{\Phi} + 2\,\Omega\,\mu\,\dot{\Phi} + \Omega^2 \Phi\right) = \mathbf{Q} \tag{5.36}$$

where $\mathbf{M}$ and $\mathbf{Q} = \mathbf{Q}_{ex} + L_{\mathbf{q}} - \mathbf{M}\dot{\mathbf{q}}$ are the mass matrix and the force vector corresponding to the system without constraints.

Remark: The second term in the LHS of equation (5.36) represents the forces that are generated by the penalty system when the constraints Φ, $\dot{\Phi}$, and $\ddot{\Phi}$ are violated. The virtual power method leads to this result directly without the need of differentiation, as is the case with the Lagrange's equations. By merely comparing equations (5.36) with (5.4), we may see that $(\alpha\Phi + 2\Omega\mu\dot{\Phi} + \Omega^2\ddot{\Phi})$ is an approximation to the true Lagrange multipliers λ. The pre-multiplication by $\Phi_{\mathbf{q}}^{\mathrm{T}}$ projects these forces unto the space of the dependent coordinates.

Substituting for $\ddot{\Phi}$ with values from equation (5.9), the following final result is obtained:

$$(\mathbf{M} + \Phi_{\mathbf{q}}^{\mathrm{T}} \, \alpha \, \Phi_{\mathbf{q}}) \, \ddot{\mathbf{q}} = \mathbf{Q} - \Phi_{\mathbf{q}}^{\mathrm{T}} \, \alpha \, (\dot{\Phi}_{\mathbf{q}} \, \dot{\mathbf{q}} + \dot{\Phi}_t + 2\,\Omega\,\mu\,\dot{\Phi} + \Omega^2 \Phi) \tag{5.37}$$

It has been demonstrated (Oden (1983)) that the solution of the modified problem coincides with that of the original problem provided that $\alpha_k \to \infty$. Numerically, this condition is achieved by merely using large penalty factors. These in turn may produce numerical ill conditioning which may be avoided by the improved technique described below. For double precision arithmetic, a factor of 10^7 times the largest term of the mass matrix gives excellent results. The coefficients ω and μ may have a stability effect similar to that produced by the α and β coefficients of the Baumgarte constraint stabilization method explained above. However, the penalty formulation of equation (5.37) does not fail near kinematic singular configurations or with redundant constraints, as the Baumgarte stabilization does.

Non-holonomic Systems. The penalty formulation also allows one to introduce, with no difficulty, non-holonomic constraints. The general form of a non-holonomic constraint is

$$\Phi_k(q_j, \dot{q}_j, t) = 0 \tag{5.38}$$

Non-holonomic constraints conditions for multibody systems typically take the form:

$$\Phi = \mathbf{A}(\mathbf{q}, t)\,\dot{\mathbf{q}} + \mathbf{B}(\mathbf{q}, t) \tag{5.39}$$

The classical Lagrange multiplier approach leads to the following equations of motion (Goldstein (1980)):

$$\mathbf{M}\,\ddot{\mathbf{q}} + \mathbf{A}^T \lambda = \mathbf{Q} \tag{5.40}$$

Considering the penalty formulation, we introduce as for the holonomic systems a set of fictitious Rayleigh's dissipative forces that are proportional to the velocities:

$$G_k^* = -\alpha_k\,\mu_k\,\phi_k \equiv -\alpha\,\mu\,\Phi \tag{5.41}$$

and of inertial forces:

$$I_k^* = -\alpha_k\,\dot{\phi}_k \equiv -\alpha\,\dot{\Phi} \tag{5.42}$$

The projection of the forces acting on the constraints over the space of dependent coordinates is given by

$$\mathbf{A}^T \alpha\,(\dot{\Phi} + \mu\,\Phi) \tag{5.43}$$

and the application of the virtual power method directly leads to

$$\mathbf{q}^{*T}\!\left(\mathbf{M}\,\ddot{\mathbf{q}} - \mathbf{Q} + \mathbf{A}^T \alpha\,(\dot{\Phi} + \mu\,\Phi)\right) = 0 \tag{5.44}$$

Since the penalty formulation makes the problem become unconstrained, the virtual velocities may be arbitrarily selected. The expression between brackets vanishes, consequently,

$$\mathbf{M}\,\ddot{\mathbf{q}} + \mathbf{A}^T \alpha\,(\dot{\Phi} + \mu\,\Phi) = \mathbf{Q} \tag{5.45}$$

Knowing that

$$\dot{\Phi}(\mathbf{q}, \dot{\mathbf{q}}, t) = \Phi_{\mathbf{q}}\,\dot{\mathbf{q}} + \Phi_{\dot{\mathbf{q}}}\,\ddot{\mathbf{q}} + \Phi_t \tag{5.46}$$

equation (5.45) becomes

$$(\mathbf{M} + \mathbf{A}^T \alpha\,\mathbf{A})\,\ddot{\mathbf{q}} = \mathbf{Q} - \mathbf{A}^T \alpha\,(\dot{\mathbf{A}}\,\dot{\mathbf{q}} + \dot{\mathbf{B}} + \mu\,\Phi) \tag{5.47}$$

As in the holonomic case, the diagonal matrix α can be substituted by a constant that multiplies the identity matrix, if the same penalty number is used for all of the constraints.

Augmented Lagrangian Formulation. Equations (5.37) and (5.47) form the modified Lagrange's equations that are obtained by virtue of a penalty formulation. Penalty methods bring forth the problem of choosing the right penalty number. While large penalty values will ensure convergence to the constraint within a tight tolerance, those values may also lead to numerical conditioning problems and develop round-off errors. It is therefore important that the analyst be supplied with a method that converges, regardless of the size of the penalty values, to the right solution within specified tolerances in the constraints. This method will have all the possible advantages of a reduced number of equations and have no need for very large penalty values to assure convergence.

To this end, we can extend the augmented Lagrangian method commonly used in optimization analysis (Vanderplaats (1984)) to improve the numerical conditioning of the proposed penalty equations. Let us consider again the classical Lagrange multipliers method as stated by equation (5.8). This method, along with the constraints (5.1), forms a system of DAEs whose solution will yield the values of the n generalized coordinates q_j as well as the m Lagrange multipliers λ_k. Instead of following this approach, we can modify equation (5.8) by adding the corresponding penalty terms, whose values will be zero if the constraints are satisfied. Therefore

$$\mathbf{M}\,\ddot{\mathbf{q}} + \Phi_q^T\,\alpha\left(\ddot{\Phi} + 2\,\Omega\,\mu\,\dot{\Phi} + \Omega^2\,\Phi\right) + \Phi_q^T\,\lambda^* = \mathbf{Q} \qquad (5.48)$$

This new equation can be viewed as a penalty method to which the Lagrange multipliers are added. In the limit, the constraint conditions are satisfied; thus $\lambda = \lambda^*$ and equations (5.8) and (5.48) become totally equivalent except for round-off errors induced by the penalty parameters and finite machine precision. In (5.48) the Lagrange multipliers λ^* play the role of correcting terms.

By merely comparing equations (5.8) and (5.48), it can be inferred that

$$\lambda \cong \lambda^* + \alpha\left(\ddot{\Phi} + 2\,\Omega\,\mu\,\dot{\Phi} + \Omega^2\,\Phi\right) \qquad (5.49)$$

We are seeking the solution of (5.48) without having to use the algebraic constraint equations (5.1). This requires that the correct values of λ^* be known, so that they can be inserted in (5.49). Since those values are not known in advance, there is a need to set up an iterative process that calculates the unknown multipliers λ^*. The iteration is easily established by taking advantage of equation (5.49):

$$\lambda_{i+1} = \lambda_i + \alpha\left(\ddot{\Phi} + 2\,\Omega\,\mu\,\dot{\Phi} + \Omega^2\,\Phi\right)_{i+1}$$
$$i = 0, 1, 2, \cdots \qquad (5.50)$$

with $\lambda_0^* = 0$ for the first iteration. Equation (5.50) physically represents the introduction at iteration $i+1$ of forces that tend to compensate the fact that the constraints are not exactly zero. It becomes obvious now that the penalty number does not need to be very large, since the resulting error in the constraint equations will be eliminated by the Lagrange's terms during the iteration procedure. The generic penalty method corresponds to the augmented Lagrangian method in which the iteration process is carried out only once.

The matrix formulation of (5.48), including the iterative process defined in (5.50), is given by the following expression:

$$\left(M + \Phi_q^T \alpha \, \Phi_q\right) \ddot{q}_{i+1} =$$
$$= M \, \ddot{q}_i - \Phi_q^T \alpha \left(\dot{\Phi}_q \dot{q} + \dot{\Phi}_t + 2 \Omega \mu \dot{\Phi} + \Omega^2 \Phi\right) \tag{5.51}$$
$$i = 0, 1, 2, \ldots$$

with $M \, \ddot{q}_0 = Q$ for the initial iteration. The subscript i represents the iteration number.

This improved formulation for the non-holonomic case leads to an iterative procedure as given by the following equation:

$$\left(M + A^T \alpha \, A\right) \ddot{q}_{i+1} = M \, \ddot{q}_i - A^T \alpha \left(\dot{A} \, \dot{q} + \dot{B} + \mu \, \Phi\right) \tag{5.52}$$
$$i = 0, 1, 2, \ldots$$

with again $M \, \ddot{q}_0 = Q$ for the initial iteration.

At first, this procedure might seem to be at a disadvantage since an iteration process and extra computation are required. However, the extra numerical effort is practically insignificant, since an iterative procedure is usually necessary to solve a system of nonlinear differential equations. A major advantage obtained in return for this additional computation is that the analyst does not have to be concerned with the value of the penalty number that simultaneously assures convergence and avoids round-off errors. The numerical integration of the equations of motion using the penalty formulation may proceed as follows:

Algorithm 5-3

1. Start at a time t, when the position q and velocity $\dot{q}$ are known.
2. Use equation (5.37) in holonomic systems, or (5.47) in non-holonomic systems, to solve for the accelerations $\ddot{q}$ at time t. If the augmented Lagrangian is desired, then use equations (5.52) and (5.53) for holonomic and non-holonomic systems, respectively.
3. Obtain the vectors q and $\dot{q}$ at time $(t - \Delta t)$ by numerical integration:

$$\dot{y}_t^T \equiv \{\ddot{q}^T, \dot{q}^T\}_t \xrightarrow{\text{n.i.s.}} y_{t+\Delta t}^T \equiv \{\dot{q}^T, q^T\}_{t+\Delta t} \tag{5.53}$$

4. Upon convergence of the n.i.s., update the time variable and go to step 2.

This numerical integration algorithm has the advantage of solving a set of n equations as compared to $(n+m)$ needed by the Lagrange multiplier method. Constraint stabilization is implicitly considered within the algorithm and is implemented more simply than the methods that use independent coordinates which are shown in the sequel. An efficient numerical implementation of this penalty method, which is more suitable for real time applications, has been proposed by Bayo et al. (1991) and will be explained in detail in Chapter 8.

Table 5.1. Maximum constraint errors for different penalty values.

Penalty	Number of iterations					
	0		1		2	
	time	error	time	error	time	error
10^1	10.5	$0.7 \ 10^{-1}$	12.8	$0.7 \ 10^{-2}$	14.1	$0.9 \ 10^{-3}$
10^3	11.7	$0.7 \ 10^{-3}$	13.3	$0.6 \ 10^{-5}$	14.6	$0.6 \ 10^{-5}$
10^5	11.7	$0.1 \ 10^{-4}$	13.3	$0.6 \ 10^{-5}$	14.6	$0.6 \ 10^{-5}$
10^7	11.7	$0.6 \ 10^{-5}$	13.3	$0.6 \ 10^{-5}$	14.6	$0.6 \ 10^{-5}$
10^9	11.7	$0.6 \ 10^{-5}$	13.3	$0.6 \ 10^{-5}$	14.6	$0.6 \ 10^{-5}$
10^{11}	17.3	$0.1 \ 10^{-4}$	19.1	$0.9 \ 10^{-5}$	21.9	$0.2 \ 10^{--4}$

Example 5.1

Given the results of a numerical simulation of the motion of a double pendulum moving in a vertical plane under gravitational forces. The pendulum has two elements of unit mass and length. Four natural coordinates with two constraints conditions are used to model the system. We use the penalty-augmented Lagrangian method with coefficients Ω and μ equal to ten and one, respectively. These provide critical damping in the stabilization process. Table 5.1 contains a comparative study of the resulting maximum constraint errors and CPU times in seconds obtained, using different penalty numbers and 0, 1 and 2 iterations. In all the cases, the integration is performed using the subroutine DGEAR (Gear (1971)) with an error tolerance equal to 10^{-4}. It may be seen how the use of only one iteration considerably widens the range of acceptable penalty values at practically no additional computational cost.

5.2 Formulations in Independent Coordinates

Some of the methods used to formulate and integrate the motion differential equations in independent coordinates will be presented below. One advantage of this type of coordinates is an important reduction in the number of equations to be integrated. Most important is the disappearance of the instability problem in the integration of the constraint equations using ODE solvers. However, this has a price in terms of computational effort since the position and velocity problems need to be solved after the function evaluations. Some of the numerical integration algorithms studied in Chapter 7, and in particular the more stable implicit algorithms are difficult to implement. In addition, the formulation and implementation of these methods become more involved than those which use dependent

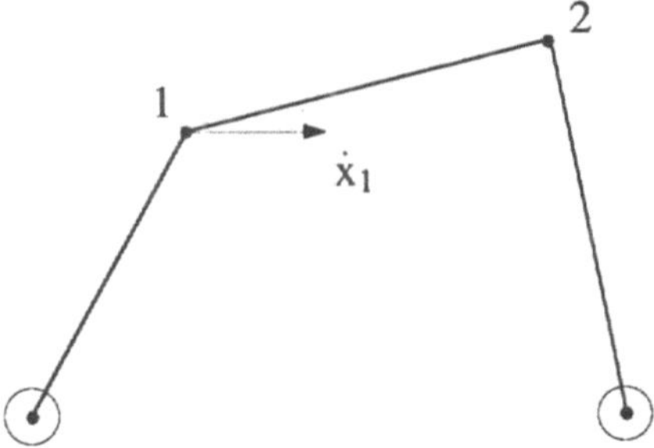

Figure 5.1. Independent velocity in a four-bar mechanism.

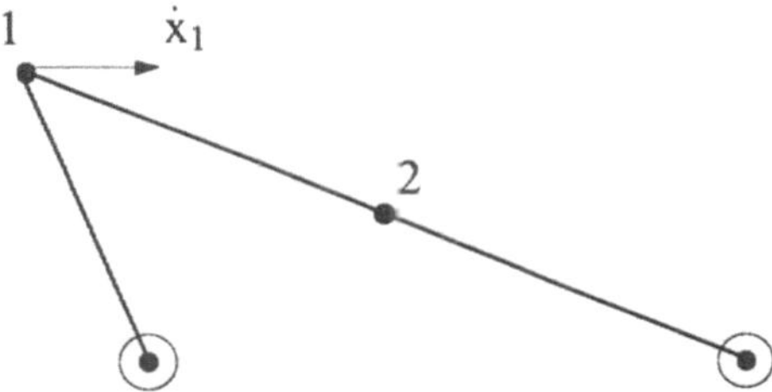

Figure 5.2. Need to change the independent velocity in a four-bar mechanism.

coordinates. One important point is the choice of the right set of independent coordinates. This point will be studied in greater detail next.

5.2.1 Determination of Independent Coordinates

This is a point of transcendental importance. The independent velocities normally
are given by the projection of the dependent velocities $\dot{\mathbf{q}}$ on certain vectors defined
by the rows of a constant matrix $\mathbf{B}$ (See Section 3.5) as:

$$\dot{\mathbf{z}} = \mathbf{B}\,\dot{\mathbf{q}} \tag{5.54}$$

The need to suitably select the independent coordinates can be illustrated from
the mechanical point of view with some very simple mechanisms. For example,
in the four-bar mechanism of Figure 5.1, the velocity $\dot{x}_1$ perfectly defines all the
mechanism's velocities. However, in the quadrilateral of Figure 5.2, this coordinate is not adequate, since in no way does it determine the velocity of point 2. In
fact, at the position of Figure 5.2, $\dot{x}_1$ will always be zero.

Figures 5.3 and 5.4 show a mechanism with five bars at two positions. At
one of these the selected coordinates are adequate, but at the other they are not.
When bars 2 and 3 are parallel, bars 3 and 4 have the possibility of being jointly
moved as a rigid body. This motion is not determined by angles ψ_1 and ψ_2.

The examples mentioned should be sufficient for understanding: first, no system of independent coordinates is adequate for the entire motion of the system;

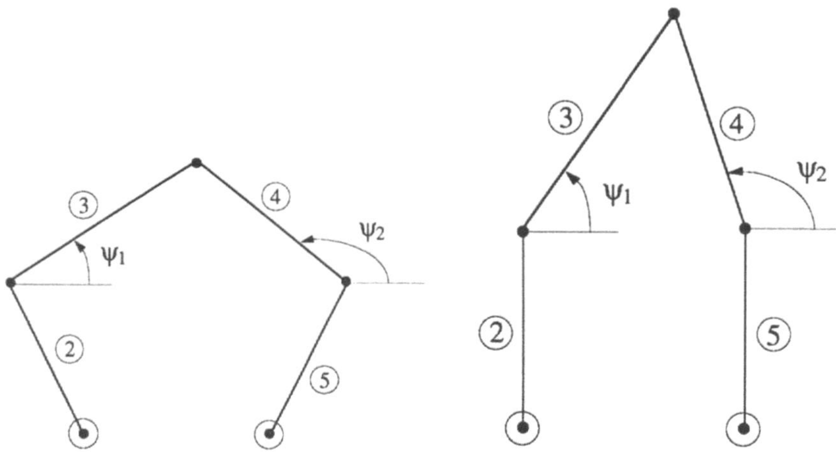

Figure 5.3. Adequate set of independent coordinates in a five-bar mechanism.

Figure 5.4. Inadequate independent coordinates in a five-bar mechanism.

and secondly, that a particular set of independent coordinates may be the most adequate at a certain position of the mechanism but not at another. Therefore, one must establish a double actuation procedure. On one hand a method must be developed that permits checking when a set of independent coordinates is becoming inadequate. On the other hand, it is necessary to establish a method for finding the most adequate new set of independent coordinates. Fortunately, there are mathematical properties of the Jacobian matrix Φ_q that permit the solution of the two problems satisfactorily. These properties will be seen later in connection with the specific methods of formulating and solving the dynamic problem with independent coordinates.

One last important point is that normally the numerical integration subroutines of ordinary differential equations are based on multistep methods (Shampine and Gordon (1975); Gear (1971)). These methods are very efficient, but they have certain limitations. They require special techniques for starting the integration process, and they are long and drawn out. Since it is necessary to change each time the independent coordinates, the numerical integration must be restarted again. One must carry out the minimum possible number of coordinate changes. On the other hand, when some determined coordinates start to be inadequate, the integration process becomes much slower. It is necessary to arrive at a compromise solution, therefore, by making the minimum number of coordinate changes that guarantee quick and accurate numerical integration. Sometimes, the speed of the numerical integration can be utilized as the criteria for the change of independent coordinates, when subroutines with automatic step size control are used.

The most important numerical integration methods for the differential equations of motion in independent coordinates are described next.

5.2.2 Extraction Methods (Coordinate Partitioning)

This method, proposed by Wehage and Haug (1982), consists of finding the dependent accelerations $\ddot{q}$ by means of equations (5.10) or (5.17). Only some of the vector's elements are integrated, specifically, those corresponding to the independent coordinates. Wehage and Haug have called this technique the *coordinate partitioning method*.

In order to choose an independent set of coordinates from vector q, it should be remembered how the numerical integration subroutine behaves in this case. Consider the vector q and a partition of dependent and independent coordinates as follows:

$$q^T = \{q_d^T, q_i^T\} \tag{5.55}$$

where there are m dependent coordinates and $(f=n-m)$ independent coordinates.

The numerical integration subroutine is applied to only the independent coordinates as follows:

$$\dot{y}_t^T \equiv \{\ddot{q}_i^T, \dot{q}_i^T\}_t \quad \xrightarrow{\text{n.i.s.}} \quad y_{t+\Delta t}^T \equiv \{\dot{q}_i^T, q_i^T\}_{t+\Delta t} \tag{5.56}$$

Since only the independent coordinates and velocities at time $(t+\Delta t)$ are known, it is necessary to calculate the remaining coordinates and velocities. This is done by solving the position problem to calculate q_d in terms of q_i after each function evaluation, and doing the same for the velocities. The latter requires the solution of the following set of linear equations which are the partitioned constraint equations for velocities:

$$\begin{bmatrix} \Phi_q^d & \Phi_q^i \end{bmatrix} \begin{Bmatrix} \dot{q}_d \\ \dot{q}_i \end{Bmatrix} = b \tag{5.57}$$

where the partition carried out on the vector $\ddot{q}$ leads to a similar partition on the Jacobian matrix Φ_q. In order to calculate the dependent variables, the inverse of matrix Φ_q^d must exist. During the Gauss triangularization process of this matrix with column pivoting to maintain the previously determined partition, all the pivots must be sufficiently different from zero. The fact that one or more of the pivots of Φ_q^d tend to zero, means that the current set of independent coordinates are becoming inadequate. More specifically, the dependent velocity corresponding to the column in which the pivot tends towards zero appears, must now be taken as a new independent coordinate.

When a system of independent coordinates becomes inappropriate in actual practice, rather than substituting one coordinate for another, it is recommended that one choose a new complete system of independent coordinates. This is done by carrying out the factorization of Φ_q with total pivoting. The $(n-m)$ columns in which the m pivots have not appeared will determine the coordinates that should be chosen as independent ones.

Although Gauss total pivoting is neither the only nor the most reliable technique that may be used for determining a decomposition of vector q into dependent and independent coordinates, it is undoubtedly the most economical. Mani et al. (1985) proposed the Singular Value decomposition (SV) and Kim and Vanderploeg (1986b) the QR decomposition. Both are, without a doubt, more reliable than the Gauss total pivoting method, but require considerably more calculation effort. These techniques can be used perhaps to choose a new set of independent coordinates at specific positions, because this process will only need to be carried out very few times in all the simulation. However, the QR or SV decompositions are completely unsuitable to detect the need for change in the set of independent coordinates. Since the detection must be carried out at each step of the numerical integration, these decompositions would unacceptably delay the integration process.

This numerical integration process requires solving the position problem and performing the velocity analysis at each iteration. The latter does not constitute an important difficulty. However, the position problem does, because it requires an iterative solution that consumes an important amount of computational time. For this reason, Paul (1975) suggested the integration of the following extended set of differential equations:

$$\dot{\mathbf{y}}_t^T \equiv \{\ddot{\mathbf{q}}_i^T, \dot{\mathbf{q}}^T\}_t \quad \xrightarrow{\text{n.i.s.}} \quad \mathbf{y}_{t+\Delta t}^T \equiv \{\dot{\mathbf{q}}_i^T, \mathbf{q}^T\}_{t+\Delta t} \tag{5.58}$$

By integrating all the velocities, and not only the independent ones, the new position of the multibody system is directly obtained as a result of the numerical integration. With numerical integration, the constraint equation stabilization problem is not so critical. The equation that is integrated is that of the velocities instead of accelerations. The general solution of the first derivative of the constraint equations is simply a constant vector. Therefore, round-off errors do not tend to increase with time, although they accumulate and slow down the integration.

Numerical integration of the extended differential equations system (5.58) is frequently used. Generally speaking it is more efficient than that of system (5.56) which, as mentioned above, entails a repeated solution of the position and velocity problem at each step. For long simulations, if the use of (5.58) leads to an accumulation of constraint errors which may even lead to numerical stiffness in the solution of equations, then the position problem is solved and the integration proceeds. Park and Haug (1986) proposed a hybrid method that combines the coordinate partitioning and Baumgarte stabilization of the constraints. Thus, when the errors in the constraint exceed a specified tolerance, the accelerations are calculated with equation (5.26) instead of (5.10). When the Baumgarte stabilization fails in the neighborhood of a kinematically singular configuration, the proposed method reverts to pure coordinate partitioning with the solution of the position and velocity problems.

To summarize, the extraction method calculates all the dependent accelerations with the same formulae used in the methods based on dependent coordinates. It then integrates only one subset of the accelerations chosen by means of Gauss triangularization with total pivoting or by the QR or SV decomposition. To determine when an independent coordinates system starts becoming inadequate, the best and most economical method is to check the pivots during the triangularization with column pivoting. This should be done at each step of the numerical integration.

5.2.3 Methods Based on the Projection Matrix $\mathbf{R}$

One should recall the velocity equation, obtained after differentiating the constraint equations with respect to time:

$$\Phi_q \, \dot{q} = - \Phi_t \equiv \mathbf{b} \tag{5.59}$$

Likewise one should consider again equation (5.54), in which the independent velocities $\dot{z}$ are defined as the projection of vector $\dot{q}$ on the rows of a constant matrix $\mathbf{B}$ of size $((n-m)\times n)$. The rows of matrix $\mathbf{B}$ satisfy the condition of being linearly independent from one another and with respect to the rows of matrix Φ_q. By jointly writing the expressions (5.59) and (5.54) one obtains

$$\begin{bmatrix} \Phi_q \\ \mathbf{B} \end{bmatrix} \dot{q} = \begin{Bmatrix} \mathbf{b} \\ \dot{z} \end{Bmatrix} \tag{5.60}$$

The matrix on the LHS of this expression is invertible and consequently,

$$\dot{q} = \begin{bmatrix} \Phi_q \\ \mathbf{B} \end{bmatrix}^{-1} \begin{Bmatrix} \mathbf{b} \\ \dot{z} \end{Bmatrix} \equiv [\mathbf{S} \quad \mathbf{R}] \begin{Bmatrix} \mathbf{b} \\ \dot{z} \end{Bmatrix} = \mathbf{S}\mathbf{b} + \mathbf{R}\,\dot{z} \tag{5.61}$$

where $\mathbf{S}$ and $\mathbf{R}$ are $(n\times m)$ and $(n\times(n-m))$ matrices, respectively. It is easy to demonstrate that term $\mathbf{R}\dot{z}$ represents the general solution of the homogeneous velocity equation, and that the term $\mathbf{S}\mathbf{b}$ represents a particular solution of the complete equation. This is a particular solution of the velocity equation for the case in which there are rheonomous constraints.

By differentiating equations (5.59) and (5.54) with respect to time, one obtains

$$\Phi_q \, \ddot{q} = -\dot{\Phi}_t - \dot{\Phi}_q \, \dot{q} \equiv \mathbf{c} \tag{5.62}$$

$$\mathbf{B} \, \ddot{q} = \ddot{z} \tag{5.63}$$

By jointly expressing these equations:

$$\begin{bmatrix} \Phi_q \\ \mathbf{B} \end{bmatrix} \ddot{q} = \begin{Bmatrix} \mathbf{c} \\ \ddot{z} \end{Bmatrix} \tag{5.64}$$

Solving for $\ddot{q}$ and introducing the matrices $\mathbf{S}$ and $\mathbf{R}$ defined in (5.61), we obtain

$$\ddot{q} = \begin{bmatrix} \Phi_q \\ \mathbf{B} \end{bmatrix}^{-1} \begin{Bmatrix} \mathbf{c} \\ \ddot{z} \end{Bmatrix} = \mathbf{S}\,\mathbf{c} + \mathbf{R}\,\ddot{z} \tag{5.65}$$

It shall be remembered that matrix $\mathbf{R}$ must be explicitly calculated with $n-m$ forward and backward substitutions, starting from the leading matrix of (5.60) factored by means of Gauss elimination. However, it is important to note that the matrix S never needs to be explicitly calculated. The terms (Sb) and (Sc) are, respectively, $\dot{q}$ and $\ddot{q}$, when $\dot{z} = 0$ and $\ddot{z} = 0$. These terms products can be directly calculated from expressions (5.60) and (5.64).

Eq. (5.16) represents the equations of motion with dependent coordinates:

$$R^T M \ddot{q} = R^T Q \qquad (5.66)$$

By introducing in this equation the equation (5.65) for the dependent accelerations in terms of the independent accelerations, we obtain

$$R^T M R \ddot{z} = R^T Q - R^T M S c \qquad (5.67)$$

which constitutes the equations of motion in terms of independent coordinates. The derivation process that starts with equation (5.59) and ends in (5.67) leads to this general form of the equations of motion in independent coordinates, which was first introduced in this context by Serna et al. (1982). Equation (5.67) represents a general matrix transformation from the vector spaces of dependent accelerations and forces to the vector space of independent accelerations and forces.

This formulation is valid for both scleronomous and rheonomous constraint equations. In addition, this layout is valid, irrespective of the matrix B chosen in equation (5.54), provided that the conditions of being constant and of having its rows linearly independent from one another and with respect to the rows of Φ_q are satisfied. The matrix B can be chosen in two different ways as explained next.

Boolean matrix. The matrix B is formed by a set of ones and zeros that extracts $(n-m)$ components of q as independent coordinates z. By partitioning equation (5.60) in a similar way to that performed in equation (5.57), one obtains

$$B = \begin{bmatrix} 0 & I \end{bmatrix} \qquad (5.68)$$

and

$$\begin{bmatrix} \Phi_q^d & \Phi_q^i \\ 0 & I \end{bmatrix} \begin{Bmatrix} \dot{q}_d \\ \dot{q}_i \end{Bmatrix} = \begin{Bmatrix} b \\ \dot{z} \end{Bmatrix} \qquad (5.69)$$

One should keep in mind that the matrix Φ_q^d should be invertible in order to express the dependent velocities in terms of the independent ones. All the pivots must be sufficiently different from zero. In this way it is assured that the chosen rows of B are independent from those of Φ_q. If during the motion of a multibody system one of the pivots of Φ_q^d becomes much smaller than the others, this independence is gradually lost. It is then necessary to choose new independent coordinates by means of a total pivoting process.

This is the simplest method of all those formulated in independent coordinates. In actual practice this method is almost always the most efficient one.

SV and QR Decompositions. As explained in Chapter 3, these methods take rows of matrix B as an orthogonal basis of the nullspace of Φ_q, calculated at a previous position of the multibody system by means of the singular value (SV) or QR decompositions. Both lead to similar results. The SVD is more stable in poor numerically conditioned problems. However, the QR decomposition is numerically more efficient, since it is a direct process that does not need iterations.

The matrix $\mathbf{R}$ is also calculated starting from equation (5.60). As before, the matrix $\mathbf{B}$ is given by an orthonormal basis of the nullspace at a previous position. Since the row subspace of matrix $\boldsymbol{\Phi}_q$ is orthogonal to this nullspace, it is expected that the rows of $\boldsymbol{\Phi}_q$ are independent from the rows of $\mathbf{B}$ for a wide range of the motion. Since matrix $\mathbf{B}$ is kept constant, one may arrive at a position of the multibody system in which this independence is lost or deteriorated. Therefore, it will be necessary to change the independent coordinates by carrying out a new SV or QR decomposition. The most efficient method to find whether the rows of $\mathbf{B}$ are becoming linearly dependent is through the monitoring of the pivots during the Gauss triangularization process of equation (5.60).

Even though this method may occasionally require fewer changes of independent coordinates than the Boolean matrix method, the latter is simpler and numerically more efficient. It is not necessary to carry out additional operations to complete on $\mathbf{B}$ the Gauss triangularization of matrix $\boldsymbol{\Phi}_q$.

Numerical integration algorithm with projection matrices R. Of all the methods based on the projection matrix $\mathbf{R}$ shown in Chapter 3, the one based on the Boolean matrix $\mathbf{B}$ with rheonomous constraints will be chosen in order to present the corresponding numerical integration algorithm. The possibility exists of solving the position problem at each step, or integrating an enlarged system of differential equations. This second option is presented first.

Algorithm 5-4

1. Start at a time when the position $\mathbf{q}$ and velocities $\dot{\mathbf{z}}$ are known.
2. Calculate a new matrix $\boldsymbol{\Phi}_q$. Triangularize this matrix with column pivoting, and verify that all the pivots are sufficiently different from zero, so that the independent coordinates continue to be valid. Otherwise, carry out a new triangularization with total pivoting and choose a new set of independent coordinates. In addition, restart the numerical integration process if using a multistep method.
3. Form matrix $\mathbf{R}$ from equation (5.60). Note that the triangularization of $\boldsymbol{\Phi}_q$ has been carried out already in step 2.
4. Calculate the new dependent velocities $\dot{\mathbf{q}}$ by means of equation (5.61).
5. Form the matrix products $(\mathbf{R}^T\mathbf{M})$, $(\mathbf{R}^T\mathbf{MR})$, and $(\mathbf{R}^T\mathbf{Q})$.
6. Calculate the terms $(\mathbf{Sb})$ and $(\mathbf{Sc})$ by making $\dot{\mathbf{z}} = 0$ and $\ddot{\mathbf{z}} = 0$ in equations (5.61) and (5.65), respectively.
7. Obtain the independent acceleration vector $\ddot{\mathbf{z}}$ from equation (5.67).
8. Obtain the vectors $\mathbf{q}$ and $\dot{\mathbf{z}}$ at time $(t-\Delta t)$ by numerical integration:

$$\dot{\mathbf{y}}_t^T \equiv \{\ddot{\mathbf{z}}^T, \dot{\mathbf{q}}^T\}_t \quad \xrightarrow{\text{n.i.s.}} \quad \mathbf{y}_{t+\Delta t}^T \equiv \{\dot{\mathbf{z}}^T, \mathbf{q}^T\}_{t+\Delta t} \tag{5.70}$$

9. Upon convergence of the n.i.s. update the time variable and go to step 2.

This algorithm constitutes an efficient and general purpose method of solving the forward dynamics. However, it is more difficult to implement than those based on dependent coordinates. By means of small modifications, this algorithm can be easily adapted to the other theoretically described projection methods. Since the integration is carried out using $\{\ddot{z}^T, \dot{q}^T\}$, errors in the constraint conditions may accumulate with the effect of slowing down the integration process. For long simulations and in order to eliminate this problem, it is necessary to solve the position problem either after a specified number of time steps or after checking at step eight the fulfillment of the position constraint equations. In what follows we also give an algorithm based on $\{\ddot{z}^T, \dot{z}^T\}$ which requires the solution of the position and velocity problems in each iteration:

Algorithm 5-5

1. Start at a time in which the independent coordinates z and $\dot{z}$ are known.
2. Solve the position and velocity problems to obtain q and $\dot{q}$. Simultaneously, do the column pivoting on the matrix Φ_q to check the validity of the current set of independent coordinates.
3. Form R from equation (5.60). Note that Φ_q has already been triangularized.
4. Form the matrix products $(R^T M)$, $(R^T M R)$, and $(R^T M)$.
5. Calculate the terms (Sb) and (Sc) by making $\dot{z} = 0$ and $\ddot{z} = 0$ in equations (5.61) and (5.65), respectively.
6. Obtain the independent acceleration $\ddot{z}$ from equation (5.67).
7. Obtain the vectors z and $\dot{z}$ at time $(t+\Delta t)$ by the numerical integration:

$$\dot{y}_t^T \equiv \{\ddot{z}^T, \dot{z}^T\}_t \quad \xrightarrow{\text{n.i.s.}} \quad y_{t+\Delta t}^T \equiv \{\dot{z}^T, z^T\}_{t+\Delta t}$$

9. Upon convergence of the n.i.s., update the time variable and go to step 2.

5.2.4 Comparative Remarks

The penalty formulation characterized by equation (5.37) has the advantage over the formulations in independent coordinates, in that the appearance or disappearance of constraints can be accommodated automatically without changing the co-ordinates. This in turn avoids the restarting procedure of the numerical integrator. The penalty formulation is also more suitable when the multibody system goes through a singular or bifurcation position. In these cases the Jacobian matrix changes its rank, and the use of independent coordinates requires a sudden change of coordinates. Unless special provisions are made, the formulation in independent coordinates and even the Lagrange's equations in dependent coordinates tends to either crash the simulation or introduce sudden large errors. However, with the penalty formulation, the term $(M + \Phi_q^T \alpha \, \Phi_q)$ of equation (5.37) is free of singularities and the integration becomes be very stable under these circumstances. This fact also makes the penalty formulation go through kinematic singular positions without problems, an advantage not shared by the classical Lagrange's method with Baumgarte stabilization (See Bayo and Avello (1993)).

The penalty formulation of Algorithm 5-3 will tend to be more efficient numerically than the Algorithms 5-4 and 5-5 in independent coordinates, because in Algorithm 5-3 the major computational burden is the formation, triangularization, and one forward reduction and backsubstitution of $(\mathbf{M} + \Phi_q^T \alpha \Phi_q)$. Since the mass matrix does not modify the sparsity of the product $(\Phi_q^T \Phi_q)$, this operation is less costly than the formation, triangularization, and f forward reductions and backsubstitutions of $(\Phi_q^T \Phi_q)$ required for the formation of the matrix $\mathbf{R}$ in a single step of Algorithms 5-4 and 5-5. These algorithms also include the formation and triangularization of $(\mathbf{R}^T\mathbf{M}\mathbf{R})$ which represents an additional computational burden of these methods.

5.3 Formulations Based on Velocity Transformations

In Section 5.2.3, a family of methods for transforming the dynamic equations from dependent to independent coordinates was presented. Equation (5.61), which defines the relation between dependent $\dot{q}$ and independent velocities $\dot{z}$, is really a particular case of the velocity transformation equations that can be introduced in the dynamic formulation. Equation (5.65) is the corresponding relation for accelerations. In this section, it will be shown how some velocity transformations can be used to improve the efficiency of the dynamic formulations described previously. These formulations, initially introduced by Jerkovsky (1978) and subsequently extended by other authors, such as: Kim and Vanderploeg (1986b), Nikravesh and Gim (1989), García de Jalón et al. (1990), and Bae and Won (1990) can be extremely efficient and simple to implement. However, they may have been presented in the literature in a rather involved way. We present these ideas in this section in a simple and yet rigorous manner, so that one can understand them easily. The concepts presented hereafter are independent of the coordinates used be they natural, reference points, or others. The efficiency of these formulations makes them be one of the best candidates for real time simulation. There will be a return to this topic in Chapter 8.

The numerical complexity in equation (5.67) comes from a double fact:

1) The computation of the matrix $\mathbf{R}$, that requires the factorization of the Jacobian matrix and as many forward and backward substitutions as columns that this matrix has; and,
2) The products of matrices that appear in equation (5.67), of which the most important and expensive to compute is the one on the left-hand side.

The relative importance of these computational tasks is problem dependent. Experience shows that very often each one of these two operations consisting of the computation of $\mathbf{R}$ and products of matrices requires around *40%* of the total computational cost involved in the numerical integration.

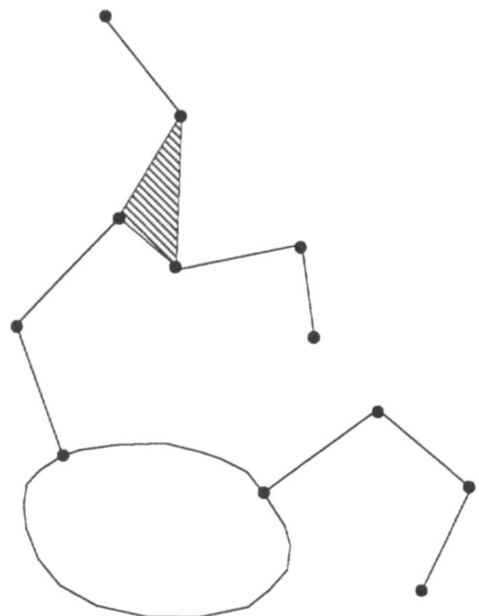

Figure 5.5. Tree-type planar multibody system.

The purpose of this section is to introduce velocity transformations similar to those of expressions (5.61) and (5.65) but with the difference of being particularly favorable in the sense of completely avoiding the Jacobian triangularization; hence allowing for an easy and efficient computation of the matrix $\mathbf{R}$ and term $(\mathbf{Sc})$. It can be seen that these velocity transformations will not necessarily represent transformations between vectors of dependent and independent velocities but transformations between different or alternative sets of dependent velocities, that are particularly suitable from the point of view of improved numerical efficiency. Open- and closed-chain configurations are considered separately.

5.3.1 Open-Chain Multibody Systems

Multibody systems that have open kinematic chain or tree configuration are most appropriate to introduce the velocity transformations described in the previous section. In the sequel, the matrix $\mathbf{R}$ that relates the natural (or mixed) velocity vector $\dot{\mathbf{q}}$ and a set of independent velocities $\dot{\mathbf{z}}$ can be constructed directly with very few arithmetic operations and avoiding the formation and factorization of the Jacobian matrix $\Phi_{\mathbf{q}}$.

An example of a tree-configured planar mechanism is presented in Figure 5.5. If the system has not gotten any fixed element, one of the elements of the system shall be defined as a *base body*. There is not a single choice for the base body, but in practice there are nearly always some natural or physical reasons

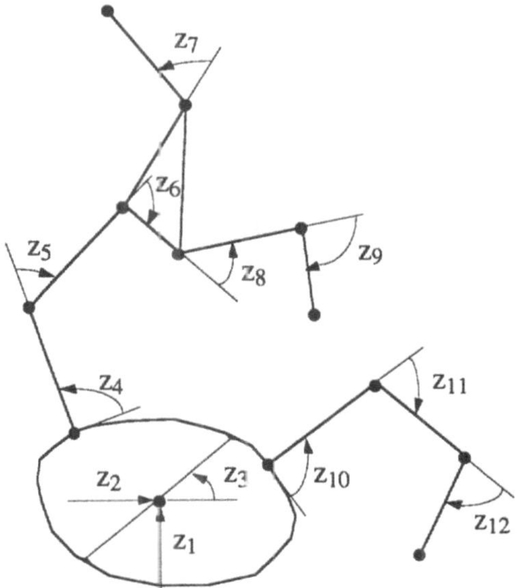

Figure 5.6. Numbering of the independent coordinates of a tree-type planar multi-body system.

that determine the selection. The base body can have several branches departing from it, and some branches can also originate other branches.

An appropriate set of independent variables or coordinates for open-chain systems, such as the one shown in Figure 5.6, is determined by the variables that describe the rigid body motion of the base body plus the relative or joint coordinates that define the motion of each body with respect to the previous one in the corresponding branch of the chain. It is quite clear that this set of relative coordinates is independent. Figure 5.6 displays the base body plus relative coordinates of the planar system of Figure 5.5. In the example shown in Figures 5.5 and 5.6, it has been assumed that all the joints are of revolute type, although prismatic or any other joint type may be considered also.

There is a very easy way of constructing a matrix $\mathbf{R}$ for the system of Figure 5.6 without any need of forming and factoring the Jacobian matrix $\mathbf{\Phi_q}$. Remember that the columns of matrix $\mathbf{R}$ are a basis of the nullspace of the Jacobian matrix or, in other words, a base of the space of allowable velocities. This means that any velocity vector can be expressed as a linear combination of the columns of matrix $\mathbf{R}$. This is exactly what expression (5.61) represents. The column (i) of matrix $\mathbf{R}$ is the velocity vector in dependent coordinates $\dot{\mathbf{q}}$ obtained with:

$$\dot{z}_i = 1$$

$$\dot{z}_j = 0 \qquad j = 1, 2, \ldots n \; (j \neq i) \qquad\qquad (5.71)$$

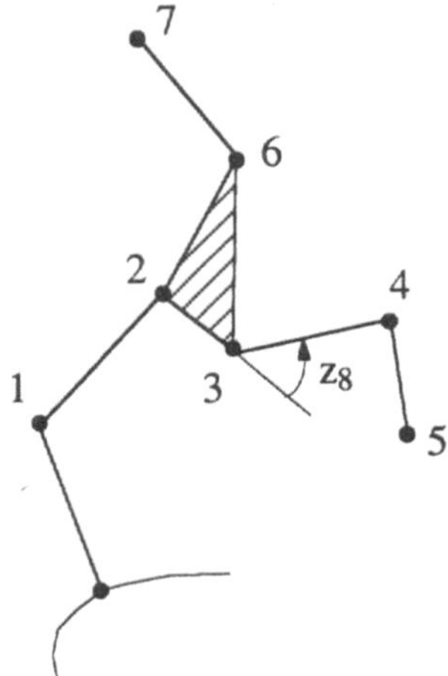

Figure 5.7. Detail numbering of one of the branches of the tree-type planar multibody system.

If the multibody system does not have closed loops, as is the case of the mechanism of Figure 5.6, this velocity vector; thus the corresponding column of matrix **R**, can be computed directly. In Figure 5.7, a branch of the mechanism of Figure 5.6 is displayed in more detail, including a possible numbering of some points.

Consider, for instance, the column of matrix **R** corresponding to the independent coordinate z_8. This column is obtained by giving a unit velocity to angle z_8, and no velocity to all the remaining coordinates z_j $(j{\neq}8)$. Only the column elements corresponding to points 4 and 5 will have a non-zero value. In general terms, only the dependent coordinates that are upwards in the branch of the independent coordinate being moved will introduce non-zero elements in the corresponding column of matrix **R**. It is very easy to take advantage of this well-defined sparsity structure on the computer implementation.

In the 3-D case, the concepts and computations are nearly as simple as in the planar case. The slightly higher complexity comes from: first, definition of the base body coordinates in 3-D; and second, the different kinds of joints that can appear in 3-D multibody systems. Only revolute, prismatic, cylindrical, universal, and spherical joints will be considered here.

5.3.1.1 Definition of Base Body Motion

There is no problem in finding six independent variables that define the velocity of the base body. Perhaps the simplest choice is determined by:

1) The three Cartesian components of the velocity of a reference point P; and,
2) The three Cartesian components of the base body angular velocity vector ω.

The difficulty arises because the velocities $\dot{z}_b^T = \{ v_P^T \; \omega^T \}$ cannot be integrated to get the corresponding position variables due to the angular velocity part. There is no problem in using this base body velocity vector to compute the corresponding columns of matrix $\mathbf{R}$. The velocity vector $\dot{z}_b$ needs to be transformed into a different one, which can be called $\dot{z}_B$, and contains only integrable variables. The most important options are:

i) *Sets of three independent parameters, such as Euler or Bryant angles.* The limitation of this option is that none of these sets are free of singular positions; that is, positions for which the angular parameters are not determined unequivocally. If these positions can be effectively reached, it is necessary to foresee the cure, perhaps in the form of a change of reference frame.

ii) *Larger sets of dependent parameters, such as Euler parameters or quaternions.* Singularities can always be avoided, but there are constraint equations that relate the variables to be integrated. This dependency can be taken into account in the integration process. It is not really a very serious problem, as seen previously in this chapter.

It will always be possible to find a position-dependent matrix $\mathbf{W}$ that relates integrable and non-integrable base body velocities:

$$\dot{z}_b = \mathbf{W}(z_B)\dot{z}_B \tag{5.72}$$

where $\dot{z}_b$ is used to construct the matrix $\mathbf{R}$, and $\dot{z}_B$ is used for the numerical integration process. This transformation shall be introduced in the vector of independent velocities $\dot{z}$ before integrating it. It is not necessary to introduce it in $\ddot{z}$ which can always be integrated once.

The first three columns of $\mathbf{R}$ related to the velocity of the reference point P can be computed as the result of applying three unit translations to the whole system on the inertial frame axes. In order to find a general expression, if $\dot{z}_1$, $\dot{z}_2$, and $\dot{z}_3$ are the related independent velocities, the velocity of a point j, and a unit vector $\mathbf{u}_j$, due to the translation of the base body can be expressed as:

$$\dot{r}_j = \dot{z}_1 \mathbf{n}_1 + \dot{z}_2 \mathbf{n}_2 + \dot{z}_3 \mathbf{n}_3 \tag{5.73}$$

$$\dot{\mathbf{u}}_j = 0 \tag{5.74}$$

where $(\mathbf{n}_1, \mathbf{n}_2, \mathbf{n}_3)$ are unit vectors on the inertial frame axes. From these equations, the elements of the columns of matrix $\mathbf{R}$, corresponding to point j and vector $\mathbf{u}_j$, can be computed in the form:

– column 1 $\qquad\qquad \dot{z}_1 = 1 \quad \dot{z}_2 = 0 \quad \dot{z}_3 = 0$

– column 2 $\qquad\qquad \dot{z}_1 = 0 \quad \dot{z}_2 = 1 \quad \dot{z}_3 = 0$

– column 3 $\qquad\qquad \dot{z}_1 = 0 \quad \dot{z}_2 = 0 \quad \dot{z}_3 = 1$

If the independent velocities are the Cartesian components of the angular velocity vector ω, the columns 4 to 6 of the matrix $\mathbf{R}$ corresponding to the base body rotation may be computed as follows:

$$\dot{\mathbf{r}}_j = \dot{z}_4\,\mathbf{n}_1 \wedge (\mathbf{r}_j - \mathbf{r}_P) + \dot{z}_5\,\mathbf{n}_2 \wedge (\mathbf{r}_j - \mathbf{r}_P) + \dot{z}_6\,\mathbf{n}_3 \wedge (\mathbf{r}_j - \mathbf{r}_P) \tag{5.75}$$

$$\dot{\mathbf{u}}_j = \dot{z}_4\,\mathbf{n}_1 \wedge \mathbf{u}_j + \dot{z}_5\,\mathbf{n}_2 \wedge \mathbf{u}_j + \dot{z}_6\,\mathbf{n}_3 \wedge \mathbf{u}_j \tag{5.76}$$

From these formulas, columns 4 to 6 can be computed by making

– column 4 $\qquad\qquad \dot{z}_4 = 1 \quad \dot{z}_5 = 0 \quad \dot{z}_6 = 0$

– column 5 $\qquad\qquad \dot{z}_4 = 0 \quad \dot{z}_5 = 1 \quad \dot{z}_6 = 0$

– column 6 $\qquad\qquad \dot{z}_4 = 0 \quad \dot{z}_5 = 0 \quad \dot{z}_6 = 1$

and this completes the information necessary to determine the part of $\mathbf{R}$ due to the base body degrees of freedom.

5.3.1.2 Different Joints in 3-D Multibody Systems

Figure 5.8 illustrates a generic joint i, a point j, and a unit vector $\mathbf{u}_j$ located upwards in the branch. The non-zero elements, corresponding to point j and vector $\mathbf{u}_j$ in the columns of matrix $\mathbf{R}$ associated with the degrees of freedom of joint i, are considered next.

Revolute joint. The joint variable is the angle z_i that defines the rotation of the joint around the axis determined by point i and vector $\mathbf{u}_i$. The velocity of point j induced by the relative velocity at joint i is

$$\dot{\mathbf{r}}_j = \dot{z}_i\,\mathbf{u}_i \wedge (\mathbf{r}_j - \mathbf{r}_i) \tag{5.77}$$

and the velocity induced in the unit vector $\mathbf{u}_j$:

$$\dot{\mathbf{u}}_j = \dot{z}_i\,\mathbf{u}_i \wedge \mathbf{u}_j \tag{5.78}$$

As done before, the corresponding values of the column (i) elements of the matrix $\mathbf{R}$ can be computed by giving a unit value to $\dot{z}_i$.

Prismatic joint. Let z_i be the translational joint variable located on a line defined by point i and vector $\mathbf{u}_i$. The induced velocities of point j and vector $\mathbf{u}_j$ are:

$$\dot{\mathbf{r}}_j = \dot{z}_i\,\mathbf{u}_i \tag{5.79}$$

$$\dot{\mathbf{u}}_j = 0 \tag{5.80}$$

These expressions allow for a very easy computation of the elements of the considered column of the matrix $\mathbf{R}$.

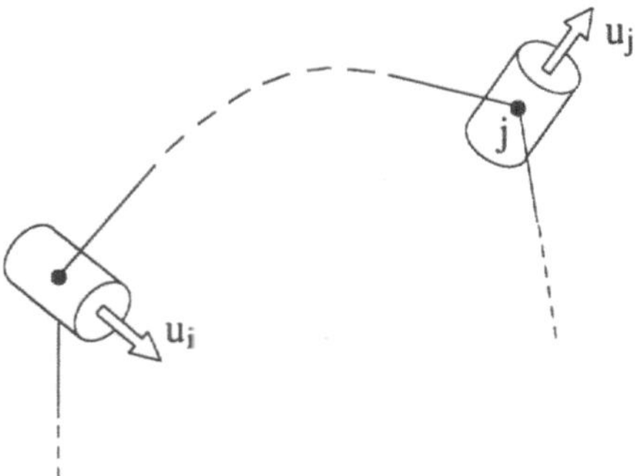

Figure 5.8. Description of a generic joint.

Cylindrical joint. This joint is different from the previous ones, since it has two degrees of freedom and it introduces two columns in matrix $\mathbf{R}$. Let z_i and z_{i+1} be respectively the relative angle and distance that constitute the joint variables. Vector $\mathbf{u}_i$ determines the direction of both the rotation axis and the translation.

The velocities of point j and vector $\mathbf{u}_j$, due to the joint variables, are in this case:

$$\dot{\mathbf{r}}_j = \dot{z}_i \mathbf{u}_i \wedge (\mathbf{r}_j - \mathbf{r}_i) + \dot{z}_{i+1}\, \mathbf{u}_i \tag{5.81}$$

$$\dot{\mathbf{u}}_j = \dot{z}_i\, \mathbf{u}_i \wedge \mathbf{u}_j \tag{5.82}$$

The non-zero elements in the corresponding matrix $\mathbf{R}$ columns, can be obtained by making, respectively:

– column i $\qquad\qquad \dot{z}_i = 1 \qquad \dot{z}_{i+1} = 0$

– column $i+1$ $\qquad\quad \dot{z}_i = 0 \qquad \dot{z}_{i+1} = 1$

Universal joint. This joint can be considered equivalent to two revolute joints, with the axes (belonging to different bodies) intersecting orthogonally at a common point. If $\dot{z}_i$ and $\dot{z}_{i+1}$ are the joint independent velocities, the elements related to point j and vector $\mathbf{u}_j$ in the columns of $\mathbf{R}$ come from the following velocity expressions:

$$\dot{\mathbf{r}}_j = \dot{z}_i\, \mathbf{u}_i \wedge (\mathbf{r}_j - \mathbf{r}_j) + \dot{z}_{i+1}\, \mathbf{v}_i \wedge (\mathbf{r}_j - \mathbf{r}_i) \tag{5.83}$$

$$\dot{\mathbf{u}}_j = \dot{z}_i\, \mathbf{u}_i \wedge \mathbf{u}_j + \dot{z}_{i+1}\, \mathbf{v}_i \wedge \mathbf{u}_j \tag{5.84}$$

From these expressions, the two columns of matrix $\mathbf{R}$ can be obtained by giving, respectively, the following values to the independent velocities:

– column i $\qquad\qquad\qquad$ $\dot{z}_i = 1$ $\qquad$ $\dot{z}_{i+1} = 0$

– column $i+1$ $\qquad\qquad$ $\dot{z}_i = 0$ $\qquad$ $\dot{z}_{i+1} = 1$

Spherical joint. The spherical joint allows for three rotations. This fact produces in the joint similar difficulties to the ones that were found at the time of defining the angular orientation of the base body. There is no problem in using three unit angular velocities on three orthogonal axes to compute the corresponding columns of matrix $\mathbf{R}$. However, these independent velocities cannot be integrated to get displacement or position variables. It is necessary to transform those independent velocities into another set of integrable velocities that are dependent or independent according to an expression similar to (5.72).

If $\dot{z}_i^b$, $\dot{z}_{i+1}^b$ and $\dot{z}_{i+2}^b$ are the Cartesian components of the relative angular velocity vector, the velocities of point j and vector $\mathbf{u}_j$ are:

$$\dot{\mathbf{r}}_j = \dot{z}_i^b\,\mathbf{n}_1 \wedge (\mathbf{r}_j - \mathbf{r}_i) + \dot{z}_{i+1}^b\,\mathbf{n}_2 \wedge (\mathbf{r}_j - \mathbf{r}_i) + \dot{z}_{i+2}^b\,\mathbf{n}_3 \wedge (\mathbf{r}_j - \mathbf{r}_i) \qquad (5.85)$$

$$\dot{\mathbf{u}}_j = \dot{z}_i^b\,\mathbf{n}_1 \wedge \mathbf{u}_j + \dot{z}_{i+1}^b\,\mathbf{n}_2 \wedge \mathbf{u}_j + \dot{z}_{i+2}^b\,\mathbf{n}_3 \wedge \mathbf{u}_j \qquad (5.86)$$

where $(\mathbf{n}_1, \mathbf{n}_2, \mathbf{n}_3)$ are orthogonal unit vectors on the inertial reference frame axes.

Using equations (5.85) and (5.86), the columns of matrix $\mathbf{R}$ (the terms corresponding to point j and vector $\mathbf{u}_j$) can be computed with the following values:

– column i $\qquad\qquad\qquad$ $\dot{z}_i^b = 1$ $\qquad$ $\dot{z}_{i+1}^b = 0$ $\qquad$ $\dot{z}_{i+2}^b = 0$

– column $i+1$ $\qquad\qquad$ $\dot{z}_i^b = 0$ $\qquad$ $\dot{z}_{i+1}^b = 1$ $\qquad$ $\dot{z}_{i+2}^b = 0$

– column $i+2$ $\qquad\qquad$ $\dot{z}_i^b = 0$ $\qquad$ $\dot{z}_{i+1}^b = 0$ $\qquad$ $\dot{z}_{i+2}^b = 1$

Once we have described how the columns of matrix $\mathbf{R}$ can be computed, the computation of the term $(\mathbf{Sc})$ that appears in equation (5.67) remains. This term represents the dependent acceleration vector $\ddot{\mathbf{q}}$ computed with the true velocities $\dot{\mathbf{q}}$ or $\dot{\mathbf{z}}$ but with zero independent accelerations $\ddot{\mathbf{z}}$.

The elements of vector $(\mathbf{Sc})$ corresponding to point j and vector $\mathbf{u}_j$ can be computed by adding to the acceleration of this point and vector the contribution of the true independent velocities of the base body and all the joints that are downwards in the branch of point j. The corresponding expressions are:

$$\ddot{\mathbf{r}}_j\big|_{\ddot{z}=0} = \sum_k \dot{z}_k \left[\mathbf{u}_k \wedge (\dot{\mathbf{r}}_j - \dot{\mathbf{r}}_i) + \dot{\mathbf{u}}_k \wedge (\mathbf{r}_j - \mathbf{r}_i) \right] \qquad (5.87)$$

$$\ddot{\mathbf{u}}_j\big|_{\ddot{z}=0} = \sum_k \dot{z}_k \left(\mathbf{u}_k \wedge \dot{\mathbf{u}}_j + \dot{\mathbf{u}}_k \wedge \mathbf{u}_i \right) \qquad (5.88)$$

Hence, all the information necessary to compute the coefficients of equation (5.67) is available.

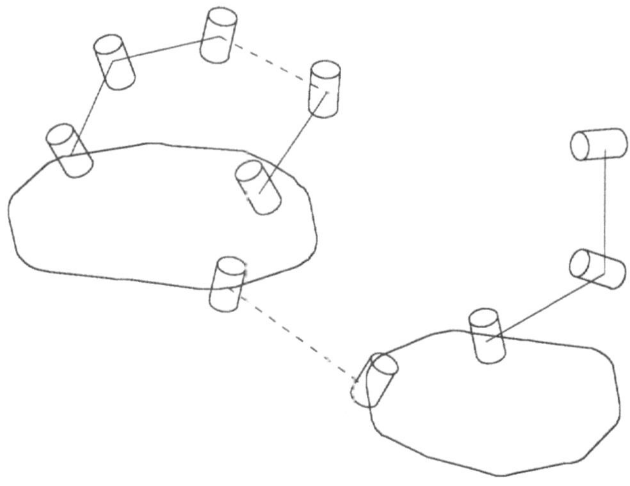

Figure 5.9. Closed-chain multibody system.

5.3.2 Closed-Chain Multibody Systems

In order to extend the previous formulation to multibody systems including closed loops, one should remember briefly the formulation of the dynamic equations using dependent coordinates and Lagrange multipliers through equation (5.10) repeated here:

$$\begin{bmatrix} \mathbf{M} & \mathbf{\Phi}_q^T \\ \mathbf{\Phi}_q & 0 \end{bmatrix} \begin{Bmatrix} \ddot{\mathbf{q}} \\ \lambda \end{Bmatrix} = \begin{Bmatrix} \mathbf{Q} \\ \mathbf{c}(\mathbf{q},\dot{\mathbf{q}}) \end{Bmatrix} \tag{5.89}$$

This equation describes the motion of the multibody system formulated with dependent coordinates. The natural coordinates can lead to a constant matrix $\mathbf{M}$, a Jacobian matrix $\mathbf{\Phi}_q$ being a linear function of coordinates $\mathbf{q}$, and to no velocity-dependent inertial forces in the right-hand side of equation (5.89). The term $\mathbf{c}(\mathbf{q}, \dot{\mathbf{q}})$ comes from the double differentiation of the constraint conditions and can be replaced by the term $\mathbf{g}$ (See equation (5.28)), if the Baumgarte constraint stabilization technique is desired.

After this brief introduction, the formulation described in the previous section can be extended to systems with closed loops and perhaps several base bodies, as in the system shown in Figure 5.9. Here, the system can be transformed into one or more open tree systems by opening the closed loops and disconnecting the base bodies. With natural coordinates, most of the constraint equations come from the rigid body conditions of the elements. The simplest way to transform the system into one or more open trees is by removing the rigid body constraints corresponding to the closure of the loops and the connection between base bodies shown by dashed lines in the system of Figure 5.9. With reference point coordinates, constraints arise mainly from the system joints. In this case, the best way to open the loops is by removing some of them.

In order to obtain the dynamic equations, it is now very convenient to distinguish between the constraint equations corresponding to the open-chain systems (superscript 1) and the constraints corresponding to the closure conditions and connection between base bodies (superscript 2). Writing the dynamic equations using the Lagrange multiplier formulation of equation (5.89) and separating the two groups of constraints, we arrive at:

$$
\begin{bmatrix} \mathbf{M} & \mathbf{\Phi}_q^{1T} & \mathbf{\Phi}_q^{2T} \\ \mathbf{\Phi}_q^1 & \mathbf{0} & \mathbf{0} \\ \mathbf{\Phi}_q^2 & \mathbf{0} & \mathbf{0} \end{bmatrix} \begin{Bmatrix} \ddot{\mathbf{q}} \\ \lambda^1 \\ \lambda^2 \end{Bmatrix} = \begin{Bmatrix} \mathbf{Q} \\ \mathbf{c}^1 \\ \mathbf{c}^2 \end{Bmatrix}
\tag{5.90}
$$

The velocity transformation corresponding to the open-chain constraints, whose Jacobian matrix nullspace is given by matrix $\mathbf{R}^1$ can now be introduced. Consequently,

$$
\dot{\mathbf{q}} = \mathbf{R}^1 \, \dot{\mathbf{z}}
\tag{5.91}
$$

and equation (5.90) becomes:

$$
\begin{bmatrix} \mathbf{R}^{1T} \mathbf{M} \mathbf{R}^1 & \mathbf{R}^{1T} \mathbf{\Phi}_q^{2T} \\ \mathbf{\Phi}_q^2 \mathbf{R}^1 & \mathbf{0} \end{bmatrix} \begin{Bmatrix} \ddot{\mathbf{z}} \\ \lambda^2 \end{Bmatrix} = \begin{Bmatrix} \mathbf{R}_1^T(\mathbf{Q} - \mathbf{M}\,\mathbf{S}^1 \mathbf{c}^1) \\ \mathbf{c}^2 - \mathbf{\Phi}_q^2 \mathbf{S}^1 \mathbf{c}^1 \end{Bmatrix}
\tag{5.92}
$$

The vector $\mathbf{z}$ in equations (5.91) and (5.92) is not a vector of independent coordinates as in Section 5.3.1, but of dependent coordinates corresponding to the base bodies plus the relative coordinates of the open tree configuration joints. These coordinates are related by the constraint equations $\mathbf{\Phi}^2$. Then equation (5.91) is a transformation between two dependent velocity vectors. Vector $\dot{\mathbf{z}}$ will contain usually far less variables than vector $\dot{\mathbf{q}}$.

This formulation is also advantageous because the matrix $\mathbf{R}^1$ which corresponds to open-chain constraints can be constructed directly without any explicit Jacobian factorization as explained in Section 5.3.1. The term $(\mathbf{\Phi}_q^2 \, \mathbf{R}^1)$ represents the projection of closure loop constraints (superscript 2) on the nullspace of the open-chain system (superscript 1).

Equation (5.92) has the form of the equations of motion in dependent coordinates with Lagrange multipliers to which Baumgarte stabilization can be directly applied. This could obviously be substituted by the penalty formulation (See Section 5.1.4), or by seeking a true independent set of coordinates. This can be done (See Section 5.2) by computing numerically the nullspace of the projected Jacobian matrix $(\mathbf{\Phi}_q^2 \, \mathbf{R}^1)$ which can be a very small matrix.

5.4 Formulations Based on the Canonical Equations

Some authors have drawn attention recently to the use of the canonical equations as a way to improve numerically the formulation of the equations of motion of mechanical systems and to perform a faster and more stable simulation; thus more suitable for real time analysis. We discuss in this section the different approaches that can be derived from the use of the canonical equations for constrained mechanical systems, and whether the use of these equations may lead to more efficient and stable numerical implementations than those coming from acceleration-based formalisms. Specifically, we will describe the canonical equations that result: first, from the use of the Lagrange multipliers; secondly, from the use of independent coordinates; and thirdly, from the penalty formulation.

5.4.1 Lagrange Multiplier Formulation

Consider again a mechanical system whose configuration is characterized by a vector $\mathbf{q}$ of n generalized coordinates that are interrelated through the m kinematic constraint conditions $\Phi(\mathbf{q}, t)=0$, of the holonomic type. The Lagrange's equations of such a system are given in equation (5.2) which along with the constraint equations (5.1) constitute a set of $(n+m)$ mixed differential algebraic equations (DAEs) of index three (See Chapter 7), with λ as the Lagrange multipliers. The *conjugate* or *canonical momenta* was defined in Chapter 4 as:

$$\mathbf{p} = \frac{\partial L}{\partial \dot{\mathbf{q}}} \tag{5.93}$$

along with the Hamiltonian:

$$H = \mathbf{p}^{\mathrm{T}} \dot{\mathbf{q}} - L \tag{5.94}$$

where the previously introduced matrix notation has been employed. Hamilton's equations for a constrained system are formulated (See Section 4.1.5) as:

$$\dot{\mathbf{q}} = \frac{\partial H}{\partial \mathbf{p}} \tag{5.95}$$

$$-\dot{\mathbf{p}} = \frac{\partial H}{\partial \mathbf{q}} - \mathbf{Q}_{ex} + \Phi_{\mathbf{q}}^{\mathrm{T}} \lambda \tag{5.96}$$

In the case of mechanical systems, the Lagrangian L is defined in terms of $\mathbf{q}$, $\dot{\mathbf{q}}$, and t and rather than following a lengthy process to form the Hamiltonian as an explicit function of $\mathbf{q}, \mathbf{p}$, and t, and then differentiating as in (5.95), the *canonical equations* can be directly obtained from (5.93) and (5.96). Since the system kinetic energy is a quadratic function of the generalized velocities, equations (5.93) and (5.96) directly lead to the following set of equations in matrix form:

$$\mathbf{p} = \mathbf{M} \dot{\mathbf{q}} \tag{5.97}$$

$$\dot{\mathbf{p}} = L_{\mathbf{q}} + \mathbf{Q}_{ex} - \Phi_{\mathbf{q}}^{T}\lambda \tag{5.98}$$

where $\mathbf{M}$ is the mass matrix, $L_{\mathbf{q}}$ is the partial derivative of the Lagrangian with respect to the coordinates, $\Phi_{\mathbf{q}}$ is the Jacobian matrix of the constraint equations, and $\mathbf{Q}_{ex}$ is the vector of applied external and dissipative forces. The combination of equations (5.97)-(5.98) and the constraints conditions constitutes a system of $(2n+m)$ differential and algebraic equations (DAE) of index two (See Chapter 7). Although there are n more equations than in equation (5.10), $\dot{\mathbf{p}}$ can be obtained explicitly by (5.98). In addition, index two DAEs are better behaved than index three DAEs (Brenan et al. (1989)). Therefore, the use of (5.97)-(5.98) may be numerically advantageous as compared to the use of (5.10), when using algorithms for the solution of the mixed differential algebraic equations.

In order to avoid the mixed differential and algebraic equations, Lankarani and Nikravesh (1988) modified the system Lagrangian to include the kinematic velocity constraints as:

$$L^{*} = L + \dot{\Phi}^{T}\sigma \tag{5.99}$$

where σ is a new set of Lagrange multipliers. It may be very easily demonstrated that $\lambda = \dot{\sigma}$. The new Hamiltonian is $H^{*} = \mathbf{p}^{T}\dot{\mathbf{q}} - L^{*}$, and the application of (5.93) and (5.95) leads to:

$$\mathbf{p} = \mathbf{M}\,\dot{\mathbf{q}} + \Phi_{\mathbf{q}}^{T}\sigma \tag{5.100}$$

$$\dot{\mathbf{p}} = L_{\mathbf{q}} + \mathbf{Q}_{ex} + \dot{\Phi}_{\mathbf{q}}^{T}\sigma \tag{5.101}$$

That, along with

$$\dot{\Phi} = \Phi_{\mathbf{q}}\,\dot{\mathbf{q}} \tag{5.102}$$

constitutes a set of $2n+m$ ordinary differential equations (ODE), with $\mathbf{p}$, $\mathbf{q}$, and σ as unknowns. If equation (5.100) is differentiated and substituted into the acceleration-based equation (5.4), the result is precisely the additional canonical equation (5.101). Thus, *the canonical equations originate from the acceleration-based equations by the mere canonical transformation* defined by equation (5.100).

Only $(n+m)$ equations need to be solved at each time step in the numerical implementation of the algorithm, which can be described as follows:

Algorithm 5-6

1. Start at time t when $\mathbf{p}$ and $\mathbf{q}$ are known.
2. Use (5.100) along with (5.102) to solve for $\dot{\mathbf{q}}$ and σ at time t, as follows:

$$\begin{bmatrix} \mathbf{M} & \Phi_{\mathbf{q}}^{T} \\ \Phi_{\mathbf{q}} & 0 \end{bmatrix} \begin{Bmatrix} \dot{\mathbf{q}} \\ -\sigma \end{Bmatrix} = \begin{Bmatrix} \mathbf{p} \\ 0 \end{Bmatrix} \tag{5.103}$$

3. Use (5.101) to compute $\dot{\mathbf{p}}$ explicitly, with no solution of equations involved.
4. Obtain the vectors $\mathbf{p}$ and $\mathbf{q}$ at time $(t+\Delta t)$ by the numerical integration:

$$\dot{s}_t^T \equiv \left\{ \dot{z}^T, \ \dot{y}^T \right\}_t \quad \xrightarrow{\text{n.i.s.}} \quad s_{t+\Delta t}^T \equiv \left\{ z^T, \ y^T \right\}_{t+\Delta t}$$

5. Upon convergence of the n.i.s., update the time variable and go to step 2.

Lankarani and Nikravesh (1988) showed in their numerical simulations that since only the first time derivative of the constraints is used, the integration of this equations is more efficient and more stable than the acceleration-based formulation. With the acceleration-based formulation to avoid the integration of the mixed differential algebraic equations, the constraint conditions need to be differentiated twice; thus leading to larger constraint violations.

5.4.2 Formulation Based on Independent Coordinates

The formulation of the canonical equations of motion can also be written as a function of a minimum set of independent coordinates. This is the approach followed by Bae and Won (1990) who used the velocity transformation method developed by Kim and Vanderploeg (1986) to transform the equations of motion from the Cartesian space to the joint space. They used an equivalence between the Lagrangian and Newton Euler formulation to derive the partial derivative of the kinetic energy with respect to the independent coordinates. We show in this section how the canonical equations in independent coordinates can be obtained very simply if one considers equations (5.97) and (5.98) as the starting point.

Given the constraint conditions (See equation (5.61)) $\Phi(q, t)=0$, one can find the two matrices R and S such that

$$\dot{q} = R\,\dot{z} + S\,b \tag{5.104}$$

where $\dot{z}$ represents a set of independent of velocities. The substitution of (5.104) into (5.97) yields

$$p = M\,R\,\dot{z} + M\,S\,b \tag{5.105}$$

and pre-multiplying both sides of equations (5.105) and (5.98) by R^T one can obtain, respectively,

$$R^T\,p = R^T\,M\,R\,\dot{z} + R^T\,M\,S\,b \tag{5.106}$$

$$R^T\,\dot{p} = R^T\,(L_q + Q_{ex}) \tag{5.107}$$

where the term containing the Jacobian matrix has been dropped, since $R^T\Phi_q^T = 0$ (See Chapter 3).

The new variable $y=R^T p$ can be defined as the projection of the canonical momenta over the subspace of allowable motions. Then

$$\dot{y} = \dot{R}^T\,p + R^T\,\dot{p} \tag{5.108}$$

and the substitution of these two expressions into equations (5.106) and (5.107) leads to

$$y = R^T\,M\,R\,\dot{z} + R^T\,M\,S\,b \tag{5.109}$$

$$\dot{\mathbf{y}} = \mathbf{R}^T (L_\mathbf{q} + \mathbf{Q}_{ex}) + \dot{\mathbf{R}}^T \mathbf{p} \qquad (5.110)$$

Substituting the value of $\mathbf{p}$ given by (5.105) into (5.110) one can arrive at the final set of canonical equations in independent coordinates:

$$\mathbf{y} = \mathbf{R}^T \mathbf{M} \mathbf{R} \dot{\mathbf{z}} + \mathbf{R}^T \mathbf{M} \mathbf{S} \mathbf{b} \qquad (5.111)$$

$$\dot{\mathbf{y}} = \mathbf{R}^T (L_\mathbf{q} + \mathbf{Q}_{ex}) + \dot{\mathbf{R}}^T (\mathbf{M} \mathbf{R} \dot{\mathbf{z}} + \mathbf{M} \mathbf{S} \mathbf{b}) \qquad (5.112)$$

These two equations may be also obtained from the acceleration-based equations (5.67) by the mere canonical transformation of (5.111). Equations (5.111) and (5.112) constitute a set of $2(n-m)$ first order ordinary differential equations. Since $\dot{\mathbf{y}}$ is given explicitly in (5.112), only $n-m$ equations need to be solved for each function evaluation in the numerical implementation of the algorithm. This can be described as follows:

Algorithm 5-7

1. Start at time t in which $\mathbf{z}$ and $\mathbf{y}$ are known.
2. Solve the position and velocity problems to get $\mathbf{q}$ and $\dot{\mathbf{q}}$.
3. Obtain the matrices $\mathbf{R}$ and $\dot{\mathbf{R}}$.
4. Use (5.111) to solve for $\dot{\mathbf{z}}$. The solution of $n-m$ equations is required.
5. Use (5.112) to solve for $\dot{\mathbf{y}}$ explicitly.
6. Obtain the vectors $\mathbf{z}$ and $\mathbf{y}$ at time $(t+\Delta t)$ by numerical integration:

$$\dot{\mathbf{s}}_t^T \equiv \left\{ \dot{\mathbf{z}}^T, \ \dot{\mathbf{y}}^T \right\}_t \quad \overset{\text{n.i.s.}}{\Longrightarrow} \quad \mathbf{s}_{t+\Delta t}^T \equiv \left\{ \mathbf{z}^T, \ \mathbf{y}^T \right\}_{t+\Delta t}$$

7. Upon convergence of the n.i.s., update the time variable and go to step 2.

This scheme can be compared to the $(n-m)$ second order ordinary differential equations resulting from the acceleration-based formulation (equation (5.67)):

$$\mathbf{R}^T \mathbf{M} \mathbf{R} \ddot{\mathbf{z}} = \mathbf{R}^T \mathbf{Q} + \mathbf{R}^T \mathbf{M} \mathbf{S} [\dot{\Phi}_t + \dot{\Phi}_\mathbf{q} (\mathbf{R} \dot{\mathbf{z}} + \mathbf{S} \mathbf{b})] \qquad (5.113)$$

where $\mathbf{Q} = \mathbf{Q}_{ex} + L_\mathbf{q} - \dot{\mathbf{M}} \dot{\mathbf{q}}$ contains the external forces plus all the inertial terms coming from the differentiation of the Lagrangian formulation. One can see that both methods require the triangularization of the same matrix $(\mathbf{R}^T \mathbf{M} \mathbf{R})$ at each function evaluation. In addition, there might not be much advantage in using the canonical equations (5.111) and (5.112). Because although equation (5.113) involves more matrix manipulations than (5.111) and (5.112) and has a more complicated forcing term, the canonical approach requires the additional evaluation of $\dot{\mathbf{R}}$ with a sizable amount of computations.

5.4.3 Augmented Lagrangian Formulation in Canonical Form

We consider in this section the penalty augmented Lagrangian formulation in its canonical form. Its better accuracy and stability properties makes this method more attractive than the generic penalty formulation.

Basic Augmented Lagrangian Formulation. Equation (5.99) can be considered as the starting point to build a modified Lagrangian formulation that will not only contain the Lagrange multipliers σ but also the penalty terms of the previous section. Accordingly

$$L^* = L + \frac{1}{2}\,\dot{\Phi}^T\,\alpha\,\dot{\Phi} - \frac{1}{2}\,\Phi^T\,\Omega^2\,\alpha\,\Phi + \dot{\Phi}^T\,\sigma^* \tag{5.114}$$

In the limit when the constraint conditions are satisfied, the penalty terms vanish, and $\sigma = \sigma^*$. This is similar to the Lagrange's formulation $\dot{\sigma}^* = \lambda^*$ and after the augmented Lagrangian iteration when the constraints are satisfied to machine precision $\dot{\sigma} = \lambda$. The differentiation of L^* with respect to $\dot{q}$ leads to the following new canonical momenta in matrix form:

$$\mathbf{p} = \frac{\partial L^*}{\partial \dot{\mathbf{q}}} = \mathbf{M}\dot{\mathbf{q}} + \Phi_{\mathbf{q}}^T\,\alpha\,\dot{\Phi} + \Phi_{\mathbf{q}}^T\,\sigma^* \tag{5.115}$$

The modified *Hamiltonian* can be written as $H^* = \mathbf{p}^T\,\dot{\mathbf{q}} - L^*$ and the use of (5.96), including the dissipative Rayleigh forces of (5.30), leads to:

$$\left[\mathbf{M} + \Phi_{\mathbf{q}}^T\,\alpha\,\Phi_{\mathbf{q}}\right]\dot{\mathbf{q}} = \mathbf{p} - \Phi_{\mathbf{q}}^T\,\alpha\,\Phi_t + \Phi_{\mathbf{q}}^T\,\sigma^* \tag{5.116a}$$

$$\dot{\mathbf{p}} = \mathbf{Q} + L_{\mathbf{q}} + \dot{\Phi}_{\mathbf{q}}^T\,\alpha\,\dot{\Phi} - \Phi_{\mathbf{q}}^T\,\alpha\,(\Omega^2\,\Phi + 2\,\Omega\,\mu\,\dot{\Phi}) + \dot{\Phi}_{\mathbf{q}}^T\,\sigma^* \tag{5.116b}$$

Equations (5.116) constitute a set of $2n$ first order ordinary differential equations. However, $\dot{\mathbf{p}}$ is given in explicit form, and therefore only n algebraic equations need be solved at each function evaluation for the numerical implementation of the algorithm. The numerical simulations have shown that equations (5.116) tend to be numerically stiff due to all the penalty terms concentrated in the RHS of (5.116b). This numerical stiffness limits the possible choices of numerical integrators. Standard ODE integrators that are based on conditionally stable predictor-corrector multi-step formulae lead to an increased number of function evaluations. A modification of (5.116) is used in the next section that circumvents this problem.

Modified Augmented Lagrangian Formulation. The canonical equation (5.116a) may be also written as:

$$\mathbf{p} = \mathbf{M}\,\dot{\mathbf{q}} + \Phi_{\mathbf{q}}^T\,\alpha\,\dot{\Phi} + \Phi_{\mathbf{q}}^T\,\sigma^* \tag{5.117}$$

which indicates that the canonical momenta is stabilized through the addition of penalty terms that are proportional to the violation of the velocity constraint equations. If equation (5.117) is differentiated and substituted into the acceleration-based augmented Lagrangian equation (5.48) the result is the additional canonical equation (5.116b).

However, we can achieve a better stabilization of the canonical momenta if we add to the RHS of (5.117) two additional penalty terms: one term proportional to the constraint violation and the other to its integral. Accordingly, we define a new momenta $\mathbf{p}$ as:

$$\mathbf{p} = \mathbf{M}\,\dot{\mathbf{q}} + \mathbf{\Phi}_\mathbf{q}^\mathsf{T}\,\alpha\left(\dot{\mathbf{\Phi}} + 2\,\mu\,\Omega\,\mathbf{\Phi} + \Omega^2\int_{t_o}^{t}\mathbf{\Phi}\,d\tau\right) - \mathbf{\Phi}_\mathbf{q}^\mathsf{T}\,\sigma^* \tag{5.118}$$

By expanding the term $\dot{\mathbf{\Phi}}$, equation (32) becomes

$$\left(\mathbf{M} + \mathbf{\Phi}_\mathbf{q}^\mathsf{T}\,\alpha\,\mathbf{\Phi}_\mathbf{q}\right)\dot{\mathbf{q}} = \mathbf{p} - \mathbf{\Phi}_\mathbf{q}^\mathsf{T}\,\alpha\left(\mathbf{\Phi}_t + 2\,\mu\,\Omega\,\mathbf{\Phi} + \Omega^2\int_{t_o}^{t}\mathbf{\Phi}\,d\tau\right) - \mathbf{\Phi}_\mathbf{q}^\mathsf{T}\,\sigma^* \tag{5.119a}$$

The differentiation of (5.119a) and substitution into (5.48) leads to the second set of modified canonical equations:

$$\dot{\mathbf{p}} = \mathbf{Q} + L_\mathbf{q} + \dot{\mathbf{\Phi}}_\mathbf{q}^\mathsf{T}\,\alpha\left(\dot{\mathbf{\Phi}} + 2\,\Omega\,\mu\,\mathbf{\Phi} + \Omega^2\int_{t_o}^{t}\mathbf{\Phi}\,d\tau\right) + \dot{\mathbf{\Phi}}_\mathbf{q}^\mathsf{T}\,\sigma^* \tag{5.119b}$$

which along with (5.119a) constitute a set of $2n$ first order ordinary differential equations in the unknowns $\mathbf{p}$, $\mathbf{q}$, and σ^*. Again, only n algebraic equations need be solved at each function evaluation for the numerical implementation of the algorithm. Contrary to equations (5.116), equations (5.119) do not become stiff. They even provide more numerical accuracy and better constraint stabilization than the acceleration-based formulation of equation (5.48).

We can compare this set of equations with the n second order ordinary differential equations resulting from the acceleration-based formulation. While both formulations require the triangularization of the same leading matrix for each function evaluation, there are advantages in the use of (5.119) as compared to (5.48). The kinematic constraint conditions are differentiated only once with the canonical procedure and twice, in the acceleration based formulation. This will lead to fewer violations of the constraints. It is shown in Bayo and Avello (1993) how this factor becomes detrimental for the acceleration-based formulation under repetitive singular positions, whereas the canonical approach leads to a much better performance.

The multipliers σ^* do not need to be solved explicitly. Following the same procedure as that used with the acceleration-based augmented Lagrangian formulation, the σ^* may be obtained in an iterative manner as:

$$\sigma_{i+1}^* = \sigma_i^* + \left(\dot{\mathbf{\Phi}} + 2\,\mu\,\Omega\,\mathbf{\Phi} + \Omega^2\int_{t_o}^{t}\mathbf{\Phi}\,d\tau\right)_{i+1} \tag{5.120}$$

$$i = 0, 1, 2, \dots$$

with $\sigma_0^* = 0$ for the first iteration. Equation (5.119a), including the iterative process of (5.120), becomes

$$\left(\mathbf{M} + \Phi_{\mathbf{q}}^{T} \alpha \, \Phi_{\mathbf{q}}\right) \dot{\mathbf{q}}_{i+1} = \mathbf{M} \, \dot{\mathbf{q}}_{i} - \Phi_{\mathbf{q}}^{T} \alpha \, (\Phi_{t} + 2 \mu \, \Omega \, \Phi + \Omega^{2}\!\! \int_{t_{0}}^{t} \Phi \, d\tau)$$

$$i = 0, 1, 2, \dots \tag{5.121}$$

with $\mathbf{M}\dot{\mathbf{q}}_{0} = \mathbf{p}$ for the first iteration. Equation (5.121) shows that the velocity calculation at each function evaluation is refined so that the weighted summation of the constraint equations (5.120) is satisfied to machine precision. After the velocity calculation equation, (5.119b) may be used to evaluate the derivative of the canonical momenta.

Algorithm 5-8

1. Start at time t in which $\mathbf{p}$ and $\mathbf{q}$ are known.
2. Use (5.121) iteratively to solve for $\dot{\mathbf{q}}$, with $\mathbf{M} \, \dot{\mathbf{q}}_{0} = \mathbf{p}$ for the first iteration. At the end of each iteration use (5.120) to calculate the Lagrange multipliers σ^{*}.
3. Use (5.119b) to compute $\dot{\mathbf{p}}$ explicitly with no solution of equations involved.
4. Call the numerical integration subroutine to compute $\mathbf{p}$ and $\mathbf{q}$ at time $t + \Delta t$.
5. Upon convergence of the n.i.s., if desired, use a differentiation scheme to obtain $\lambda = \dot{\sigma}$.
6. Update the time variable and go to step 2.

This algorithm is as efficient numerically as Algorithm 5-3, but much more stable under repetitive singular positions.

Canonical Augmented Lagrangian Formulation for Non-Holonomic Systems. The modified augmented Lagrangian formulation described above may also be extended to non-holonomic systems with constraints of the form:

$$\Phi(\dot{\mathbf{q}}, \mathbf{q}, t) = 0 \tag{5.122}$$

Typically, non-holonomic constraint conditions for multibody systems are such, that

$$\Phi = \mathbf{A}(\mathbf{q}, t) \, \dot{\mathbf{q}} + \mathbf{B}(\mathbf{q}, t) \tag{5.123}$$

The acceleration-based augmented Lagrangian formulation for this type of constraints is

$$\mathbf{M} \, \ddot{\mathbf{q}} = \mathbf{Q} + L_{\mathbf{q}} - \dot{\mathbf{M}} \, \dot{\mathbf{q}} - \mathbf{A}^{T} \alpha \left(\dot{\Phi} + \mu \, \Phi \right) - \mathbf{A}^{T} \lambda^{*} \tag{5.124}$$

In order to obtain the canonical equations, we follow a procedure similar to that used for the holonomic case and establish the following canonical transformation

$$\mathbf{p} = \mathbf{M} \, \dot{\mathbf{q}} + \mathbf{A}^{T} \alpha \, (\Phi + \mu \!\! \int_{t_{0}}^{t} \Phi \, d\tau) + \mathbf{A}^{T} \sigma^{*} \tag{5.125a}$$

which indicates that a better stabilization of the canonical momenta may be achieved by considering one penalty term proportional to the constraint violation and the other term proportional to its integral. The differentiation of (5.125a) and posterior substitution into (5.124) leads to the second set of canonical equations:

$$\dot{p} = Q + L_q + \dot{A}^T \alpha (\Phi + \mu \int_{t_o}^{t} \Phi \, d\tau) + \dot{A}^T \sigma^* \qquad (5.125b)$$

which along with (5.125a) constitutes a set of $2n$ first order ordinary differential equations in the unknowns p, q, and σ^*. Again, only n equations need to be solved at each function's evaluation. The multipliers are given by

$$\sigma_{i+1}^* = \sigma_i^* + (\Phi + \mu \int_{t_o}^{t} \Phi \, d\tau)_{i+1} , \quad i = 0, 1, 2, \dots \qquad (5.126)$$

with $\sigma_0^* = 0$ for the first iteration.

References

Bae, D.S. and Won, Y.S., "A Hamiltonian Equation of Motion for Real-time Vehicle Simulation", *Advances in Design and Automation 1990*, Vol. 2, pp. 151-157, ASME Press, (1990).

Bae, D.S. and Yang, S.M., "A Stabilization Method for Kinematic and Kinetic Constraint Equations", *Real-Time Integration Methods for Mechanical System Simulation*, NATO ASI Series, Vol. 69, pp. 209-232, Springer-Verlag, (1990).

Baumgarte, J., "Stabilization of Constraints and Integrals of Motion in Dynamical Systems", *Computer Methods in Applied Mechanics and Engineering*, Vol. 1, pp. 1-16, (1972).

Bayo, E., García de Jalón, J., and Serna, M.A., "A Modified Lagrangian Formulation for the Dynamic Analysis of Constrained Mechanical Systems", *Computer Methods in Applied Mechanics and Engineering*, Vol. 71, pp. 183-195, (1988).

Bayo, E., García de Jalón, J., Avello, A., and Cuadrado, J., "An Efficient Computational Method for Real Time Multibody Dynamic Simulation in Fully Cartesian Coordinates", *Computer Methods in Applied Mechanics and Engineering*, Vol. 92, pp. 377-395, (1991).

Bayo, E. and Avello, A., "Singularity Free Augmented Lagrangian Algorithms for Constraint Multibody Dynamics", to appear in the *Journal of Nonlinear Dynamics*, (1993).

Brenan, K.E, Campbell, S.L., and Petzold, L.R., *The Numerical Solution of Initial Value Problems in Differential-Algebraic Equations*, Elsevier Science Publishing, (1989).

Chang, C.O. and Nikravesh, P.E., "An Adaptive Constraint Violation Stabilization Method for Dynamic Analysis of Mechanical Systems", *ASME Journal of Mechanisms, Transmissions and Automation in Design*, Vol. 107, pp. 488-492, (1985).

García de Jalón, J., Avello, A., Jiménez, J.M., Martín, F., and Cuadrado, J., "Real Time Simulation of Complex 3-D Multibody Systems with Realistic Graphics", *Real-Time Integration Methods for Mechanical System Simulation*, NATO ASI Series, Vol. 69, pp. 265-292, Springer-Verlag, (1990).

Gear, C.W., *Numerical Initial Value Problems in Ordinary Differential Equations*, Prentice-Hall, (1971).

Goldstein H., *Classical Mechanics*, 2nd edition, Addison-Wesley, (1980).

Haug, E.J., *Computer–Aided Kinematics and Dynamics of Mechanical Systems, Volume I: Basic Methods*, Allyn and Bacon, (1989).

Jerkovsky, W., "The Structure of Multibody Dynamic Equations", *Journal of Guidance and Control*, Vol. 1, pp. 173-182, (1978).

Kamman, J.W. and Huston, R.L., "Dynamics of Constrained Multibody Systems", *ASME Journal of Applied Mechanics*, Vol. 51, pp. 899-903, (1984).

Kim, S.S. and Vanderploeg, M.J., "A General and Efficient Method for Dynamic Analysis of Mechanical Systems Using Velocity Transformations", *ASME Journal of Mechanisms, Transmissions and Automation in Design*, Vol. 108, pp. 176-182, (1986a).

Kim, S.S. and Vanderploeg, M.J., "QR Decomposition for State Space Representation of Constrained Mechanical Dynamic Systems", *ASME Journal on Mechanisms, Transmissions and Automation in Design*, Vol. 108, pp. 183-188, (1986b).

Kurdila, A. J. and Narcowich F.J., "Sufficient Conditions for Penalty Formulation Methods in Analytical Dynamics", to appear in *Computational Mechanics*, (1993).

Lankarani, H.M. and Nikravesh, P.E., "Application of the Canonical Equations of Motion in Problems of Constrained Multibody Systems with Intermittent Motion", *Advances in Design Automation 1988*, DE-Vol. 14, pp. 417-423, edited by S.S. Rao, ASME Press, (1988).

Mani, N.K., Haug, E.J., and Atkinson, K.E., "Application of Singular Value Decomposition for Analysis of Mechanical System Dynamics", *ASME Journal on Mechanisms, Transmissions and Automation in Design*, Vol. 107, pp. 82-87, (1985).

Nikravesh, P.E., "Some Methods for Dynamic Analysis of Constrained Mechanical Systems: A Survey", *Computer-Aided Analysis and Optimization of Mechanical System Dynamics*, ed. by E.J. Haug, Springer-Verlag, pp. 351-368, (1984).

Nikravesh, P.E. and Gim, G., "Systematic Construction of the Equations of Motion for Multibody Systems Containing Closed Kinematic Loops", *Advances in Design Automation 1989*, Vol. 3, pp. 27-33, ASME Press, (1989).

Oden, J.T., *Finite Elements. A Second Course, Volume II*, Chapter 3, Prentice-Hall, (1983).

Park, T.W. and Haug, E.J., "A Hybrid Numerical Integration Method for Machine Dynamic Simulation", *ASME Journal of Mechanisms, Transmissions and Automation in Design*, Vol. 108, pp. 211-216, (1986).

Pars, L.A., *A Treatise of Analytical Dynamics*, William Heineman Ltd., (1965).

Paul, B., "Analytical Dynamics of Mechanisms - A Computer-Oriented Overview", *Mechanism and Machine Theory*, Vol. 10, pp. 481-507, (1975).

Serna, M.A., Avilés, R., and García de Jalón, J., "Dynamic Analysis of Planar Mechanisms with Lower-Pairs in Basic Coordinates", *Mechanism and Machine Theory*, Vol. 17, pp. 397-403, (1982).

Shampine, L. and Gordon, M., *Computer Solution of ODE. The Initial Value Problem*, Freeman, (1975).

Steigerwald, M.F., "BDF Methods for DAEs in Multibody Dynamics: Shortcomings and Improvements", in *Real-Time Integration Methods for Mechanical System Simulation*, NATO ASI Series, Vol. 69, pp. 345-352, Springer-Verlag, (1990).

Unda, J., García de Jalón, J., Losantos, F., and Enparantza, R., "A Comparative Study of Some Different Formulations of the Dynamic Equations of Constrained Mechanical Systems", *ASME Journal of Mechanisms, Transmissions and Automation in Design*, Vol. 109, pp. 466-474, (1987).

Vanderplaats, G.N., *Numerical Optimization Techniques for Engineering Design: with Applications*, McGraw-Hill, (1984).

Wehage, R.A. and Haug, E.J., "Generalized Coordinate Partitioning for Dimension Reduction in Analysis of Constrained Dynamic Systems", *ASME Journal of Mechanical Design*, Vol. 104, pp. 247-255, (1982).

Problems

5/1   Using natural coordinates, write the equations of motion of the slider-crank mechanism of the figure with: a) Lagrange multipliers b) Penalty formulation. Assume that the mass is uniformly distributed and the center is located at the middle of each element. Also $L_2=L_3/2=L$, $m_2=1$, $m_3=2$, and $m_4=1$.

5/2   Form the matrix **R** of the mechanism of Problem 5/1 and find the equations of motion: a) in dependent coordinates using $q^T=(x_1,y_1,x_2)$ and equation (5.17); and b) in independent coordinates using $z_1=x_2$ and equation (5.67). For case a) discuss the regularization process to avoid the singular position when $L_3=L_2$ and both slider and crank are in the vertical position.

5/3   Repeat Problem 5/2 using the canonical formulation (equations (5.111) and (5.112)) in independent coordinates using $z_1=x_2$.

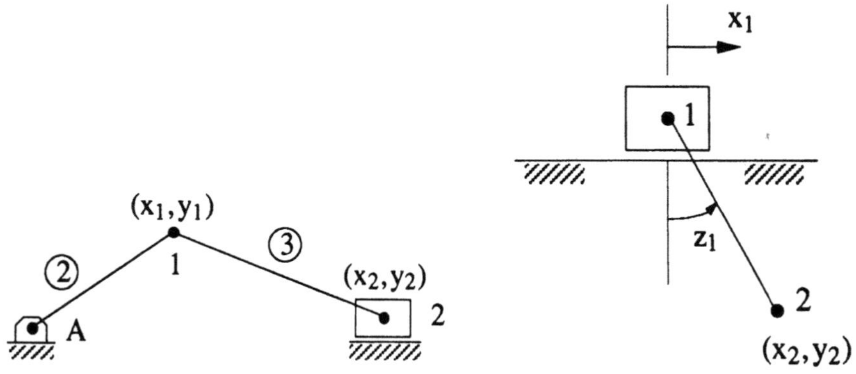

Figure P5/1.   Figure P5/4.

5/4 Find the equations of motion of the mechanism of the figure when the coordinate x_1 is kinematically imposed. Use: a) Dependent coordinates with Lagrange multipliers; b) independent coordinates choosing the angle z_1 as the independent variable.

5/5 Solve Problem 5/3 using the canonical formulation of Section 5.4 with dependent coordinates and the penalty formulation for the constraint equations.

5/6 Solve for the equations of motion of the mechanism shown in the figure, using the velocity transformation methods of Section 5.3. Open the loop at joint 2 where indicated and then apply the closure conditions using the Lagrange multiplier approach with Baumgarte stabilization.

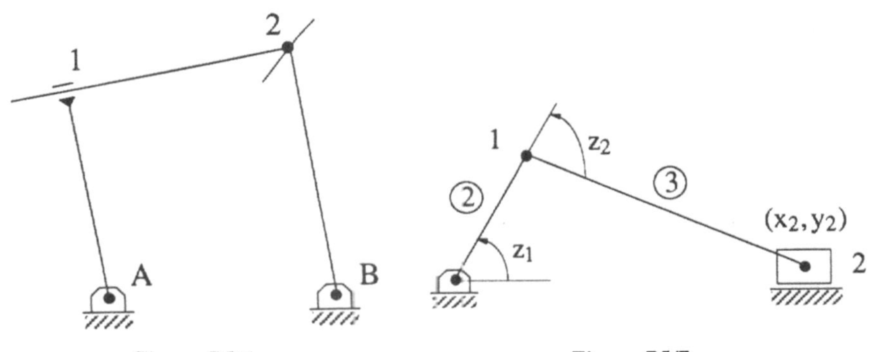

Figure P5/6.   Figure P5/7.

5/7 Form the differential equations of motion of the mechanism of the figure, using velocity transformations and the penalty formulation for the closure condition $y_2=0$.

5/8 Derive the equations of motion of a coin that rolls over a flat surface with no slipping, using the natural coordinates that are shown in the figure. The nonholonomic constraint condition is given by $v_P=0$ (3 equations).

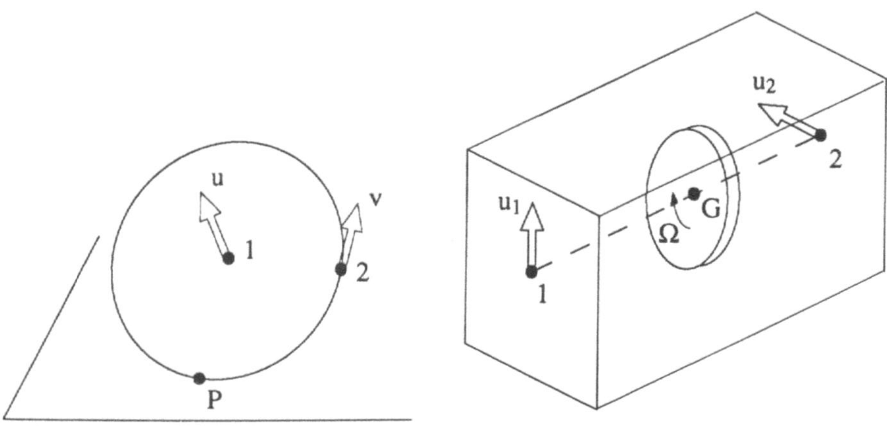

Figure P5/8.   Figure P5/9.

5/9   Consider a satellite that is modeled with two points 1 and 2, and two unit vectors $\mathbf{u}_1$ and $\mathbf{u}_2$. This satellite contains a high speed rotor that rotates at a constant relative angular velocity Ω. Model the effect of the rotor by means of an equivalent set of forces.

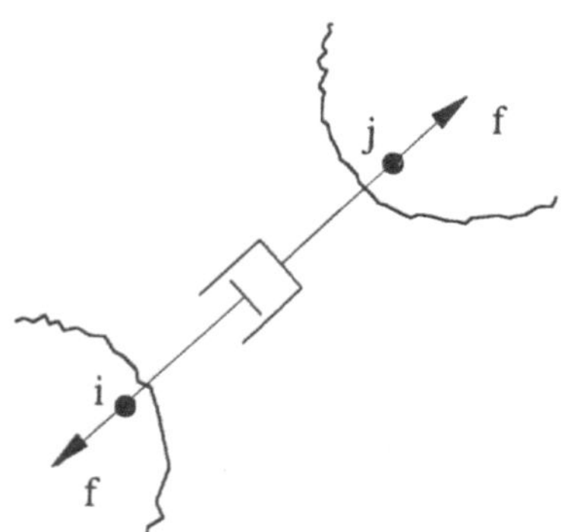

Figure P5/10.

5/10   The controlled mechanical system in the figure consists of two rods with two revolute joints that move on a vertical plane. Joint A is torque free. Joint 2 has an actuator that applies two opposite torques on both bars, so as to keep the system in the vertical position ($Y_1 = Y_2 = 0$). Write the differential equations of motion using Y_1 and Y_2 as independent coordinates.

6
Static Equilibrium Position and Inverse Dynamics

This chapter deals with two important multibody problems related to forces: the determination of the static equilibrium position and the solution of the inverse dynamics. In both cases, it is assumed that the motion, that is, velocities and accelerations, is known, and, in the former case also, that the motion does not exist. At least, there is not relative motion with respect to the reference frame on which the problem is to be solved.

In Section 6.1 we will consider the static equilibrium position problem. This problem consists in determining the position of the multibody system, when all the acting balanced forces (gravitational and external, forces in the springs, and external reactions) are known. The static equilibrium condition requires that the total potential energy for the system be at a minimum; that is, the sum of the gravitational potential energy of the elements, the elastic potential energy of the springs, and the potential energy of external forces has to satisfy a minimum condition. It is not always easy to determine the static equilibrium position by inspection or by means of simple calculations. Generally, the solution of this problem leads to a system of nonlinear equations which need to be solved iteratively.

Section 6.2 deals with the inverse dynamics that solves for the driving forces and joint reactions necessary to produce a specified motion. The inverse dynamics requires a previous knowledge of positions, velocities, accelerations, and external forces such as weight and forces in springs and dampers and involves the finding of an unknown driving force for each one of the kinematically guided input elements. It involves finding an unknown reaction force also for each one of the degrees of freedom constrained by the kinematic joints.

The solution of the inverse dynamics has different important applications. It permits the determination of the forces to which the multibody system is subjected in both the dynamic and the kinematic simulation problems. The inverse dynamics is also of special importance when trying to control a multibody system, so that it follows a specified trajectory in either the Cartesian or joint space. The inverse dynamics solves for the forces that the motors must apply in order to achieve this movement.

6.1 Static Equilibrium Position

The *static equilibrium position problem* consists of determining the equilibrium position of a multibody system which may contain elastic, rectilinear or torsion, linear or nonlinear springs. The multibody system is subjected to the action of different external forces such as: its own weight, centrifugal forces, or, in general, any other type of inertial forces corresponding to a known field of accelerations.

The static position problem is typically nonlinear, since the final equilibrium position is not known with sufficient accuracy to formulate the equilibrium equations about this position. At the final equilibrium position, not only should the external forces and reactions be in equilibrium at each element and at the whole system but also the kinematic constraint equations be satisfied. Therefore, all the methods for solving this problem should simultaneously impose both types of conditions: the equilibrium of forces and the fulfillment of constraint equations.

The method of computing the generalized external and spring introduced forces **Q**, from the applied forces and/or torques have been seen in Section 4.3. Included in this chapter are some detailed descriptions for evaluating the *potential energy* and the *work* of these forces, an essential condition for establishing the solution methods, that are based on the minimum potential energy (two methods) and on the theorem of virtual power (one method).

6.1.1 Computation of Derivatives of Potential Energy

We will develop in the next sections the solution methods for the static equilibrium position problem. First, we will study two methods based on the minimum condition for the total potential energy of the system. The total potential energy minimum condition is a necessary and sufficient condition for the stability of the solution.

We first consider the formulation of the minimum potential energy condition, introducing the kinematic constraints by means of Lagrange multipliers. The solution of the resulting system of nonlinear equations by Newton-Raphson iteration requires the first and second derivatives of the total potential energy which was obtained from the expressions developed in Section 4.3. This total potential energy contains terms that can originate from different sources:

a) Potential of concentrated external forces.
b) Potential of external torques.
c) Potential energy of translational springs.
d) Potential energy of rotational springs.
e) Potential of gravitational forces and forces coming from known acceleration fields.

The way of calculating the first and second derivatives with respect to the position variables of each one of these terms will be developed below.

6.1.1.1 Derivatives of the Potential of External Forces

The derivative of the potential of a force applied at point P (equation (4.95)) is

$$\frac{\partial V}{\partial \mathbf{q}} = - \mathbf{C}_P^T \, \mathbf{f}(\mathbf{q}) \tag{6.1}$$

If the external forces are constant, the second derivative of its potential function is also zero, provided the element has at least two points and two non-coplanar unit vectors; so as to guarantee that $\mathbf{C}_P$ exists and is constant. If the forces depend on the position, it is necessary for one to carry out the differentiation taking the specific case in question into account.

Example   6.1

Determine the first derivative of the expression $\mathbf{r}_P = \mathbf{C}_P \, \mathbf{q}^e$, when the element has only two points and one unit vector; thus matrix $\mathbf{C}_P$ is not being constant.

If the element has two points and one unit vector (or an equivalent structure), the derivative of $\mathbf{C}_P$ is not zero, but can be calculated as indicated below. The position of point P can be expressed in this case as:

$$\mathbf{r}_P = \mathbf{r}_i + \mathbf{A} \, (\bar{\mathbf{r}}_P - \bar{\mathbf{r}}_i) \tag{i}$$

The rotation matrix $\mathbf{A}$ may be formed from the two points $\mathbf{r}_i$ and $\mathbf{r}_j$, and a non-coplanar vector defined by the cross product of $(\mathbf{r}_i - \mathbf{r}_j)$ and $\mathbf{u}$ in the form:

$$\mathbf{A} = \left[\mathbf{r}_i - \mathbf{r}_j \,|\, \mathbf{u} \,|\, (\mathbf{r}_i - \mathbf{r}_j) \wedge \mathbf{u}\right] \bar{\mathbf{X}}^{-1} \tag{ii}$$

where $\bar{\mathbf{X}}$ is a constant matrix defined as follows:

$$\bar{\mathbf{X}} = \left[\bar{\mathbf{r}}_i - \bar{\mathbf{r}}_j \,|\, \bar{\mathbf{u}} \,|\, (\bar{\mathbf{r}}_i - \bar{\mathbf{r}}_j) \wedge \bar{\mathbf{u}}\right] \tag{iii}$$

It may be seen that this matrix is defined by the points and vectors of the element in local coordinates. Introducing the vector $\mathbf{a}$ as

$$\mathbf{a} = \bar{\mathbf{X}}^{-1} \, (\bar{\mathbf{r}}_P - \bar{\mathbf{r}}_i) \tag{iv}$$

expression (i) becomes

$$\mathbf{r}_P = \mathbf{r}_i + a_1 \, (\mathbf{r}_i - \mathbf{r}_j) + a_2 \, \mathbf{u} + a_3 \, (\mathbf{r}_i - \mathbf{r}_j) \wedge \mathbf{u} =$$

$$= \left[(1 + a_1) \, \mathbf{I}_3 \,|\, -a_1 \, \mathbf{I}_3 \,|\, a_2 \, \mathbf{I}_3 + a_3 \, \widetilde{\mathbf{R}}_{ij}\right] \begin{Bmatrix} \mathbf{r}_i \\ \mathbf{r}_j \\ \mathbf{u} \end{Bmatrix} \equiv \mathbf{C}_P(\mathbf{q}) \, \mathbf{q}^e \tag{v}$$

where $\widetilde{\mathbf{R}}_{ij}$ is the skew-symmetric matrix related with the cross product of $(\mathbf{r}_i - \mathbf{r}_j)$, and $\mathbf{I}_3$ is the (3×3) unit matrix. The derivative of expression (v) with respect to vector $\mathbf{q}$ is

$$\frac{\partial \mathbf{r}_P}{\partial \mathbf{q}} = \left[(1 + a_1) \, \mathbf{I}_3 \,|\, -a_1 \, \mathbf{I}_3 \,|\, a_2 \, \mathbf{I}_3 + a_3 \, \widetilde{\mathbf{R}}_{ij}\right] + a_3 \left[-\widetilde{\mathbf{u}} \,|\, \widetilde{\mathbf{u}} \,|\, 0\right] \tag{vi}$$

6.1.1.2 Derivatives of the Potential of External Torques

Where one is dealing with an external torque applied perpendicularly to a unit vector $\mathbf{u}_f$, the use of (4.101) directly leads to

$$\frac{\partial V}{\partial \mathbf{q}} = -\,(\mathbf{C}_i^T - \mathbf{C}_{i+\mathbf{u}\,f}^T)\,\mathbf{f} \tag{6.2}$$

If the element has at least two points and two non coplanar unit vectors and, if the external torques are constant, the second derivative of its potential function is zero. If the torques are position-dependent, it is necessary to take into account their specific dependency expression.

6.1.1.3 Derivatives of the Potential Energy of Translational Springs

a) Spring connecting two basic points. When the spring connects two basic points, the potential is given by equations (4.102), (4.104), and (4.106) and its derivative with respect to the position vector $\mathbf{q}$ is:

$$\frac{\partial V}{\partial \mathbf{q}} = \mathbf{Q} = \frac{k(L)\,(L - L_o)}{L}\begin{pmatrix} x_i - x_j \\ y_i - y_j \\ z_i - z_j \\ x_j - x_i \\ y_j - y_i \\ z_j - z_i \end{pmatrix} \tag{6.3}$$

That can be written as:

$$\frac{\partial V}{\partial \mathbf{q}} = k(L)\left(1 - \frac{L_o}{L}\right)\begin{Bmatrix} \mathbf{r}_i - \mathbf{r}_j \\ \mathbf{r}_j - \mathbf{r}_i \end{Bmatrix} =$$

$$= k(L)\left(1 - \frac{L_o}{L}\right)\begin{bmatrix} 1 & 0 & 0 & -1 & 0 & 0 \\ 0 & 1 & 0 & 0 & -1 & 0 \\ 0 & 0 & 1 & 0 & 0 & -1 \\ -1 & 0 & 0 & 1 & 0 & 0 \\ 0 & -1 & 0 & 0 & 1 & 0 \\ 0 & 0 & -1 & 0 & 0 & 1 \end{bmatrix}\begin{Bmatrix} x_i \\ y_i \\ z_i \\ x_j \\ y_j \\ z_j \end{Bmatrix} \tag{6.4}$$

Before differentiating the previous expression with respect to $\mathbf{q}$, it is convenient to differentiate the scalar term that appears in it:

$$\frac{\partial}{\partial \mathbf{q}}\left[k(L)\left(1 - \frac{L_o}{L}\right)\right] = \frac{\partial}{\partial L}\left[k(L)\left(1 - \frac{L_o}{L}\right)\right]\frac{\partial L}{\partial \mathbf{q}} =$$

$$= \left[\frac{\partial k(L)}{\partial L}\left(1 - \frac{L_o}{L}\right) + k(L)\frac{L_o}{L^2}\right]\frac{\partial L}{\partial \mathbf{q}} \tag{6.5}$$

Taking into account that L is

$$L = \left((x_i-x_j)^2 + (y_i-y_j)^2 + (z_i-z_j)^2\right)^{1/2} \tag{6.6}$$

and it is easy to see that its derivative with respect to the dependent coordinates is

$$\frac{\partial L}{\partial \mathbf{q}} = \frac{1}{L}
\begin{bmatrix}
1 & 0 & 0 & -1 & 0 & 0 \\
0 & 1 & 0 & 0 & -1 & 0 \\
0 & 0 & 1 & 0 & 0 & -1 \\
-1 & 0 & 0 & 1 & 0 & 0 \\
0 & -1 & 0 & 0 & 1 & 0 \\
0 & 0 & -1 & 0 & 0 & 1
\end{bmatrix}
\begin{pmatrix} x_i \\ y_i \\ z_i \\ x_j \\ y_j \\ z_j \end{pmatrix} \tag{6.7}$$

Now, we differentiate equation (6.4) with respect to $\mathbf{q}$. Taking into account equations (6.5) and (6.7), we obtain

$$\frac{\partial^2 V}{\partial \mathbf{q}^2} = k(L)\left(1 - \frac{L_o}{L}\right)\mathbf{J} +$$

$$+ \left[\frac{\partial k(L)}{\partial L}\left(1 - \frac{L_o}{L}\right) + k(L)\frac{L_o}{L^2}\right]\mathbf{J}
\begin{pmatrix} x_i \\ y_i \\ z_i \\ x_j \\ y_j \\ z_j \end{pmatrix}
\left\{ x_i\, y_i\, z_i\, x_j\, y_j\, z_j \right\}\mathbf{J} \tag{6.8}$$

where $\mathbf{J}$ is the matrix defined in equation (6.7).

b) Spring with its length defined as a dependent coordinate. When the distance between the extremes of the spring is defined as generalized coordinate s, the potential energy is given by equation (4.110). Its derivative with respect to the distance s is:

$$\frac{\partial V}{\partial \mathbf{q}} = \frac{\partial V}{\partial s} = k(s)\,(s-s_o) \tag{6.9}$$

and its second derivative with respect to the distance s is

$$\frac{\partial^2 V}{\partial \mathbf{q}^2} = \frac{\partial^2 V}{\partial s^2} = k(s) + \frac{\partial k(s)}{\partial s}\,(s-s_o) \tag{6.10}$$

The simplicity of this expression resulting from the use of mixed coordinates is noteworthy, as compared with the expressions developed previously for translational springs connecting points without mixed coordinates.

c) Spring connecting any two points. Finally, when the spring connects two points that are not basic points, the potential energy is given by equation (4.122). Its derivative with respect to the generalized coordinates vector $\mathbf{q}$ is:

$$\frac{\partial V}{\partial \mathbf{q}} = k(L)\left(1 - \frac{L_o}{L}\right)\begin{bmatrix} \mathbf{C}_1^T \mathbf{C}_1 & -\mathbf{C}_1^T \mathbf{C}_2 \\ -\mathbf{C}_2^T \mathbf{C}_1 & \mathbf{C}_2^T \mathbf{C}_2 \end{bmatrix}\begin{Bmatrix} \mathbf{q}_1 \\ \mathbf{q}_2 \end{Bmatrix} =$$
$$= k(L)\left(1 - \frac{L_o}{L}\right)\mathbf{C}_{12}\begin{Bmatrix} \mathbf{q}_1 \\ \mathbf{q}_2 \end{Bmatrix} \tag{6.11}$$

In accordance with expression (6.11), it can be verified that

$$L^2 = \begin{Bmatrix} \mathbf{q}_1^T & \mathbf{q}_2^T \end{Bmatrix}\mathbf{C}_{12}\begin{Bmatrix} \mathbf{q}_1 \\ \mathbf{q}_2 \end{Bmatrix} \tag{6.12}$$

Differentiation of this expression with respect to vector $\mathbf{q}$ yields

$$2L\frac{\partial L}{\partial \mathbf{q}} = 2\,\mathbf{C}_{12}\begin{Bmatrix} \mathbf{q}_1 \\ \mathbf{q}_2 \end{Bmatrix} \tag{6.13}$$

and, differentiating expression (6.11) with respect to $\mathbf{q}$ and substituting (6.13):

$$\frac{\partial^2 V}{\partial \mathbf{q}^2} = k(L)\left(1 - \frac{L_o}{L}\right)\mathbf{C}_{12} +$$
$$+ \left(\frac{\partial k}{\partial L}\left(1 - \frac{L_o}{L}\right) + k(L)\frac{L_o}{L^3}\right)\mathbf{C}_{12}\begin{Bmatrix} \mathbf{q}_1 \\ \mathbf{q}_2 \end{Bmatrix}\begin{Bmatrix} \mathbf{q}_1^T & \mathbf{q}_2^T \end{Bmatrix}\mathbf{C}_{12} \tag{6.14}$$

which is the final expression of the second derivative. If the elements to which points $\mathbf{r}_1$ and $\mathbf{r}_2$ belong have less than two unit vectors, or they are coplanar, matrix $\mathbf{C}_{12}$ is not constant, and the previous derivation process will be different.

6.1.1.4 Derivatives of the Potential Energy of Rotational Springs

The potential energy of a rotational spring corresponding to a revolute joint, whose axis is defined by a unit vector $\mathbf{u}$ and whose angle is defined as generalized coordinate Ψ, is defined by equation (4.128). Its derivative with respect to the angle variable is

$$\frac{\partial V}{\partial \mathbf{q}} = \frac{\partial V}{\partial \Psi} = M(\Psi) = k(\Psi)(\Psi - \Psi_0) \tag{6.15}$$

where M is the torque and Ψ is the angle formed by the two elements, calculated by means of the scalar and cross products of vectors.

Again, differentiating this equation with respect to Ψ yields

$$\frac{\partial^2 V}{\partial \mathbf{q}^2} = \frac{\partial^2 V}{\partial \Psi^2} = \frac{\partial k(\Psi)}{\partial \Psi}(\Psi - \Psi_0) + k(\Psi) \tag{6.16}$$

6.1.1.5 Derivatives of the Potential Energy of Gravitational Forces

a) Weight forces. The derivative of the potential gravitational energy can be obtained from equation (4.132) as follows:

$$\frac{\partial V}{\partial \mathbf{q}} = - m \; \mathbf{C}_{\mathrm{G}}^{\mathrm{T}} \, \mathbf{g} \tag{6.17}$$

where $\mathbf{C}_{\mathrm{G}}$ is a constant matrix if the element has two points and two non-coplanar unit vectors. Here, all the terms of this expression are constant, and the second derivative is zero.

b) Forces due to a known field of accelerations. When the known inertia forces originate from a known field of accelerations, the potential is given by equation (4.138), and its derivative is:

$$\frac{\partial V}{\partial \mathbf{q}} = - \mathbf{Q}_{\mathrm{in}}^{e} = - \mathbf{M}^{e} \, \ddot{\mathbf{q}}^{e} \tag{6.18}$$

The second derivative depends on the particular expression of the known acceleration field.

6.1.2 Method of the Lagrange Multipliers

Once the different terms of the potential function and their derivatives have been calculated, the total value of the potential can be determined as the sum of the potentials due to the springs, the gravitational, and the external forces. In order to apply the Lagrange multiplier method, and before imposing the minimum condition to the total potential energy, it is necessary for one to add the virtual work performed by the Lagrange multipliers in the form of an energy term as:

$$V^{*} = \Sigma V + \Phi(\mathbf{q})^{\mathrm{T}} \lambda \tag{6.19}$$

The minimum condition of this function is obtained by making its first derivative with respect to the natural coordinates $\mathbf{q}$ equal to zero:

$$\frac{\partial V^{*}}{\partial \mathbf{q}} = \Sigma \frac{\partial V}{\partial \mathbf{q}} + \Phi_{\mathbf{q}}^{\mathrm{T}}(\mathbf{q}) \, \lambda = 0 \tag{6.20}$$

where $\Phi_{\mathbf{q}}$ is the Jacobian matrix of the constraint equations. The derivatives of the different potential terms have been calculated previously. Equation (6.20), that is, the mathematical condition of the minimum of the total potential energy, leads to a set of nonlinear equations in the dependent coordinates $\mathbf{q}$ and in the Lagrange multipliers λ. One can symbolically represent it as

$$\Psi(\mathbf{q}, \lambda) = 0 \tag{6.21}$$

In order for this system to have a number of equations equal to all the unknowns $\mathbf{q}$ and λ, it is necessary to add the constraint equations, resulting in:

$$\left\{ \begin{array}{c} \Psi(\mathbf{q}, \lambda) \\ \Phi(\mathbf{q}) \end{array} \right\} = 0 \tag{6.22}$$

By applying the Newton-Raphson method to the nonlinear equations system (6.22):

$$\begin{bmatrix} \Psi_q & \Psi_\lambda \\ \Phi_q & 0 \end{bmatrix}_i \left(\begin{Bmatrix} q \\ \lambda \end{Bmatrix}_{i+1} - \begin{Bmatrix} q \\ \lambda \end{Bmatrix}_i \right) = - \begin{Bmatrix} \Psi \\ \Phi \end{Bmatrix}_i \tag{6.23}$$

In accordance with equation (6.19),

$$\Psi_\lambda = \Phi_q^T \tag{6.24}$$

and, therefore, it results in the following set of symmetric linear equations:

$$\begin{bmatrix} \Psi_q & \Phi_q^T \\ \Phi_q & 0 \end{bmatrix}_i \left(\begin{Bmatrix} q \\ \lambda \end{Bmatrix}_{i+1} - \begin{Bmatrix} q \\ \lambda \end{Bmatrix}_i \right) = - \begin{Bmatrix} \Psi \\ \Phi \end{Bmatrix}_i \tag{6.25}$$

6.1.3 Penalty Formulation

This method can be considered as a variation of the method introduced in the previous section, with the difference that the constraint equations are introduced by means of *penalty functions* instead of the Lagrange multipliers. The advantages of this method are that it reduces the number of unknowns in the problem, it has less convergence problems, and it also permits an easy calculation of the reactions at the joints.

Expression (6.19), which gives the potential function to be minimized by using the Lagrange multipliers method, is now replaced by the expression

$$V^{**} = \Sigma V + \frac{1}{2} \Phi(q)^T \alpha \, \Phi(q) \tag{6.26}$$

where V represents the potentials of the applied external forces, springs, and the gravitational forces. Matrix α is a diagonal one, whose elements are the penalty coefficients of each one of the constraint equations.

In order to minimize the potential function defined in expression (6.26), it is necessary to cancel out the first derivative with respect to the dependent coordinates q:

$$\frac{\partial V^{**}}{\partial q} = \Sigma \frac{\partial V}{\partial q} + \Phi_q^T \alpha \, \Phi(q) = 0 \tag{6.27}$$

This yields a system of nonlinear equations that may be solved by Newton-Raphson method. It is again necessary to differentiate the potential V^{**} with respect to q. The derivatives of the potential terms in equation (6.27) have already been calculated in the previous sections. Therefore, special attention must now be paid to the derivative with respect to q of the last term of this equation. Consequently,

$$\frac{\partial}{\partial \mathbf{q}}\left(\Phi_{\mathbf{q}}^{T} \alpha \, \Phi(\mathbf{q})\right) = \Phi_{\mathbf{q}}^{T} \alpha \, \Phi_{\mathbf{q}} + \frac{\partial}{\partial \mathbf{q}} \, \Phi_{\mathbf{q}}^{T} \alpha \, \Phi(\mathbf{q}) \tag{6.28}$$

The derivative of $\Phi_{\mathbf{q}}^{T}$ with respect to $\mathbf{q}$ can be calculated column by column, as constraint equation to constraint equation. The column corresponding to a constant angle condition between segment $(i\text{-}j)$ and unit vector $\mathbf{u}$ can be considered here:

$$\left(\mathbf{r}_{j} - \mathbf{r}_{i}\right)^{T} \mathbf{u} - constant = 0 \tag{6.29}$$

This equation can also be written as:

$$\frac{1}{2}\left\{\mathbf{r}_{i}^{T}\ \mathbf{r}_{j}^{T}\ \mathbf{u}^{T}\right\}\begin{bmatrix} 0 & 0 & -I_{3} \\ 0 & 0 & I_{3} \\ -I_{3} & I_{3} & 0 \end{bmatrix}\begin{Bmatrix} \mathbf{r}_{i} \\ \mathbf{r}_{j} \\ \mathbf{u} \end{Bmatrix} - constant = 0 \tag{6.30}$$

By differentiating with respect to $\mathbf{q}$, we obtain

$$\Phi_{iq}^{T} = \begin{bmatrix} 0 & 0 & -I_{3} \\ 0 & 0 & I_{3} \\ -I_{3} & I_{3} & 0 \end{bmatrix}\begin{Bmatrix} \mathbf{r}_{i} \\ \mathbf{r}_{j} \\ \mathbf{u} \end{Bmatrix} \tag{6.31}$$

By differentiating again with respect to $\mathbf{q}$, we obtain the contribution to the last term of expression (6.28) equal to

$$\frac{\partial \Phi_{iq}^{T}}{\partial \mathbf{q}} \alpha_{i}\, \Phi_{i}(\mathbf{q}) = \alpha_{i}\, \Phi_{i}(\mathbf{q}) \begin{bmatrix} 0 & 0 & -I_{3} \\ 0 & 0 & I_{3} \\ -I_{3} & I_{3} & 0 \end{bmatrix} \tag{6.32}$$

The contributions of the other constraint equations could be calculated in a similar way.

Once the second derivatives of the potential function have been evaluated, the process continues with the typical scheme of the Newton-Raphson method. This is similar to expression (6.23),

$$\Psi_{\mathbf{q}\,i}\,(\mathbf{q}_{i+1} - \mathbf{q}_{i}) = -\,\Psi(\mathbf{q})_{i} \tag{6.33}$$

where

$$\Psi_{\mathbf{q}} \equiv \left[\frac{\partial^{2} V^{**}}{\partial \mathbf{q}^{2}}\right] \tag{6.34}$$

6.1.4 Virtual Power Method

6.1.4.1 Theoretical Development

The application of the virtual power method is a third alternative for solving the static equilibrium position problem. This method is based on the fact that at the final equilibrium position, the virtual power of all the forces that act on the multibody system should be zero. To set out this method, it is necessary for one to start by formulating $\mathbf{Q}$ with all the forces acting on the system and which generally will be position-dependent.

Let $\dot{\mathbf{q}}^*$ be a virtual dependent velocity vector related to the virtual independent velocities by means of the expression:

$$\dot{\mathbf{q}}^* = \mathbf{R}\,\dot{\mathbf{z}}^* \tag{6.35}$$

By imposing the null virtual power condition to the set of all the applied forces and taking the arbitrary nature of virtual independent velocities into account, we obtain

$$\Psi(\mathbf{q}) \equiv \mathbf{R}(\mathbf{q})^{\mathrm{T}}\,\mathbf{Q}(\mathbf{q}) = 0 \tag{6.36}$$

which constitutes the system of nonlinear equilibrium equations. The static equilibrium position is defined by the vector $\mathbf{q}$ that satisfies equation (6.36).

In order to solve the nonlinear equations system (6.36), one can resort to Newton-Raphson method. One should remember that the number of equations of the system (6.36) is equal to the number of degrees of freedom of the multibody system and not equal to the number of dependent coordinates. Equation (6.36) must be modified to put it in terms of independent coordinates. For this purpose, the following expressions are to be used later:

$$\delta\mathbf{q} = \mathbf{R}\,\delta\mathbf{z} \tag{6.37}$$

$$\dot{\mathbf{q}} = \mathbf{R}\,\dot{\mathbf{z}} \tag{6.38}$$

To apply the Newton-Raphson method to (6.36), it is necessary for one to calculate the Jacobian matrix of the system, which will be given by:

$$\frac{\partial\Psi}{\partial\mathbf{z}} = \frac{\partial}{\partial\mathbf{z}}(\mathbf{R}^{\mathrm{T}}\mathbf{Q}) = \frac{\partial}{\partial\mathbf{q}}(\mathbf{R}^{\mathrm{T}}\mathbf{Q})\frac{\partial\mathbf{q}}{\partial\mathbf{z}} = \left(\frac{\partial\mathbf{R}^{\mathrm{T}}}{\partial\mathbf{q}}\mathbf{Q} + \mathbf{R}^{\mathrm{T}}\frac{\partial\mathbf{Q}}{\partial\mathbf{z}}\right)\frac{\partial\mathbf{q}}{\partial\mathbf{z}} \tag{6.39}$$

The direct calculation of the derivatives appearing in this expression will be included as an exercise at the end of this section.

Using the result of equation (6.38) in equation (6.39) yields

$$\frac{\partial\Psi}{\partial\mathbf{z}} = \frac{\partial\mathbf{R}^{\mathrm{T}}}{\partial\mathbf{z}}\mathbf{Q} + \mathbf{R}^{\mathrm{T}}\frac{\partial\mathbf{Q}}{\partial\mathbf{q}}\mathbf{R} \tag{6.40}$$

By transposing the equation that defines matrix $\mathbf{R}$ (See Section 3.5):

$$\mathbf{R}^{\mathrm{T}}\left[\boldsymbol{\Phi}_{\mathbf{q}}^{\mathrm{T}}\ \ \mathbf{B}^{\mathrm{T}}\right] = \left[\,0\ \mid\ \mathbf{I}\,\right] \tag{6.41}$$

and differentiating with respect to $\mathbf{z}$,

$$\frac{\partial\mathbf{R}^{\mathrm{T}}}{\partial\mathbf{z}}\left[\boldsymbol{\Phi}_{\mathbf{q}}^{\mathrm{T}}\ \ \mathbf{B}^{\mathrm{T}}\right] + \mathbf{R}^{\mathrm{T}}\left[\frac{\partial\boldsymbol{\Phi}_{\mathbf{q}}^{\mathrm{T}}}{\partial\mathbf{q}}\ \ 0\right]\mathbf{R} = 0 \tag{6.42}$$

By again transposing this equation to recover the habitual form:

$$\begin{bmatrix} \Phi_q \\ B \end{bmatrix} \left[\left[\frac{\partial R^T}{\partial z} \right] + R^T \left[\left[\begin{array}{c} \dfrac{\partial \Phi_q}{\partial q} \\ \\ 0 \end{array} \right] \right] \right] R = [[0]] \qquad (6.43)$$

The three-dimensional hyper-matrices have been shown with double square brackets.

Example 6.2

Determine direct expressions for the derivatives that appear in equation (6.39).
- $\partial q/\partial z = R$, by virtue of equation (6.38).
- $\partial Q/\partial q$ can be calculated in accordance with the dependence that Q has in re lation to vector q.
- $\partial R^T/\partial q$

This term, which is the variation of the basis of the nullspace of Φ_q with respect to the dependent coordinates q, constitutes a third order tensor or a three-dimensional hyper-matrix. To calculate it, one must start from the equation that defines R (See Section 3.5),

$$\begin{bmatrix} \Phi_q \\ B \end{bmatrix} R = \begin{bmatrix} 0 \\ I \end{bmatrix} \qquad (i)$$

By differentiating with respect to coordinate q_i,

$$\begin{bmatrix} \dfrac{\partial \Phi_q}{\partial q_i} \\ \\ 0 \end{bmatrix} R + \begin{bmatrix} \Phi_q \\ B \end{bmatrix} \begin{bmatrix} \dfrac{\partial R}{\partial q_i} \end{bmatrix} = \begin{bmatrix} 0 \\ 0 \end{bmatrix} \qquad (ii)$$

one can obtain the desired derivative of R with respect to coordinate q_i. Even though the leading matrix of (ii) remains constant, the number of forward reductions and back substitutions to be performed can be very high.

6.1.4.2 Practical Computation of Derivatives

An alternative way of finding the derivatives with certain interesting mechanical interpretations can be obtained from equation (6.43). We start with equation (6.38) in summation form:

$$\dot{q} = \sum_j r^j \dot{z}_j \qquad (6.44)$$

By again differentiating with respect to time:

$$\ddot{q} = \sum_j \sum_k \frac{\partial r^j}{\partial z_k} \dot{z}_k \dot{z}_j + \sum_j r^j \ddot{z}_j \qquad (6.45)$$

This expression permits calculating the derivatives with respect to the independent coordinates of the columns of matrix $\mathbf{R}$.

By making $\ddot{z}_j = 0$ ($\forall\ j$), and $\dot{z}_k = \delta_{kj}$, one obtains

$$\ddot{\mathbf{q}} \equiv \frac{\partial \mathbf{r}^k}{\partial z_k} \tag{6.46}$$

This indicates that the derivative of the column k of $\mathbf{R}$ with respect to the independent coordinate k is the acceleration vector of the multibody system, when all the independent accelerations and all the independent velocities are made zero. An exception is that corresponding to the degree of freedom k, which is made equal to unity.

By making $\ddot{z}_j = 0$ ($\forall\ j$), $\dot{z}_k = \dot{z}_j = 1$, $\dot{z}_i = 0$ ($i = j, k$), and substituting in expression (6.45), we obtain

$$\ddot{\mathbf{q}} = \frac{\partial \mathbf{r}^j}{\partial z_k} + \frac{\partial \mathbf{r}^k}{\partial z_j} + \frac{\partial \mathbf{r}^j}{\partial z_j} + \frac{\partial \mathbf{r}^k}{\partial z_k} \tag{6.47}$$

From this expression, taking into account the equality of crossed derivatives one obtains

$$\frac{\partial \mathbf{r}^j}{\partial z_k} = \frac{1}{2}\left(\ddot{\mathbf{q}} - \frac{\partial \mathbf{r}^j}{\partial z_j} - \frac{\partial \mathbf{r}^k}{\partial z_k} \right) \tag{6.48}$$

Expression (6.48) assumes that the derivatives of columns j and k with respect to the corresponding independent variables have been calculated previously.

Once all the derivatives appearing in expression (6.40) have been evaluated, the Newton-Raphson iteration proceeds in accordance with the habitual scheme:

$$\left[\frac{\partial \Psi}{\partial \mathbf{z}} \right]_i (\mathbf{z}_{i+1} - \mathbf{z}_i) = -\Psi_i \tag{6.49}$$

6.1.5 Dynamic Relaxation

A last and most appropriate method to solve the static position problem is to use the dynamic simulation procedures explained in Chapter 5. To apply this method, it is enough to start a dynamic simulation from any position, applying the external static forces and introducing dampers so as to dissipate the kinetic energy. At a certain time of the simulation, the system will converge to the static equilibrium position. The largest difficulty in applying this method is how to define the values of the dampers. It shall be large enough to avoid oscillations and small enough not to introduce stiffness in the motion differential equations. A trial and error procedure seems to be the simplest solution for most practical cases.

The main advantages of dynamic relaxation are that computation of derivatives is not necessary at all, and that the user can have better control over the convergence process as compared to the Newton-Raphson iteration.

6.2 Inverse Dynamics

The inverse dynamics consists in the determination of the driving or motor forces and reactions at the joints of a multibody system once the movement (inertial forces and forces produced by the springs and dampers) and the external forces (weight, point forces, torques, etc.) are known. The inverse dynamics can be applied either to an isolated position of the system or to the entire kinematic and dynamic simulations or a static equilibrium position.

Several methods exist for solving the inverse dynamic problem. The basic theories and practices of the Newton method, of the Lagrange multiplier method and the virtual power method will be developed in this section. Whether one method is better than another will depend on the type of dependent coordinates used, the calculation of all or some motor forces and reactions, and also on the specific multibody system being considered to a certain extent. We will consider in the last subsection the particular case of inverse dynamics for open-chain systems.

6.2.1 Newton's Method

Newton's method basically consists of laying out the simultaneous equilibrium for all the multibody system and considering the motor forces and the reactions at the joints as unknowns. A simple recount of the number of unknowns and equations permits checking that both are equal and, therefore, the problem has only one solution in multibody systems that follow the Grübler criterion. The Grübler criterion for three-dimensional multibody systems establishes that:

$$G = 6\left(N - 1\right) - 5\,P_I - 4\,P_{II} - 3\,P_{III} - 2\,P_{IV} - P_V \qquad (6.50)$$

where G is the number of degrees of freedom, N is the number of elements, and P_i is the number of class I joints consisting of joints that allow I degrees of freedom of relative motion between the elements. In expression (6.50), $6(N-1)$ is the total number of equations (six for each moving element) and the term $(G+5P_I+5P_{II}+5P_{III}+2P_{IV}+P_V)$ is the number of unknowns with G motor forces and as many reactions per joint as degrees of freedom that restrict the joint.

In order for Newton's method to be applicable, it is necessary to clearly define how the unknowns are going to be considered and how the equilibrium equations of each element are going to be set. Reference will be made to these two problems further on.

There are two possibilities for defining the problem inputs: one with regard to motor forces that will normally be produced by translation and/or rotation motors or actuators, and one by kinematically driven degrees of freedom.

Normally, both translational and rotational motors will be associated with distance or angle coordinates (relative coordinates), whose variation is known as

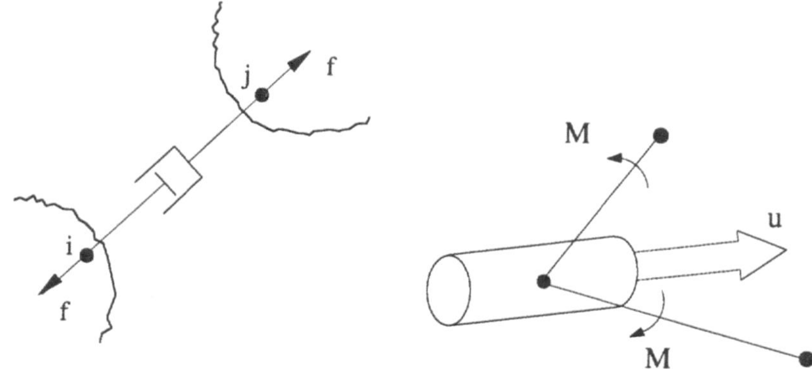

Figure 6.1. Forces produced by a hydraulic actuator.

Figure 6.2. Torques produced by a rotational actuator.

functions of time. With a distance between two points that varies in a predefined
manner, the associated motor force is constituted by two equal and opposed
forces that act on these points in the direction determined by them (Figure 6.1).
In the case of the known angle between two elements, the motor forces will be
two equal and opposite torques in the direction of the axis joint, acting on the elements. The analytic expressions of these forces and torques for both translational and for rotational motions are included in the following pages.

In the case of the translational motors of Figure 6.1, the value of forces acting on the basic points i and j are:

$$\mathbf{Q}_i^\mathrm{T} = \frac{f}{d_{ij}} \left\{ \ (x_i - x_j) \ \ (y_i - y_j) \ \ (z_i - z_j) \ \right\} \tag{6.51}$$

$$\mathbf{Q}_j^\mathrm{T} = \frac{f}{d_{ij}} \left\{ \ (x_j - x_i) \ \ (y_j - y_i) \ \ (z_j - z_i) \ \right\} \tag{6.52}$$

In the case of rotational motors between the two elements as seen in Figure
6.2, the torque acting on each one of them has the direction of the joint axis and
may be written as

$$\mathbf{M} = M \left\{ \begin{matrix} u_x \\ u_y \\ u_z \end{matrix} \right\} \tag{6.53}$$

There is a similar expression for the torques produced by the rotational
springs and dampers. The difference lies in the fact that in the motor torques the
magnitude M is unknown; whereas in the case of the springs and dampers, it is
perfectly known.

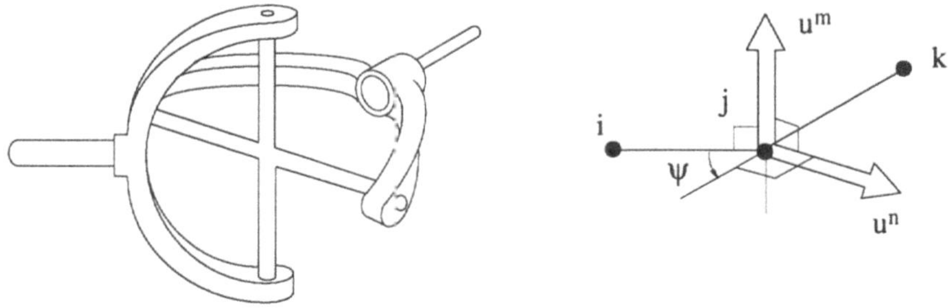

Figure 6.3. Universal joint with natural coordinates.

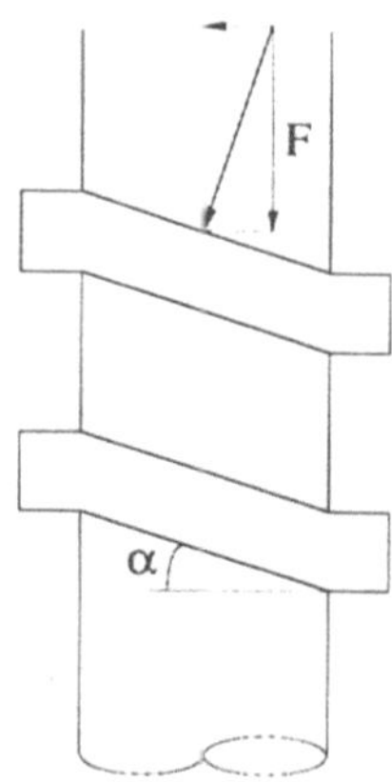

Figure 6.4. Helical joint with natural coordinates.

The other unknowns of the problem, not including the motor forces and torques, are the reactions at the joints. At each joint, there are as many unknown reactions as there are degrees of freedom that the joint constraints. For example: at a ball joint or spherical joint, there are three unknown reactions; at a revolute joint, there are three reaction forces and two torques (the torques perpendicular to the axis); at a cylindrical joint, there are two reaction forces and two torques (both also perpendicular to the axis), and so forth. The consideration of only the reactions that really exist at a joint is possible, although somewhat complicated, because it requires the establishing of certain local coordinate axes at each joint in which the components are cancelled in accordance with the degrees of freedom. At a cylindrical joint, for example, the axis should be taken as one of coordinate axes.

At the time of choosing the unknowns, it is better to choose six unknowns per joint, three forces and three torques in the direction of the general coordinate axes, and formulate some additional equations that compensate for the excess of unknowns. These additional equations are very easy to establish, since they pose the inability of the joint to transmit forces in the direction of its degrees of free-

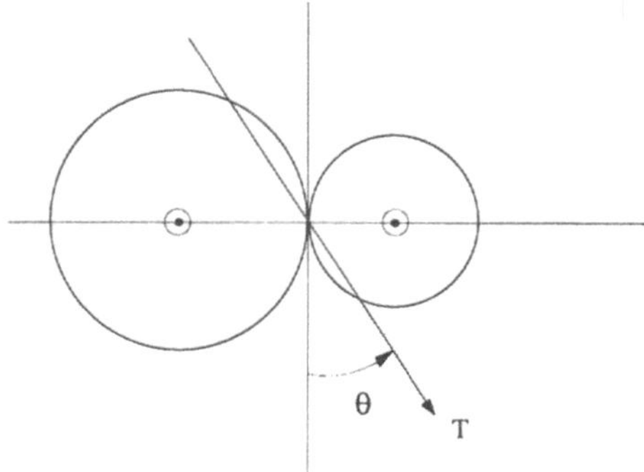

Figure 6.5. Gear joint with natural coordinates.

dom. At a cylindrical joint, the three components of the force and the three components of the reaction torque would be considered as unknowns. The additional equations would be taken as those resulting from imposing the condition that the component of the force and torque in the direction of the joint axis be zero. This condition could be imposed by means of the scalar or dot product with the unit vector in the direction of the joint axis. The spherical, revolute, and prismatic joints generate similar types of equations.

In other types of joints, the generation of these additional equations is somewhat more complicated. The universal, helical, and gear joints are briefly reviewed below.

a) Universal joint. The universal joint is formed by sharing a point with two perpendicular revolute joints R. It is a joint with two degrees of freedom. The two additional equations originate by imposing that the components of the torques in the direction of the rotational joints be zero. This direction is determined by the corresponding unit vectors (See Figure 6.3).

b) Helical joint. Consider a helical joint without friction, as shown in Figure 6.4. The contact force is perpendicular to the thread of the screw. This force is decomposed into an axial force F and into a tangential force ($F \, tan \, \alpha$). This tangential component gives a torque in relation to the screw axis that is equal to

$$M = F \, tan \, \alpha \, r \tag{6.54}$$

where r is the average radius of the thread. Equation (6.81) indicates that there is a constant ratio between the axial force and the torque with respect to the joint axis. This is the additional equation that relates the three components of the force and the three components of the reaction torque.

c) Gear joint. Consider the planar gear joint shown in Figure 6.5. The constraint equations corresponding to this joint are established by imposing the condition

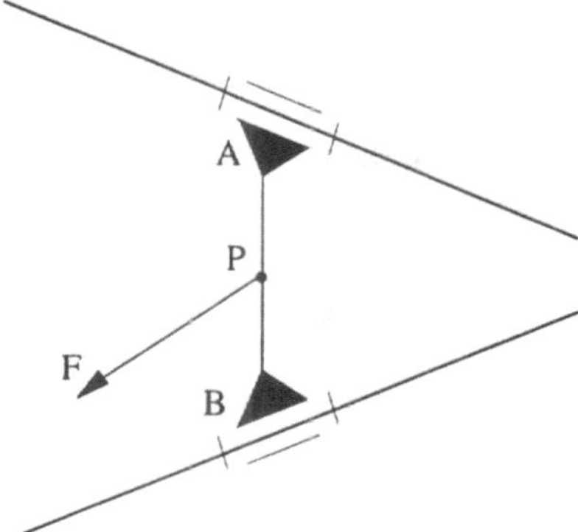

Figure 6.6. 3-D Gear joint.

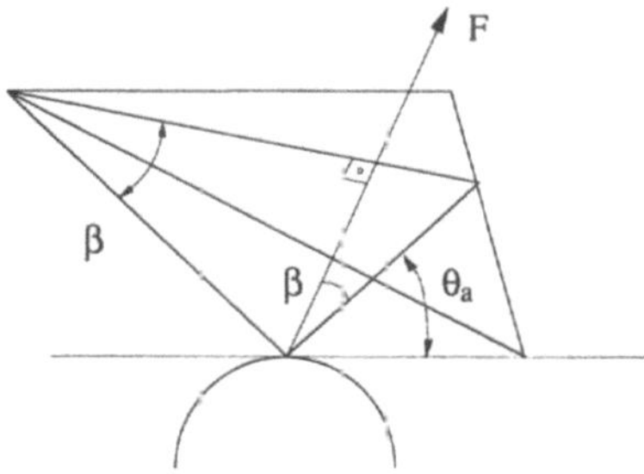

Figure 6.7. Contact force in helical joints.

of constant distance between the centers of both wheels, and of a ratio between the angles turned by both wheels and considered as mixed or relative coordinates. Normally, the element that links the centers of the wheels will truly exist. If nonexistent, there is no way of guaranteeing that the centers remain at a constant distance. This element will have its own inertial forces and equilibrium equations. The problem is reduced to establishing the equilibrium equations of each wheel. This is accomplished by entering the contact force T that forms an angle θ (pressure angle) with the tangent to both wheels or by entering two components in the direction of the absolute coordinates and imposing the condition that the resultant force is on the contact line (force transmission line).

Now consider the three-dimensional gear joint of Figure 6.6. The relative position between the axes of both wheels is guaranteed by an element with two joints R, assuming that it is located on the normal line common to both axes. In the case of helical gears, the direction of the transmission force is defined by the helix inclination angle β and by the apparent pressure angle θ_a (Figure 6.7).

These angles can be calculated starting from those of inclination β_a and normal pressure θ by means of well-known formulae (Shigley (1969)). To determine the contact force that acts on each one of the wheels, a force F can be assumed. This force F consists of three unknown components expressed in accordance with axes of the general coordinates, acting at a point P of the common normal calculated in accordance with the distances to the points A and B and dependent on the contact ratio (See Figure 6.6). At the components of the force F, it is necessary to impose two ratios in order to make the force have the direction shown in Figure 6.7. These ratios can be two scalar products with two directions contained in a plane perpendicular to F or two components of a vector product with a vector in the direction that F should have been.

One can see briefly how the unknowns of the inverse dynamics are introduced and how the additional equations are calculated, when more unknowns than necessary are defined. The only remaining problem is to determine how to set out the force equilibrium equations for each element. The simplest way is to carry out this setting in general coordinates.

To set out the equilibrium of an element, it is necessary to consider the following forces:

a) inertial forces of the element.
b) external forces that act on the element, including the forces produced by the springs and dampers.
c) reaction forces at the joints of the element.

The sum of all these forces and their resulting torque with respect to a point should be zero. This point can be any one, but the two simplest options are the center of gravity and the origin of the system of absolute coordinates. The latter possibility appears to be the most adequate.

The inertial forces can be calculated by multiplying the inertial matrix of the element by the accelerations of its basic points and unit vectors, in accordance with the expression:

$$Q_{in}^e = - M^e \ \ddot{q}^e \tag{6.55}$$

The inertial forces vector Q_{in}^e contains point forces corresponding to basic points and couples of forces corresponding to unit vectors. The components of the inertial forces conjugated with the unit vectors are force couples that act on the ends of the vectors, as seen in Figure 6.8. These force couples have a null resultant and a resultant torque equal to the cross product of the force by the corresponding unit vector. Thus, the components of Q_{in}^e associated with basic points are forces that intervene in the force equilibrium. These yield a torque about the origin of coordinates; that is, the components of Q_{in}^e, conjugated with the unit vectors, are cross multiplied by the associated unit vector and constitute torques that are added up in the torque's equilibrium equation of the element.

With respect to the velocity-dependent inertial forces, the external forces, and those produced by the springs and dampers, when known, are directly incorpo-

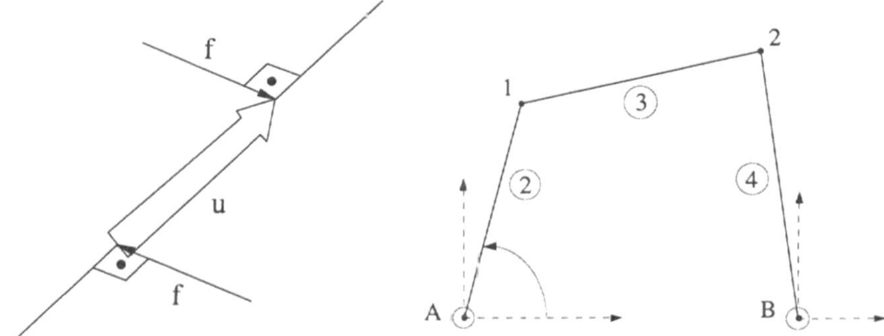

Figure 6.8. Couple of forces conjugated with a unit vector.

Figure 6.9. Inverse dynamics in a four-bar mechanism.

rated without further difficulties into the equilibrium equations of the elements on which they act.

Lastly, the reactions at the joints intervene as unknowns in the equilibrium equations of each element.

To avoid the calculation of all the reactions at all joints is also possible, by formulating the equilibrium of groups of bodies or even of all the multibody system. For example, in the four-bar mechanism of Figure 6.9 there are five overall unknowns: the reactions at joints A and B, and the motor torque M. It is necessary to establish five equations, which will be the three static equilibrium equations of the entire mechanism, and two additional equations. These two additional equations could be the null moment condition of all the forces acting on bar 2 with respect to point 1, and the null moment condition with respect to point 2 of all the forces acting on elements 2 and 3. This equation could be substituted by the null moment condition with respect to two of the forces acting on bar 4.

An interesting case, with regard to solving the inverse dynamics by means of Newton's method, is that of the *over-determined multibody systems* that do not satisfy the Grübler criterion. They are systems that in theory should not move, since Grübler predicts a null or negative number of degrees of freedom for them. But in practice, due to their specific dimensions and the orientation of the axes, these systems do in fact move. A typical example is the planar four-bar mechanism with four revolute joints. When considered as a three-dimensional mechanism, the Grübler criterion predicts –2 degrees of freedom for it. However the mechanism moves when the four revolute joints have parallel axes. If this condition is not satisfied, it would be locked.

The inverse dynamics presents additional difficulties in over-determined multibody systems. From the forces equilibrium point of view, these systems constitute undetermined problems in which there are more unknowns than equations. The reactions at the joints cannot be calculated without resorting to additional hypotheses. For example, if at a planar articulated quadrilateral there is an

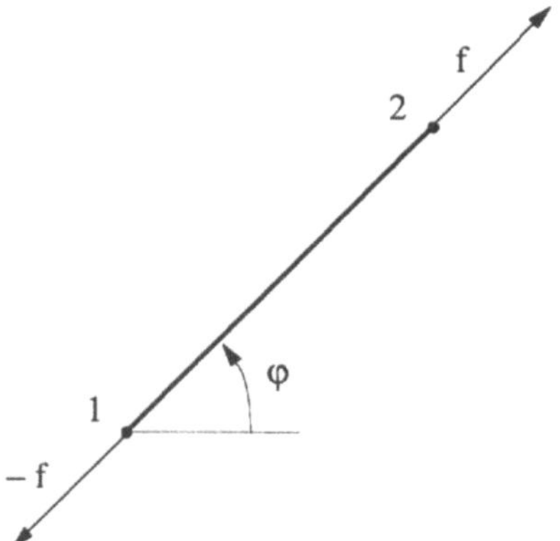

Figure 6.10. Constraint force corresponding to a constant distance condition.

external force applied to it in a direction perpendicular to the motion plane, there are infinite systems of external reactions and torques that compensate this force. It can even have external reactions that are self-compensated with one another, without affecting the motion.

To solve this difficulty, the additional condition of the external reactions at the joints having a minimum norm should be applied. It makes sense physically since it presupposes that the multibody system is well-constructed or installed without initial self-compensating forces, and permits finding a mathematical solution to the problem without excessive calculation effort.

6.2.2 *Method of the Lagrange Multipliers*

The Lagrange multiplier method permits determining the motor forces and the reactions that appear at the joints of a multibody system and makes optimum use of the relationship that exists between the Lagrange multipliers and the forces associated with the constraint equations. The equations of motion developed in Chapter 5 establish that

$$\mathbf{M}\ddot{\mathbf{q}} + \mathbf{\Phi}_\mathbf{q}^\mathrm{T}\,\lambda - \mathbf{Q} = 0 \tag{6.56}$$

where the first term represents the inertial forces; the second, the forces produced by the constraints; and the third, the external forces plus additional velocity-dependent inertial forces. Each column of matrix $\mathbf{\Phi}_\mathbf{q}^\mathrm{T}$, multiplied by the corresponding λ, represents the vector of forces associated with the constraint.

The forces introduced by some of the most common constraint equations in planar and three-dimensional multibody systems will be studied in greater detail below.

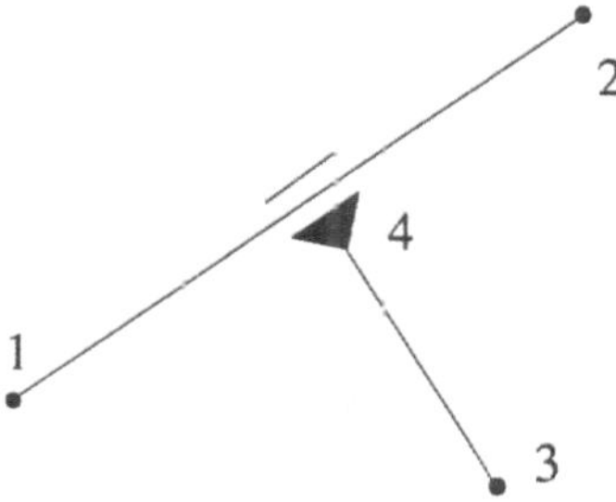

Figure 6.11. Planar prismatic joint.

6.2.2.1 Constraint Forces in Planar Multibody Systems

The most commonly used constraint equations in planar multibody systems are
those arising from the constant distance, and those in which the prismatic joint
is introduced. These are the ones in which three points are kept aligned, and the
scalar product between the segments determined by two couples of basic points
are constant.

Figure 6.10 shows two points between which there is a constant distance
condition and the force system f necessary to maintain this condition. This force
system will consist of two equal and opposite forces in the direction of the bar.
These forces will amount to

$$Q = \frac{f}{L} \begin{Bmatrix} x_1 - x_2 \\ y_1 - y_2 \\ x_2 - x_1 \\ y_2 - y_1 \end{Bmatrix} \tag{6.57}$$

On the other hand, the constant distance constraint is

$$\Phi_i(\mathbf{q}) \equiv (x_2 - x_1)^2 + (y_2 - y_1)^2 - L^2 = 0 \tag{6.58}$$

The derivative of this equation with respect to $\mathbf{q}$ is

$$\frac{\partial \Phi_i}{\partial \mathbf{q}} = \left\{ (x_1 - x_2) \quad (y_1 - y_2) \quad -(x_1 - x_2) \quad -(y_1 - y_2) \right\} \tag{6.59}$$

By the Lagrange multipliers theory, the constraint force will be this vector in
column form, multiplied by the corresponding Lagrange multiplier λ:

$$\mathbf{f} = \lambda \begin{Bmatrix} x_1 - x_2 \\ y_1 - y_2 \\ -(x_1 - x_2) \\ -(y_1 - y_2) \end{Bmatrix} \tag{6.60}$$

By identifying expressions (6.57) and (6.60), it can be concluded that the
force associated with the constraint has the following value:

$$f = \lambda L \tag{6.61}$$

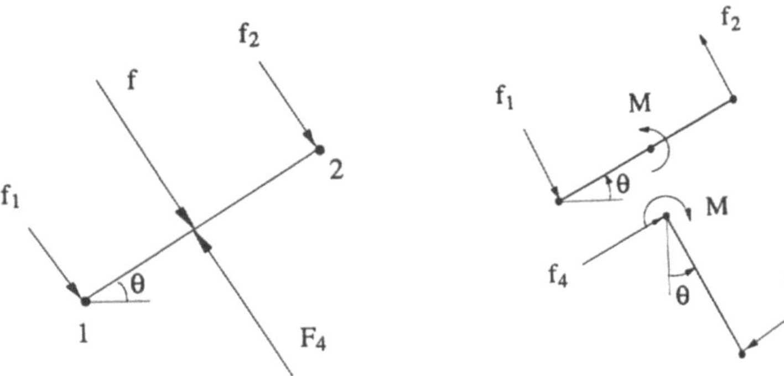

Figure 6.12. Constraint forces corresponding to the alignment condition.

Figure 6.13. Constraint forces corresponding to the perpendicular condition.

Figure 6.11 shows a prismatic joint corresponding to a planar mechanism. The joint generates two constraint equations. The forces associated with the alignment condition are shown in Figure 6.12. There is a normal reaction f that is balanced by forces f_1 and f_2 acting on the basic points.

In this case the alignment constraint equation is

$$\Phi_i(\mathbf{q}) \equiv \{(x_2 - x_1)\ (y_4 - y_1)\ -(x_4 - x_1)\ (y_2 - y_1)\} \tag{6.62}$$

and the corresponding force, in accordance with the Lagrange multiplier method is

$$\mathbf{f} = \lambda\ \frac{\partial \Phi_i}{\partial \mathbf{q}} = \lambda \begin{Bmatrix} y_2 - y_4 \\ x_4 - x_2 \\ y_4 - y_1 \\ x_1 - x_4 \\ 0 \\ 0 \\ y_1 - y_2 \\ x_2 - x_1 \end{Bmatrix} \tag{6.63}$$

On the other hand, based on Figure 6.12, the reaction forces can be calculated in terms of the force f:

$$Q_{1x} = f\ ((y_2 - y_4)/(y_2 - y_1))\ \sin\theta = f\ (y_2 - y_4)/L_{12} \tag{6.64}$$

$$Q_{1y} = -f\ ((x_2 - x_4)/(x_2 - x_1))\ \cos\theta = -f\ (x_2 - x_4)/L_{12} \tag{6.65}$$

$$Q_{2x} = f\ ((y_4 - y_1)/(y_2 - y_1))\ \sin\theta = f\ (y_4 - y_1)/L_{12} \tag{6.66}$$

$$Q_{2y} = -f \left((x_4 - x_1)/(x_2 - x_1)\right) \cos \theta = -f \ (x_4 - x_1)/L_{12} \qquad (6.67)$$

$$Q_{4x} = -f \sin \theta = -f \ (y_2 - y_1)/L_{12} \qquad (6.68)$$

$$Q_{4y} = f \cos \theta = f \ (x_2 - x_1)/L_{12} \qquad (6.69)$$

By comparing these expressions with the vector of reaction forces defined by equation (6.63), one can conclude, in this case, that

$$f = \lambda \cdot L_{12} \qquad (6.70)$$

Figure 6.13 shows the forces associated with the perpendicular condition between segments (1-2) and (3-4). If M is the torque that keeps this angle condition constant, one must formulate the forces at the basic points in terms of the value of M. It is easy to see that:

$$Q_{1x} = (M \ / \ L_{12}) \sin \theta = (M \ /(L_{12} \ L_{34})) \ (x_3 - x_4) \qquad (6.71)$$

$$Q_{1y} = (M \ / \ L_{12}) \cos \theta = - (M \ /(L_{12} \ L_{34})) \ (y_4 - y_3) \qquad (6.72)$$

$$Q_{2x} = - Q_{1x} = - (M \ /(L_{12} \ L_{34})) \ (x_3 - x_4) \qquad (6.73)$$

$$Q_{2y} = - Q_{1y} = + (M \ /(L_{12} \ L_{34})) \ (y_4 - y_3) \qquad (6.74)$$

$$Q_{3x} = - (M \ / \ L_{34}) \cos \theta = - (M \ /(L_{12} \ L_{34})) \ (x_2 - x_1) \qquad (6.75)$$

$$Q_{3y} = - (M \ / \ L_{34}) \sin \theta = - (M \ /(L_{12} \ L_{34})) \ (y_2 - y_1) \qquad (6.76)$$

$$Q_{4x} = - Q_{3x} = (M \ /(L_{12} \ L_{34})) \ (x_2 - x_1) \qquad (6.77)$$

$$Q_{4y} = - Q_{3y} = (M \ /(L_{12} \ L_{34})) \ (y_2 - y_1) \qquad (6.78)$$

On the other hand the perpendicular constraint equation is in this case:

$$(x_2 - x_1) \ (x_4 - x_3) + (y_2 - y_1) \ (y_4 - y_3) = 0 \qquad (6.79)$$

The associated constraint force vector will be

$$\mathbf{Q} = \lambda \frac{\partial \Phi_i}{\partial \mathbf{q}} = \lambda \begin{Bmatrix} x_3 - x_4 \\ y_3 - y_4 \\ x_4 - x_3 \\ y_4 - y_3 \\ x_1 - x_2 \\ y_1 - y_2 \\ x_2 - x_1 \\ y_2 - y_1 \end{Bmatrix} \qquad (6.80)$$

By identifying the coefficients of equations (6.71)-(6.78) with the vector equation (6.80), one can conclude that the Lagrange multiplier λ is equivalent to:

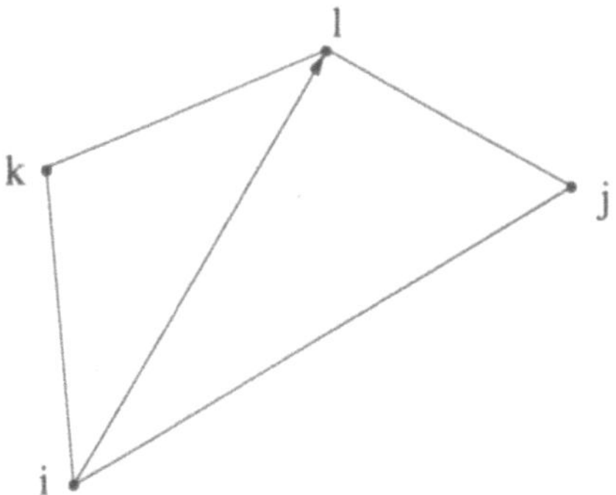

Figure 6.14. Linear combination of planar vectors condition.

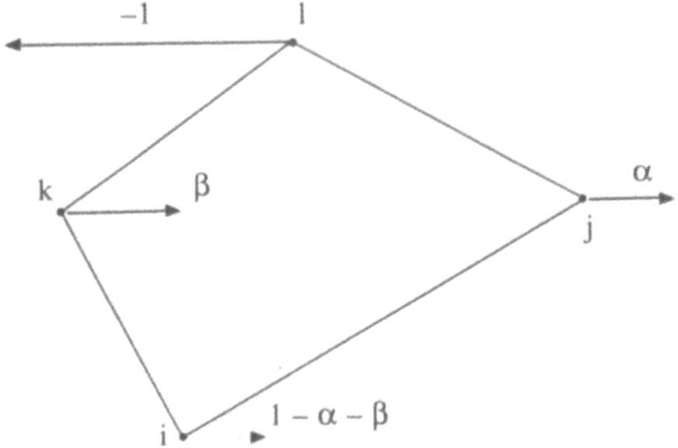

Figure 6.15. Constraint forces corresponding to a linear combination condition.

$$\lambda = \frac{M}{L_{12}\,L_{34}} \tag{6.81}$$

In order to carry out this identification of the Lagrange multiplier, it is generally not necessary to calculate all the terms of the forces vector. Only one must be calculated and the identification carried out with that term.

Figure 6.14 shows a different type of constraint equation that may appear in planar multibody systems and also in three-dimensional ones more frequently. In this case, the position vector of one point in relation to another is expressed as a linear combination of the position vectors of other points of the same multibody system. In the planar multibody system shown in Figure 6.14, the constraint equations that establish the position of point l in relation to the other three points are

$$(\mathbf{r}_l - \mathbf{r}_i) - \alpha\,(\mathbf{r}_j - \mathbf{r}_i) - \beta\,(\mathbf{r}_k - \mathbf{r}_i) = 0 \tag{6.82}$$

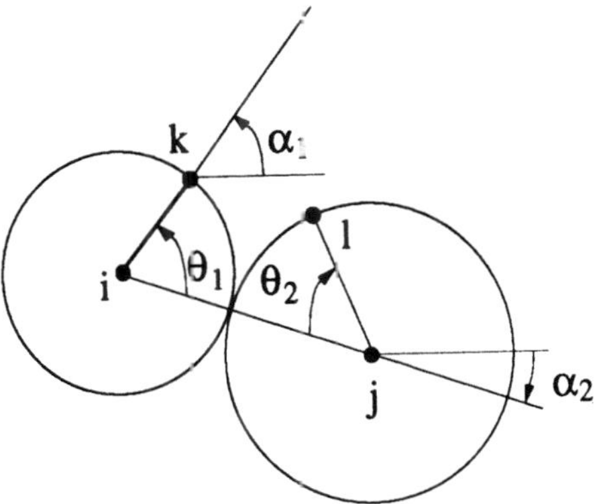

Figure 6.16. Gear joint.

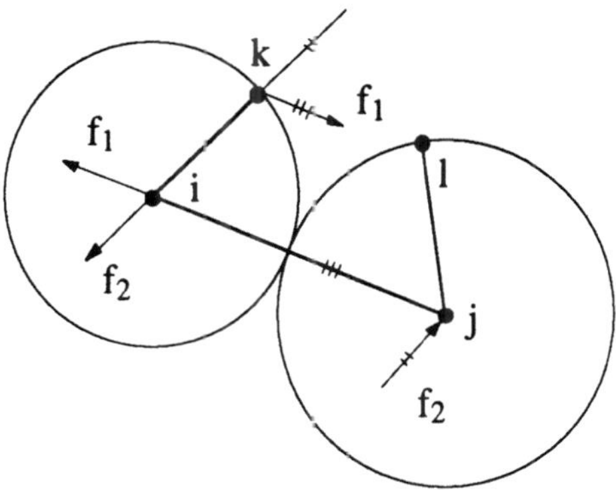

Figure 6.17. Constraint forces corresponding to a gear joint.

where α and β are known constants. The vector equation (6.82) represents two constraint equations. The derivative of the first of them with respect to the dependent coordinate vector $\mathbf{q}$ results in:

$$\Phi_{\mathbf{q}i} \equiv \{(1-\alpha-\beta), 0, \alpha, 0, \beta, 0, -1, 0\} \tag{6.83}$$

Multiplied by the Lagrange multiplier λ, the vector of equation (6.83) represents a system of four forces in the direction x acting on the four basic points of the element, as can be seen in Figure 6.15. It is evident that these forces are in equilibrium, because their sum is zero. It is also easy to demonstrate that they give a zero torque about point i. The torque's equilibrium equation in accordance with axis z is

$$M_z \equiv \alpha\,(y_j - y_i) + \beta\,(y_k - y_i) - (y_l - y_i) \tag{6.84}$$

This expression is cancelled because it is nothing more than component y of equation (6.83). Therefore, once the Lagrange multiplier is known, each linear combination of constraint equations produces a system of resultant forces and a null resultant torque acting on the basic point of the element. These forces should be equivalent to those that would have been obtained if point l had been determined in relation to the other three points with three constant distance conditions.

The only case left to be seen is the *gear joint*.

Figure 6.16 shows a gear joint determined by points i and k, and j and l, respectively. The constraint equation that enters the angular coordinate θ_1 for wheel 1, is

$$(x_k - x_i)\,(x_j - x_i) + (y_k - y_i)\,(y_j - y_i) - L_{ik}\,L_{ij}\,\cos\theta_1 = 0 \qquad (6.85)$$

For the angular coordinate θ_2:

$$(x_l - x_j)\,(x_i - x_j) + (y_l - y_j)\,(y_i - y_j) - L_{jl}\,L_{ij}\,\cos\theta_2 = 0 \qquad (6.86)$$

The ratio between both angular coordinates is

$$\theta_1 + m\,\theta_2 = 0 \qquad (6.87)$$

To see the significance of the constraint forces associated with equation (6.85), one should differentiate this equation with respect to the dependent variables vector $\mathbf{q}$. We exclusively consider the following terms

$$\frac{\partial \Phi}{\partial x_k} = x_j - x_i\,; \qquad \frac{\partial \Phi}{\partial y_k} = y_j - y_i \qquad (6.88)$$

$$\frac{\partial \Phi}{\partial \theta_1} = L_{ij}\,L_{ik}\,\sin\theta_1 \qquad (6.89)$$

The two equations (6.88) indicate that at point k there is a force f_1 that is parallel to the segment $(i\text{-}j)$, as shown in Figure 6.17. This force is compensated with another one, equal and opposite, applied at point i. Likewise, other forces of magnitude f_2, equal, opposite, and parallel to the segment $(i\text{-}k)$ appear at points i and j. The magnitude of the forces f_1, applied at points i and k, will be

$$M_1 = f_1\,L_{ik}\,\sin\theta_1 \qquad (6.90)$$

The component x of force at point k can be used to identify the Lagrange multiplier value λ:

$$\lambda\,(x_j - x_i) = f_1\,\cos\alpha_2 = f_1\,\frac{x_j - x_i}{L_{ij}} \qquad (6.91)$$

From expressions (6.90) and (6.91), one can arrive at the following result:

$$\lambda = \frac{f_1}{L_{ij}} = \frac{M}{L_{ij}\,L_{ik}\,\sin\theta_1} \qquad (6.92)$$

which shows how one can calculate the torque acting on wheel 1, once the value of λ is known. From expressions (6.89) and (6.92), it can be concluded that the value of the force associated with the angular coordinate θ_1 is

Figure 6.18.   Constant angle condition.

$$\lambda \frac{\partial \Phi}{\partial \theta_1} = \lambda L_{ij} L_{ik} \sin\theta_1 = M_1 \tag{6.93}$$

This is the torque M_1 that acts on each of the elements. The force f_2 could be calculated by means of an expression similar to (6.118). Constraint equation (6.113), however, leads to certain forces similar to those studied and which produce a torque M_2 at wheel 2. The angle's constraint equation (6.114) establishes the relationship existing between torques M_1 and M_2.

6.2.2.2 Constraint Forces in Three-Dimensional Multibody Systems

The determination of the constraint forces in three-dimensional multibody systems by means of Lagrange multipliers follows the same basic steps as in the planar case. The only difference is that the mathematical formulation is more complicated. The development of the constraint forces corresponding to some of the most normal constraint equations is shown in an example included below.

a) Constant distance constraint. The constant distance constraint between two points is also very easy to consider. This constraint equation can be formulated as follows:

$$\Phi \equiv (x_i - x_j)^2 + (y_i - y_j)^2 + (z_i - z_j)^2 - L_{ij}^2 = 0 \tag{6.94}$$

The constraint forces vector will have the following form:

$$\lambda \frac{\partial \Phi}{\partial \mathbf{q}} \equiv \lambda \left\{ (x_i - x_j) \ \ (y_i - y_j) \ \ (z_i - z_j) \ \ (x_j - x_i) \ \ (y_j - y_i) \ \ (z_j - z_i) \right\} \tag{6.95}$$

from which it is inferred that the force f acting at the ends of the bar is in the same direction of the bar. The component of this force, at node i and in the direction x, is

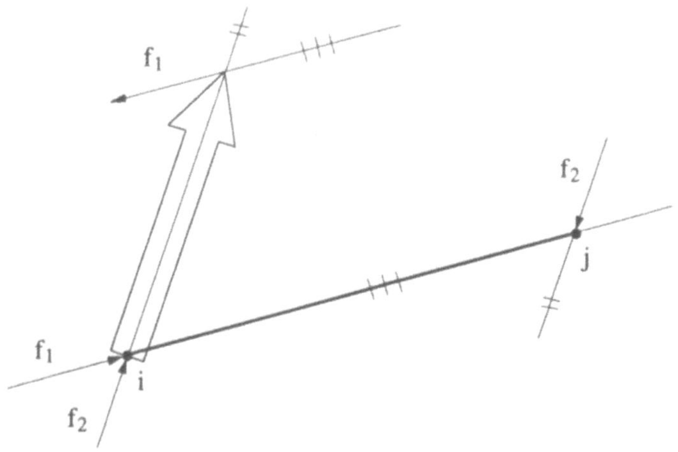

Figure 6.19. Constraint forces corresponding to a constant angle condition.

$$\lambda\,(x_i - x_j) = f\,\frac{x_i - x_j}{L_{ij}} \tag{6.96}$$

from which the value of the constraint force can be identified depending on the Lagrange multiplier, or vice versa.

b) Constant angle constraint between a segment determined by two points and one unit vector. Figure 6.18 shows an element with two points and one unit vector, between which a constant angle θ should be maintained. The corresponding constraint equation of the constant scalar product can be formulated as:

$$\Phi \equiv \mathbf{u}^T\left(\mathbf{r}_j - \mathbf{r}_i\right) = 0 \tag{6.97}$$

The derivative of this equation with respect to $\mathbf{q}$, and multiplied by the Lagrange multiplier λ, will yield the constraint force:

$$\lambda\left(\frac{\partial\Phi}{\partial\mathbf{q}}\right)^T = \lambda\left\{\frac{\partial\Phi}{\partial\mathbf{r}_i}\ \frac{\partial\Phi}{\partial\mathbf{r}_j}\ \frac{\partial\Phi}{\partial\mathbf{u}}\right\}^T = \lambda\left\{\begin{array}{c} -\mathbf{u} \\ \mathbf{u} \\ \mathbf{r}_j - \mathbf{r}_i \end{array}\right\} \tag{6.98}$$

This expression indicates that there are certain equal and opposite reaction forces at i and j in the direction of vector $\mathbf{u}$. There is a reaction force associated with vector $\mathbf{u}$ that has the same direction as the bar. Remember that the reaction forces associated with the unit vectors are equal and opposed pairs of forces acting at its ends. Consequently, the reaction forces that appear as a result of this constraint equation are those shown in Figure 6.19.

These forces have zero resultant. It is easy to see that the torque of the pair of forces acting on the unit vector is equal to the pair of forces acting on segment $(i\text{-}j)$. It is verified that:

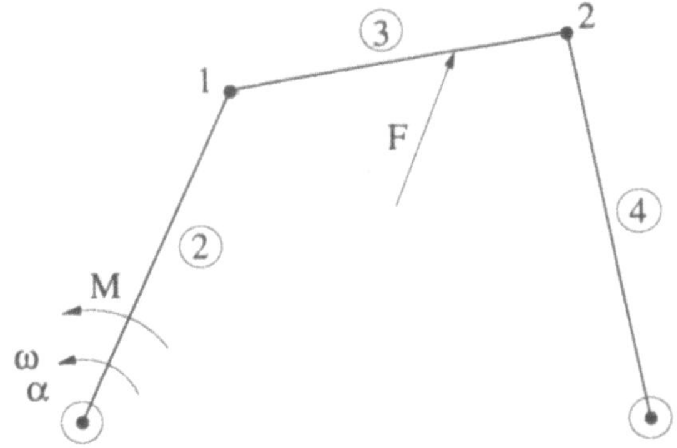

Figure 6.20. Inverse dynamics in a four-bar mechanism.

$$\mathbf{u} \wedge \mathbf{f}_1 = \mathbf{u} \wedge \left(\mathbf{r}_i - \mathbf{r}_j\right) \lambda \qquad (6.99)$$

$$\mathbf{f}_2 \wedge \left(\mathbf{r}_i - \mathbf{r}_j\right) = - \lambda \, \mathbf{u} \wedge \left(\mathbf{r}_i - \mathbf{r}_j\right) \qquad (6.100)$$

which are equal and opposite torques.

c) Unit vector constraint. The unit module vector equation is a variant of the constant distance constraint equation. This constraint equation generates two equal and opposite forces on the extremes of the unit vector and on its same direction.

d) Other constraint equations. Other types of three-dimensional constraint equations such as: a vector that is a linear combination of others, a constraint equation corresponding to the entry of an angular coordinate, and the three-dimensional gear joint, generate the corresponding constraint forces similar to those studied in the planar multibody system case. Their mathematical development follows the same rules as stated previously.

6.2.2.3 Calculation of Reaction Forces at the Joints

With reference point coordinates and joint constraints, the Lagrange multipliers directly provide the constraint forces at the joints. With natural coordinates, the Lagrange multipliers do not, in general, directly provide the forces at the joints. The joints are almost always considered without constraint equations by means of sharing of points and/or unit vectors or with a number of constraint equations that is smaller than the number of degrees of freedom prevented by the joint. The element's constraints, which are always present, give rise to forces on the elements and not on the joints.

Figure 6.20 shows a four-bar mechanism whose motion is perfectly known. Assume that it is wished to calculate the forces at joints 1 and 2. The equilibrium condition of element 3 must be established. Four types of forces act on the element: external forces $\mathbf{Q}_{ex}$, inertial forces, forces associated with the element's constraint equations, and the reactions $\mathbf{Q}_r$ at the joints. The following equilibrium equation can be established:

$$\mathbf{Q}_r - \mathbf{M}^e \, \ddot{\mathbf{q}}^e - \mathbf{\Phi_q^e}^T \lambda^e + \mathbf{Q}_{ex} = 0 \tag{6.101}$$

This is an equation in which all the terms are known, except for the unknown reactions $\mathbf{Q}_r$ produced at the joints by adjacent elements. In expression (6.101), the mass matrix, the dependent accelerations, the element's constraint equations and the Lagrange multipliers only intervene at the level of the element being studied. This has been indicated with the superscript (^{e}). In addition, equation (6.101) is general and can be applied to any element of any multibody system, once the Lagrange multipliers have been calculated.

The Lagrange multiplier method is attractive especially when using the Lagrange multipliers for the numerical integration of the differential equations of motion. An important part of the previously carried out numerical effort is then used to the outmost. If other numerical integration methods are used, it will be necessary to carry out a specific calculation of Lagrange multipliers term ($\mathbf{\Phi_q^T} \lambda$) by means of the equation:

$$\mathbf{M} \, \ddot{\mathbf{q}} + \mathbf{\Phi_q^T} \lambda = \mathbf{Q}_{ex} \tag{6.102}$$

in which everything, except the second term, is supposedly known. The Lagrange multiplier method is more useful with reference point coordinates and joint constraints than with natural coordinates and element constraints, since in the latter case, the Lagrange multipliers do not directly give the reactions at the joints.

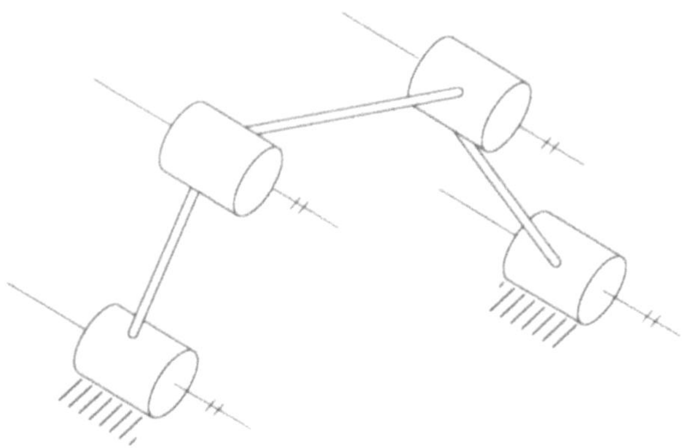

Figure 6.21. 3-D four-bar mechanism with parallel axes.

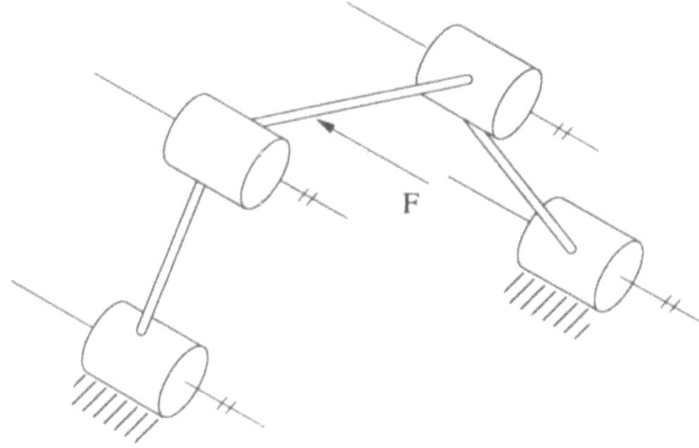

Figure 6.22. External force perpendicular to the four-bar plane.

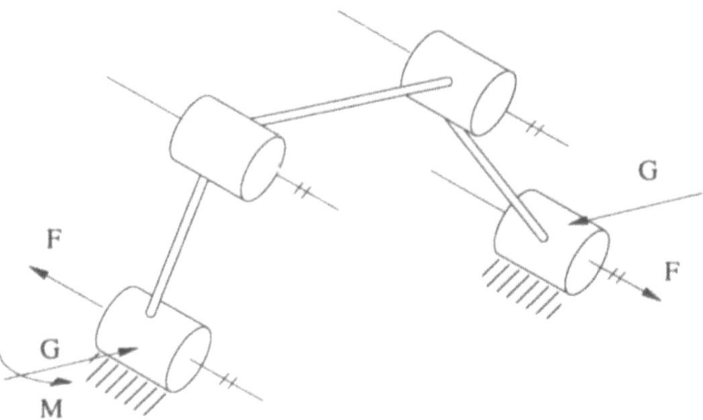

Figure 6.23. Self-equilibrating external forces.

6.2.3 *Method of the Lagrange Multipliers with Redundant Constraints*

If the multibody system is over-constrained or if there are redundant constraints in the formulation (See Section 3.4.), Φ_q^T has more columns than rows ($m > n{-}f$). There is an infinite set of solutions for vector λ in the following system of linear equations:

$$\Phi_q^T \, \lambda = Q_{ex} - M \, \ddot{q} \tag{6.103}$$

This difficulty can be solved as is indicated above. The infinite number of solutions for vector λ has a physical meaning. There are many possible sets of applied external and reaction forces that produce or arise from the same motion. Difficulties arise in finding a unique mathematical solution, because physically

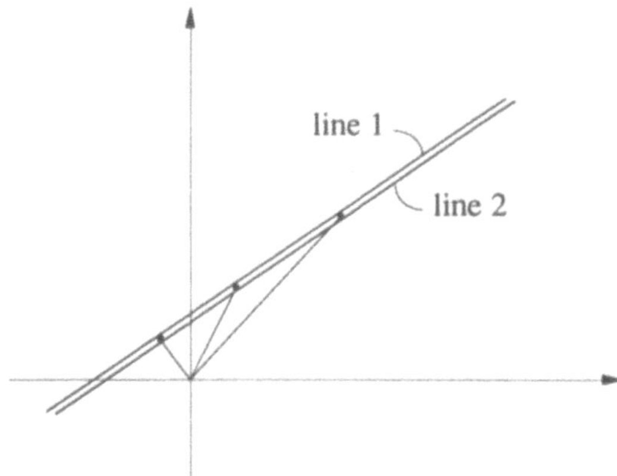

Figure 6.24. Geometric meaning of the minimum norm condition.

there are many solutions that are possible. A very simple example will help to explain this point. Figure 6.21 shows a four-bar mechanism with four revolute joints with parallel axes. This is a 1 degree of freedom system, but when considered as a three-dimensional mechanism, the Grübler formula predicts –2 degrees of freedom; so this mechanism should not move, but in fact it may move only when the four joint axes are parallel. For this system, using reference point coordinates, the Jacobian matrix has twenty rows, with five constraint equations for each joint, and eighteen columns with six coordinates for each one of the three moving links. Thus, there is not a unique solution for the system of linear equations (6.103). It is possible to have, as indicated in Figure 6.22, external forces that are perpendicular to the plane motion and that do not produce virtual work. These forces can be balanced externally with forces and torques in the fixed joints in many different ways. It is even possible, as indicated in Figure 6.23, to have self-equilibrating external reaction forces and torques that could be due, for instance, to assembly or manufacturing errors and that do not appear in any term of the equations of motion (6.103).

In order to find a single solution for equation (6.103), one must add some information or introduce a new hypothesis. A very reasonable addition is to impose to the solution λ the minimum norm condition (See Figure 6.24). To fulfill this condition, λ shall be orthogonal to the null-space of Φ_q^T. It shall belong to its orthogonal complement that is the column space of Φ_q (Strang (1980)):

$$\lambda = \Phi_q \, \sigma \tag{6.104}$$

where σ is a vector of n components, which are the coefficients of vector λ expressed as a linear combination of the columns of Φ_q. Substituting this expression in equation (6.103):

$$\Phi_q^T \, \Phi_q \, \sigma = Q_{ex} - M \, \ddot{q} \tag{6.105}$$

This is an expression that can be solved for σ without major difficulties, because $(\Phi_q^T \Phi_q)$ is a ($n \times n$) matrix of rank n. Vector λ is computed from equation (6.104) and the inverse dynamics problem continues in the standard way.

6.2.4 Penalty Formulation

Equation (6.102) allows the calculation of the Lagrange multipliers for those cases in which they have not been explicitly obtained during the simulation process. The penalty formulation provides an equivalent alternative to equation (6.102) in the form:

$$\mathbf{M}\,\ddot{\mathbf{q}} + \Phi_q^T \, \alpha \, (\ddot{\Phi} + 2\,\mu\,\Omega\,\dot{\Phi} + \Omega^2 \Phi) = \mathbf{Q} \tag{6.106}$$

It may be seen by comparing equations (6.102) and (6.106) that the penalty method approximates the Lagrange multipliers by the term:

$$\lambda^* = \alpha\,(\ddot{\Phi} + 2\,\mu\,\Omega\,\dot{\Phi} + \Omega^2\,\Phi) \tag{6.107}$$

The accuracy of λ^* will depend on the method of integration and the value of α. For $\alpha = 10^7$ and working in double precision arithmetic, one should expect an accuracy in λ^* up to the sixth digit. The augmented Lagrangian formulation described in Chapter 5 will improve this accuracy.

6.2.5 Virtual Power Method

The Newton and Lagrange multiplier methods are by themselves global methods, that is to say, methods used to calculate all the motor forces and all the reactions at the joints. There are variations of these methods that permit saving a certain amount of work when it is not really necessary to calculate all the reactions and all the motor forces.

The virtual power method, which is going to be described below, is a method

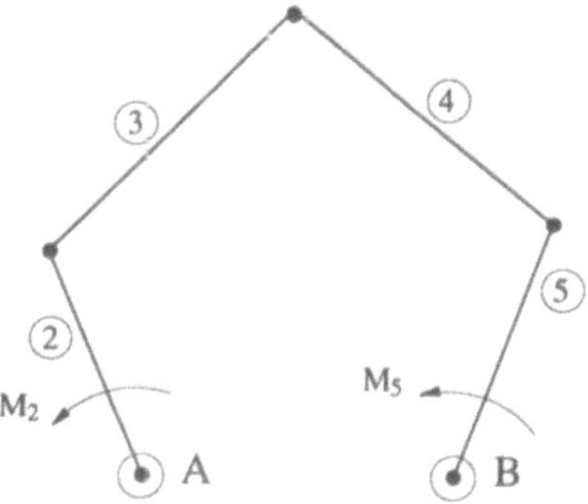

Figure 6.25. Five-bar mechanism with two input torques.

specifically used to calculate a reaction or a motor force. In this case it is much more efficient than the previous methods. However, the said methods must be used whenever it is required to calculate a large number of forces and/or reactions. The virtual power method will be shown in the following sections. The case for evaluation the motor forces will be distinguished from the case for evaluation of reactions at the joints. In principle these cases use natural coordinates with minimum adaptations which are mostly simplifications. These cases can be applied to the other types of coordinates.

6.2.5.1 Calculation of Motor Forces

Motor forces are very easy to calculate with the virtual power method. The idea of the method is very simple and can best be explained by means of an example, such as the planar mechanism with five bars and two degrees of freedom, shown in Figure 6.25. It can be assumed that bars two and five are the mechanism's input elements on which torques M_2 and M_5 are applied.

To determine M_2, one can evaluate the virtual power of the forces that act on the mechanism with the virtual velocities $\dot{\mathbf{q}}_1^*$ calculated by giving a unit angular velocity at element 2 and a zero angular velocity at element 5. The following equation will be obtained:

$$M_2 \cdot 1 + M_5 \cdot 0 = \dot{\mathbf{q}}_1^{*T} \left(\mathbf{M}\,\ddot{\mathbf{q}} - \mathbf{Q}_{ex} \right) \tag{6.108}$$

In the previous equation, the RHS represents the virtual power of the inertial forces and of the external forces (sign changed). If the mechanism's movement is known, its dependent accelerations vector $\ddot{\mathbf{q}}$ will also be known. The only calculation that is required to obtain the moment M_2 is that of the virtual velocities vector $\dot{\mathbf{q}}_1^*$. This vector is easy to calculate, since normally the triangularized matrix $\mathbf{\Phi}_q$, or even matrix $\mathbf{R}$, will be already available.

To calculate the torque M_5, one can determine a virtual velocities vector $\dot{\mathbf{q}}_2^*$ based on a unit angular velocity at element 5 and null velocity at element 2. The resulting equation will be

$$M_2 \cdot 0 + M_5 \cdot 1 = \dot{\mathbf{q}}_2^{*T} \left(\mathbf{M}\,\ddot{\mathbf{q}} - \mathbf{Q}_{ex} \right) \tag{6.109}$$

The virtual velocity vectors $\dot{\mathbf{q}}_1^*$ and $\dot{\mathbf{q}}_2^*$, used in expressions (6.108) and (6.109) are nothing more than the columns of matrix $\mathbf{R}$, when angles θ_2 and θ_5 are taken as independent coordinates. In this case, expressions (6.108) and (6.109) can be jointly represented as

$$\begin{Bmatrix} M_2 \\ M_5 \end{Bmatrix} = \mathbf{R}^T \left(\mathbf{M}\,\ddot{\mathbf{q}} - \mathbf{Q}_{ex} \right) \tag{6.110}$$

Expression (6.110) or its other variants, formed from it with slight modifications, is general, and represents the optimum way to determine the motor forces.

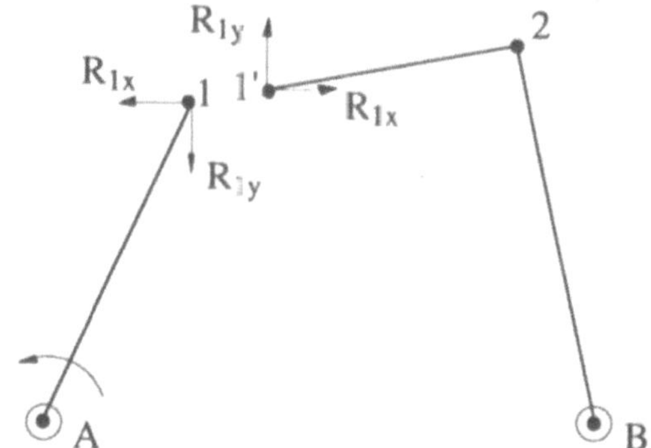

Figure 6.26. Cutting a joint in a four-bar mechanism.

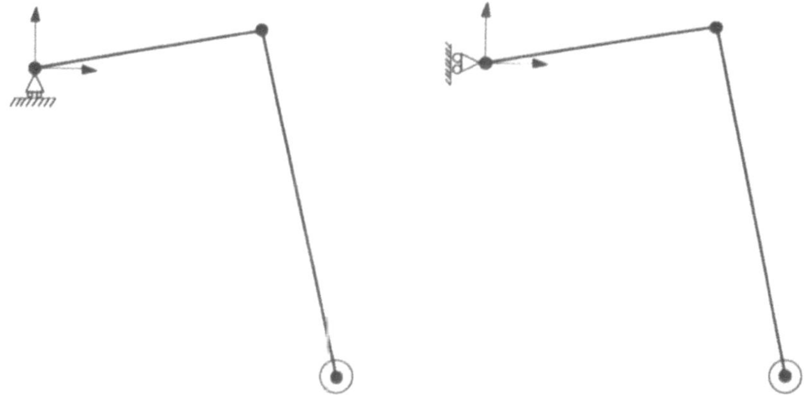

Figure 6.27. Vertical constraint in the cut joint.

Figure 6.28. Horizontal constraint in the cut joint.

6.2.5.2 Calculation of Reactions at the Joints

The calculation of reactions at the joints by means of the virtual power method is slightly more complicated than the calculation of the motor forces. There are several possibilities for tackling this problem, and not all of them are equally simple and efficient. The most important variants are described below:

a) Method of elimination of a joint. As its name implies, this method consists of eliminating the joint from the multibody system in which it is wished to calculate the reactions by substituting the said joint for the corresponding reaction forces. Figure 6.26 shows this method applied to a four-bar mechanism. When eliminating a joint that constrains a certain number of degrees of freedom (two, in the example of Figure 6.26), the total number of degrees of freedom of the mechanism is increased by the same number. Sometimes, such as happens in the

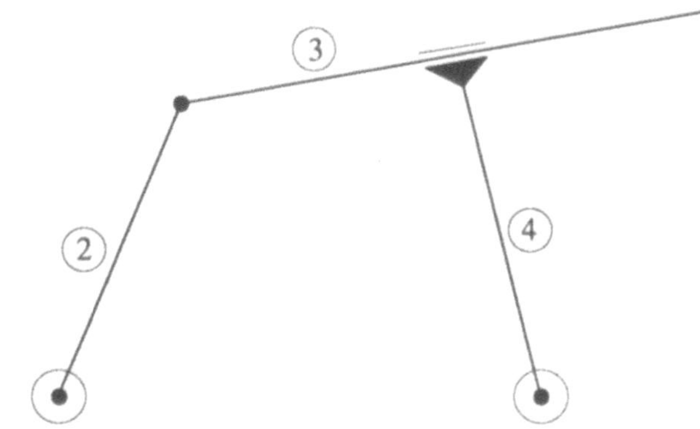

Figure 6.29. Four-bar mechanism with a prismatic joint.

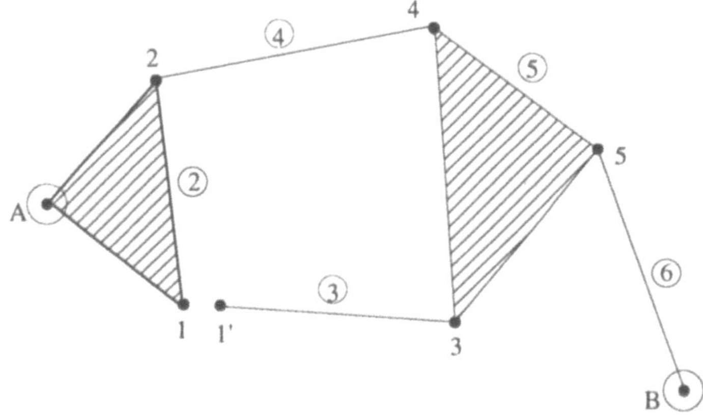

Figure 6.30. Cutting a joint in a six-bar mechanism.

example being considered, the multibody system is divided into two that share the total degrees of freedom.

It is now necessary to enter some virtual velocities in the mechanism that are compatible with the other constraints which permit determining the reaction forces at the joint. In the example of Figure 6.26, this can be done by means of the method shown in Figures 6.27 and 6.28; in the first case, the zero virtual velocity at point 1 and the unit horizontal virtual velocity at point 1' is given; in the second case, the unit vertical velocity at point 1' is given. In the first case, the horizontal reaction Q_{r1x} is obtained; in the second case the vertical reaction Q_{r1y} is obtained.

If one wishes to calculate the reactions at other joints in the example above, the method could be used in a similar way. If one wishes to calculate the reactions at point 2, there is no difference with the methods explained previously. If the reactions at a fixed point (for example B) are required, a mechanism with three degrees of freedom is obtained when the joint is eliminated. The virtual velocities that permit calculating Q_{rBx} and Q_{rBy} could be the following: $\dot{q}_1^*$ is obtained by giving a zero angular velocity to element 2, a unit horizontal velocity to point B, and a null vertical velocity to the said point; $\dot{q}_2^*$ is obtained by also giving zero velocity to element 2 and a zero horizontal velocity and unit vertical velocity to point B.

If one wishes to calculate the reactions at a prismatic joint as shown in Figure 6.29, one should keep in mind that the reactions will be the normal forces at the bar and one torque. To solve this case, it would be necessary to bisect at the prismatic joint and use virtual velocities that immobilize bar 4 at the same time so they give virtual power with one and only one of the contact forces. In order for only the force to give power, it would be necessary to force bar 3 to move in a direction parallel to itself. In order for only the element to give power, it must turn without the point of contact having velocity in the direction of bar 4.

As a last example of this method, consider the six-bar mechanism of Figure 6.30 in which one can calculate the reaction forces at joint 1. In this case, the virtual velocities could be chosen as follows: to determine the horizontal reaction, select a virtual velocity vector based on giving a zero angular velocity at element 2 and a unit horizontal velocity at point 1'; the vertical reaction would be likewise calculated, with a virtual velocity obtained by giving null angular velocity at element 2 and unit vertical velocity at point 1'.

The method for eliminating a joint, described by means of examples for planar multibody systems, is perfectly adaptable to three-dimensional multibody systems and all types of joints, but there will be a slight increase in practical difficulties. The only thing required is the means for imposing any linear velocity or any angular velocity on the elements adjacent to the joint that has been eliminated.

In the case of natural coordinates, the main practical inconvenience of the method just described lies in the fact that when the joint is eliminated, new dependent coordinates appear, and, sometimes, constraint equations disappear. This means that the kinematic problem to be solved for calculating the virtual velocities is significantly different from the original, and very little of the work carried out can be used. This results in a higher cost for the calculation of the reactions by this method and also complicates the implementation.

b) Method of elimination of an element. This method can be considered as a variant of the previous one. In order to avoid adding dependent coordinates at the point for which it is wished to calculate the reactions, one of the elements connected at the joint is eliminated. The multibody system increases the number of total degrees of freedom by a number equal to the constraint equations of the eliminated element and joints. In either case, the inertial forces and external forces that act on the eliminated element must be transmitted to the points and vectors of the adjacent elements.

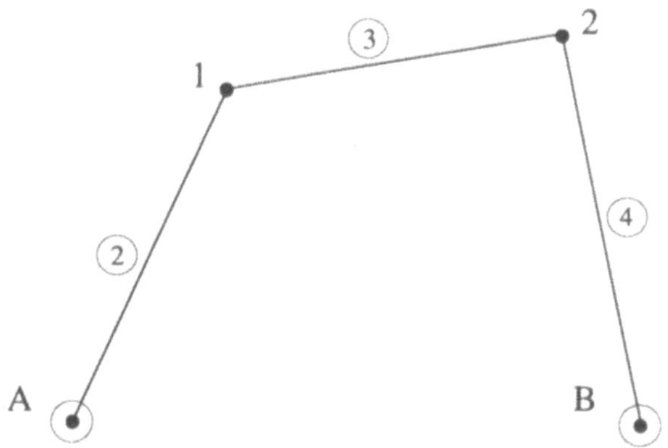

Figure 6.31. Four-bar mechanism.

The implementation of this method is complicated, since a large number of specific cases and special situations should be taken into account. The main advantage lies in the fact that it can use, to calculate the virtual velocities, the calculations carried out for determining the true velocities and accelerations.

c) Method of variation of the constraints. The method described below is oriented to calculating the force associated with each constraint equation in a way similar to that done with the Lagrange multipliers. The difference is that with the multiplier method, all of the constraint forces are calculated at the same time; whereas with this method, only the desired force or forces are calculated.

The key point of this method is that of imposing a unit variation in the corresponding constraint equation. Once again one can consider the example of the four-bar mechanism in Figure 6.31. Let's assume that it is wished to calculate the constraint force associated with the constant length condition of bar 3. The corresponding constraint equation is

$$(x_2 - x_1)^2 + (y_2 - y_1)^2 - L_{12}^2 = 0 \quad (6.111)$$

In order to calculate the force associated with this constraint equation, it is necessary to eliminate the constant distance condition and replace it with a distance condition that is variable with the time (rheonomic). Therefore, in order to calculate the virtual velocities, equation (6.111) must be replaced by

$$(x_2 - x_1)^2 + (y_2 - y_1)^2 - L_{12}(t)^2 = 0 \quad (6.112)$$

which, differentiated with respect to the time, becomes

$$(x_2 - x_1)(\dot{x}_2 - \dot{x}_1) + (y_2 - y_1)(\dot{y}_2 - \dot{y}_1) = L_{12}(t)\,\dot{L}_{12}(t) \quad (6.113)$$

Thus, to calculate the virtual velocities that permit finding the constraint force, one can give a zero value to the velocities of all the mechanism's input el-

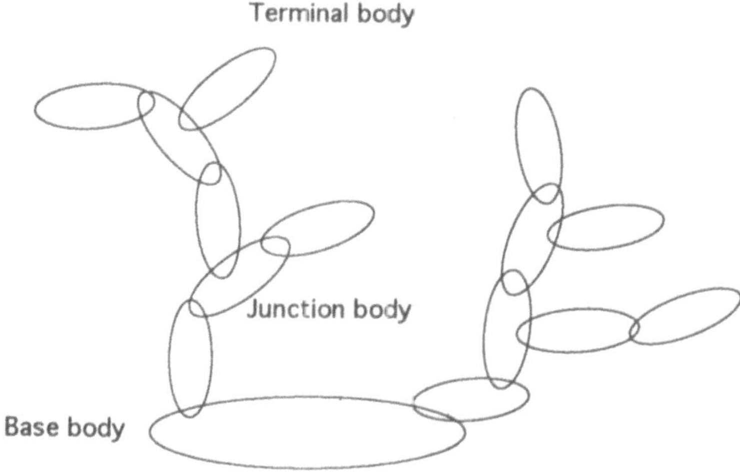

Figure 6.32. Open-chain system with branched tree structure

ements and value 1 to the derivative $\dot{L}_{12}(t)$. The equation which results from applying the theorem of virtual power will be

$$f \cdot 1 = \dot{\mathbf{q}}^{*T} \left(\mathbf{M}\, \ddot{\mathbf{q}} - \mathbf{Q}_{ex} \right) \tag{6.114}$$

where f is the force associated with the constant length constraint of bar 3.

The main advantage of this method lies in the fact that the left-hand member of the velocity constraint equation (6.113) is identical to the left-hand member of the equation which it has replaced. Since the triangularization of matrix $\boldsymbol{\Phi}_{\mathbf{q}}$ is already carried out, it is only necessary for one to do the corresponding forward and backward substitutions.

If the forces $\mathbf{Q}_c$ associated with all the constraints of an element are calculated by this method, the corresponding equilibrium equations can be set out in accordance with the natural coordinates

$$\mathbf{Q}_r + \mathbf{Q}_c - \mathbf{M}\, \ddot{\mathbf{q}} + \mathbf{Q}_{ex} = 0 \tag{6.115}$$

where $\mathbf{Q}_r$ are the forces at the joints looked for, and $\mathbf{Q}_{ex}$ are the external forces.

When one wishes to calculate all the reactions at the joints, using this method is probably the most efficient alternative to the global methods.

6.2.6 Inverse Dynamics of Open-Chain Systems

A very important particular case will be considered in the last section of this chapter: the inverse dynamics problem of open-chain systems, such as the branched tree shown in Figure 6.32.

Inverse dynamics for open-chain multibody systems can be carried out recursively in an extremely efficient way. In Chapter 8 it will be seen how to take ad-

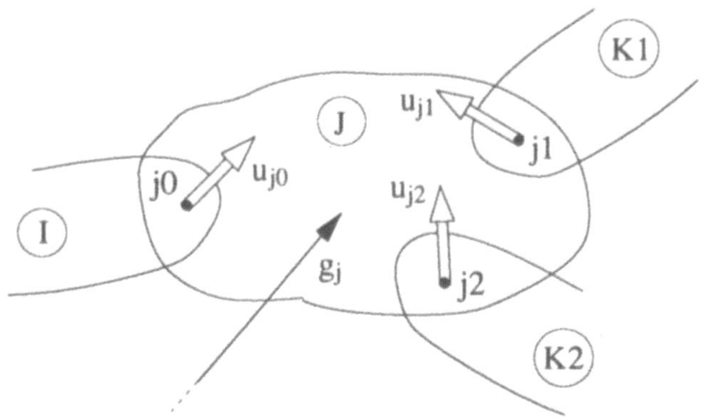

Figure 6.33. Force equilibrium of a junction body

vantage of the system topology (open-chain, for instance) so as to improve the efficiency in the direct dynamic simulation. We will concentrate now on the inverse dynamics.

It is possible to distinguish in a branched tree multibody system, such as the one shown in Figure 6.32, three kinds of bodies:

a) The *base body*, that is the body from which all the main branches originate. If there is in the system a fixed or non-moving body, it shall be chosen as the base body; otherwise, the base body can be chosen rather arbitrarily. It is common to select the base body according to physical considerations or using some criteria, for instance trying to equilibrate the lengths of the different branches.

b) The *junction bodies*, that are bifurcation bodies with one input joint connected to the *father* body, and two or more output joints connected to the *filial* bodies.

c) The *terminal bodies*, that are the last bodies in a branch.

In order to solve for the inverse dynamics in open-chain systems, it will be assumed that the system's motion, including positions, velocities and accelerations, as well as the external forces are known. We will see how to obtain the driving forces acting on the system's degrees of freedom and the constraint reaction forces at the joints.

The method that we will describe next is a variant of the well-known recursive Newton-Euler method widely used in robotics (Luh et al. (1980)). Let us consider the general case of the force and torque equilibrium of a *junction body* J, such as the one shown in Figure 6.33. Terminal bodies, simple two-joint bodies, and even the base body, can be considered as particular cases of this general junction body. In order to make this analysis as simple as possible, we will consider only multibody systems with revolute R and prismatic P joints.

Link J originates from link I, and links K1, K2, etc., originate from J; whereas $\mathbf{g}_j$ is the position vector for the link J center of gravity. Point $j0$ and unit vector $\mathbf{u}_{j0}$ are the *input* point and vector (reference point and direction in J for the joint with link I). Points $j1, j2$, etc., and vectors $\mathbf{u}_{j1}$, $\mathbf{u}_{j2}$, etc., are the *output* points and vectors (reference points and directions in J, for the joints with links K1, K2, etc.). All the joint constraint or driving forces and torques will be referred to these points and vectors.

In order to formulate the equilibrium equations, the following sets of forces will be considered:

a) *External forces and torques.* It is assumed that $\mathbf{f}_j^{ext}$ and $\mathbf{n}_j^{ext}$ are the force and torque resulting from all the externally applied loads, respectively, applied at point $\mathbf{g}_j$.

b) *Inertial forces.* They are also applied at $\mathbf{g}_j$. According to the Newton and Euler expressions they are:

$$\mathbf{f}_j^{in} = - m_j \ddot{\mathbf{g}}_j \tag{6.116}$$

$$\mathbf{n}_j^{in} = - \mathbf{I}_j \dot{\boldsymbol{\omega}}_j - \widetilde{\boldsymbol{\omega}}_j \mathbf{I}_j \boldsymbol{\omega}_j \tag{6.117}$$

where m_j and $\mathbf{I}_j$ are the mass and inertia tensor of link (J). The inertial tensor is expressed on a reference frame with axes parallel to the global or inertial frame and with its origin in the center of gravity $\mathbf{g}_j$.

c) *Joint forces.* We will consider that $\mathbf{f}_{jkn}$ and $\mathbf{n}_{jkn}$ are the total (driving and reaction) force and torque that link Kn exerts on link J at point kn. Vectors $\mathbf{f}_{ji}$ and $\mathbf{n}_{ji}$ are the reaction force and torque between bodies I and J.

The equilibrium of forces can be expressed as:

$$\mathbf{f}_{ji} + \mathbf{f}_j^{ext} + \sum_n \mathbf{f}_{jkn} - m_j \ddot{\mathbf{g}}_j = 0 \tag{6.118}$$

The equilibrium of torques with respect to the input point $j0$ becomes:

$$\mathbf{n}_{ji} + \mathbf{n}_j^{ext} + \sum_n \mathbf{n}_{jkn} + \mathbf{n}_j^{in} +$$
$$+ \sum_n (\mathbf{r}_j - \mathbf{r}_{j0}) \wedge \mathbf{f}_j^{ext} + \sum_n (\mathbf{r}_{jkn} - \mathbf{r}_{j0}) \wedge \mathbf{f}_{jkn} + \sum_n (\mathbf{r}_j - \mathbf{r}_{j0}) \wedge \mathbf{f}_j^{in} = 0 \tag{6.119}$$

Equations (6.118) and (6.119) allow one to solve for the inverse dynamics recursively. It is assumed that these equations apply to a terminal body. In this case there is no summation extended to the output joints, so equations (6.118) and (6.119) can be solved respectively for the force $\mathbf{f}_{ji}$ and the torque $\mathbf{n}_{ji}$ in the input joint. If one goes backwards, applying equations (6.118) and (6.119), one will always know the forces and torques in the summation extended to the output points, since all of them have been computed previously. One should remember that $\mathbf{f}_{ij} = - \mathbf{f}_{ji}$ and $\mathbf{n}_{ij} = - \mathbf{n}_{ji}$ according to the Newton's Third Law.

The force $\mathbf{f}_{ji}$ and the torque $\mathbf{n}_{ji}$ represent the total action in joint $j0$. It is easy to separate by taking into account the nature of the joint and the driving and reaction components of the total force and torque that link I transmits to link J. One can call τ the scalar value of the driving component on a revolute joint. τ will be the projection of torque $\mathbf{n}_{ji}$ on the joint axis $\mathbf{u}_{j0}$:

$$\tau = \mathbf{n}_{ji} \cdot \mathbf{u}_{j0} \qquad (6.120)$$

For a prismatic joint, the driving force f is the projection of force $\mathbf{f}_{ji}$ on the joint axis $\mathbf{u}_{j0}$:

$$f = \mathbf{f}_{ji} \cdot \mathbf{u}_{j0} \qquad (6.121)$$

The constraint reaction force or torque can be computed accordingly, subtracting the driving component from the total force and torque acting at the input joint. This recursive procedure is far more efficient than the general or particular methods previously explained. Although its application is restricted to open-chain systems, it is always possible in a closed-chain system to compute the reaction forces at a joint using the virtual power method. Then it is possible to compute recursively the remaining reactions using the concepts explained in this section.

References

Luh, J.Y.S., Walker, M.W., and Paul, R.P.C., "On Line Computational Scheme for Mechanical Manipulators", *ASME Journal of Dynamic Systems, Measurements, and Control*, Vol. 102, pp. 69-76, (1980).

Shigley, J.E., *Kinematic Analysis of Mechanisms*, McGraw-Hill, (1969).

Strang, G., *Linear Algebra and its Applications*, 2nd edition, Academic Press, (1980).

7
Numerical Integration of the Equations of Motion

It was shown in Chapter 5 how the application of the laws of dynamics to constrained multibody systems leads to a set of differential algebraic equations (DAE). These can be transformed to second order ordinary differential equations (ODE) by proper differentiation of the kinematic constraint equations, by use of an independent set of coordinates, or by penalty formulations. A stable and accurate integration of both DAE and ODE is of great importance for the solution of the equations of motion. Although analytical solutions may be found for some simple cases, the number and complexity of the equations resulting from the majority of multibody systems requires numerical solutions. Because the theory of ordinary differential equations has been known for a long time, the stability, convergence, and accuracy of many methods have been studied in great detail. This has led to a wide use of these methods as compared to the differential algebraic equations, not so thoroughly known at this stage. As a consequence, many of the computer programs currently available for the computer-aided analysis and design of multibody systems rely on well-established methods for the solution of ODE.

In this chapter, we first present a brief introduction to the concepts involved in the solution of ODE with particular emphasis on the notion of stability, which is of fundamental importance for the integration of constrained multibody systems. This is followed by a description of the most widely-used methods from the viewpoint of their application to the numerical integration of the equations of motion of multibody systems. Finally, we also describe some of the methods used for the direct integration of DAE and draw some conclusions for real time analysis.

7.1 Integration of Ordinary Differential Equations

As shown in Chapter 5, the mixed differential algebraic equations of motion can be written after differentiation of the kinematic constraint equations as a set of n

second order ordinary differential equations. They can be expressed in a general way as

$$\ddot{\mathbf{q}} = \mathbf{F}(t, \mathbf{q}, \dot{\mathbf{q}}) \tag{7.1}$$

The solution $\mathbf{q}$ to this equation must satisfy also the initial conditions $\mathbf{q}(t_0){=}\mathbf{q}_0$ and $\dot{\mathbf{q}}(t_0){=}\dot{\mathbf{q}}_0$. The majority of general purpose computer programs for the solution of ODE integrate *first order* equations. For this purpose, the second order system of equation (7.1) may be written as a system of *2n* first order equations by defining the additional set of variables $\mathbf{s} = \dot{\mathbf{q}}$. Consequently (7.1) becomes:

$$\dot{\mathbf{s}} = \mathbf{F}(t, \mathbf{q}, \mathbf{s}) \tag{7.2}$$

$$\dot{\mathbf{q}} = \mathbf{s} \tag{7.3}$$

Introducing $\mathbf{y}^T \equiv \{\mathbf{q}^T, \mathbf{s}^T\}$ as a new vector variable that includes $\mathbf{q}$ and $\mathbf{s}$, one can write equation (7.2) in the form commonly used in textbooks dealing with ODE:

$$\dot{\mathbf{y}} = \mathbf{f}(t, \mathbf{y}) \tag{7.4}$$

Function evaluation is the name of the process by which, given t and $\mathbf{y}$, the value of $\dot{\mathbf{y}}$ or $\mathbf{f}$ is computed. This obviously entitles the calculation of the accelerations as a function of the positions and velocities. Equation (5.67) may be used to perform the function evaluation with independent coordinates; equation (5.10) with Lagrange multipliers; and equation (5.37) for the penalty method.

7.1.1 General Background

Since in a general setting it is not possible to find closed form (analytic) solutions to the set of equations (7.4) describing the motion of a multibody system, we seek numerical methods (time-marching schemes) to approximate the solution at discrete times t_1, t_2, ... t_n. We will define the time step Δt as the difference $(t_{n+1}-t_n)$, and we will consider it constant during the integration process unless it is otherwise specified. If the function $\mathbf{f}$ is continuously differentiable with respect to t and $\mathbf{y}$ over the interval of interest, then there is a unique solution to (7.3) which also satisfies the initial conditions $\mathbf{y}_0 = \mathbf{y}(t_0)$. A proof of this theorem can be found in Ince (1956).

Taylor's Series Method. A first approach to the discrete approximate solution of (7.4), assuming that $\mathbf{f}$ is sufficiently differentiable with respect to t and $\mathbf{y}$, is to expand $\mathbf{y}$ in Taylor series as

$$\mathbf{y}(t) = \mathbf{y}_0 + \Delta t\, \mathbf{y}^{\mathrm{I}}(t_0) + \frac{\Delta t^2}{2!} \mathbf{y}^{\mathrm{II}}(t_0) + \dots \dots \tag{7.5}$$

where the total time derivatives of $\mathbf{y}$ (denoted by $(^{\mathrm{I}})$) can be found by differentiating $\mathbf{f}$ as:

$$\mathbf{y}^{\mathrm{I}} = \mathbf{f}(t, \mathbf{y}) \tag{7.6}$$

$$y^{II} = f^I = f_t + f_y f \tag{7.7}$$

Therefore Taylor's expansion (7.4) can be written in terms of the function f and its derivatives. The difficulties involved in obtaining high order derivatives of f make this method unsuitable except for low order approximations resulting from the truncation of the higher order terms of equation (7.5). One such approximation is the well-known *Euler's method* which, given the solution at time step n, approximates the solution at step $n+1$ as follows:

$$y_{n+1} = y_n + \Delta t\, f(t_n, y_n) \tag{7.8}$$

which is an *explicit* method, because the RHS does not depend on y_{n+1}.

One can infer from Taylor's series expansion of (7.5) that the local truncation error resulting from Euler's method is $(\Delta t^2 f'(\xi)/2)$. It has been demonstrated (Gear (1971)), that Euler's method converges to the true solution as Δt decreases and that it is first order accurate. Although the local truncation error is proportional to the square of Δt, or $O(\Delta t^2)$, the global error, consisting of error accumulated as the integration proceeds in time, depends linearly on Δt, or $O(\Delta t)$. It is also demonstrated in Gear (1971) how an rth order method in which its global error is $O(\Delta t^r)$ has a local truncation error equal to $O(\Delta t^{r+1})$.

Accuracy is a very important aspect that needs to be considered when choosing a method to integrate the equations of motion. Euler's method is extremely simple and easy to implement in a computer program, but it yields low accuracy and requires rather small time steps. In some cases, the time step Δt may need to be so small that the round-off errors become important and render the method useless. The search for higher order methods that will produce more accurate results as the time step is decreased would lead us to include more terms in Taylor's series (7.5) and approximate the solution at each time step with a more accurate polynomial interpolation. This is not a simple task since the high order derivatives of f are not readily available. One could use numerical or even symbolic differentiation; but the equations of motion of multibody systems are so involved that these tasks would not be of practical implementation in a general purpose computer program.

Stability. In addition to accuracy, another important aspect to be considered for the integration of ODE is *stability*. One can loosely define stability as the property of an integration method to keep the errors resulting in the integration process of a given equation bounded at subsequent time steps. An unstable method will make the integration errors increase exponentially, and an arithmetic overflow can be expected even after just a few time steps. Since stability depends not only on the given method but also on the type of problem, the test equation $y' = \lambda y$, where λ is a complex valued constant, is customarily used to characterize the stability properties of a given method. This characterization is performed by defining the set of values of λ and Δt for which the corresponding method is stable.

Mathematical literature has defined a large number of different kinds of stability. We only address three kinds of stability which are of use for the analysis of multibody systems and which will be referred to repeatedly later in the text.

1. Algorithms that are stable for some restricted range of values $(\lambda \Delta t)$ are called *conditionally stable*. When using these methods, the time step should be chosen depending on the characteristics of the problem as defined by λ (or a set of λ). In the case of a nonlinear problem for which the value of λ changes with time, the algorithm may be stable for some part of the integration and unstable for another. Consequently, it is very important when using conditionally stable algorithms to know in advance the range of values $(\lambda \Delta t)$ for which the method is stable and to compare it with the possible range of λ values of the given problem. For this purpose the *region of absolute stability* of a method is defined as that set of values $(\lambda \Delta t)$ (area of the complex plane C) for which a perturbation in the solution y_n will produce a change in subsequent values which does not increase from step to step. The region of absolute stability is an intrinsic characteristic of the method which should be considered prior to the use of conditionally stable algorithms. As an example, Euler's method described above is conditionally stable and Δt must be less than $|\lambda|/2$ to assure stability.

2. Algorithms whose regions of absolute stability are $Re(\lambda \Delta t)<-\eta$ and where η is a positive constant, are call *stiffly stable* methods. These methods are important when dealing with *stiff systems* of ordinary differential equations. The *stiff problems* contain a very large dispersion (several orders of magnitude) on the values of λ and arise either as a consequence of the type of formulation such as the penalty formulations, or simply because of the physical characteristics of the multibody system. The integration of these systems by conditionally stable algorithms should be avoided, because it would require such small time steps that the integration would become exceedingly expensive and even inaccurate due to round-off errors. Stiffly stable algorithms do not include the imaginary axis of the complex plane as part of their region of absolute stability. Multibody systems may have pure vibration modes whose λ may lie in the imaginary axis. For these cases the stiff stable methods are inadequate and a more stable family of algorithms, such as *A-stable* methods, should be required for the integration.

3. An algorithm is said to be *A-stable* if the solution to $y'=\lambda y$ tends to zero as $n \rightarrow \infty$ when the $Re(\lambda)<0$, which means that the numerical solution decays to zero whenever the corresponding exact solution decays to zero. An A-stable algorithm may be also defined as an algorithm whose region of absolute stability is the complete left half complex plane including the imaginary axis. The most important consequence of the A-stability property is that there is no limitation on the size of Δt for the stability of the integration process. A-stable algorithms have also been called *unconditionally stable* in the linear setting. It is apparent that this property is very important and generally desired in

the integration of multibody and other engineering systems, since the analyst would only have to be concerned with the step size for accuracy purposes and not for stability. The use of A-stable algorithms may become strictly necessary for stiff problems, where the imaginary axis must be part of the region of absolute stability. An important subclass of the A-stable methods is the *L-stable* ones. A method is said to be *L-stable* (also called *stiff A-stable*), if it is A-stable and when applied to the solution of $y'=\lambda y$ with $\mathrm{Re}(\lambda)<0$, it gives $y_{n+1}=A(\lambda\Delta t)y_n$ where $A(\lambda\Delta t)$ tends to zero as $\mathrm{Re}(\lambda)\to\infty$. The difference between L-stability and A-stability is that the former damps out the response of the stiff components (that is equations with high λ) very rapidly, almost in only one time step. This property is interesting and very applicable in those cases with spurious stiff equations that may arise as a consequence of the formulation or modeling process. However, caution should be practiced when using these methods since they tend to artificially damp out part of the response of interest corresponding to values of λ that are not that high.

7.1.2 Runge-Kutta Methods

Euler's method is not useful because of its low accuracy. Higher order Taylor's series are also impractical because of the difficulty involved in obtaining the derivatives of $\mathbf{f}(t, \mathbf{y})$. It is possible, however, to develop one step methods that match the accuracy of the higher order Taylor's series methods by sequentially computing the function $\mathbf{f}(t, \mathbf{y})$ at several points within the time interval. The task of computing higher order derivatives is replaced by function evaluations at a given number of points. Such methods are called the *Runge-Kutta* methods. One can refer to classical books in numerical analysis (Carnahan et al. (1969), Conte and Boor (1972)) for a detailed description and derivation of these methods.

One of these algorithms which is commonly used in engineering applications is the second order explicit Runge-Kutta method defined by:

$$\mathbf{y}_{n+1} = \mathbf{y}_n + \frac{\Delta t}{2}\,(\mathbf{k}_1 + \mathbf{k}_2) \tag{7.9}$$

$$\mathbf{k}_1 = \mathbf{f}(t_n, \mathbf{y}_n) \tag{7.10}$$

$$\mathbf{k}_2 = \mathbf{f}(t_n + \Delta t, \mathbf{y}_n + \Delta t\,\mathbf{k}_1) \tag{7.11}$$

which is also called the *improved Euler's method, modified trapezoidal method,* or *Heun's method*. Note that two function evaluations are required per time step, which in the case of multibody systems implies the solution of the equations of motion to obtain the accelerations twice at the given time step. The method is also explicit because $\mathbf{k}_1$ does not depend on $\mathbf{k}_2$, and neither one depends on $\mathbf{y}_{n+1}$.

One of the most widely used Runge-Kutta methods is the fourth order method which requires four function evaluations per time step:

$$\mathbf{y}_{n+1} = \mathbf{y}_n + \frac{\Delta t}{6}\left(\mathbf{k}_1 + \mathbf{k}_2 + \mathbf{k}_3 + \mathbf{k}_4\right) \tag{7.12}$$

$$k_1 = f(t_n, y_n) \tag{7.13}$$

$$k_2 = f\left(t_n + \frac{\Delta t}{2}, \, y_n + \frac{\Delta t}{2} k_1\right) \tag{7.14}$$

$$k_3 = f\left(t_n + \frac{\Delta t}{2}, \, y_n + \frac{\Delta t}{2} k_2\right) \tag{7.15}$$

$$k_4 = f(t_n + \Delta t, \, y_n + \Delta t \, k_3) \tag{7.16}$$

This method is *explicit*, because all the k_i depend on previous values already calculated. Otherwise, the method is said to be *implicit*. Solving for y_{n+1} at each time step requires an iterative process for the solution of a set of nonlinear equations. A general r-stage implicit Runge-Kutta method requires r function evaluations and has the following general form:

$$k_i = f\left(t_n + c_i \Delta t, \, y_n + \Delta t \sum_{j=1}^{r} a_{ij} k_j\right) \tag{7.17}$$

$$y_{n+1} = y_n + \Delta t \sum_{i=1}^{r} b_i \, k_i \tag{7.18}$$

If $a_{ij}=0$ for $j{\geq}i$, then the values k_i can be computed explicitly from the preceding values $k_1, k_2,....k_{i-1}$, and the method is *explicit*. If $a_{ij}=0$ for $j>i$ and some $a_{ii}{\neq}0$, the method is called *semi-explicit*. If all $a_{ii}{\neq}0$ the method is *diagonally implicit*. Butcher (1964) has shown that it is possible to achieve order $2r$ for an implicit r-stage method. The two-stage method with $a_{11}=a_{12}=0$, $a_{21}=a_{22}=1/2$, $c_1=0$, $c_2=1$, and $b_1=b_2=1/2$ corresponds to the *trapezoidal rule* which according to Dalhquist (1963) is the most accurate second order A-stable method. This method can also be expressed in finite difference form as

$$y_{n+1} = y_n + \frac{\Delta t}{2}\left[f(t_n, y_n) + f(t_{n+1}, y_{n+1})\right] \tag{7.19}$$

Norsett (1974) proposed a diagonally implicit two-stage Runge-Kutta family of methods for which $a_{11}=a_{22}=\alpha$, $a_{21}=1-2\alpha$, $a_{12}=0$; $b_1=b_2=1/2$ and $c_1=\alpha$, and $c_2=1-\alpha$. For general values of α, the method is second order accurate. For $\alpha=1\pm\sqrt{2}/2$ the method is L-stable. For $\alpha=(3+\sqrt{6})/2$ the method is third order and A-stable and corresponds to the approximation used by Calahan (1968). Good results of the Norsett algorithms have been reported by Smith (1975) who used them for the solution of first as well as second order ODEs.

The explicit Runge-Kutta methods are easy to implement because they only require function evaluations and are self-starting, meaning that they do not need any other algorithm or technique to start the integration process. However, they are only conditionally stable. Figure 7.1 shows the area of absolute stability of the fourth order explicit Runge-Kutta method of equation (7.12). In addition, they require several function evaluations per time step. This turns out to be computationally expensive when considering large systems of nonlinear equations, such as those arising in the analysis of multibody systems.

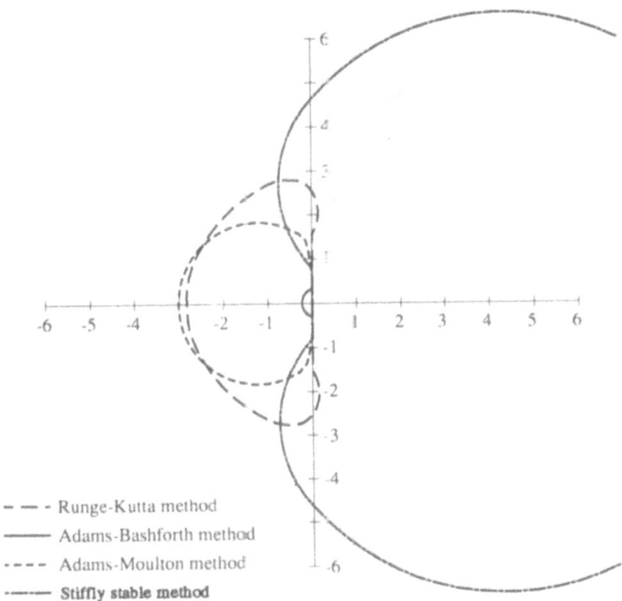

Figure 7.1. Areas of absolute stability of some fourth order methods.

Implicit Runge-Kutta methods are more stable (the r-stage methods of order $2r$ are A-stable), and much more accurate than the explicit ones. However, except for the simplest trapezoidal rule, they are more difficult to implement and much more expensive to use. A set of nonlinear equations that involve repeated function evaluations needs to be solved at each time step. A disadvantage of both the implicit and explicit methods is that it is very difficult to evaluate bounds for the accumulated or propagated error. This makes error control and time-step adjustments rather involved with this kind of method. An algorithm that uses error control and time step adjustment for the Runge-Kutta method of order four is due to Felhberg (1970). The implementation is explained in detail by Burden et al. (1989).

7.1.3 Explicit and Implicit Multistep Methods

The Taylor's series and the Runge-Kutta Methods are called *single step* methods because they only require information on the time step $(n, n+1)$ to advance to the next step. When information of other previous steps is also used, the resulting method is called *multistep*. These methods are very simply derived by integrating the differential equation $y'=f(t, y)$ from t_{n-p} to t_{n+1}, where p is a positive integer.

$$\mathbf{y}_{n+1} = \mathbf{y}_{n-p} + \int_{t_{n-p}}^{t_{n+1}} \mathbf{f}(t, \mathbf{y})\, dt \tag{7.20}$$

In order to carry out the integration, one can approximate the function $\mathbf{f}(t, \mathbf{y})$ by using backward finite differences, with the values $\mathbf{f}(t, \mathbf{y})$ calculated at previous time steps. One is referred to Carnahan et al. (1969) and Conte and Boor (1972) for a detailed description of this process. The end result is a family of methods that depends on the order of approximation of $\mathbf{f}(t, \mathbf{y})$ and on the value of p. If the approximation for $\mathbf{f}(t, \mathbf{y})$ includes the value at $n+1$, the method is called implicit, and explicit otherwise. Implicit methods are much more accurate and stable than the explicit ones. However, they are also more difficult to use, since $\mathbf{y}_{n+1}$ can not be solved for explicitly and an iteration process is required.

The general form of the multistep methods is given by the following expression:

$$\sum_{i=0}^{p+1} \alpha_i\, \mathbf{y}_{n+1-i} + \sum_{i=0}^{k} \Delta t\, \beta_i\, \mathbf{f}(t_{n+1-i},\, \mathbf{y}_{n+1-i}) = 0 \tag{7.21}$$

The α_i and β_i are parameters that define the method. If $\beta_0=0$ the term $\mathbf{f}(t_{n+1}, \mathbf{y}_{n-1})$ does not appear in the difference equation and the method is explicit. If $\beta_0 \neq 0$ the method is implicit. The methods resulting from $\beta_i=0$ for all $i \geq 1$ are called *backward-difference methods*.

One can readily see that the trapezoidal rule defined by (7.19) corresponds to $\alpha_0=-\alpha_1=1$ and $\beta_0=\beta_1=-1/2$. In addition, the more terms involved in (7.21) the better the approximation and the accuracy of the method will be. However, increasing the number of terms also leads to a larger amount of information from previous steps that needs to be stored. This can be a serious disadvantage in the analysis of multibody systems with a large number of equations, since it may lead to swapping of information from the in-core to out-of-core computer storage.

Three important conclusions known as Dahlquist's theorem (1963) may be mentioned at this stage, which can be summarized as follows:

1. There is not an *A*-stable explicit linear multistep method.

Table 7.1. Coefficients of the Adams-Bashforth Methods.

k	α_0	α_1	β_0	β_1	β_2	β_3	β_4	β_5
1	1	1	0	1				
2	2	2	0	−3	1			
3	12	12	0	−23	16	−5		
4	24	24	0	−55	59	−37	9	
5	720	720	0	−1901	2774	−2616	1274	−251

Table 7.2. Coefficients of the Adams-Moulton Methods.

k	α_0	α_1	β_0	β_1	β_2	β_3	β_4
1	1	1	-1				
2	2	2	-1	-1			
3	12	12	-5	-8	1		
4	24	24	-9	-19	5	-1	
5	720	720	-251	-646	264	-106	19

2. The order of accuracy of a linear multistep A-stable method cannot exceed two.
3. The second order accurate A-stable linear multistep method with the smallest error constant is the trapezoidal rule.

The widely known Adams-Bashforth methods are explicit with $p=0$. Table 7.1 shows the parameters corresponding to different orders of accuracy. The most commonly used Adams-Bashforth method is the fourth order method that takes the following form:

$$y_{n+1} = y_n + \frac{\Delta t}{24} (55\, f_n - 59\, f_{n-1} + 37\, f_{n-2} - 9\, f_{n-3}) \qquad (7.22)$$

with local truncation error $E = (251/720)\,\Delta t^5\, f^{IV}(\xi)$. Being explicit, these methods are conditionally stable, and the area of absolute stability for the fourth order method is illustrated in Figure 7.1.

The family of implicit methods obtained for $p=0$ are called the *Adams-Moulton* methods. Table 7.2 depicts the coefficients of several of the methods. The widely used fourth order method is

$$y_{n+1} = y_n + \frac{\Delta t}{24} (9\, f_{n+1} + 19\, f_n - 5\, f_{n-1} + f_{n-2}) \qquad (7.23)$$

with the local truncation error $E = -(19/720)\,\Delta t^5\, f^{IV}(\xi)$.

The use of (7.23) requires an iteration process that is usually initiated using (7.22) to obtain an estimate or prediction of the value of f_{n+1}. It is worth paying this price, since the Adams-Moulton method is much more accurate than the Adams-Bashforth as seen by the respective constants of the error terms. Note also that the (7.22) is a fourth order method with only one function evaluation per time step as compared to the four evaluations required by the explicit fourth order Runge-Kutta Method. As a consequence of Dahlquist's theorem (1975), the Adams-Moulton method is also conditionally stable. The area of absolute stability for the fourth order method is also illustrated in Figure 7.1.

Predictor-Corrector Iteration. The use of an implicit multistep method leads to much more accurate results than the explicit ones. Since $f(t_{n+1}, y_{n+1})$ is nonlinear for the case of multibody systems, a set of nonlinear equations is required for the solution of y_{n+1}. The necessary iteration process can be initiated by using an explicit method of the same order to obtain a predictor. For the case of the Adams-Moulton method of (7.23) a good predictor is the Adams-Bashforth method of equation (7.22). One can obtain a predicted value y_{n+1}^0 with which f_{n+1} of (7.23) can be computed and from these a new approximation y_{n+1}^1. This value can be reentered again in (7.23) to obtain a new estimate y_{n+1}^2, and so on until the difference of two consecutive values is smaller than a prescribed tolerance. This process can be summarized in the following algorithm:

Algorithm   7-1

1. Use the explicit method (predictor) to obtain y_{n+1}^0
2. Use the implicit method (corrector) for the successive y_{n+1}^k, $k = 1, \ 2, \ ...$
3. Check $\dfrac{y_{n+1}^k - y_{n+1}^{k-1}}{y_{n+1}^k} <$ tolerance
4. If 3 is true, go to the next step. Otherwise go to step 2.

The predictor-corrector algorithm outlined above is equivalent to performing a fixed point iteration for the solution of the nonlinear equations on y_{n+1}. This iteration is rather slow, since the rate of convergence is only linear in the neighborhood of the solution. It can also be shown (Carnahan et al. (1969)) that the time step required for this iteration to converge has to satisfy the following condition:

$$\Delta t < \frac{C}{\partial f / \partial y} \tag{7.24}$$

where C is a constant that depends on the type of explicit method being used. One may see how Δt is limited not only by the stability criteria but also by the convergence of the iteration process that is required at each time step.

It is also possible to use a Newton-Raphson or quasi Newton-Raphson iteration by providing a derivative of the function $f(t, y)$ with respect to y. In this way the convergence rate increases from linear to quadratic. Some integrators have the option of computing f_y by finite difference approximations or allowing the user to provide it. In this last case the calculation of f_y is entirely equivalent to linearize the equations of motion as explained in Chapter 9.

Error Estimates. A very positive advantage of the use of predictor-corrector algorithms is the possibility of finding an estimate of the error at each time step. This can be accomplished using the local truncation errors of both predictor-corrector methods. For example, if $y(t_{n+1})$ represents the exact value of y at time t_{n+1} the errors produced by the predictor and corrector of (7.22) and (7.23) will respectively be:

$$y(t_{n+1}) - y_{n+1}^0 = \frac{251}{720} \Delta t^5 \, f^{IV}(\xi_1) \tag{7.25}$$

$$y(t_{n+1}) - y_{n+1}^1 = -\frac{19}{720} \Delta t^5 \, f^{IV}(\xi_2) \tag{7.26}$$

For a sufficiently small time step one can assume that f^{IV} remains constant and by subtracting (7.25) and (7.26), one can get the following estimate of the error after one iteration:

$$y(t_{n+1}) - y_{n+1}^1 \cong \frac{1}{14} (y_{n+1}^1 - y_{n+1}^0) \tag{7.27}$$

This error estimate allows for the possibility of general purpose integrators with time step adjustments. If the error is larger than a pre-specified tolerance, it may be more convenient to reduce the time step rather than continue the fixed point iteration. However, if the error is small, the time step can be enlarged to achieve a faster integration. The most expensive part of the integration process is the function evaluation $f(t, y)$ which requires the solution of the dynamic equations. These function evaluations can be composed in the case of multibody systems of a large number of coupled nonlinear equations.

7.1.4 Comparison Between the Runge-Kutta and the Multistep Methods

The Runge-Kutta and multistep methods can be used for the solution of the equations of motion of multibody systems. The Runge-Kutta methods are single-step and therefore self-starting. In addition they need a minimum amount of storage requirements. However, they require a larger number of function evaluations (four for the fourth order method). Due to the difficulty of estimating their local truncation errors, the time step adjustment can only be performed by integrating with two different time steps. This is computationally expensive.

The multistep methods require a smaller amount of function evaluations, particularly if the time step is chosen so that the number of predictor-corrector iterations per step is kept below two or three. Error estimates are easily provided, and step-size adjustments can be performed with no difficulties. Being a multistep method, they are not self-starting, requiring the help of a single-step method to start the integration. In the case of multibody systems, the forcing terms may have jumps or discontinuities. While this does not affect a single-step method, a multistep method will need to be reinitialized. The necessary historical data pool is larger for these methods thus requiring a larger amount of storage.

Regardless of these difficulties and since the total cost of integration is mostly due to the number of function evaluations, the general purpose integrators that are based on predictor-corrector Adams-Moulton-Bashforth methods (Gear (1971), and Shampine and Gordon (1975)) have been customarily used for the integration of the equations of motion of multibody systems. Both explicit Runge-Kutta methods and multistep methods suffer from being only condition-

Table 7.3. Parameter values of the Gear's Stiffly Stable Methods.

p	β_0	α_0	α_1	α_2	α_3	α_4
1	-1	1	-1			
2	-2	3	-4	1		
3	-6	11	-18	9	-2	
4	-12	25	-48	36	-16	3

ally stable. This poses serious limitations in the size of the time steps in those systems with a large dispersion on the values of λ, the so-called *stiff systems*, and also in those cases in which the forcing terms contain high frequency components.

Stiff Systems of Equations. A stiff system of differential equations is characterized by a large dispersion of the values λ of each of the individual equations. The stiffness can be produced by the physical characteristics of the multibody system (components with large differences in their masses, stiffness and/or damping). However in many other instances, stiffness is numerically induced due to either the discretization process, the large number of components and equations of motion, or sudden or accumulated violations in the constraint conditions. Gear (1971) developed a family of variable order *stiffly-stable* algorithms for the solution of stiff problems. Table 7.3 illustrates the parameter values of such methods. Each method is a backward-difference multistep algorithm. The second order method is A-stable (it actually is L-stable) but much less accurate than the trapezoidal rule, which is the most accurate second order A-stable method. For orders larger than two, the regions of stability do not include the complete imaginary axis. This feature does not appear appropriate for typical multibody systems with oscillatory motion. In second order systems, the backward difference methods may introduce an excess of artificial damping in the interesting part of the response. They tend to yield better results in first order problems with exponentially decaying responses; such as those in heat conduction, mass transport, and ground water flow (Wood (1990)).

In the case of constrained multibody systems with stiff equations, the analyst may benefit by the choice of an A-stable method (unconditionally stable) and not have to worry about the stiff effects produced by either the physical characteristics of the multibody system, the forcing terms, or the large number of equations of motion. Unconditionally stable methods are widely used in the field of structural dynamics because of the high frequencies produced by the finite element discretization, even to such an extent that A-stability is considered as a necessary condition for any new method to gain acceptance within the structural engineering community. The next section reviews a few of the mostly used A-stable methods in structural dynamics. Some of them have also been successfully used in the integration of the equations of motion of multibody systems (See Bayo et al. (1991), and Chapter 8).

7.1.5 Newmark Method and Related Algorithms

Newmark (1959) proposed what has become one of the most popular family of algorithms for the solution of problems in structural dynamics. His method relies on the following interpolations that relate positions, velocities, and accelerations from step n to $n+1$:

$$\mathbf{v}_{n+1} = \mathbf{v}_n + \Delta t\left[(1-\gamma)\,\mathbf{a}_n + \gamma\,\mathbf{a}_{n+1}\right] \tag{7.28}$$

$$\mathbf{x}_{n+1} = \mathbf{x}_n + \Delta t\,\mathbf{v}_n + \frac{\Delta t^2}{2}\left[(1-2\beta)\,\mathbf{a}_n + 2\beta\,\mathbf{a}_{n+1}\right] \tag{7.29}$$

where $\mathbf{x}_n$, $\mathbf{v}_n$, and $\mathbf{a}_n$ are approximations to the position, velocity, and acceleration vectors at time step n; and β and γ are the parameters that define the method. The method is implicit, and A-stability is guaranteed for $2\beta \geq \gamma \geq 1/2$ (Bathe (1982)). The trapezoidal rule is a particular case of this family for which $\beta=1/4$ and $\gamma=1/2$. This case also corresponds to the assumption that the acceleration is constant over the time interval $[t_n,\ t_{n+1}]$ and equal to $(\mathbf{a}_n+\mathbf{a}_{n+1})/2$. This method is also known as the *average acceleration method*. For any other set of β and γ values within the range of A-stability, the degree of accuracy of the method degrades to order one. The linear acceleration method, in which a linear variation of the acceleration in the time interval $[t_n,t_{n+1}]$ is assumed, corresponds to the case $\beta=1/6$ and $\gamma=1/2$. This method is conditionally stable and has little practical importance; however, it was used as the basis for another important method known as the *Wilson-θ method* (Bathe (1982)).

Similar to the multistep methods, the implicit algorithm of equations (7.28) and (7.29) can be used in a predictor-corrector fashion with fixed point iteration. This is not the way it is customarily applied in structural dynamics. Rather, the interpolations of (7.28) and (7.29) are directly introduced into the equations of motion. This leads to a set of algebraic equations which can be linear or nonlinear depending on the type of problem, with $\mathbf{a}_{n+1}$ as the resulting unknowns. The algebraic equations can alternatively be solved with $\mathbf{x}_{n+1}$ as primary unknowns, by substituting $\mathbf{a}_{n+1}$ and $\mathbf{v}_{n+1}$ in terms of $\mathbf{x}_n$, $\mathbf{v}_n$, $\mathbf{a}_n$ and $\mathbf{x}_{n+1}$. Equations (7.28) and (7.29) can then be written as:

$$\mathbf{a}_{n+1} = \frac{1}{\beta\,\Delta t^2}\,(\mathbf{x}_{n+1} - \mathbf{x}_n) - \frac{1}{\beta\,\Delta t}\,\mathbf{v}_n - (1 - \frac{1}{2\beta})\,\mathbf{a}_n \tag{7.30}$$

$$\mathbf{v}_{n+1} = \frac{\gamma}{\beta\,\Delta t}\,(\mathbf{x}_{n+1} - \mathbf{x}_n) - (\frac{\gamma}{\beta} - 1)\,\mathbf{v}_n - (\frac{\gamma}{2\beta} - 1)\,\Delta t\,\mathbf{a}_n \tag{7.31}$$

The procedure can be explained by applying these equations to a linear structural dynamics problem which has the following form (Bathe (1982) or Hughes (1987)):

$$\mathbf{M}\,\mathbf{a}_{n+1} + \mathbf{C}\,\mathbf{v}_{n+1} + \mathbf{K}\,\mathbf{x}_{n+1} = \mathbf{F}(t) \tag{7.32}$$

with $\mathbf{M}$, $\mathbf{K}$, and $\mathbf{C}$ as the mass, stiffness, and damping matrices, respectively; and $\mathbf{F}$ as the vector of externally applied forces. The substitution of equations (7.30) and (7.31) into (7.32) yields

$$\left[\frac{1}{\beta\,\Delta t^{2}}\,\mathbf{M} + \frac{\gamma}{\beta\,\Delta t}\,\mathbf{C} + \mathbf{K}\right]\mathbf{x}_{n+1} =$$

$$= \mathbf{F}(t) + \mathbf{M}\left[\frac{1}{\beta\,\Delta t^{2}}\,\mathbf{x}_n + \frac{1}{\gamma\,\Delta t}\,\mathbf{v}_n + (1 - \frac{1}{2\beta})\,\mathbf{a}_n\right] + \tag{7.33}$$

$$+ \mathbf{C}\left[\frac{\gamma}{\beta\,\Delta t}\,\mathbf{x}_n + (\frac{\gamma}{\beta} - 1)\,\mathbf{v}_n + (\frac{\gamma}{2\beta} - 1)\,\Delta t\,\mathbf{a}_n\right]$$

Once the constant matrix that multiplies $\mathbf{x}_{n+1}$ on the left-hand side has been triangularized, the solution for the displacements at each time step only requires the formation of the right-hand side of (7.33) plus a forward reduction and a backwards substitution. This implementation is by far more efficient than the iterative process required with a fixed point iteration which would require a function evaluation per iteration with several iterations per time step. This substitution has also been carried out for nonlinear problems in structural dynamics. The resulting set of nonlinear algebraic equations in $\mathbf{x}_{n+1}$ is customarily solved by either Newton-Raphson iteration, which has a quadratic convergence in the neighborhood of the solution, secant methods, or quasi-Newton methods (Bathe (1982)).

The same idea can be applied in principle to the equations of motion of multibody systems which take the following general form:

$$\mathbf{M}(\mathbf{q})\,\ddot{\mathbf{q}} + \mathbf{P}(\mathbf{q}, \dot{\mathbf{q}}) = \mathbf{F}(t) \tag{7.34}$$

However, the substitution of (7.30) and (7.31) in (7.34) usually yields a highly nonlinear set of equations in the unknowns $\mathbf{q}_{n+1}$. The tangent or quasi-tangent matrices necessary for the solution of the resulting set of equations through Newton-Raphson iteration may be of such complexity, that one may end up being forced to use the simpler but less efficient predictor-corrector algorithms with the fixed point iteration outlined above. For those cases in which the formation of the tangent or quasi-tangent matrix is possible, this way of integrating the equations of motion tends to be much more efficient than the predictor-corrector schemes which only have linear convergence in the neighborhood of the solution. Such cases and special implementations will be dealt with in Chapter 8.

The most accurate A-stable algorithm of the Newmark family is the *trapezoidal rule* ($\beta=1/4$ and $\gamma=1/2$) which is energy preserving for linear systems. It does not damp out any of the frequency content of the system during the integration process (for a detailed analysis of the accuracy and stability of the Newmark method refer to Bathe (1982) and Hughes (1987)). For $\gamma > 1/2$, the A-stability is preserved and artificial damping is introduced. However, the accuracy is reduced to first order. This artificial damping is in some instances necessary because the mathematical model may contain spurious high frequency content that needs to be damped out.

There is another reason why artificial damping may be necessary. The trapezoidal rule is unconditionally stable for linear problems. However, this property is not maintained in the nonlinear regime. As examples of instabilities, Hughes (1976) reported pathological energy growth in structural dynamic problems with bilinear softening material. Cardona and Geradin (1989) and Bayo et al. (1991) also reported instability of the trapezoidal rule in the integration of constrained multibody systems. Gourlay (1970) considered the nonlinear scalar equation $\dot{y}+\lambda(y)y = 0$. He reported that when $\lambda_n>\lambda_{n+1}$, the trapezoidal rule becomes conditionally stable with critical time step $\Delta t \leq 4/(\lambda_n-\lambda_{n+1})$ (See example below). One way of circumventing this problem is to use the *midpoint rule* and related algorithms (Simo and Wong (1991)) that preserve unconditional stability in the nonlinear regime. The midpoint rule also uses the same interpolations defined by the equations (7.28) and (7.29), with the difference that equilibrium is computed at the middle of the time step rather than at the end. Accordingly equation (7.34) becomes

$$\mathbf{M}(\mathbf{q}_{n+1/2})\,\ddot{\mathbf{q}}_{n+1/2} + \mathbf{P}(\mathbf{q}_{n+1/2}, \dot{\mathbf{q}}_{n+1/2}) = \mathbf{F}(t_{n+1/2}) \tag{7.35}$$

with $t_{n+1/2} = t_n+\Delta t/2$, $\mathbf{q}_{n+1/2} = (\mathbf{q}_{n+1}+\mathbf{q}_n)/2$, and the same for $\dot{\mathbf{q}}_{n+1/2}$ and for $\ddot{\mathbf{q}}_{n+1/2}$.

The instability of the trapezoidal rule in nonlinear problems can also be avoided by: first, introducing the energy constraints (Hughes (1983)) or the kinematic velocity constraints (Bayo et al. (1991) or Section 8.5); and secondly, by adding some artificial damping through the numerical integration scheme. We discuss in the next sections a couple of these methods.

Example 7.1

Demonstrate that the trapezoidal rule is conditionally stable in the nonlinear setting for $\lambda_n>\lambda_{n+1}$ and that the critical time step becomes $\Delta t \leq 4/(\lambda_n-\lambda_{n+1})$.

Consider the equation of motion $\dot{y}+\lambda(y)y = 0$. $\qquad\qquad$ (i)

The trapezoidal rule is given by

$$y_{n+1} = y_n+\frac{\Delta t}{2}\left[\dot{y}_n + \dot{y}_{n+1}\right] \tag{ii}$$

Substituting (i) into (ii) and grouping terms together one obtains

$$y_{n+1} (2 + \lambda_{n+1}\,\Delta t) = y_n (2 - \lambda_n\,\Delta t) \tag{iii}$$

and the condition for stability becomes

$$\frac{y_{n+1}}{y_n} = \frac{\left|2 - \lambda_n\,\Delta t\right|}{\left|2 + \lambda_{n+1}\,\Delta t\right|} \leq 1 \tag{iv}$$

which leads to the critical time step $\Delta t \leq 4/(\lambda_n-\lambda_{n+1})$.

α-Method (Hilber, Hughes and Taylor). Since numerical damping can not be introduced with the Newmark family unless the order of accuracy is degraded, Hilber, Hughes, and Taylor (1977) proposed the α-method which can still maintain second order accuracy and A-stability and in addition introduce variable damping depending on the value of the parameter α. This method uses the same finite difference expressions of the Newmark method (7.28) and (7.29), however the equations of motion of the type (7.34) are modified in the following form:

$$\mathbf{M}(\mathbf{q}_{n+1})\,\ddot{\mathbf{q}}_{n+1} + (1+\alpha)\,\mathbf{P}(\mathbf{q}_{n+1}, \dot{\mathbf{q}}_{n+1}) - \alpha\,\mathbf{P}(\mathbf{q}_n, \dot{\mathbf{q}}_n) = \mathbf{F}(t_{n+\alpha}) \qquad (7.36)$$

where $t_{n+\alpha} = (1+\alpha)t_{n+1} - \alpha t_n$. For $\alpha=0$ this method reduces to the Newmark family. The best choices for the parameter α lie in the interval $[-1/3, 0]$. One can reduce this three parameter family of methods to only one parameter by choosing $\gamma = (1-2\alpha)/2$ and $\beta = (1-\alpha)^2/4$. This choice of parameters results in a second order accurate, A-stable algorithm with variable artificial damping, depending on the value of α. The smaller the value of α the larger the damping is. When $\alpha=0$, there is no damping. The resulting method is the trapezoidal rule. A comparison of the accuracy of the Newmark method and the α-method as measured by the period elongation and algorithmic damping ratios can be seen in Hughes (1987) and Hilber (1976). Cardona and Geradin (1989) used the α-method successfully in the integration of constrained multibody systems.

Wilson-θ and Collocation Methods. In order to make the linear acceleration method (Newmark's method with $\beta=1/6$ and $\gamma=1/2$) unconditionally stable, Wilson (1968) proposed the idea of applying the equations of motion not at time $t+\Delta t$ but at $t+\theta\Delta t$, where $\theta > 1.37$ for unconditional stability. This method, when applied to the type of equations of multibody systems (7.22), yields the following expressions:

$$\mathbf{M}(\mathbf{q}_{n+\theta})\,\ddot{\mathbf{q}}_{n+\theta} + \mathbf{P}(\mathbf{q}_{n+\theta}, \dot{\mathbf{q}}_{n+\theta}) = \mathbf{F}(t_{n+\theta}) \qquad (7.37)$$

$$\ddot{\mathbf{q}}_{n+\theta} = (1-\theta)\,\ddot{\mathbf{q}}_{n+\theta} + \theta\,\ddot{\mathbf{q}}_n \qquad (7.38)$$

$$\mathbf{F}_{n+\theta} = (1-\theta)\,\mathbf{F}_{n+\theta} + \theta\,\mathbf{F}_n \qquad (7.39)$$

with the displacement and velocity expressions given by the Newmark interpolation (7.18) with $\beta=1/6$, and $\gamma=1/2$.

The *collocation methods* were proposed by Hilber (1976) as a three-parameter family of methods resulting from the combination of the Newmark interpolation with the Wilson-θ method. The resulting equations are composed of equations (7.37)-(7.39) plus the Newmark finite difference equations that now include the θ parameter (also called the collocation parameter), as follows:

$$\dot{\mathbf{q}}_{n+\theta} = \dot{\mathbf{q}}_n + \theta\,\Delta t \left[(1-\gamma)\,\ddot{\mathbf{q}}_n + \gamma\,\ddot{\mathbf{q}}_{n+\theta}\right] \qquad (7.40)$$

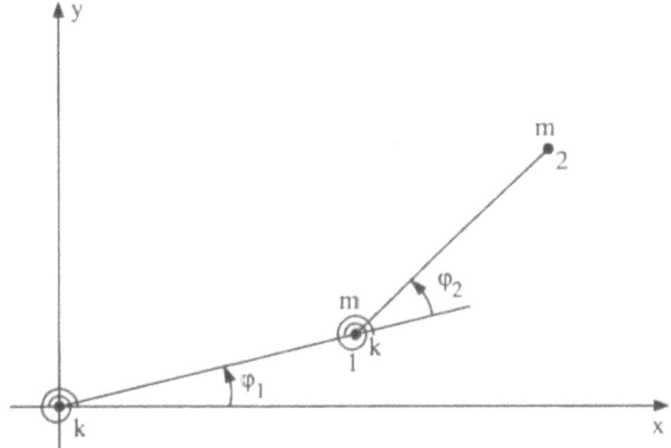

Figure 7.2. Double pendulum with rotational springs.

$$\mathbf{q}_{n+\theta} = \mathbf{q}_n + \theta \, \Delta t \, \dot{\mathbf{q}}_n + \frac{\left(\theta \, \Delta t\right)^2}{2}\left[(1 - 2\,\beta)\,\ddot{\mathbf{q}}_n + 2\,\beta\,\ddot{\mathbf{q}}_{n+\theta}\right] \qquad (7.41)$$

This family of methods now encompasses the Newmark method for which $\theta=1$ and the Wilson-θ for which $\beta=1/6$ and $\gamma=1/2$. Second order accuracy and unconditional stability are assured if the following conditions on γ, β, and θ are met:

$$\gamma = 1/2, \quad \theta \geq 1, \quad \frac{\theta}{2(\theta+1)} \geq \beta \geq \frac{2\,\theta^2 - 1}{4\,(2\theta^3 - 1)} \qquad (7.42)$$

Similar to the α-method, the collocation method also allows for controllable algorithm damping. A good comparison on accuracy and damping characteristics of both methods is included in Hughes (1987) and Hilber (1976). However, as reported by Goudreau and Taylor (1973), the method suffers from a tendency to a spurious *overshoot* of the response in the first few steps of integration. Such disadvantage is not present in either the Newmark or the α-methods.

Example 7.2

Double pendulum with rotational springs. With this example, we show the stability and convergence properties of the trapezoidal rule when its difference equations are introduced in the equations of motion as explained in Section 7.1.4. The positions are the primary variables, and the resulting set of equations are solved by means of Newton-Raphson iteration. Figure 7.2 illustrates a double pendulum that has two elements of unit mass $m = 1$ and unit length with rotational springs at the joints of value k. The joints are subjected to equal initial velocities of value 100 rad/sec. The system is then analyzed using the two independent joint rotations φ_1 and φ_2 as coordinates for a total time of 20 seconds.

The kinetic energy of this system is

$$T = \frac{1}{2} \left\{ \dot{\varphi}_1 \quad \dot{\varphi}_2 \right\} \begin{bmatrix} 3 + 2\,cos\varphi_2 & 1 + cos\varphi_2 \\ 1 + cos\varphi_2 & 1 \end{bmatrix} \left\{ \begin{matrix} \dot{\varphi}_1 \\ \dot{\varphi}_2 \end{matrix} \right\}$$

and the potential energy,

$$V = \frac{1}{2} k \left(\varphi_1^2 + \varphi_2^2 \right)$$

The application of the Lagrange's equations leads to the following equations of motion:

$$\begin{bmatrix} 3 + 2\,cos\varphi_2 & 1 + cos\varphi_2 \\ 1 + cos\varphi_2 & 1 \end{bmatrix} \left\{ \begin{matrix} \ddot{\varphi}_1 \\ \ddot{\varphi}_2 \end{matrix} \right\} = - \left\{ \begin{matrix} k\,\varphi_1 \\ k\,\varphi_2 \end{matrix} \right\} + sin\varphi_2 \begin{bmatrix} 2\,\dot{\varphi}_2 & \dot{\varphi}_2 \\ (\dot{\varphi}_2 - \dot{\varphi}_1) & -\dot{\varphi}_1 \end{bmatrix} \left\{ \begin{matrix} \dot{\varphi}_1 \\ \dot{\varphi}_2 \end{matrix} \right\}$$

Since independent joint coordinates are being used, the mass matrix is not constant, and there are velocity-dependent terms in the RHS of the equations of motion which make the integration more critical. Also, the terms coming from the springs constants k in the RHS of the equation add additional numerical stiffness to the above equations of motion. These are integrated for different values of the spring constants k: first, in the predictor-corrector fashion (Algorithm 7-1) by using the subroutine DE (Shampine and Gordon (1975)); and secondly, by substituting the difference equations corresponding to the trapezoidal rule (equations (7.28) and (7.29) with $\beta = 1/4$ and $\gamma = 1/2$) and solving the resulting set of nonlinear algebraic equations by means of the Newton-Raphson iteration. Although it is rather involved, the expression for the tangent matrix can be obtained for this particular case in closed form and its elements are:

$$t_{11} = \frac{4}{\Delta t^2}(3 + 2\,cos\varphi_2) + k - \frac{4}{\Delta t}\dot{\varphi}_2\,sin\varphi_2$$

$$t_{12} = \frac{4}{\Delta t^2}(1 + cos\varphi_2) - \frac{4}{\Delta t}(\dot{\varphi}_1 + \dot{\varphi}_2)\,sin\varphi_2 - (2\,\dot{\varphi}_1\dot{\varphi}_2 + \dot{\varphi}_2^2)\,cos\varphi_2 - (2\,\ddot{\varphi}_1 + \ddot{\varphi}_2)\,sin\varphi_2$$

$$t_{21} = \frac{4}{\Delta t^2}(1 + cos\varphi_2) - \frac{4}{\Delta t}(\dot{\varphi}_1 + \dot{\varphi}_2)\,sin\varphi_2$$

$$t_{22} = \frac{4}{\Delta t^2} + k + \dot{\varphi}_1^2\,cos\varphi_2 - \ddot{\varphi}_1\,sin\varphi_2$$

where the velocities and accelerations can be expressed in terms of the positions using the difference equations (7.28 and 7.29). Even in this simple case, the elements of the tangent matrix are quite involved. One can conclude that for a general case the formulation of the equations of motion in independent coordinates makes the task of obtaining the tangent matrix exceedingly complicated. On the other hand, the use of dependent coordinates and in particular the natural coordinates leads to a much simpler tangent matrix at the expense of increasing the number of equations and adding constraint conditions. This can be done as an exercise (See problem at the end of the chapter). These concepts are further developed in Chapter 8 (Section 8.5) in the context of real time analysis.

Another very important point is that the spring terms k appear in the tangent matrix and do not produce numerical instabilities. On the contrary, these terms help in the convergence process with increasing values of k, thus coping very nicely with the problem of numerical stiffness.

Table 7.4 contains a comparative study of both methods, by showing the resulting maximum energy errors along with the processing times for a time step of 0.001 seconds and different values of the spring constant k. The constant is changed with the idea of adding numerical stiffness to the problem. One can see how the integration with the trapezoidal rule, although not as accurate, compares favorably with the subroutine DE for small values of k (low numerical stiffness). When stiffness is present due to large values of k, the implementation of the trapezoidal rule with the positions as primary variables becomes much faster and accurate than DE which behaves very slowly in some cases and does not converge in others. With this implementation of the trapezoidal rule, the integration becomes faster as the system gets stiffer.

7.2 Integration of Differential-Algebraic Equations

7.2.1 Preliminaries

We present in this section a brief discussion on the direct integration of the differential algebraic equations (DAEs) with special emphasis in its application to the integration of the equations of motion of multibody systems. The intention is to give a basic notion of where the state of the art in this field currently is and of how these new methods are applied to the problems at hand. One is referred to specialized works in this area (Brenan et al. (1989), and Haug and Deyo (1990)) for a more comprehensive study on these topics.

Nonlinear DAEs are classified into two major groups: *implicit* and *semi-explicit*. Implicit equations take the following form:

Table 7.4. Maximum errors and total CPU time in the integration of the double pendulum.

k_1	k_2	h	Error		CPU time	
			Trapez.	DE	Trapez.	DE
0	0	0.001	$4.6 \; 10^{-4}$	$-3.4 \; 10^{-5}$	14.6	11.2
100	100	0.001	$2.7 \; 10^{-3}$	$7.3 \; 10^{-3}$	17.2	14.2
10^4	10^4	0.001	$2.3 \; 10^{-3}$	$3.8 \; 10^{-3}$	14.7	18.0
10^6	10^6	0.001	$1.2 \; 10^{-3}$	$-2.0 \; 10^{-3}$	12.1	48.4
10^8	10^8	0.001	$7.7 \; 10^{-5}$	$-4.9 \; 10^{-3}$	9.2	342.6
0	100	0.001	$4.0 \; 10^{-4}$	$-1.7 \; 10^{-5}$	14.2	10.9
0	10^4	0.001	$6.0 \; 10^{-5}$	$6.0 \; 10^{-5}$	14.0	12.1
0	10^6	0.001	$1.0 \; 10^{-3}$	$3.5 \; 10^{-2}$	11.7	47.3
0	10^8	0.001	$3.7 \; 10^{-3}$	No conv.	8.8	No conv.

$$F(t, \mathbf{y}, \mathbf{y'}) = 0 \tag{7.43}$$

with initial conditions $\mathbf{y}(t_o) = \mathbf{y}_o$, and where $\partial F/\partial \mathbf{y'}$ may be singular. This type of equation arises in problems related to electrical circuits. Semi-explicit equations can be written as:

$$\mathbf{y'} = \mathbf{f}(t, \mathbf{z}, \mathbf{y}) \tag{7.44}$$

$$\mathbf{0} = \mathbf{g}(t, \mathbf{z}, \mathbf{y}) \tag{7.45}$$

with initial conditions $\mathbf{y}(t_o) = \mathbf{y}_o$ and $\mathbf{z}(t_o) = \mathbf{z}_o$. This type of equation arises commonly in constrained multibody systems, optimal control, and trajectory prescribed path problems.

Whereas the theory of existence and uniqueness of ordinary differential equations (ODEs) is complete, that of the DAEs is still incomplete. It is also more difficult to establish than that of the ODEs. Solutions to DAEs may not always exist. If they do, they may not be unique. Other important issues that pertain to the integration of DAE are their theoretical as well as numerical solvability; that is, the identification of analytical as well as numerical solutions. Such topics are very important when considering the development of both special and general purpose DAE solvers. The amount of research dedicated to this important mathematical problem has been steadily increasing in recent years, as measured by the amount of related literature appearing in specialized journals and conferences. One may identify such published works in Brenan et al. (1989), and Haug and Deyo (1990).

Differential algebraic equations are classified according to their *differential index* or simply *index*, defined as the number of times that the DAE has to be differentiated to obtain a standard set of ODE. This ODE is also called the *underlying ODE*, and is satisfied by all the components of the solution. The differential index can also be defined as the number of differentiations necessary to solve for $\mathbf{y'}$ uniquely in terms of $\mathbf{y}$ and t. The higher the index the more complex the integration becomes. As an example, one can verify that the equations of constrained multibody systems of the form:

$$\mathbf{M}(\mathbf{q})\,\ddot{\mathbf{q}} = \mathbf{Q}(t, \mathbf{q}, \dot{\mathbf{q}}) - \boldsymbol{\Phi}_\mathbf{q}^{\mathrm{T}}(\mathbf{q})\,\boldsymbol{\lambda} \tag{7.46}$$

$$\boldsymbol{\Phi}(t, \mathbf{q}) = 0 \tag{7.47}$$

is semi-explicit of index three, with $\mathbf{q}$ as the positions, $\dot{\mathbf{q}}$ the velocities, $\ddot{\mathbf{q}}$ the accelerations, and $\mathbf{Q}$ as all the forcing terms. Equations (7.46) and (7.47) can be transformed to first order form by the following transformation $\dot{\mathbf{q}} = \mathbf{s}$, yielding:

$$\mathbf{M}(\mathbf{q})\,\dot{\mathbf{s}} = \mathbf{Q}(t, \mathbf{q}, \mathbf{s}) - \boldsymbol{\Phi}_\mathbf{q}^{\mathrm{T}}(\mathbf{q})\,\boldsymbol{\lambda} \tag{7.48}$$

$$\dot{\mathbf{q}} = \mathbf{s} \tag{7.49}$$

$$\boldsymbol{\Phi}(t, \mathbf{q}) = 0 \tag{7.50}$$

Two families of methods used for the integration of ODEs have also been utilized for DAEs: the backward difference formulae (BDF), and the implicit Runge-Kutta methods (IRK). This does not necessarily mean that all DAE problems are solvable by ODE methods. In fact this does not hold true in all the cases. However, success has been reported when using these methods for certain classes of DAE. Special care must be exercised at the time of using these algorithms since their implementation is not as simple as in the case of ODEs.

The way of solving a differential algebraic equation in essence is to approximate $\mathbf{y}'$ in (7.43) or (7.44), and (7.45) by a finite difference formula or implicit Runge-Kutta method, and solve the resulting set of nonlinear algebraic equations by some iterative procedure for an approximation to $\mathbf{y}$. For example, the simple backward Euler method ($\mathbf{y}_{n+1} = \mathbf{y}_n + \Delta t\, \dot{\mathbf{y}}_{n+1}$) can be substituted into the implicit equation (7.28) to yield

$$\mathbf{F}\left[t_{n+1}, \mathbf{y}_{n+1}, \frac{1}{\Delta t}(\mathbf{y}_{n+1} - \mathbf{y}_n)\right] = 0 \qquad (7.51)$$

The procedure can be applied in the same manner for the semi-explicit equations (7.44) and (7.45). Equation (7.51) constitutes a set of nonlinear equations with $\mathbf{y}_{n+1}$ as unknowns that can be solved by standard methods such as Newton-Raphson iteration. This way of proceeding is similar to that outlined above for the solution of the equations in structural dynamics and used in the numerical example of the double pendulum, in which the finite difference equations of the integrator (the Newmark family, collocation, or α-methods, etc.) that approximate the derivatives (accelerations and velocities) are directly introduced in the equations of motion. More recently, the implicit Runge-Kutta methods have also been used in a similar fashion, resulting in a set of nonlinear equations with $\mathbf{y}_{n+1}$ as unknowns.

7.2.2 Solutions by Backward Difference Formulae

The BDF methods were first successfully applied to the solution of DAEs. The theory of stability and convergence has been almost completed, and there is even public domain software currently available (Petzold (1982)). The method consists of directly applying the backward difference formulae to the system equations. This approach is one of the options used in commercial programs for the integration of the equations of motion of constrained mechanical systems (Chace (1984)). The procedure consists of substituting the backward difference approximations:

$$\dot{\mathbf{q}}_{n+1} = \frac{1}{\Delta t\, \beta_o}\left(\mathbf{q}_{n+1} - \sum_{i=0}^{p} \alpha_i\, \mathbf{q}_{n-i}\right) \qquad (7.52)$$

$$\dot{\mathbf{s}}_{n+1} = \frac{1}{\Delta t\, \beta_o}\left(\mathbf{s}_{n+1} - \sum_{i=0}^{p} \alpha_i\, \mathbf{s}_{n-i}\right) \qquad (7.53)$$

into the equations of motion (7.48)-(7.50) to form the following set of nonlinear algebraic equations with $\mathbf{q}_{n+1}$ and $\mathbf{s}_{n+1}$ as unknowns:

$$\mathbf{M}(\mathbf{q}_{n+1})\frac{1}{\Delta t\,\beta_o}\left(\mathbf{s}_{n+1} - \sum_{i=0}^{p} \alpha_i\,\mathbf{s}_{n-i}\right) = \mathbf{Q}_{n+1} - \Phi_{\mathbf{q}}^{T}(\mathbf{q}_{n+1})\,\lambda_{n+1} \qquad (7.54)$$

$$\frac{1}{\Delta t\,\beta_o}\left(\mathbf{q}_{n+1} - \sum_{i=0}^{p} \alpha_i\,\mathbf{q}_{n-i}\right) = \mathbf{s}_{n+1} \qquad (7.55)$$

$$\Phi(t,\,\mathbf{q}_{n+1}) = 0 \qquad (7.56)$$

The solution of this system of equations can be carried out by Newton-Raphson iteration provided a tangent matrix is available. Otherwise, a less accurate secant method or simplest fixed point iteration would have to be used. An approach that consists of substituting the backward difference equations (7.52) and (7.53) into the equations of motion written in the form of the generalized coordinate partitioning method has been proposed by Haug and Yen (1990).

It has been demonstrated by Gear and Petzold (1984) that for all index one DAEs, the k-step BDF with fixed step size are stable and convergent to order $O(\Delta t^k)$, if all initial values are sufficiently accurate. An extension of this result to variable step size is shown by Gear et al. (1985). General purpose DAE solvers designed to solve this kind of problems are not free of some implementation problems that are described below. For higher index systems, instability may be present. Encouraging results have been obtained for the particular case of semi-explicit systems which are those in which the multibody systems are classified. It has been demonstrated (Lotstedt and Petzold (1986)) that the k-step BDF with constant step-size are also convergent to order $O(\Delta t^k)$ for semi-explicit index two DAEs and even index three of the type of equations (7.46) and (7.47), when the initial conditions are defined within sufficient accuracy. This result has been generalized (Gear et al. (1985)) for variable step-size and index two problems.

While the BDF seem to achieve convergence and yield satisfactory results for a wide variety of DAEs, the actual implementation of the algorithms in general purpose solvers is not free from serious numerical difficulties. This becomes more acute for index three problems such as constrained multibody systems with equations (7.46) and (7.47). Such difficulties stem from the following points:

- It can be shown that for an index m DAE, the tangent or quasi-tangent matrix used in the Newton-Raphson iteration has a condition number of order $O(1/\Delta t^m)$. Consequently, the practical implementation of the method is bound to have large round-off errors for small sizes of the time step.

- Instabilities may result for sudden changes in the system variables and constraints, such as impacts or sudden appearances or disappearances of constraints. Any time there is a situation of discontinuity in the response, the multistep BDF tries to fit a polynomial through such discontinuity, and the step size must be severely reduced. This results in an ill-condition iteration matrix, and the Newton-Raphson iteration may end up near a solution and yet be unable to converge. These problems can be circumvented as explained by Steigerwald (1990) at the expense of reinitializing the integration with consistent initial conditions which have to be obtained from the derivatives of the constraint equations. This reinitialization produces serious delays in the integration process.

- The multistep methods are not self-starting. A k-step method requires sufficiently accurate $(k-1)$ starting values which have to be obtained by other methods which may render the method sensitive to the starting values. This problem also arises when either the time step or order of the BDF method is changed during the integration process.

7.2.3 Solutions by Implicit Runge-Kutta Methods

The IRK methods have been used as an alternative to the BDF methods for the integration of DAEs, offering some important advantages over them for the integration of multibody systems. Being single step, the IRK methods do not suffer from systems discontinuities and changes in the order or time step as the BDF do. The IRK are also self-starting which means that the only starting values required are the initial conditions. These methods have the disadvantage of leading after the time discretization to larger and more complex sets of nonlinear algebraic equations. In addition, the convergence and stability analysis is still in early stages and no public domain software is yet available.

The method consists of substituting the expression of the IRK (7.17) into the system equations. In the case of the implicit type of equations (7.43), this substitution leads to

$$\mathbf{F}(t_n + c_i\,\Delta t,\ \mathbf{y}_n + \Delta t\sum_{j=1}^{r} a_{ij}\,\mathbf{k}_j,\ \mathbf{k}_i) = 0\ ,\quad i = 1,2,...,r \tag{7.57}$$

Once the system (7.57) is solved for the values of the stage derivatives $\mathbf{k}_i$, the solution at step $n+1$ is given by equation (7.18). The application of the IRK to the solution of the equations of motion of constrained multibody systems of the form (7.48)-(7.50) becomes:

$$\mathbf{M}(\mathbf{q}_n + \Delta t\sum_{j=1}^{r} a_{ij}\,\mathbf{k}_j)\,\mathbf{L}_i = \mathbf{Q}_{n+1} + \Phi_{\mathbf{q}}^{\mathrm{T}}(\mathbf{q}_n + \Delta t\sum_{j=1}^{r} a_{ij}\,\mathbf{k}_j)\,\lambda_{n+1}$$
$$i = 1, 2, ...,r \tag{7.58}$$

$$\mathbf{k}_i = \mathbf{s}_n + \Delta t\sum_{j=1}^{r} a_{ij}\,\mathbf{L}_j \qquad i = 1, 2, ..., r \tag{7.59}$$

$$\Phi\,(t_n + c_i\,\Delta t,\ \mathbf{q}_n + \Delta t\sum_{j=1}^{r} a_{ij}\,\mathbf{k}_j) = 0 \tag{7.60}$$

where $\mathbf{k}_i$ and $\mathbf{L}_i$ are the stage derivatives for the approximations to $\mathbf{q}$ and $\mathbf{s}$, respectively. Equations (7.58) to (7.60) have to be solved for the unknowns $\mathbf{k}_i$, $\mathbf{L}_i$, and Lagrange multipliers λ. The resulting set of equations can be seen becoming larger and more involved than that resulting from the application of the BDF. Obtaining the tangent matrix of equation (7.58) for a Newton-Raphson iteration becomes an exceedingly complicated task. The use of the IRK method still defies implementation in a general purpose DAE solver.

Stability conditions for index one and semi-explicit index two DAE are given by Burrage (1982), and Brenan and Petzold (1989).

7.3 Considerations for Real-Time Simulation

Real time simulation of multibody systems requires an analysis time (integration time plus time for graphical display) smaller than the physical time taken by the actual motion of the multibody system. This has an important influence on the type of method used for the integration of the equations of motion. Commonly used integration routines in multibody dynamics are based on a variable order, variable step size multistep methods with error control. The user specifies the maximum allowable error, and the routine adapts the order and the step to fulfill the error conditions. With this kind of integrators, it is not possible to predict the computer time necessary to integrate the equations on a determined period of time. Multistep methods adjust very poorly to force and/or system discontinuities such as those arising from sudden forces and appearances or disappearances of constraints, leading to very small time steps and ill-conditioning of the Jacobian matrices. Special formulas are required to restart the integration after system or force discontinuities. These restarting procedures do not fit well into the real time condition, since the requirement for a fixed time of integration per time step will not be met.

Explicit multistep methods can be inexpensive and accurate for real time analysis provided the time step is chosen sufficiently small. However, they do not present good stability conditions. This poses a serious limiting factor for real time integration. Sometimes, the equations of motion may become stiff, whereby the solution has components whose time constants can differ in several orders of magnitude. It is convenient that the chosen integrator performs well under such conditions. This makes the implicit stiffly-stable or A-stable methods more suitable.

It seems that for real time integration, it would be more convenient to use an implicit single step integration formula with fixed time step and order that will have the same computational cost in each integration step (fixed number of iterations) and will be capable of adjusting to system discontinuities. Since the computational time needs to be smaller than the step size, the selected integration method must be computationally inexpensive, with few function evaluations and iterations in each step, and must allow for large step sizes without introducing excessive loss of accuracy and, most importantly, must not become unstable.

The stated preference towards implicit A-stable methods does not necessarily mean that the explicit methods cannot be used for real time integration. This would require, on the analyst's part, a previous knowledge of the range of the values $\lambda(t)$ of the given problem, so that the necessary Δt may be chosen. As soon as the problem has a certain degree of stiffness as most practical cases do, the necessary time step will be so small that the use of an implicit method will be more than justified.

The suitable methods for real time simulation of multibody systems should be implicit, simple step, and inexpensive in terms of computational power required, with good stability properties (A-stability, if possible), and sufficiently accurate. Accuracy contradicts other characteristics such as stability, economy, and single step integration. For real time applications, accuracy is the feature that must be sacrificed in conflicts with other properties. It is better to obtain a solution with some small error than not be able to obtain it at all in the allowed time. Moreover, many real time applications incorporate a feedback control either embedded in the dynamic formulation or carried out visually as in the case of a teleoperator training system. Feedback control helps to compensate errors and disturbances, including integration errors.

According to these considerations and in view of the theory and methods explained above, it seems that the methods used in structural dynamics (Newmark family, α-method, collocation methods), the midpoint rule, and the implicit Runge-Kutta methods are suitable choices for the integration of the equations of motion in their ODE form. In particular, the α-method which adds artificial damping to the trapezoidal rule without losing second order accuracy, the generalized trapezoidal rule with energy or velocity constraints, and midpoint rule seem to be good choices and somehow preferable over the more expensive implicit Runge-Kutta methods. However, in order to make these methods computationally inexpensive, it is necessary to limit the number of iterations necessary to solve the resulting system of algebraic nonlinear equations. We will show in Chapter 8 how the trapezoidal rule with velocity constraints performs very satisfactorily in real time analysis when it is directly introduced in the equations of motion of multibody systems. We will also show how this integration scheme fits quite well in the framework of the natural Cartesian coordinates and the penalty formulation, because it allows for a very simple expression of the tangent matrix necessary for the Newton-Raphson iteration.

Regarding the integration of the equations of motion in their DAE form, it seems that currently available DAE solvers are not capable of competing against their ODE counterparts in speed of integration.

References

Bathe, K.-J., *Finite Element Procedures in Engineering Analysis*, Prentice-Hall, (1982).

Bayo, E., García de Jalón, J., Avello, A., and Cuadrado, J., "An Efficient Computational Method for Real Time Multibody Dynamic Simulation in Fully Cartesian Coordinates", *Computer Methods in Applied Mechanics and Engineering*, Vol. 92, pp. 377-395, (1991).

Brenan, K.E, Campbell, S.L., and Petzold, L.R., *The Numerical Solution of Initial Value Problems in Differential-Algebraic Equations*, Elsevier, (1989).

Brenan, K.E. and Petzold, L.R., "The Numerical Solution of Higher Index Differential-Algebraic Equations by Implicit Runge-Kutta Methods", *SIAM Journal on Numerical Analysis*, Vol. 26, pp. 976-996, (1989).

Burden, R.L., Faires, J.D., and Reynolds, A.C., *Numerical Analysis*, 4th edition, Prindle, Weber, and Schmidt, Boston, Massachusetts, (1989).

Burrage, K., "Efficiently Implementable Algebraically Stable Runge-Kutta Methods", *SIAM Journal on Numerical Analysis*, Vol. 19, pp. 245-258, (1982).

Butcher, J.C., "Implicit Runge-Kutta Processes", *Mathematical Computation*, Vol. 18, pp. 50-64, (1964).

Calahan, D.A., "A-stable Accurate Method of Numerical Integration for Nonlinear Systems", *Proceedings of IEEE*, Vol. 56, pp. 744-751, (1968).

Cardona, A. and Geradin, M., "Time Integration of the Equations of Motion in Mechanism Analysis", *Computer & Structures*, Vol. 33, pp. 801-820, (1989).

Carnahan, B., Luther, H.A., and Wilkes, J. O., *Applied Numerical Methods*, Wiley, (1969).

Chace, M.A, "Methods and Experience in Computer-Aided Design of Large-Displacement Mechanical Systems", *Computer-Aided Analysis and Optimization of Mechanical System Dynamics*, NATO ASI Series, Vol. F9, pp. 233-259, Springer-Verlag, (1984)

Conte, S.D and Boor, C., *Elementary Numerical Analysis*, 2nd edition, McGraw-Hill, (1972).

Dahlquist, G., "A Special Stability Problem for Linear Multistep Methods", *BIT*, Vol. 3, pp. 27-43, (1963).

Fehlberg, E., "Klassische Runge-Kutta Formeln Vierter und Niedrigerer Ordnung mit Schrittweiten-Kontrolle und ihre Anwendung auf Warmeleitungs-Probleme", *Computing*, Vol. 6, pp. 61-71, (1970).

Gear, C.W., *Numerical Initial Value Problems in Ordinary Differential Equations*, Prentice-Hall, (1971).

Gear, C.W. and Petzold, L.R., "ODE Methods for the Solution of Differential/Algebraic Systems", *SIAM Journal on Numerical Analysis*, Vol. 21, pp. 367-384, (1984).

Gear, C.W., Leimkuhler, B., and Gupta, G.K., "Automatic Integration of Euler-Lagrange Equations with Constraints", *Journal of Computational and Applied Mathematics*, Vol. 12, pp. 77-90, (1985).

Goudreau, G.L. and Taylor, R.L., "Evaluation of Numerical Methods in Elastodynamics", *Computer Methods in Applied Mechanics and Engineering*, Vol. 2, pp. 69-97, (1973).

Gourley, A.R., "A Note on Trapezoidal Methods for the Solution of Initial Value Problems", *Mathematics of Computation*, Vol. 24, pp. 629-633, (1970).

Haug, E.J. and Deyo, R.C. (editors), *Real-Time Integration Methods for Mechanical System Simulation*, NATO ASI Series, Vol. F 69, Springer-Verlag, (1990).

Haug, E.J and Yen, J., "Generalized Coordinate Partitioning Methods for Numerical Integration of Differential Algebraic Equations of Dynamics", in *Real-Time Integration Methods for Mechanical System Simulation*, NATO ASI Series, Vol. F 69, pp. 97-114, Springer-Verlag, (1990).

Hilber, H.M., "Analysis and Design of Numerical Integration Methods in Structural Dynamics", *EERC Report No. 76-29*, Earthquake Engineering Research Center, University of California, Berkeley, (1976).

Hilber, H.M., Hughes, T.J.R., and Taylor, R.L., "Improved Numerical Dissipation for Time Integration Algorithms in Structural Dynamics", *Earthquake Engineering and Structural Dynamics*, Vol. 5, pp. 283-292, (1977).

Hughes, T.J.R., "Stability, Convergence, and Growth and Decay of Energy of the Average Acceleration Method in Nonlinear Structural Dynamics", *Computer and Structures*, Vol. 6, pp. 313-324, (1976).

Hughes, T.J.R., *Computational Methods for Transient Analysis* (eds. T. Belytschko and T.J.R. Hughes), Chapter 2, North-Holland, (1983).

Hughes, T.J.R., *The Finite Element Method: Linear Static and Dynamic Analysis*, Prentice-Hall, (1987).

Ince, E.L., *Ordinary Differential Equations*, Dover, New York, (1956).

Lodsted, P. and Petzold, L., "Numerical Solution of Nonlinear Differential Equations with Algebraic Constraints: Convergence Results for Backward Differentiation Formulae", *Mathematics of Computation*, Vol. 46, pp. 491-516, (1986).

Newmark, N.M., "A Method of Computation for Structural Dynamics", *Journal of the Engineering Mechanics Division, ASCE*, pp. 67-94, (1959).

Norsett, S.P., "One Step Methods of Hermite Type for the Numerical Solution of Stiff Systems", *BIT*, Vol. 14, pp. 63-77, (1974).

Petzold, L.R., "A Description of DASSL: A Differential/Algebraic System Solver", *IMACS Transactions on Scientific Computation*. Vol. 1, R.S. Stepleman (ed.), (1982).

Shampine, L. and Gordon, M., *Computer Solution of Ordinary Differential Equations: The Initial Value Problem*, Freeman, San Francisco, (1975).

Simo, J.C. and Wong, S., "Unconditionally Stable Algorithms for Rigid Body Dynamics that Exactly Preserve Energy and Momentum", *International Journal for Numerical Methods in Engineering*, Vol. 31, pp. 19-52, (1991).

Smith, I.M., "Some Time Dependent Soil-Structure Interaction Problems", *In Finite Elements in Geomechanics* (ed. Gudehus), Chapter 8, Wiley, (1975).

Steigerwald, M.F., "BDF Methods for DAEs in Multibody Dynamics: Shortcomings and Improvements", in *Real-Time Integration Methods for Mechanical System Simulation*, NATO ASI Series, Vol. F 69, pp. 345-352, Springer-Verlag, (1990).

Wilson, E.L., "A Computer Program for the Dynamic Stress Analysis of Underground Structures," *SESM Report No. 68-1*, Division of Structural Engineering Structural Mechanics, University of California, Berkeley, (1968).

Wood, W.L., *Practical Time-Stepping Schemes*, Oxford University Press, Chapter 7, (1990).

Problems

7/1   Integrate the second order differential equation: $\ddot{y}+y=0$ with initial conditions $y(t=0)=1$ and $\dot{y}(t=0)=0$, during the time interval $[0,10]$ with $\Delta t=0.1$, using the following methods:
a) The trapezoidal rule (Newmark with $\gamma=1/2$ and $\beta=1/4$).
b) The a-method with $a=-0.1$.
c) The backward difference method with p=1 of Table 7.3 equivalent to Newmark with $\gamma=1$ and $\beta=1/2$.
In all cases, introduce the difference equations of the corresponding algorithm into the equation of motion and solve the resulting set of linear equations for the unknown y_{n+1}. Plot the results of each of the methods and the exact solution and compare the period elongation and amplitude decay. Draw some conclusions about the accuracy and artificial damping introduced by each of the methods.

7/2   Write a subroutine for the integration of the first order differential equations using the four order Adams-Bashforth-Moulton predictor-corrector algorithm (Eqs. 7.22 and 7.23). Solve Problem 7/1 by transforming the second order equation into a set of two first order ones and then using the new subroutine. Compare your results with those obtained in Problem 7/1.

7/3   Formulate the equations of motion of the double pendulum used in the numerical example of Section 7/1 including rotational dash pots and using the two independent joint coordinates. Then use the trapezoidal rule and the Adams-Bashforth-Moulton predictor-corrector algorithm developed in Problem 7/2 (or any other multistep method available in the software library of your computer). Write up a table with the energy errors and CPU times obtained by both methods for different values of the springs and dash pots.

7/4   Repeat Problem 7/3 using the Cartesian coordinates of the ends of the two links with additional constraints to introduce the joint angles. Then use the penalty formulation (Chapter 5) to form the equations of motion (take a penalty value equal to 10^7). Compare the results with those obtained in Problem 7/3.

7/5   Repeat Problem 7/4 with the same coordinates and the Lagrange multiplier formulation to form the equations of motion of the type of equation (7.46). Then use the backward difference formula with $p=1$ (Table 7.3) to solve the resulting set of differential and algebraic equations. Compare your results with those obtained in Problems 7/4 and 7/3.

8
Improved Formulations for Real-Time Dynamics

The general purpose dynamic formulations described in Chapter 5 are simple and efficient, but they are not suitable for real time dynamic simulation. Real time performance requires faster formulations. These can be developed by taking into account the system's kinematic configuration or topology. In the last two decades, a big effort has been dedicated to developing very efficient dynamic formulations for serial robots or manipulators. These formulations have been extended later on to general open and closed chain configurations.

In the first section of this chapter, a survey of two of the most efficient available formulations that need $O(N^3)$ and $O(N)$ arithmetic operations is presented. It follows, a detailed description of a formulation based on *velocity transformations* that can be parallelized at body level. In the next section, a description of how the penalty formulation can be used for improved performance is also included. Finally, two complex examples: a model of the human body and a heavy truck, with theoretical count of arithmetic floating-point operations, and some numerical results are presented.

The methods and results presented in this chapter are contributions coming mainly from Jiménez (1993), and Avello et al. (1993).

8.1 Survey of Improved Dynamic Formulations

Most of the improvements in multibody dynamic formulations that have been developed in the last 25 years come from the robotics field. This field was very active in the 60's and 70's with scientists trying to solve very hard problems of simulation and control with the limited computational resources available at that time. Anyone familiar with robot control algorithms based on dynamic models knows the very competitive race that took place in order to decrease the number of arithmetic operations required for the inverse dynamics in serial robots. This research was quite important and led to the solution of the inverse dynamic problem two or three hundred times per second with a DEC-PDP 11 processor. The recursive Newton-Euler formulation seems to have been the winner of the race

(Luh, Walker, and Paul (1980)), at least for general geometries of serial manipulators with about six degrees of freedom.

Inverse dynamics, which consists in computing the motor torques and/or forces required to produce a desired motion, was directly interrelated to control purposes. However, it was realized that the high efficiency reached in solving the inverse dynamics could also benefit the formulation of the simulation problem, that is, the computation of the accelerations from the state variables (position and velocities) and the external and driving forces. As it may be seen in Walker and Orin (1982), the solution of the inverse dynamics allows for a formulation of the equations of motion much more efficient than conventional or even recursive Euler-Lagrange formulations.

Other authors, such as Armstrong (1979) and Featherstone (1983, 1987) have developed fully recursive $O(N)$ algorithms for open-chain systems, that is, algorithms whose number of floating-point arithmetic operations grows linearly with the number of degrees of freedom or bodies in the open-chain. The algorithms of Walker and Orin (1982) conclude with the solution of a system of N linear equations. They are of order $O(N^3)$, if Gaussian elimination is used.

Although it has been demonstrated (Featherstone (1987)) that the best $O(N^3)$ algorithms are better and faster than the best $O(N)$ algorithms for $N<10$ (that is, for most of the practical cases), the elegance and attractiveness of Featherstone's $O(N)$ formulation has exerted a strong influence on later developments that have generalized these ideas for non-serial (tree-configuration) systems (Bae and Haug (1987)) and closed-loop systems (Bae and Haug (1987-88), Rodriguez et al. (1991)). A limitation arises when closed-chain multibodies are analyzed, since for these cases special provisions must be made to account for the reaction forces between the different loops. More interest has been placed recently in looking for improved efficiency using $O(N^3)$ methods in cases with small values of N per chain (Bae, Hwang and Haug (1988), Bae and Won (1990)).

The foundations of the $O(N^3)$ formulation of Walker and Orin (1982), the $O(N)$ formulation of Featherstone (1987), and the generalizations of Bae and Haug (1987-88) will be described in the following subsections . Since it is not possible to reproduce these formulations in detail, the fundamental ideas on which each formulation is based will be outlined. Those interested in further details are referred to the original references given at the end of the chapter.

The number of floating-point arithmetic operations has been the criterion followed to compare the efficiency of different dynamic algorithms. From the *60's* to the late *80's*, this seemed to be the most reasonable criterion, since the average computers required over 20 times more *CPU* time to carry out a *SP* or a *DP* floating-point operation than an integer operation. Thus, it made full sense to neglect integer or logical operations and to draw comparative conclusions only from the theoretical account of floating-point operations. This procedure does not make sense any longer, since the RISC computers of the 90's have similar *CPU* cost for integer and floating-point operations. The new RISC processors can carry out two double precision floating-point operations (a product and an accu-

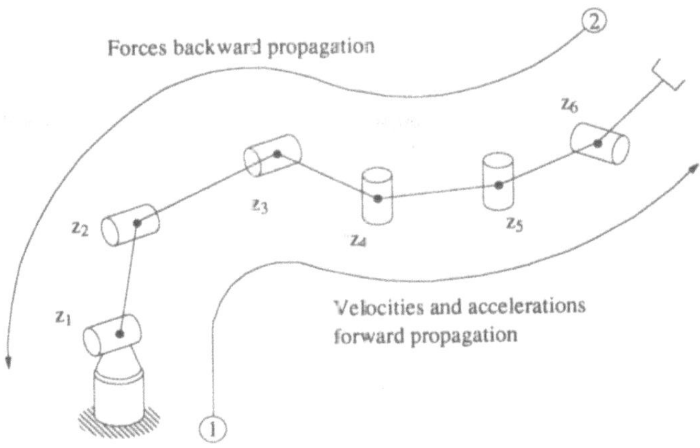

Figure 8.1. Recursive Newton-Euler method.

mulative addition) for each clock cycle. In these new machines, a clear and neat logic of the algorithm and an efficient use of the registers and of the cache memory can be more important than the number of floating-point operations. It will be seen that the methods presented in Sections 8.2 and 8.3 are simpler than other methods found in the literature. Moreover, these methods offer excellent opportunities to get very efficient computer implementations on modern computers using high level languages.

8.1.1 Formulations $O(N^3)$: Composite Inertia

One of the most efficient dynamic formulation for serial robots with $N<10$ is the one based on the recursive Newton-Euler solution of the inverse dynamic problem (Walker and Orin (1982)). This formulation proceeds as indicated below.

The inverse dynamics consists in determining the vector τ of the N motor torques and/or forces from the external forces Q_{ex} and the joint positions z, velocities $\dot{z}$, and accelerations $\ddot{z}$. Symbolically, it can be written

$$\tau = \mathbf{tau}(Q_{ex}, z, \dot{z}, \ddot{z}) \tag{8.1}$$

where $\mathbf{tau}(-)$ represents a function that solves the inverse dynamics for the values of the input parameters.

The recursive Newton-Euler method proceeds as indicated in Figure 8.1. In the first step, the joint positions, velocities, and accelerations are recursively propagated forward to compute the absolute link position, velocity, and acceleration and from them the inertial forces. In the second step, inertia and external forces are recursively propagated backwards so as to compute the equilibrating actuator forces and torques. If local reference frames are properly chosen on each body and used for the recursive equations, one can further reduce the computational cost.

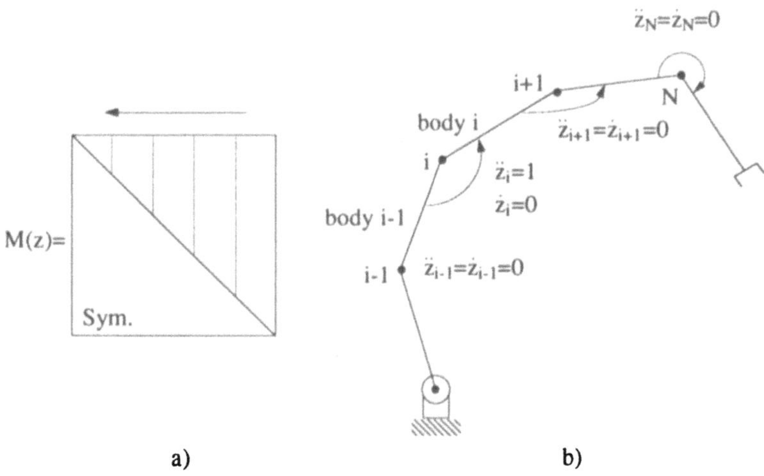

Figure 8.2.   Composite-Inertia method.

Now it will be described how the inverse dynamic solution given by equation (8.1) can be used for direct dynamic simulation following Walker and Orin (1982). The equations of motion can be written in the form,

$$M(z)\,\ddot{z} + C(z,\dot{z})\,\dot{z} + G(z) + Q(z, Q_{ex}) = \tau \qquad (8.2)$$

where $M(z)$ is the position-dependent inertial matrix, $C(z, \dot{z})\,\dot{z}$ is a term representing velocity-dependent (centrifugal and Coriolis) inertial forces, $G(z)$ represents the effect of gravitational forces, and $Q(z, Q_{ex})$ represents the vector of generalized external forces. Remember that in the direct dynamic problem, everything is known except the accelerations $\ddot{z}$.

If a vector b is defined as

$$b \equiv C(z,\dot{z})\,\dot{z} + G(z) + Q(z, Q_{ex}) \qquad (8.3)$$

equation (8.2) can be written as

$$M(z)\,\ddot{z} = \tau - b \qquad (8.4)$$

Now, all the terms in equation (8.4) can be computed with the inverse dynamics function **tau(–)** defined in (8.1):

– From equations (8.2) and (8.3), b can be computed by introducing a null acceleration in equation (8.1)

$$b = \mathbf{tau}(Q_{ex}, z, \dot{z}, 0) \qquad (8.5)$$

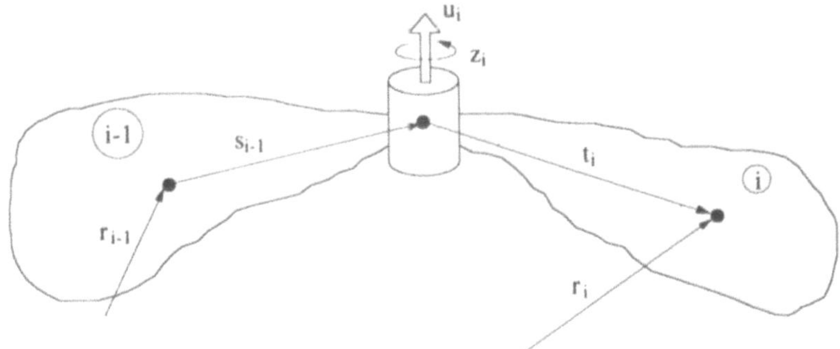

Figure 8.3. Revolute joint between contiguous bodies.

– From equation (8.4) it can be seen that the column i of $\mathbf{M}(\mathbf{z})$ can be obtained with an acceleration vector $\mathbf{e}^i$ (all elements zeroes, except for a unit value on component i),

$$\mathbf{m}_i(\mathbf{z}) = \mathbf{tau}(0, \mathbf{z}, 0, \mathbf{e}^i) \qquad i = 1, 2, ..., N \tag{8.6}$$

– Once all the columns of $\mathbf{M}(\mathbf{z})$ and vector $\mathbf{b}$ have been computed, the actual acceleration vector $\ddot{\mathbf{z}}$ can be obtained by solving the linear system of equations (8.4).

This algorithm can be improved if one takes into account that the matrix $\mathbf{M}(\mathbf{z})$ is symmetric and only the upper (or the lower) part needs to be calculated. The most efficient algorithm is the so-called *composite inertia* method (Walker and Orin (1982)). This method computes only the terms of $\mathbf{M}(\mathbf{z})$ located over the diagonal, starting from column N and proceeding to column 1 (See Figure 8.2a). In order to compute column i use equation (8.6). This equation computes the torques and/or forces in joints $(1, 2, ..., i)$ with null velocities and null accelerations in all the joints, except in joint i, where $\ddot{z}_i = 1$ (See Figure 8.2b).

At the time of computing column i, the joints from $(i+1)$ to N have zero velocity and zero acceleration. Furthermore, for all $j<i$, that is, in all subsequent calls to function $\mathbf{tau}(-)$, both the velocity and acceleration of joints $(i+1)$, $(i+2)$, ... , N remain zero. Therefore, bodies i to N do not exhibit relative velocities and accelerations and move as a single rigid body. Walker and Orin's idea (1982) makes use of this property and computes the *composite inertia*, or inertia of a fictitious rigid body, with the same center of gravity and inertial tensor as the set of rigid bodies i, $(i+1)$, ... , N. The composite inertia is updated before each call to function $\mathbf{tau}(-)$. For instance, before computing column $(i-1)$, the inertia of body i is added to the composite inertia. The complexity of expressions in this method can be considerably reduced by choosing appropriate local (body-fixed) reference frames. So far, the composite inertial method is the most efficient general purpose algorithm for serial manipulators with $N<10$ which includes most practical cases.

8.1.2 Formulations O(N): Articulated Inertia

This method has been fully developed by Featherstone (1983, 1987) and has had a major influence on later works, in spite of being less efficient than the composite inertial method for $N<9$. Featherstone described this method using *spatial vector* notation for both the kinematic and dynamic equations. This notation yields simpler, more compact, and more efficient expressions, but many engineers and students are not familiar with it. The main ideas behind Featherstone's formulation will be explained employing a more conventional notation used by Bae and Haug (1987).

Two contiguous elements, $(i-1)$ and i, linked by joint i can be seen in Figure 8.3. It can be assumed that joint i is of revolute type. The following kinematic relationships can be written for the rotation matrices:

$$\mathbf{A}_i = \mathbf{A}_{i,i-1}(\mathbf{u}_i, \mathbf{z}_i) \cdot \mathbf{A}_{i-1} \tag{8.7}$$

for angular velocities:

$$\boldsymbol{\omega}_i = \boldsymbol{\omega}_{i-1} + \dot{z}_i \, \mathbf{u}_i \tag{8.8}$$

and for the velocities of the centers of gravity:

$$
\begin{aligned}
\dot{\mathbf{r}}_i &= \dot{\mathbf{r}}_{i-1} + \tilde{\boldsymbol{\omega}}_{i-1} \, \mathbf{s}_{i-1} + \tilde{\boldsymbol{\omega}}_i \, \mathbf{t}_i = \\
&= \dot{\mathbf{r}}_{i-1} + \tilde{\boldsymbol{\omega}}_{i-1} \, \mathbf{s}_{i-1} + \left(\tilde{\boldsymbol{\omega}}_{i-1} + \dot{z}_i \, \tilde{\mathbf{u}}_i \right) \mathbf{t}_i = \\
&= \dot{\mathbf{r}}_{i-1} + \tilde{\boldsymbol{\omega}}_{i-1} \left(\mathbf{s}_{i-1} + \mathbf{t}_i \right) + \dot{z}_i \, \tilde{\mathbf{u}}_i \, \mathbf{t}_i = \\
&= \dot{\mathbf{r}}_{i-1} + \tilde{\boldsymbol{\omega}}_{i-1} \left(\mathbf{r}_i - \mathbf{r}_{i-1} \right) + \dot{z}_i \, \tilde{\mathbf{u}}_i \, \mathbf{t}_i
\end{aligned}
\tag{8.9}
$$

Equations (8.8) and (8.9) can be written together in the following matrix form:

$$
\{\mathbf{Y}_i\} = \left\{ \begin{array}{c} \dot{\mathbf{r}}_i \\ \boldsymbol{\omega}_i \end{array} \right\} = \begin{bmatrix} \mathbf{I} & \tilde{\mathbf{r}}_{i-1} - \tilde{\mathbf{r}}_i \\ \mathbf{0} & \mathbf{I} \end{bmatrix} \left\{ \begin{array}{c} \dot{\mathbf{r}}_{i-1} \\ \boldsymbol{\omega}_{i-1} \end{array} \right\} + \left\{ \begin{array}{c} \tilde{\mathbf{u}}_i \, \mathbf{t}_i \\ \mathbf{u}_i \end{array} \right\} \dot{z}_i \tag{8.10}
$$

or, in compact form:

$$\mathbf{Y}_i = \mathbf{B}_i \, \mathbf{Y}_{i-1} + \mathbf{b}_i \, \dot{z}_i \tag{8.11}$$

Equation (8.11) is a recursive relation between the velocities of two consecutive bodies, in terms of the relative joint velocity. Matrix $\mathbf{B}_i$ and vector $\mathbf{b}_i$ depend on the position variables. Differentiating this equation one obtains

$$\dot{\mathbf{Y}}_i = \mathbf{B}_i \, \dot{\mathbf{Y}}_{i-1} + \mathbf{b}_i \, \ddot{z}_i + \mathbf{d}_i \tag{8.12}$$

where $\mathbf{d}_i$ is a vector that groups the velocity-dependent terms:

$$\mathbf{d}_i = \dot{\mathbf{B}}_i \, \mathbf{Y}_{i-1} + \dot{\mathbf{b}}_i \, \dot{z}_i \tag{8.13}$$

Now, for an N-link serial manipulator, it is possible to formulate the principle of virtual power in the form

$$\sum_{i=1}^{N} \mathbf{Y}_i^{*T}\left(\mathbf{M}_i\,\dot{\mathbf{Y}}_i - \mathbf{Q}_i\right) = 0 \tag{8.14}$$

where

$$\mathbf{M}_i = \begin{bmatrix} \mathbf{m}_i\,\mathbf{I} & 0 \\ 0 & \mathbf{J}_i \end{bmatrix} \tag{8.15}$$

$$\mathbf{Q}_i = \begin{bmatrix} \mathbf{f}_i \\ \mathbf{n}_i - \tilde{\omega}_i\,\mathbf{J}_i\,\omega_i \end{bmatrix} \tag{8.16}$$

and where $\mathbf{n}_i$ and $\mathbf{f}_i$ are the external torques and forces acting at the center of gravity of link i. Vector $\mathbf{Y}_i^{*}$ represents the virtual velocities of link i. In equation (8.14), the virtual velocities cannot be eliminated, because they are not independent. Only the relative velocities $\dot{z}$ are independent.

One can transform expression (8.14) and write explicitly the last two terms,

$$\sum_{i=1}^{N-2} \mathbf{Y}_i^{*T}\left(\mathbf{M}_i\,\dot{\mathbf{Y}}_i - \mathbf{Q}_i\right) +$$
$$+ \mathbf{Y}_{N-1}^{*T}\left(\mathbf{M}_{N-1}\,\dot{\mathbf{Y}}_{N-1} - \mathbf{Q}_{N-1}\right) + \mathbf{Y}_N^{*T}\left(\mathbf{M}_N\,\dot{\mathbf{Y}}_N - \mathbf{Q}_N\right) = 0 \tag{8.17}$$

The virtual velocities must satisfy the compatibility equation (8.11). Substituting $\mathbf{Y}_N^{*}$ and $\dot{\mathbf{Y}}_N$ in equation (8.17)

$$\sum_{i=1}^{N-2} \mathbf{Y}_i^{*T}\left(\mathbf{M}_i\,\dot{\mathbf{Y}}_i - \mathbf{Q}_i\right) + \mathbf{Y}_{N-1}^{*T}\left(\mathbf{M}_{N-1}\,\dot{\mathbf{Y}}_{N-1} - \mathbf{Q}_{N-1}\right) +$$
$$+ \left(\mathbf{Y}_{N-1}^{*T}\,\mathbf{B}_N^T + \dot{z}_N^{*}\,\mathbf{b}_N^T\right)\left[\mathbf{M}_N\left(\mathbf{B}_N\,\dot{\mathbf{Y}}_{N-1} + \mathbf{b}_N\,\ddot{z}_N + \mathbf{d}_N\right) - \mathbf{Q}_N\right] = 0 \tag{8.18}$$

Reordering and grouping the terms yields

$$\sum_{i=1}^{N-2} \mathbf{Y}_i^{*T}\left(\mathbf{M}_i\,\dot{\mathbf{Y}}_i - \mathbf{Q}_i\right) + \mathbf{Y}_{N-1}^{*T}\Big[\ \left(\mathbf{M}_{N-1} + \mathbf{B}_N^T\,\mathbf{M}_N\,\mathbf{B}_N\right)\dot{\mathbf{Y}}_{N-1} -$$
$$-\left(\mathbf{Q}_{N-1} + \mathbf{B}_N^T\,\mathbf{Q}_N - \mathbf{B}_N^T\,\mathbf{M}_N\,\mathbf{d}_N\right) + \mathbf{B}_N^T\,\mathbf{M}_N\,\mathbf{b}_N\,\ddot{z}_N\Big] + \tag{8.19}$$
$$+ \dot{z}_N^{*}\left(\mathbf{b}_N^T\,\mathbf{M}_N\,\mathbf{B}_N\,\dot{\mathbf{Y}}_{N-1} + \mathbf{b}_N^T\,\mathbf{M}_N\,\mathbf{b}_N\,\ddot{z}_N + \mathbf{b}_N^T\,\mathbf{M}_N\,\mathbf{d}_N - \mathbf{b}_N^T\,\mathbf{Q}_N\right) = 0$$

The joint virtual velocity $\dot{z}_N^{*}$ appears in this equation. This virtual velocity is independent of the remaining virtual velocities; thus the parenthesis multiplying $\dot{z}_N^{*}$ in equation (8.19) must be zero,

$$\mathbf{b}_N^T\,\mathbf{M}_N\,\mathbf{B}_N\,\dot{\mathbf{Y}}_{N-1} + \mathbf{b}_N^T\,\mathbf{M}_N\,\mathbf{b}_N\,\ddot{z}_N + \mathbf{b}_N^T\,\mathbf{M}_N\,\mathbf{d}_N - \mathbf{b}_N^T\,\mathbf{Q}_N = 0 \tag{8.20}$$

From this equation the independent acceleration $\ddot{z}_N$ can be computed as

$$\ddot{z}_N = \left(b_N^T M_N b_N\right)^{-1} \left(b_N^T Q_N - b_N^T M_N B_N \dot{Y}_{N-1} - b_N^T M_N d_N\right) \qquad (8.21)$$

On the other hand, defining

$$\overline{M}_{N-1} \equiv M_{N-1} + B_N^T \left(M_N - M_N b_N \left(b_N^T M_N b_N\right)^{-1} b_N^T M_N\right) B_N \qquad (8.22)$$

$$\overline{Q}_{N-1} \equiv Q_{N-1} + \left(B_N^T - B_N^T M_N b_N \left(b_N^T M_N b_N\right)^{-1} b_N^T\right) \left(Q_N - M_N d_N\right) \qquad (8.23)$$

Equation (8.19) can be written as

$$\sum_{i=1}^{N-2} Y_i^{*T} \left(M_i \dot{Y}_i - Q_i\right) + Y_{N-1}^{*T} \left(\overline{M}_{N-1} \dot{Y}_{N-1} - \overline{Q}_N\right) = 0 \qquad (8.24)$$

Equation (8.24) is similar to equation (8.17) but with two important differences:

- There are only $(N-1)$ terms.
- The inertial matrix and the vector of forces corresponding to link $(N-1)$ have been modified according to equations (8.22) and (8.23) in order to incorporate the effects of link N. $\overline{M}_{N-1}$ is the *articulated inertia* of links $(N-1)$ and N.

The process of eliminating the last element in the virtual power expression can continue in the same manner using the recursive relations,

$$\overline{M}_{i-1} \equiv M_{i-1} + B_i^T \left(\overline{M}_i - \overline{M}_i b_i \left(b_i^T \overline{M}_i b_i\right)^{-1} b_i^T \overline{M}_i\right) B_i \qquad (8.25)$$

$$\overline{Q}_{i-1} \equiv Q_{i-1} + \left(B_i^T - B_i^T \overline{M}_i b_i \left(b_i^T \overline{M}_i b_i\right)^{-1} b_i^T\right) \left(\overline{Q}_i - \overline{M}_i d_i\right) \qquad (8.26)$$

$$\ddot{z}_i = \left(b_i^T \overline{M}_i b_i\right)^{-1} \left(b_i^T \overline{Q}_i - b_i^T \overline{M}_i B_i \dot{Y}_{i-1} - b_i^T \overline{M}_i d_i\right) \qquad (8.27)$$

$$i = N, N-1, ..., 2, 1$$

Finally, one arrives at the equation of the first link. If it is a floating link, one can write,

$$Y_1^{*T} \left(\overline{M}_1 \dot{Y}_1 - \overline{Q}_1\right) = 0 \qquad (8.28)$$

But now the virtual velocities are independent, so $\dot{Y}_1$ can be computed as,

$$\dot{Y}_1 = \overline{M}_1^{-1} \overline{Q}_1 \qquad (8.29)$$

Summarizing: the method of articulated inertia proceeds with a triple recursion in the following way:

1. Knowing the position and velocity of the base body and the joint relative positions z and velocities $\dot{z}$, one can compute recursively forward the Cartesian position and velocity of all the links from $i=1$ to $i=N$.

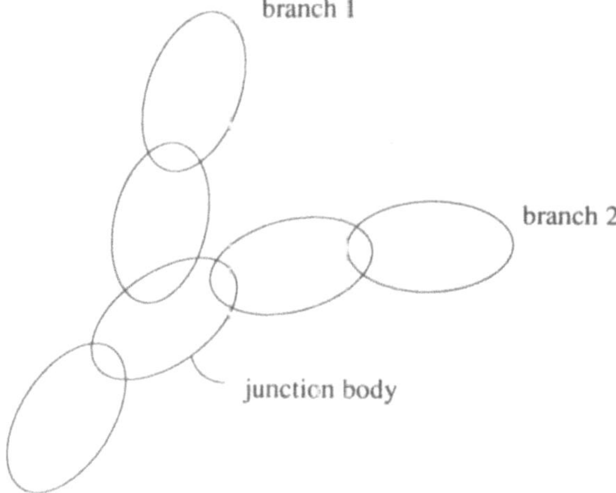

Figure 8.4. Kinematic chain with two branches.

2. The articulated inertias $\overline{M}$, the forces $\overline{Q}$, and the coefficients of equation (8.27) are then computed recursively backwards from $i=N$ to $i=1$, using equations (8.25)-(8.27).
3. Finally, the acceleration of the base body is computed from equation (8.29) and then the relative accelerations $\ddot{z}_i$ are computed recursively forward from $i=1$ to $i=N$, using equation (8.27).

The Featherstone version of this algorithm is probably much more efficient from the computational point of view. Bae, Hwang, and Haug (1988) presented an improved version of this algorithm, using as reference point the point of the moving body that instantaneously coincides with the origin of the inertial reference frame. Looking at the triple recursive procedure, one can see that the number of arithmetic operations grows proportionally with the number of degrees of freedom of the open-chain; thus it is an $O(N)$ method.

8.1.3 Extension to Branched and Closed-Chain Configurations

The ideas explained in the two previous sections can be extended to multibody systems with any kinematic configuration. There are three main directions in which this formulation can be generalized: a) include any kind of joints, b) extend it to multibody systems with many branches on a tree-configuration, and c) generalize it to systems with closed loops. The extension of this formulation to multibody systems with joints of any type is a straightforward task. It can be found in the original references.

The consideration of branches in the kinematic chain (See Figure 8.4) is also a simple task. Both the composite inertial and the articulated inertial methods can be easily accommodated to include junction bodies, that is, links with more

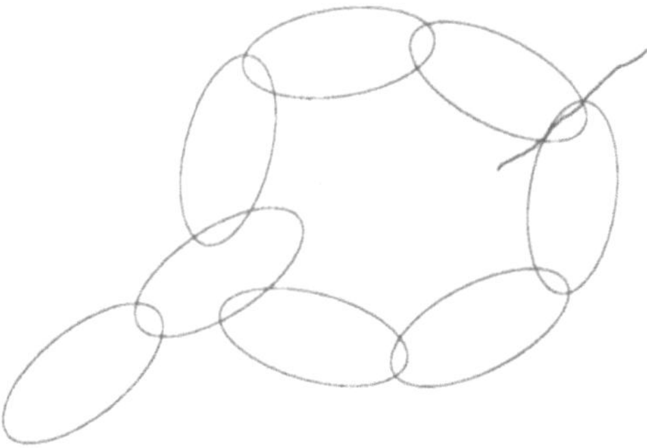

Figure 8.5. Cut-joint method to open a closed-chain multibody system.

than two joints. In the junction bodies, remember that the forward recursive computations must be split into two separate procedures that move independently along each branch. In the backward recursive computations, the two separate procedures along each branch meet at the junction body and yield a single procedure. From the theoretical point of view, the formulation can be extended to include tree-structured multibody systems without difficulty. The main difficulties arise in the practical implementation, when one tries to compute in parallel terms corresponding to different branches, since it is necessary to set up synchronization points at the junction bodies.

More difficulties can be found when the above formulations are modified to tackle closed-chain multibody systems. These can be transformed into open-chain systems through the *cut-joint* method which eliminates or *cuts* a joint in each loop, as seen in Figure 8.5. This method is described by Bae et al. (1987-88, 1988).

Where a joint is removed, the corresponding constraint forces, formulated through the Lagrange multipliers method as $(\Phi_q^T \lambda)$ and $(-\Phi_q^T \lambda)$ on both links, can be propagated backwards on both branches just as equation (8.23) indicates. All joint accelerations in the loop, given by equation (8.27), depend on the Lagrange multipliers vector λ.

In order to have enough equations to compute the relative accelerations $\ddot{z}$ and the Lagrange multipliers λ, it is necessary to differentiate twice the constraint equations corresponding to the cut joint.

Initially, these constraints are formulated using the Cartesian coordinates of the adjacent bodies. One must to substitute backwards on both branches the Cartesian velocities and accelerations by the corresponding relative variables using equations (8.11) and (8.12). Finally, as many equations as unknowns are available, and the details of this fully recursive formulation are described in Bae and Haug (1987-88). In a later work Bae, Hwang, and Haug (1988) introduced a modification addressed to compute all the relative accelerations $\ddot{z}$ at once by solving a system of linear equations, thus becoming an $O(N^3)$ method. More recently, García de Jalón et al. (1989) and Bae and Won (1990) presented other formulations better suited for real time analysis, which are based on *velocity transformations* (Jerkovsky (1978), Kim and Vanderploeg (1986)).

The recursive methods, so eagerly accepted at the beginning, have been steadily losing ground and interest in favor of the methods based on velocity transformations which seem simpler to formulate and easier to parallelize.

8.2 Velocity Transformations for Open-Chain Systems

Some of the most efficient formulations for the forward dynamic analysis that have been developed in the last decade have been dealt with in the preceding sections. These methods were classified, according to the number of floating-point operations that they require, as methods of order $O(N)$ or $O(N^3)$. Although there are important theoretical and practical differences among them, there are also common aspects that are worth pointing out.

The first point in common is their *origin*. These formulations arose from the study of the dynamics of serial robots, were posteriorly extended to other more general open-chain and tree-type multibody systems, and to those with a closed-chain configuration. Therefore, the *topology* of the multibody system has played a key role in the development history of these recursive formulations. This is a point to be contrasted with kinematic and dynamic methods exposed in Chapters 3 and 5 which can be really considered as *configuration independent*.

The second point in common between the $O(N)$ and $O(N^3)$ families is the extensive use, with a different emphasis, that those methods make of the concept of *recursion* along the kinematic chain. The recursive formulations have meant an important contribution to the dynamics of robots in particular, and to the dynamics of multibody systems in general, since they have substantially reduced the number of required floating-point operations. This does not mean that they do not suffer from some limitations that make the other non-recursive formulations compete advantageously against them.

From a computational viewpoint, vectorization has an enormous importance in the solution of large computational problems with thousands of degrees of freedom, such as those arising in fluid and solid mechanics. It may not influence the analysis of multibody systems very much, because these problems lead to systems of equations of small or moderate size. Parallelization, however, may have a very important effect in the dynamics of multibody systems, provided the dynamic method is such that different parts of the computational process may be executed simultaneously.

A general and simple method that formulates the dynamic equations of any open- or closed-chain multibody system, and which can be parallelized even to the body (or element) level will be studied in this and the next sections. This formulation is based upon the *velocity transformations* considered in Section 5.3.

The general purpose dynamic formulations described in Sections 5.1 and 5.2, are simple and may be applied to any multibody system regardless of their con-

Figure 8.6. Open-chain multibody system with tree structure.

figuration. They treat all systems in the same way, regardless of their topology and particular characteristics which make those methods insufficiently efficient for real time applications. A way to improve the efficiency of these formulations is to take advantage of the open-chain configurations that the multibody systems may have, or in which they may be transformed.

Open-chain multibody systems are less constrained than the closed-chain ones. This is a fact that when conveniently considered can be used to improve the efficiency of the dynamic formulation. Open-chain multibody systems are not only frequently found in robotics, aerospace, and biomechanics, but in *any closed-chain multibody system that may also be transformed into an open-chain configuration by simply opening its loops* (removing certain constraint equations, which can be considered separately). Consequently, all the benefits that may be drawn from the formulation of open-chain systems can be posteriorly extended to closed kinematic chains as well. The study of the open-chain multibody systems is next.

8.2.1 Dependent and Independent Coordinates

An open-chain multibody system will be considered that consists of one or several *branches*, forming a *tree structure* and connected to a *base element*, as shown in Figure 8.6. The total number of degrees of freedom f (or independent coordinates) will be equal to the six degrees of freedom of the base body plus all the relative degrees of freedom allowed by the kinematic joints along the branches of the multibody system. In order to be consistent with the formulations previously written in this book, the vector of independent coordinates will be denoted as z. This vector can be composed for instance of the three translations and three rotations of the base body, the rotations of the revolute joints, the translations of the prismatic joints the two rotations of an universal joint, and so forth. The vector z, thus formed and its derivatives, will completely and univocally define the position and motion of the open-chain system.

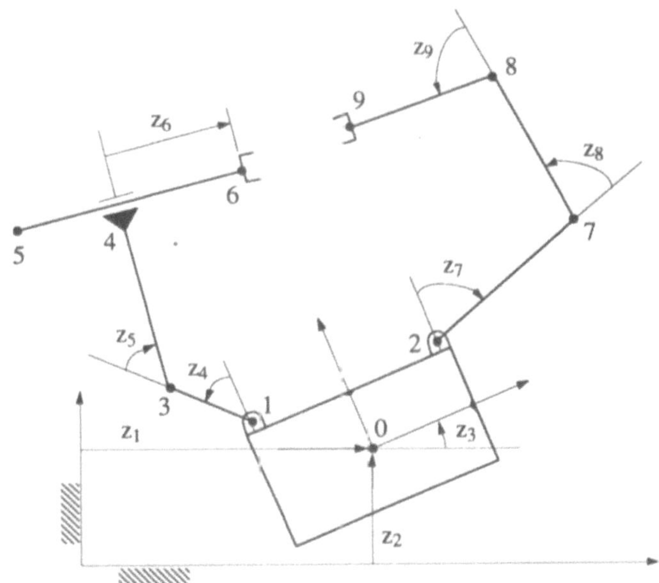

Figure 8.7. Planar system with floating base body and two robot arms.

It is also possible to describe the motion of the open-chain multibody system by means of an augmented set of coordinates q with n components, where $n > f$. These coordinates will no longer be independent, but interrelated through $(n-f)$ constraint equations of the form:

$$\Phi(q) = 0 \tag{8.30}$$

where it has been assumed, without important loss of generality, that the constraints are scleronomous.

Whereas there is a *natural choice* for the independent coordinates z (degrees of freedom of the base body plus the relative coordinates at the joints), the choice of dependent coordinates q is not so restrictive. One may make a choice from among the different sets of dependent coordinates based on convenience of implementation or even on personal preference. The following example clarifies this point.

Example 8.1

Figure 8.7 illustrates a planar multibody system with a base element that can move freely and to which two robots with three degrees of freedom each are attached. Therefore, there is a total number of nine degrees of freedom, corresponding to the independent coordinates z_1 to z_9. Among the many possible sets of dependent coordinates, the following set q of natural coordinates can be chosen,

$$q^T = \{r_1^T, r_2^T, r_3^T, r_4^T, r_5^T, r_6^T, r_7^T, r_8^T, r_9^T\}$$

where $\mathbf{r}_i^T = \{x_i, y_i\}$. These 18 Cartesian coordinates will be interrelated by means of nine constraint conditions, that the reader may formulate as an exercise following the rules given in Chapter 2.

Another possibility is to form a vector of dependent coordinates q composed of the previous set plus the coordinates of the center of gravity of the base body, those that describe its orientation, and the relative coordinates of the joints. Thus, the following set of 27 mixed coordinates is obtained,

$$\mathbf{q}^T = \{\mathbf{r}_0^T, z_3, \mathbf{r}_1^T, z_4, \mathbf{r}_2^T, z_7, \mathbf{r}_3^T, z_5, \mathbf{r}_4^T, z_6, \mathbf{r}_5^T, \mathbf{r}_6^T, \mathbf{r}_7^T, z_8, \mathbf{r}_8^T, z_9, \mathbf{r}_9^T\}$$

which will require 18 constraint equations.

A third possibility is to form $\mathbf{q}$ from the reference point coordinates of all the elements, including the coordinates of the center of gravity and the angular orientation of each element with respect to the inertial frame, which will lead to the following 21 dependent coordinates

$$\mathbf{q}^T = \{\mathbf{g}_1^T, \varphi_1, \mathbf{g}_2^T, \varphi_2, \mathbf{g}_3^T, \varphi_3, \mathbf{g}_4^T, \varphi_4, \mathbf{g}_5^T, \varphi_5, \mathbf{g}_6^T, \varphi_6, \mathbf{g}_7^T, \varphi_7\}$$

where $\mathbf{g}_i^T = \{g_{ix}, g_{iy}\}$ is the position vector of the center of gravity of body (i). These coordinates can also be augmented by the addition of all or part of the relative coordinates at the kinematic pairs.

The last possibility is to form a vector $\mathbf{q}$ that includes *all* the available information in terms of the position of the basic points, position of the center of gravity, and the rotation matrix that relates the orientation of the element with respect to the inertial frame. Accordingly,

$$\mathbf{q}^T = \{\mathbf{r}_1^T, \mathbf{r}_2^T, \mathbf{g}_1^T, \mathbf{A}_1, \mathbf{r}_3^T, \mathbf{g}_2^T, \mathbf{A}_2, \mathbf{r}_4^T, \mathbf{g}_3^T, \mathbf{A}_3, \ldots\}$$

where $\mathbf{r}_i$ is the position vector of the point i, and $\mathbf{g}_j$ and $\mathbf{A}_j$ contain respectively the coordinates of the center of gravity and the rotation matrix of element (j). As above, this vector $\mathbf{q}$ can be augmented by the relative coordinates of the joints and by the Cartesian components of the unit vectors in the case of three-dimensional multibody systems.

It has been seen in the previous example that there are many different sets of dependent coordinates that can conveniently represent the position of open-chain multibody systems. One should choose the most favorable one from the viewpoint of practical implementation, and the velocity and acceleration vectors need not be the derivatives, term by term, of the dependent position coordinates. For example, for the dependent position coordinates for a rigid body, one can chose those of the center of gravity and the Euler angles. As dependent velocities, one can chose those of the center of gravity plus the vector of angular velocities ω, which is not obtained through the direct differentiation of the Euler angles. The Euler angles could be replaced by the Euler parameters or by the nine components of the rotation matrix. They would still use as dependent velocities those of the center of gravity and the vector of angular velocities ω.

The reason for this peculiar choice comes from the non-integrability of the vector of angular velocities ω. One is forced to look for other sets of position variables (obviously more complicated than ω) that will enable representation of the angular orientation. In order to take into account the different character between the position vector and the velocity or acceleration vectors, we introduce the following notation: $\overline{q}$ will represent the vector of position dependent coordinates; whereas $\dot{q}$ and $\ddot{q}$ will represent the dependent velocity and acceleration vectors. Accordingly, the following relations will hold:

$$\dot{q} \neq \dot{\overline{q}} \qquad (8.31)$$

$$\dot{q} = V(\overline{q}) \, \dot{\overline{q}} \qquad (8.32)$$

$$\dot{\overline{q}} = U(\overline{q}) \, \dot{q} \qquad (8.33)$$

where $V(\overline{q})$ is a position-dependent matrix that transforms the derivatives of the position vector $\overline{q}$ into the velocities $\dot{q}$. The matrix $U(\overline{q})$ represents the velocity inverse transformation between $\dot{\overline{q}}$ and $\dot{q}$.

8.2.2 Dependent and Independent Velocities: Matrix R

The *velocity transformations* introduced in Chapters 3 and 5 will now be considered. The corresponding equations will be rewritten here for convenience as,

$$\dot{q} = R(\overline{q}) \, \dot{z} + Sb(\overline{q}) \qquad (8.34)$$

$$\ddot{q} = R(\overline{q}) \, \ddot{z} + Sc(\overline{q}, \dot{\overline{q}}) \qquad (8.35)$$

where the dependency of the matrix R and the terms (Sb) and (Sc) on $\overline{q}$ has been explicitly indicated in (8.34) and (8.35). The term (Sb) will be zero if the constraints are scleronomous.

The columns of the matrix R constitute a basis for the vector space of all the possible velocity vectors $\dot{q}$. Any vector $\dot{q}$ that satisfies the velocity constraint equations may be expressed as a linear combination of the f columns of the matrix R. As shown in equation (8.34), the components of the vector $\dot{z}$ of independent velocities are the coefficients of such linear combination.

Equation (8.34) also indicates the physical significance of the columns of the matrix R. Column i represents the dependent velocities $\dot{q}$ obtained by giving a unit value to the component $\dot{z}_i$ of the vector $\dot{z}$, and zero to the rest of the components. Thus,

$$r^i = \dot{q} \big|_{\dot{z}_i = 1} \text{ and } \dot{z}_j = 0 \text{ for } j \neq i \qquad (8.36)$$

where r^i represents the column i of the matrix R, as compared with r_i which represents the position vector of point i.

The definition made in equation (8.36) of the columns of the matrix R leads for the general case of open- and closed-chain multibody systems, to the methods explained in Chapter 5, such as the LU factorization of the Jacobian Φ_q, Singular Value decomposition, and so forth. In the particular case of open-chain multibody systems that are represented by the independent coordinates mentioned above, *the columns of the matrix* $\mathbf{R}$ *can be obtained directly* through velocity computations, without the need of forming and factoring the Jacobian matrix; thus leading to a greatly reduced numerical effort. In addition, the following advantages can also be obtained:

a) The *sparsity* pattern of the matrix $\mathbf{R}$ becomes apparent and can be used in subsequent matrix operations.

b) The part of the matrix $\mathbf{R}$ that affects a particular link or element of the multibody system can be formed independently of the rest of the elements. This property leads to an *element-by-element* treatment of the equations of motion.

These advantages become apparent when considering the physical significance of each of the columns of the matrix $\mathbf{R}$. Column i represents the dependent velocities resulting from a unit value of the independent velocity $\dot{z}_i$ and zero values of the rest of independent velocities. Taking into account that vector z contains the position coordinates of the base body plus the relative joint coordinates, three possibilities can be considered:

1) $\dot{z}_i$ corresponds to a translational velocity of the base body along one of the inertial axis. All the elements of the multibody system will have a unit translational velocity along the same inertial axis. Only the Cartesian coordinates of the basic points will be affected by this translation. The rest of the dependent velocities, including unit vectors and relative joint coordinates, will take zero values.

2) $\dot{z}_i$ corresponds to a rotational velocity of the base body along one of the inertial axis. The corresponding column of the matrix $\mathbf{R}$ will contain the rotation velocities of all the points and unit vectors about an axis parallel to the inertial one. This axis goes through the reference point of the base body.

3) $\dot{z}_i$ corresponds to the relative velocity of one of the kinematic joints. Only the distal elements (those after the corresponding joint) of the kinematic chain will be affected by the relative joint velocity.

The following example will make use of these considerations to show the ease by which the different columns of the matrix $\mathbf{R}$ can be calculated in an open-chain multibody system, when the proposed sets of coordinates are used.

Example 8.2

Consider the open-chain multibody system illustrated in Figure 8.7 which is represented by a set of mixed coordinates. Derive the columns of the matrix $\mathbf{R}$ that correspond to the following independent velocities: $\dot{z}_1$, $\dot{z}_3$, and $\dot{z}_7$.

The vector of dependent mixed coordinates is, in this case,

$$\mathbf{q}^T = \{\mathbf{r}_0^T, z_3, \mathbf{r}_1^T, z_4, \mathbf{r}_3^T, z_5, \mathbf{r}_4^T, \mathbf{r}_5^T, z_6, \mathbf{r}_6^T, \mathbf{r}_2^T, z_7, \mathbf{r}_7^T, z_8, \mathbf{r}_8^T, z_9, \mathbf{r}_9^T\}$$

where $\mathbf{r}_i$ is the position vector of point i, and z_j represents the relative coordinates.

a) When $\dot{z}_1=1$, all the points will have the same velocity $\mathbf{n}_1^T=\{1,0\}$, and the relative coordinates will have zero velocity. Accordingly,

$$\mathbf{r}^{1T} = \{\mathbf{n}_1^T, 0, \mathbf{n}_1^T, 0, \mathbf{n}_1^T, 0, \mathbf{n}_1^T, \mathbf{n}_1^T, 0, \mathbf{n}_1^T, \mathbf{n}_1^T, 0, \mathbf{n}_1^T, 0, \mathbf{n}_1^T, 0, \mathbf{n}_1^T\}$$

b) When $\dot{z}_3 = 1$ (unit rotation of the base element), each point will have the following velocity: $\mathbf{v}_i=\dot{\mathbf{r}}_i=\mathbf{k}\wedge(\mathbf{r}_i-\mathbf{r}_0)$, where $\mathbf{k}$ is the unit vector perpendicular to the plane of the multibody system. Therefore

$$\mathbf{r}^{3T} = \{0^T, 1, \mathbf{v}_1^T, 0, \mathbf{v}_3^T, 0, \mathbf{v}_4^T, \mathbf{v}_5^T, 0, \mathbf{v}_6^T, \mathbf{v}_2^T, 0, \mathbf{v}_7^T, 0, \mathbf{v}_8^T, 0, \mathbf{v}_9^T\}$$

c) Finally, if a unit relative velocity is introduced in the joint 7, $\dot{z}_7 = 1$. Only those points that belong to the distal elements after joint 7 will have a non-zero velocity, which will have the following value: $\mathbf{v}_i=\dot{\mathbf{r}}_i=\mathbf{k}\wedge(\mathbf{r}_i-\mathbf{r}_2)$, ($i=7,8,9$). Consequently, the seventh column of $\mathbf{R}$ becomes

$$\mathbf{r}^{7T} = \{0^T, 0, 0^T, 0, 0^T, 0, 0^T, 0^T, 0, 0^T, 0^T, 1, \mathbf{v}_7^T, 0, \mathbf{v}_8^T, 0, \mathbf{v}_9^T\}$$

d) In order to complete this example, the part of the matrix $\mathbf{R}$ that corresponds to the element that joins the points 3 and 4 will be calculated separately. This part of the matrix $\mathbf{R}$ is composed of the columns that correspond to the degrees of freedom of the base element (z_1, z_2, z_3) and those of the joints (z_4, z_5) that are before the element within the same kinematic chain. The result for the element is

$$\mathbf{R}_{3\text{-}4} = \begin{bmatrix} 1 & 0 & \mathbf{k}\wedge(\mathbf{r}_3-\mathbf{r}_0) & \mathbf{k}\wedge(\mathbf{r}_3-\mathbf{r}_1) & 0 \\ 0 & 1 & & & \\ 1 & 0 & \mathbf{k}\wedge(\mathbf{r}_4-\mathbf{r}_0) & \mathbf{k}\wedge(\mathbf{r}_4-\mathbf{r}_1) & \mathbf{k}\wedge(\mathbf{r}_4-\mathbf{r}_3) \\ 0 & 1 & & & \end{bmatrix}$$

$$\begin{array}{ccccc} \dot{z}_1 & \dot{z}_2 & \dot{z}_3 & \dot{z}_4 & \dot{z}_5 \end{array}$$

The previous example clearly illustrates two very important advantages of this method: a) the way in which the matrix $\mathbf{R}$ can be calculated is direct, systematic, and general; and, b) the procedure can be carried out for each body independently in a *body-by-body* or *element-by-element* basis (rather than recursively). Therefore, the method can take full advantage of parallel computer architectures.

Finally, the columns of matrix $\mathbf{R}$, thus calculated, constitute a basis for the nullspace (subspace of possible motions) of the Jacobian matrix Φ_q. Although the constraint equations are not explicitly calculated, the following relation will always hold:

$$\Phi_q\,\mathbf{R} = 0 \tag{8.37}$$

Table 8.1. Algorithm to formulate and integrate the equations of motion of an open-chain system.

Step	Data	Result	Mode
1	z	$\overline{q}$	recursive
2	$\overline{q}$	R	e-by-e or rec.
3	$\dot{z}, \overline{q}$	$\dot{q}$	recursive
4	$\dot{z}, \overline{q}, \dot{q}$	(Sc)	e-by-e
5	R, M	$R^T M R$	e-by-e
6	R, Q	$R^T Q$	e-by-e
7	R, Sc, M	$R^T M Sc$	e-by-e
8	Linear equations	$\ddot{z}$	global
9	$(\ddot{z}, \dot{z})_t$	$(\dot{z}, z)_{t+\Delta t}$	global
10	GO TO 1		

8.2.3 Equations of Motion

Once the matrix R is known, the methods presented in Chapter 5 for the formulation of the equations of motion in independent coordinates can be used. We will use equation (5.67) that is written again for convenience as:

$$R^T M R \, \ddot{z} = R^T Q - R^T M S c \tag{8.38}$$

where Q is the vector containing the external forces, those coming from a potential and the velocity-dependent ones. The term (Sc) contains the dependent accelerations $\ddot{q}$ that are calculated from the true velocities $\dot{q}$ by equating to zero the independent accelerations $\ddot{z}$ in equation (5.65).

The algorithm that allows the formulation and numerical integration of the motion differential equations of an open-chain system, according to the proposed method and to equation (8.38), is given in detail in Table 8.1. The integration is carried out in independent coordinates; and therefore no constraint violation stabilization is considered. Table 8.1 also indicates the way each step can be calculated in an optimal way. Some steps can be carried out in an *element-by-element* basis; thus susceptible for being parallelized. Other steps are carried out recursively, and others in a global manner.

The steps included in Table 8.1 will be described more in detail, starting from the position problem.

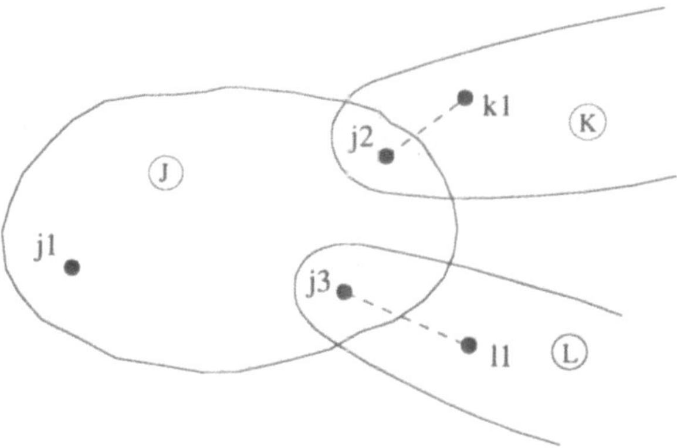

Figure 8.8. Body with one input point and two output points.

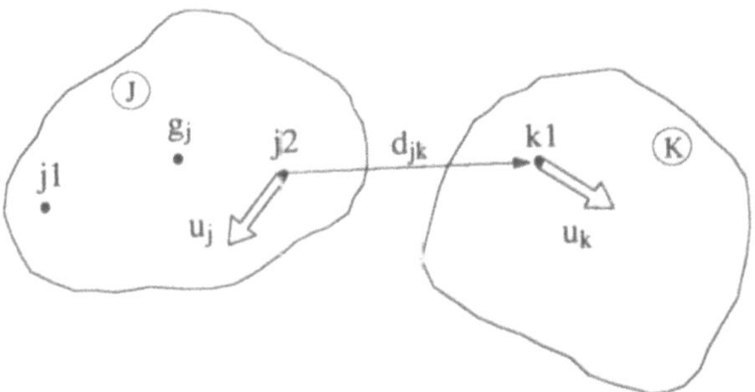

Figure 8.9. Model of a generalized joint.

8.2.4 Position Problem

A recursive solution to the position problem of a spatial open-chain multibody system will be shown in this section. We assume that the multibody system is composed of a base element and a series of branches with arbitrary size and distribution composed of rigid bodies interconnected by any of the following joints: revolute (R), prismatic (P), cylindrical (C), universal (U), or spherical (S). The formulation can also be extended without difficulty to other types of joints.

The position problem, required in Table 8.1, consists in finding the dependent coordinates $\bar{q}$ that define the position of the individual elements of the multibody system from the independent coordinates z, composed of the degrees of freedom of the base body plus the joint coordinates. Later on, in order to avoid possible singular positions, dependent angular orientation coordinates will be included for the base body and for the spherical joints. The vector of dependent coordinates $\bar{q}$ includes the natural coordinates of the elements plus the following additional variables:

- Cartesian coordinates of the points $\mathbf{r}_i$ referred to the inertial frame.
- Cartesian components of the unit vectors $\mathbf{u}_i$ referred to the inertial frame.
- Rotation matrices $\mathbf{A}_i$, that relate the current position of every element with respect to a known reference or initial position.
- Cartesian components of the center of gravity $\mathbf{g}_i$ of each body expressed in the inertial frame.

The aim of this section is to calculate the components of the position vector $\bar{\mathbf{q}}$ in a recursive manner (avoiding the expensive Newton-Raphson iterations described in Chapter 3), starting from the base body and moving forward towards the distal elements in the different branches. We will assume that the vector $\bar{\mathbf{q}}_0$ with the initial or reference positions is known.

Figure 8.8 shows a general element J that has an input point $j1$, whose position is already known, and one or more output points $j2$, $j3$, and so forth. These output joints are related to the pairs that join the element J with the posterior elements in the chain. Figure 8.9 illustrates the model of the *generalized joint* that joins elements J and K. The points $j2$ and $k1$ are the output point of element J and input point of K, respectively. The vectors $\mathbf{u}_j$ and $\mathbf{u}_k$ belong to J and K, respectively, and must be chosen according to the type of the kinematic pair that joins both elements. The relative degrees of freedom allowed by this joint will be represented by a vector of relative joint coordinates $\mathbf{z}_{jk}$. The vector $\mathbf{d}_{jk}$ corresponds to the position vector between points $j2$ and $k1$.

Assuming that the position vector $\mathbf{r}_{j1}$ and rotation matrix $\mathbf{A}_j$ corresponding to the element J are known, the solution of the position problem gives the remaining position variables of J (such as the center of gravity $\mathbf{g}_j$, output point $\mathbf{r}_{j2}$ and unit vector $\mathbf{u}_j$), plus the input position variables of element K (position vector $\mathbf{r}_{k1}$ and rotation matrix $\mathbf{A}_k$). The necessary calculations are represented schematically in Figure 8.10. The corresponding analytical expressions are:

$$\mathbf{g}_j = \mathbf{r}_{j1} + \mathbf{A}_j \left(\mathbf{g}_j^o - \mathbf{r}_{j1}^o \right) \tag{8.39}$$

$$\mathbf{u}_j = \mathbf{A}_j \, \mathbf{u}_j^o \tag{8.40}$$

$$\mathbf{r}_{k1} = \mathbf{r}_{j1} + \mathbf{A}_j \left(\mathbf{r}_{j2}^o - \mathbf{r}_{j1}^o \right) + \mathbf{d}_{jk} \tag{8.41}$$

$$\mathbf{A}_k = \mathbf{A}_{jk}(\mathbf{z}_{jk}) \, \mathbf{A}_j \tag{8.42}$$

where the superscript $(-)^o$ refers to the corresponding variable to the known initial or reference position defined in the inertial frame. The matrix $\mathbf{A}_{jk}(\mathbf{z}_{jk})$ is the relative rotation matrix of the kinematic joint between elements J and K that depends on the relative rotation. Note that equations (8.39) to (8.42) apply to any output variable and joint of the element J.

Equations (8.39) through (8.42) will be particularized to each of the different joints.

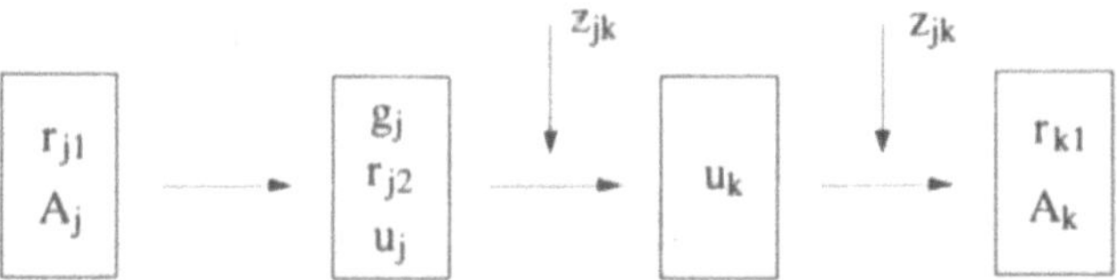

Figure 8.10. Scheme of recursive position calculations.

Revolute Joint R. Equations (8.41) and (8.42) become in this case:

$$\mathbf{d}_{jk} = 0 \tag{8.43}$$

$$\mathbf{A}_{jk} = \mathbf{A}(z_{jk}, \mathbf{u}_j) \tag{8.44}$$

where z_{jk} is the angle rotated by the joint with respect to the reference configuration $\bar{\mathbf{q}}_0$, and $\mathbf{u}_j$ is the unit vector in the direction of the joint axis. The rotation matrix in (8.44) is defined in terms of the angular rotation z_{jk} about the axis defined by the unit vector $\mathbf{u}_j$. It can be demonstrated (See Argyris (1982)) that

$$\mathbf{A}(z_{jk}, \mathbf{u}) = \mathbf{I} + \tilde{\mathbf{u}}\,\tilde{\mathbf{u}}\left(1 - \cos z_{jk}\right) + \tilde{\mathbf{u}}\,\sin z_{jk} \tag{8.45}$$

where $\tilde{\mathbf{u}}$ is the skew-symmetric matrix commonly used to perform the cross product of vectors.

Prismatic Joint P. Considering that the vectors $\mathbf{u}_j$ and $\mathbf{d}_{jk}$ have the direction of the relative translation allowed by the prismatic joint, the following equations describe the relative configuration of the elements:

$$\mathbf{d}_{jk} = z_{jk}\,\mathbf{u}_j \tag{8.46}$$

$$\mathbf{A}_k = \mathbf{A}_j \tag{8.47}$$

Cylindrical Joint C. The vectors $\mathbf{u}_j$ and $\mathbf{d}_{jk}$ are considered to have the direction of the joint axis. Consequently,

$$\mathbf{d}_{jk} = z_{jk}^{t}\,\mathbf{u}_j \tag{8.48}$$

$$\mathbf{A}_k = \mathbf{A}_{jk}\left(z_{jk}^{t}, \mathbf{u}_j\right)\mathbf{A}_j \tag{8.49}$$

where z_{jk}^{r} and z_{jk}^{t} are the relative coordinates that represent the rotation and translation of the joint, respectively.

Spherical Joint S. The spherical joint allows an arbitrary rotation that may be defined either by the Euler angles, the Bryant angles, or the Euler parameters $\mathbf{p}$ which are interrelated by the following condition $p_0^2 + p_1^2 + p_2^2 + p_3^2 = 1$. Equations (8.48) and (8.49) become:

$$\mathbf{d}_{jk} = 0 \tag{8.50}$$

$$\mathbf{A}_k = \mathbf{A}_{jk}(z_{jk} \equiv \mathbf{p}) \cdot \mathbf{A}_j \tag{8.51}$$

The rotation matrix in terms of the Euler parameters is defined as (Nikravesh (1988) and Haug (1989))

$$\mathbf{A}(\mathbf{p}) = \left(2\, e_o^2 - 1\right)\mathbf{I}_3 + 2\left(\mathbf{e}\, \mathbf{e}^T + e_o\, \tilde{\mathbf{e}}\right) \tag{8.52}$$

where $e_o = p_o$, and $\mathbf{e}^T = (p_1, p_2, p_3)$.

Universal Joint U. This joint allows two rotations z_{jk}^1 and z_{jk}^2 with respect to two perpendicular axes along the unit vectors $\mathbf{u}_j$ and $\mathbf{u}_k$. The axes intersect at the common point between both bodies J and K. The equations that correspond to this joint become

$$\mathbf{d}_{jk} = 0 \tag{8.53}$$

$$\mathbf{u}_j = \mathbf{A}_j\, \mathbf{u}_j^o \tag{8.54}$$

$$\mathbf{u}_k = \mathbf{A}_{jk}^1(z_{jk}^1, \mathbf{u}_j)\, \mathbf{A}_j\, \mathbf{u}_k^o \tag{8.55}$$

$$\mathbf{A}_k = \mathbf{A}_{jk}^2(z_{jk}^2, \mathbf{u}_k)\, \mathbf{A}_{jk}^1(z_{jk}^1, \mathbf{u}_j)\, \mathbf{A}_j \tag{8.56}$$

Base body. Let the independent coordinate vector of the base element z_b be defined by the coordinates of a reference point i and by the Euler parameters. Consequently,

$$\mathbf{z}_b^T \equiv \left(\Delta\mathbf{r}_i^T\, \mathbf{p}^T\right) \tag{8.57}$$

and the following two relations will hold for the reference point and rotation matrix:

$$\mathbf{r}_i = \mathbf{r}_i^o + \Delta\mathbf{r}_i \tag{8.58}$$

$$\mathbf{A} = \mathbf{A}(\mathbf{p}) \tag{8.59}$$

Equations (8.39) to (8.59) allow one to solve recursively the position problem for a wide range of open-chain multibody systems.

8.2.5 Velocity and Acceleration Problems

A recursive process addressed to obtain the dependent velocity and acceleration vectors of a multibody system can be similar and even simpler than that used in the preceding section to solve the position problem.

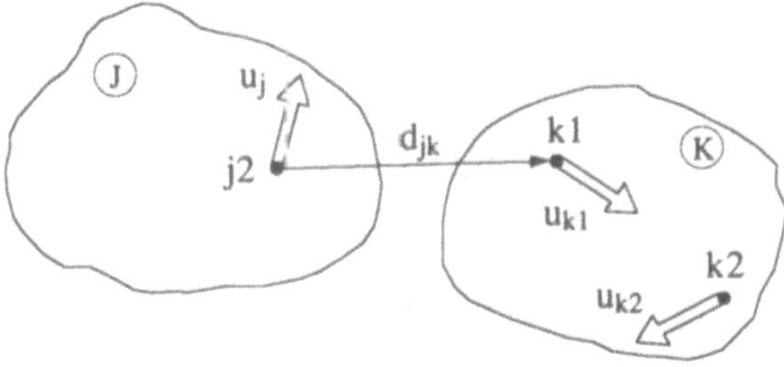

Figure 8.11. Consecutive bodies for recursive velocity calculation (Form. A)

It is possible and even convenient to use a set of dependent velocities and accelerations that does not correspond to the derivatives of the dependent position
vector $\bar{q}$. It has been shown in the previous section that the vector $\bar{q}$ may be
composed of the Cartesian components of basic points, those of the centers of
gravity, and the components of unit vectors and the rotation matrices. Two different sets of dependent velocities and accelerations will be considered, that will
lead to the following formulations:

a) *Formulation A,* based on the Cartesian components of basic points and unit
   vectors. This formulation follows the concept of natural coordinates so much
   used throughout this book. If an element is defined by two basic points and
   two unit vectors the corresponding mass matrix is constant and no velocity-
   dependent inertial terms are involved in the forcing function. It is possible to
   take advantage of these important facts at the time of implementing the algorithm of Table 8.1.

b) *Formulation B,* based on the reference point coordinates. This formulation
   uses the velocity and acceleration vectors of the center of gravity and the angular velocity and acceleration vectors of each element. This leads to some
   important advantages at the time of calculating the matrix **R** in an *element-
   by-element* basis, because contrary to the Formulation A, there are no variables shared by the different elements. These advantages materialize in an ease
   of parallelization and in computational savings.

The analytical expressions necessary to calculate the dependent velocities and
accelerations in both formulations will be described next.

8.2.5.1 Formulation A

In this case, it is necessary to compute the velocities of the basic points and unit
vectors. Figure 8.11 illustrates two consecutive elements J and K whose positions are known. In addition, it will be assumed that the velocity of J, as well as
the relative velocity of the joint $\dot{z}_{jk}$, are also known.

The aim is to compute the velocity of the basic points *k1* and *k2*, and those of the unit vectors $\mathbf{u}_{k1}$ and $\mathbf{u}_{k2}$. These are given by the following velocity equations:

$$\dot{\mathbf{r}}_{k1} = \dot{\mathbf{r}}_{j2} + \dot{\mathbf{d}}_{jk} \tag{8.60}$$

$$\boldsymbol{\omega}_k = \boldsymbol{\omega}_j + \boldsymbol{\omega}_{jk} \tag{8.61}$$

$$\dot{\mathbf{r}}_{k2} = \dot{\mathbf{r}}_{k1} + \boldsymbol{\omega}_k \wedge (\mathbf{r}_{k2} - \mathbf{r}_{k1}) \tag{8.62}$$

$$\dot{\mathbf{u}}_{k1} = \boldsymbol{\omega}_k \wedge \mathbf{u}_{k1} \tag{8.63}$$

$$\dot{\mathbf{u}}_{k2} = \boldsymbol{\omega}_k \wedge \mathbf{u}_{k2} \tag{8.64}$$

Similarly, the accelerations can be obtained from the following equations:

$$\ddot{\mathbf{r}}_{k1} = \ddot{\mathbf{r}}_{j2} + \ddot{\mathbf{d}}_{jk} \tag{8.65}$$

$$\dot{\boldsymbol{\omega}}_k = \dot{\boldsymbol{\omega}}_j + \dot{\boldsymbol{\omega}}_{kj} \tag{8.66}$$

$$\ddot{\mathbf{r}}_{k2} = \ddot{\mathbf{r}}_{k1} + \dot{\boldsymbol{\omega}}_k \wedge (\mathbf{r}_{k2} - \mathbf{r}_{k1}) + \boldsymbol{\omega}_k \wedge (\boldsymbol{\omega}_k \wedge (\mathbf{r}_{k2} - \mathbf{r}_{k1})) \tag{8.67}$$

$$\ddot{\mathbf{u}}_{k1} = \dot{\boldsymbol{\omega}}_k \wedge \mathbf{u}_{k1} + \boldsymbol{\omega}_k \wedge (\boldsymbol{\omega}_k \wedge \mathbf{u}_{k1}) \tag{8.68}$$

$$\ddot{\mathbf{u}}_{k2} = \dot{\boldsymbol{\omega}}_k \wedge \mathbf{u}_{k2} + \boldsymbol{\omega}_k \wedge (\boldsymbol{\omega}_k \wedge \mathbf{u}_{k2}) \tag{8.69}$$

These equations will now be particularized to the different types of kinematic joints.

Revolute joint R. In this case, *k1* coincides with *j2*, and the unit vector $\mathbf{u}_{k1}$ coincides with $\mathbf{u}_j$ and the joint axis. Consequently, the following expressions can be easily obtained:

$$\mathbf{d}_{jk} = \dot{\mathbf{d}}_{jk} = \ddot{\mathbf{d}}_{jk} = 0 \tag{8.70}$$

$$\boldsymbol{\omega}_{jk} = \dot{z}_{jk}\, \mathbf{u}_j \tag{8.71}$$

$$\dot{\boldsymbol{\omega}}_{jk} = \ddot{z}_{jk}\, \mathbf{u}_j + \dot{z}_{jk}\, \boldsymbol{\omega}_j \wedge \mathbf{u}_j \tag{8.72}$$

Prismatic joint P. This joint can be modeled by considering $\mathbf{u}_{k1} \equiv \mathbf{u}_j$ in the direction of the joint which is defined by the points *j2* and *k1*. After this consideration, the following equations define the required velocities and accelerations:

$$\dot{\mathbf{d}}_{jk} = \dot{z}_{jk}\mathbf{u}_j + z_{jk}\omega_j \wedge \mathbf{u}_j \tag{8.73}$$

$$\omega_{jk} = \dot{\omega}_{jk} = 0 \tag{8.74}$$

$$\ddot{\mathbf{d}}_{jk} = \ddot{z}_{jk}\,\mathbf{u}_j + 2\dot{z}_{jk}\,\omega_j \wedge \mathbf{u}_j + z_{jk}\,\dot{\omega}_j \wedge \mathbf{u}_j + z_{jk}\,\omega_j \wedge \left(\omega_j \wedge \mathbf{u}_j\right) \tag{8.75}$$

Cylindrical joint C. This joint can be considered as a combination of a revolute and a prismatic joint. Therefore the following equations yield the required velocities and accelerations:

$$\dot{\mathbf{d}}_{jk} = \dot{z}_{jk}\,\mathbf{u}_j + z^{t}_{jk}\,\omega_j \wedge \mathbf{u}_j \tag{8.76}$$

$$\omega_{jk} = \dot{z}^{r}_{jk}\,\mathbf{u}_j \tag{8.77}$$

$$\ddot{\mathbf{d}}_{jk} = \ddot{z}^{t}_{jk}\,\mathbf{u}_j + 2\dot{z}^{t}_{jk}\,\omega_j \wedge \mathbf{u}_j + z^{t}_{jk}\,\dot{\omega}_j \wedge \mathbf{u}_j + z^{t}_{jk}\,\omega_j \wedge \left(\omega_j \wedge \mathbf{u}_j\right) \tag{8.78}$$

$$\dot{\omega}_{jk} = \ddot{z}^{t}_{jk}\,\mathbf{u}_j + \dot{z}^{t}_{jk}\,\omega_j \wedge \mathbf{u}_j \tag{8.79}$$

Spherical joint S. This joint can be modeled by joining points *j2* and *k1* in the system of Figure 8.11. The relative velocities and accelerations of the joint will be defined by the first and second derivatives of the Euler parameters. This is not strictly necessary for the accelerations, since $\dot{\omega}$ is integrable. The expressions that define the motion of the joint are (Nikravesh (1988) and Haug (1989)):

$$\dot{\mathbf{d}}_{jk} = \ddot{\mathbf{d}}_{jk} = 0 \tag{8.80}$$

$$\omega_{jk} = 2\,\mathbf{G}\,\dot{\mathbf{p}} \tag{8.81}$$

$$\mathbf{G} \equiv \begin{bmatrix} -p_1 & p_0 & -p_3 & p_2 \\ -p_2 & p_3 & p_0 & -p_1 \\ -p_3 & -p_2 & p_1 & p_0 \end{bmatrix} \tag{8.82}$$

$$\dot{\omega}_{jk} = 2\,\mathbf{G}\,\ddot{\mathbf{p}} \tag{8.83}$$

Universal joint U. The points *j2* and *k1* that belong to this joint are joined together, and the unit vectors $\mathbf{u}_j$ and $\mathbf{u}_{k1}$ that belong to elements *J* and *K*, respectively, are perpendicular to each other. The axes of the two relative rotations z^{1}_{jk} and z^{2}_{jk}, are $\mathbf{u}_j$ and $\mathbf{u}_{k1}$, respectively. The equations that define the motion of the joint become:

$$\dot{\mathbf{d}}_{jk} = \ddot{\mathbf{d}}_{jk} = 0 \tag{8.84}$$

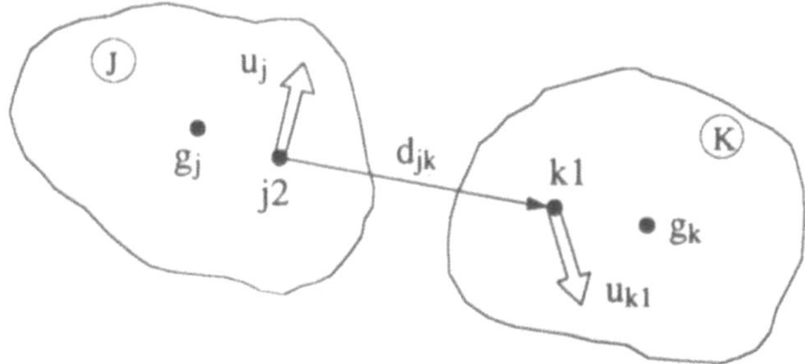

Figure 8.12. Consecutive bodies for recursive velocity calculation (Form. B)

$$\dot{\mathbf{u}}_{k1} = \left(\boldsymbol{\omega}_j + \dot{z}_{jk}^{1}\,\mathbf{u}_j\right) \wedge \mathbf{u}_{k1} \tag{8.85}$$

$$\boldsymbol{\omega}_k = \boldsymbol{\omega}_j + \dot{z}_{jk}^{1}\,\mathbf{u}_j + \dot{z}_{jk}^{2}\,\mathbf{u}_{k1} \tag{8.86}$$

$$\dot{\boldsymbol{\omega}}_k = \dot{\boldsymbol{\omega}}_j + \ddot{z}_{jk}^{1}\,\mathbf{u}_j + \ddot{z}_{jk}^{2}\,\mathbf{u}_{k1} + \dot{z}_{jk}^{1}\,\dot{\mathbf{u}}_j + \dot{z}_{jk}^{2}\,\dot{\mathbf{u}}_{k1} \tag{8.87}$$

8.2.5.2 Formulation B

The main difference between this formulation and the previous one is the kind of dependent variables used to derive the equations of motion. In this formulation, the velocity of the center of gravity and the angular velocity of each body are taken as velocity variables instead of the velocities of points and unit vectors that were used in Formulation A. Using these variables, one can obtain a similar expression of the equations of motion as the one presented in equation (8.38). However, the mass matrix is of size (6x6) with the following expression:

$$\mathbf{M} = \begin{bmatrix} m\,\mathbf{I}_3 & 0 \\ 0 & \mathbf{A}_i\mathbf{J}_i\mathbf{A}_i^{\mathrm{T}} \end{bmatrix} \tag{8.88}$$

An additional term appears in the right-hand side of equation (8.38), to account for the centrifugal terms giving $\mathbf{R}^{\mathrm{T}}(\mathbf{Q}-\mathbf{C})$ instead of $\mathbf{R}^{\mathrm{T}}\mathbf{Q}$ alone.

Consider the general joint of Figure 8.12. The following data is assumed known:

- the position of all the points and the unit vectors,
- the velocity $\dot{\mathbf{g}}_j$ and acceleration $\ddot{\mathbf{g}}_j$ of the center of gravity of body J,
- the angular velocity $\boldsymbol{\omega}_j$ and acceleration $\dot{\boldsymbol{\omega}}_j$ of body J, and,
- the relative velocities $\dot{z}_{jk}$ and accelerations $\ddot{z}_{jk}$ of the joint that links bodies J and K.

The velocities of the output points of body J and the input point of body K will be given by:

$$\dot{\mathbf{r}}_{j2} = \dot{\mathbf{g}}_j + \omega_j \wedge \left(\mathbf{r}_{j2} - \mathbf{g}_j\right) \tag{8.89}$$

$$\dot{\mathbf{u}}_j = \omega_j \wedge \mathbf{u}_j \tag{8.90}$$

$$\dot{\mathbf{r}}_{k1} = \dot{\mathbf{r}}_{j2} + \dot{\mathbf{d}}_{jk} \tag{8.91}$$

$$\omega_k = \omega_j + \omega_{kj} \tag{8.92}$$

$$\dot{\mathbf{u}}_{k1} = \omega_k \wedge \mathbf{u}_{k1} \tag{8.93}$$

$$\dot{\mathbf{g}}_k = \dot{\mathbf{r}}_{k1} + \omega_k \wedge \left(\mathbf{g}_k - \mathbf{r}_{k1}\right) \tag{8.94}$$

The accelerations will be given by:

$$\ddot{\mathbf{r}}_{j2} = \ddot{\mathbf{g}}_j + \dot{\omega}_j \wedge \left(\mathbf{r}_{j2} - \mathbf{g}_j\right) + \omega_j \wedge \left(\omega_j \wedge \left(\mathbf{r}_{j2} - \mathbf{g}_j\right)\right) \tag{8.95}$$

$$\dot{\omega}_k = \dot{\omega}_j + \dot{\omega}_{kj} \tag{8.96}$$

$$\ddot{\mathbf{r}}_{k1} = \ddot{\mathbf{r}}_{j2} + \ddot{\mathbf{d}}_{jk} \tag{8.97}$$

$$\ddot{\mathbf{g}}_k = \ddot{\mathbf{r}}_{k1} + \dot{\omega}_k \wedge \left(\mathbf{g}_k - \mathbf{r}_{k1}\right) + \omega_k \wedge \left(\omega_k \wedge \left(\mathbf{g}_k - \mathbf{r}_{k1}\right)\right) \tag{8.98}$$

In equations (8.89)-(8.98), the particular values of the variables that define the relative motion of each kind of joint are given by the same equations developed for the Formulation A: equations (8.70) to (8.87).

8.2.6 Element-by-Element Computation of Matrix **R**

The matrix **R** may be obtained *globally* (for all the multibody system) or in an *element-by-element* basis. Each column of matrix **R** represents the result of velocity analysis, in which one computes the dependent velocities corresponding to a unit relative independent velocity (rigid body motion of the base body or relative joint velocity). On the other hand, one may consider separately the rows of matrix **R** that correspond to the dependent velocities of a particular body. In these rows, only the columns corresponding to the independent velocities that affect this body, that is, base body velocities and relative velocities in joints that are located backward in the branch of the body, will contain non-zero elements. This presents very good opportunities for carrying out the computations in parallel.

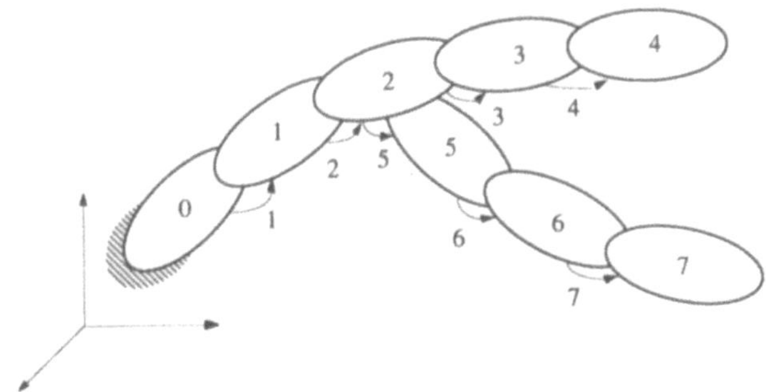

Figure 8.13. Open-chain system with two branches and seven bodies.

Since **R** is a transformation matrix from independent to dependent velocities, **R** may be obtained by either one of the two Formulations A or B, seen in the previous section. The topic of how to obtain **R** using velocity analysis should not present any conceptual difficulty. Formulation B, which seems to be the most efficient method, will be dealt with here. The basic ideas of how to obtain the matrix **R** using Formulation A will be treated through some exercises.

In Section 8.2.2, the physical meaning of the velocity transformation matrix **R** of order ($6n_b{\times}f$) was explained. In this section, a procedure for the fast parallel computation of matrix **R** will be presented. Recall from Section 8.2.5 that in the Formulation B, the *i-th* column of matrix **R** represents the velocity of the center of gravity and the angular velocity of all bodies, that is vector $\dot{q}$, when all the relative joint velocities $\dot{z}$ are zero except the *i-th* component that takes a unit value. Consequently, the columns of matrix **R** can be computed, one at a time, by solving f times the velocity problem. The parallelization of the algorithm becomes easier, if an *element-by-element* computational scheme is adopted. This scheme is based on computing 'separately' the rows of **R** corresponding to each body. In this way, most of the computations can be carried out independently for each body and can be computed concurrently.

Example 8.3

Consider the example in Figure 8.13, with seven elements or rigid bodies forming an open-chain with two branches. It is assumed that all the joints have a single degree of freedom, and that the base body is fixed.

In this example, the size of matrix **R** is of (42×7). Suppose that it is necessary to compute the rows of **R** corresponding to body 5. First of all, columns 3, 4, 6, and 7 will be zero, because the unit velocities of joints 3, 4, 6 and 7 will not produce motion in the bodies that are backward in the kinematic chain. Hence, labeling $\mathbf{R}_5$ as the six rows-matrix **R** corresponding to body 5, omitting the zero columns, one can write it as

$$\mathbf{R}_5 = \left[\mathbf{R}_5^1\, \mathbf{R}_5^2\, \mathbf{R}_5^5\right]$$

where $\mathbf{R}_5^i$ represents a (6×1) vector obtained by extracting from $\mathbf{R}$ the rows of column i that correspond to body 5. Therefore, matrix $\mathbf{R}_5$ is of (6×3) size. The number of columns of $\mathbf{R}_j$ in the general case coincides with the number of degrees of freedom found in the path from body J to the base body. Vector $\mathbf{R}_5^i$ can be written in the following form:

$$\mathbf{R}_5^i = \left\{ \begin{array}{c} \dot{\mathbf{g}}_5^i \\ \boldsymbol{\omega}_5^i \end{array} \right\}$$

where $\dot{\mathbf{g}}_5^i$ and $\boldsymbol{\omega}_5^i$ are, respectively, the velocity of the center of gravity and the angular velocity of body 5, when all the joint relative velocities are zero, except the i-th component, which has unit relative velocity.

Generally, the rows of matrix $\mathbf{R}$ corresponding to body b are contained in matrix $\mathbf{R}_b$. The size of matrix $\mathbf{R}_b$ is (6×f_b), where f_b is the number of degrees of freedom found in the path that goes from body b to the base body including the base body degrees of freedom. Matrix $\mathbf{R}_b$ can be written in the following way:

$$\mathbf{R}_b = \left[\mathbf{R}_b^1 \ \mathbf{R}_b^2 \cdots \mathbf{R}_b^{p_b} \right] \tag{8.99}$$

where p_b is the number of joints in the path from body b to the base body. Subscripts are used for bodies and superscripts for joints. Each submatrix $\mathbf{R}_b^i$ is of size (6×d^i), where d^i is the number of degrees of freedom of joint i. In the example of Figure 8.13, d^i takes unit value for all the joints, since all of them were assumed to allow a single degree of freedom.

The computation of matrices $\mathbf{R}_b^i$ for the different types of joints are presented below:

Revolute joint R. Since a revolute joint allows a single degree of freedom, $d^i=1$ in this case. Assigning $\mathbf{u}_i$ to the unit vector that points in direction of the revolute axis, $\mathbf{R}_b^i$ is calculated as

$$\mathbf{R}_b^i = \left\{ \begin{array}{c} \mathbf{u}_i \wedge \left(\mathbf{g}_b - \mathbf{r}_i \right) \\ \mathbf{u}_i \end{array} \right\} \tag{8.100}$$

where $(\mathbf{g}_b - \mathbf{r}_i)$ is the vector that goes from the revolute joint i to the center of gravity of body b.

Prismatic joint P. Again $d^i=1$. Using the same notation, one can obtain

$$\mathbf{R}_b^i = \left\{ \begin{array}{c} \mathbf{u}_i \\ 0 \end{array} \right\} \tag{8.101}$$

Cylindrical joint C. Since a cylindrical joint allows two degrees of freedom, $d^i=2$; and hence $\mathbf{R}_b^i$ has two columns, that can be written as follows:

$$\mathbf{R}_b^i = \left[\begin{array}{cc} \mathbf{u}_i \wedge (\mathbf{g}_b - \mathbf{r}_i) & \mathbf{u}_i \\ \mathbf{u}_i & 0 \end{array} \right] \tag{8.102}$$

Spherical joint S. In this case, $d^i=3$. Since the relative angular velocity vector is taken as relative joint velocity, the expression of $\mathbf{R}_b^i$ is

$$\mathbf{R}_b^i = \begin{bmatrix} \mathbf{i} \wedge (\mathbf{g}_b - \mathbf{r}_i) & \mathbf{j} \wedge (\mathbf{g}_b - \mathbf{r}_i) & \mathbf{k} \wedge (\mathbf{g}_b - \mathbf{r}_i) \\ \mathbf{i} & \mathbf{j} & \mathbf{k} \end{bmatrix} \qquad (8.103)$$

where $\mathbf{i}$, $\mathbf{j}$, and $\mathbf{k}$ are three orthonormal unit vectors parallel to the inertial axes.

Universal joint U. As the cylindrical joint, the universal joint has $d^i=2$. Vectors $\mathbf{u}_{i1}$ and $\mathbf{u}_{i2}$ are designated to the two orthogonal unit vectors that point in the directions of the axes. Hence,

$$\mathbf{R}_b^i = \begin{bmatrix} \mathbf{u}_{i1} \wedge (\mathbf{g}_b - \mathbf{r}_i) & \mathbf{u}_{i2} \wedge (\mathbf{g}_b - \mathbf{r}_i) \\ \mathbf{u}_{i1} & \mathbf{u}_{i2} \end{bmatrix} \qquad (8.104)$$

Base body. An unconstrained floating base body has six rigid body degrees of freedom; thus $d^i=6$. As in the spherical joint, the rotational part of matrix $\mathbf{R}$ is defined from three unit angular velocities in the directions of the inertial axes. Vector $\mathbf{g}_b^i$ is the vector that goes from the center of gravity of the base body (reference point) to the center of gravity of body b. Matrix $\mathbf{R}_b^i$ is thus calculated in the following way:

$$\mathbf{R}_b^i = \begin{bmatrix} \mathbf{i} & \mathbf{j} & \mathbf{k} & \mathbf{i} \wedge \mathbf{g}_b & \mathbf{j} \wedge \mathbf{g}_b & \mathbf{k} \wedge \mathbf{g}_b \\ 0 & 0 & 0 & \mathbf{i} & \mathbf{j} & \mathbf{k} \end{bmatrix} \qquad (8.105)$$

where columns one to three correspond to the translational degrees of freedom, and columns four to six to the rotational ones.

The procedure described in this section for the computation of matrix $\mathbf{R}$ works independently for each of the bodies. Thus, matrices $\mathbf{R}_b$ can be computed concurrently.

8.2.7 Computation of Mass Matrices $\mathbf{M}_b$

As it can be seen in Nikravesh (1988) and Haug (1989), with reference point coordinates the multibody mass matrix $\mathbf{M}$ can be obtained by assembling the body mass matrices $\mathbf{M}_b$. It is not necessary to obtain explicitly matrix $\mathbf{M}$, but only to compute $\mathbf{M}_b$ and perform the required product with matrix $\mathbf{R}_b$.

The body mass matrix $\mathbf{M}_b$ is a (6×6) matrix formed by two (3×3) submatrices on its diagonal. The leading (3×3) submatrix is the unit matrix $\mathbf{I}$ times the mass of the element. The second (3×3) submatrix $\mathbf{J}_b$, which represents the inertial tensor of body b expressed in the inertial reference frame, requires further computation because it is position dependent. Using the well-known tensor transformation expression between two reference frames, one can obtain

$$\mathbf{J}_b = \mathbf{A}_b \, \bar{\mathbf{J}}_b \, \mathbf{A}_b^T \qquad (8.106)$$

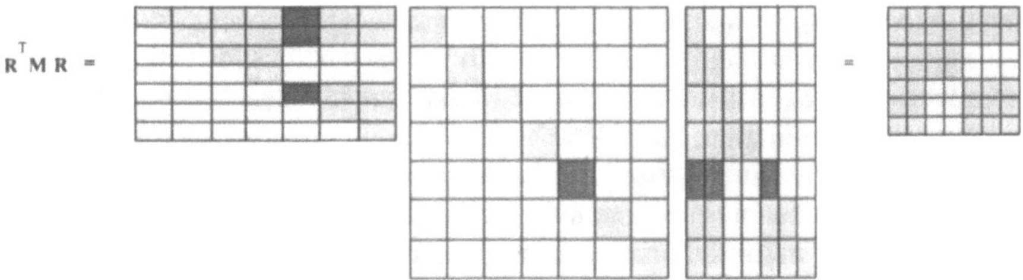

Figure 8.14. Non-zero pattern for the product $\mathbf{R}^T\mathbf{M}\mathbf{R}$ in Example 8.3.

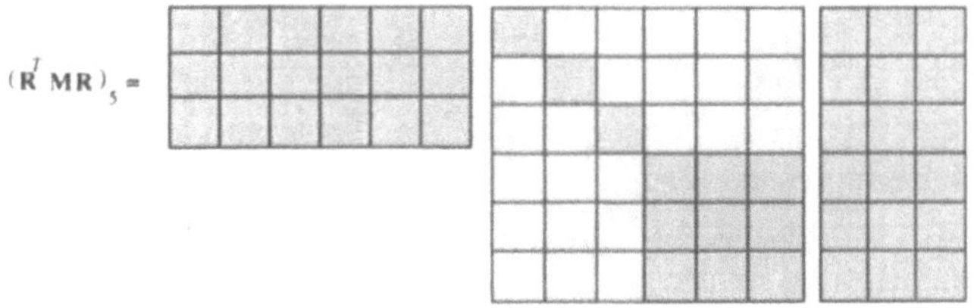

Figure 8.15. Non-zero pattern for the product $\mathbf{R}^T\mathbf{M}\mathbf{R}$ for body 5 in Example 8.3.

where matrix $\overline{\mathbf{J}}_b$ is the constant inertial tensor expressed in the moving frame of body b.

8.2.8 Computation of the Matrix Product $\mathbf{R}^T\mathbf{M}\mathbf{R}$

One of the most attractive features of the choice of dependent coordinates that has been made (reference point coordinates) is that the multibody system mass matrix $\mathbf{M}$ has a block-diagonal structure without coupling terms between contiguous bodies. The immediate application of this property is that the triple matrix product $(\mathbf{R}^T\mathbf{M}\mathbf{R})$ that appears in the equations of motion (8.38) can be computed efficiently on an *element-by-element* basis. This means this triple product can be computed as

$$\mathbf{R}^T\,\mathbf{M}\,\mathbf{R} = \sum_{i=1}^{N_b}{}^{*}\ \mathbf{R}_i^T\,\mathbf{M}_i\,\mathbf{R}_i \tag{8.107}$$

where the symbol Σ^* represents the combined action of summation and assembly of the resulting matrix. Due to the fact that there are no coupling terms in matrix $\mathbf{M}$, equation (8.107) shows that the triple product can be performed *ele-*

ment-by-element without increasing the total number of arithmetic operations. This product can be computed independently for each body which allows an easy parallelization that fits perfectly into the whole computational scheme.

With this formulation, there is no need to treat in a special way the *junction bodies*, or bodies that are linked to more than two other bodies. In the example of Figure 8.13, the only junction body is body 2, which is linked to bodies 1, 3, and 5. The term in the summation corresponding to it can be computed exactly in the same way as all the other terms.

If one returns again to the multibody system of Example 8.3 (which is depicted in Figure 8.13), one can symbolically represent the product $\mathbf{R}^T\mathbf{M}\mathbf{R}$ a s shown in Figure 8.14. Submatrices that contain non-zero terms have been represented with a shaded pattern. When the product $\mathbf{M}\mathbf{R}$ is calculated, the distribution of zero and non-zero terms in the resulting matrix coincides with the one in matrix $\mathbf{R}$. This means that the product between $\mathbf{R}^T$ and $\mathbf{M}\mathbf{R}$ involves two matrices with exactly the same structure of matrix $\mathbf{R}$. Then it is possible to obtain the whole product as a summation of individual terms corresponding to each body. The terms that involve body 5, $(\mathbf{R}^T\mathbf{M}\mathbf{R})_5$ have been outlined with a dark shaded pattern and are represented separately in Figure 8.15, once the zero columns of matrix $\mathbf{R}$ have been removed. Again, shaded terms correspond to non-zero elements in the different matrices. The result of this small triple product is a symmetric matrix. Only the terms above the diagonal need to be computed. Finally, the partial result corresponding to body 5 has to be assembled in the final matrix on rows and columns 1, 2, and 5. This computational scheme can be applied to all the bodies in the system, so as to obtain the matrix $\mathbf{R}^T\mathbf{M}\mathbf{R}$ in a very efficient way.

8.2.9 *Computation of the Matrix Product* $\mathbf{R}^T\mathbf{M}\mathbf{S}\mathbf{c}$

Looking at equation (5.65) in Chapter 5, it can be seen that the product $(\mathbf{S}\mathbf{c})$ has a clear physical meaning, that makes it very easy to compute. If one makes the independent accelerations $\ddot{\mathbf{z}}$ equal to zero and then computes the dependent accelerations $\ddot{\mathbf{q}}$, the resulting vector turns out to be $(\mathbf{S}\mathbf{c})$. Therefore, $(\mathbf{S}\mathbf{c})$ represents the vector of dependent accelerations that depends only on the independent velocities. The recursive computation of accelerations has been described in Section 8.2.5. It can be applied to this particular case.

Once the term $(\mathbf{S}\mathbf{c})$ has been computed, the final product $\mathbf{R}^T\mathbf{M}\mathbf{S}\mathbf{c}$ can be computed again in an *element-by-element* basis, recovering the partial product $\mathbf{R}^T\mathbf{M}$ from the previous section.

8.2.10 *Computation of the Term* $\mathbf{R}^T(\mathbf{Q} - \mathbf{C})$

Once more, this product corresponds to the projected external and velocity-dependent inertial forces and does not offer any special difficulty. This product can be computed on an *element-by-element* basis and then conveniently assembled to produce the corresponding forcing term in the equations of motion.

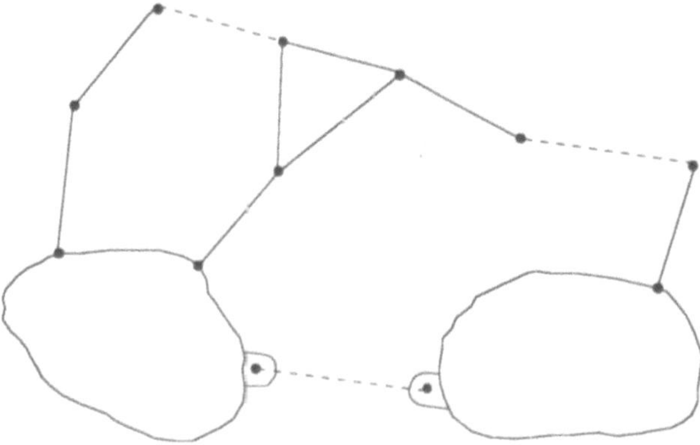

Figure 8.16. Opening a closed-chain system by removing rigid body constraints.

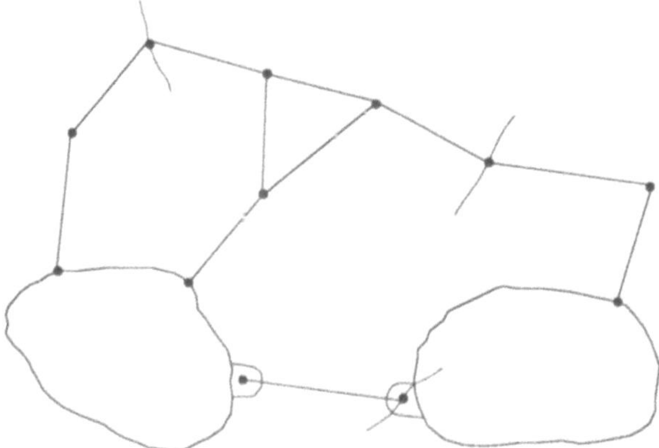

Figure 8.17. Opening a closed-chain system by removing joint constraints.

8.3 Velocity Transformations for Closed-Chain Systems

It was shown in Section 5.3 how a closed-chain multibody system can be transformed to an open-chain by simply eliminating certain constraint equations that enforce the closure of the loops.

Two different formulations, A and B, have been introduced in Section 8.2.5. Which one is used depends on how the velocities and accelerations are represented. Formulation A considers as variables the *natural coordinates*, that is, the Cartesian components of basic points and unit vectors. The constraint equations that these coordinates generate (See Chapter 2) are primarily due to the rigid body conditions. Therefore, the way to proceed in this case to open the kinematic chain is by *eliminating certain rigid body conditions* of some of the elements of the multibody system. Figure 8.16 illustrates how this procedure may be applied to a complex system using the concepts just explained.

Formulation B considers that the dependent velocities of each element are characterized by the velocity of the center of gravity and by the angular velocity of the element. In this case, the constraint equations (See Chapter 2) are originated not at the elements but at the kinematic joints. Therefore, in order to open a loop, one needs to *cut a joint* by removing the corresponding constraints. Figure 8.17 illustrates this procedure.

Regardless of what formulation is used to represent the dependent velocities and accelerations of the multibody system, it is always possible to divide the constraint equations into two major groups: the first, that will be represented by the superscript 1, is formed by the constraints of the open-chain multibody system that result from the opening of the kinematic loops; the second, that in the sequel will be represented by the superscript 2, will be formed by those constraints needed to close the loops previously opened. Consequently, the velocity constraint equations become

$$\begin{bmatrix} \Phi_q^1 \\ \Phi_q^2 \end{bmatrix} \{\dot{q}\} = \{b\} \equiv -\begin{Bmatrix} \Phi_t^1 \\ \Phi_t^2 \end{Bmatrix} \tag{8.108}$$

where $b=0$, if the constraints are scleronomous. Let m_1 and m_2 be the number of rows of Φ_q^1 and Φ_q^2, respectively. Assume that $m_1 >> m_2$ (m_1 much greater than m_2), since the opening of the loops will be done by removing only a few constraint equations.

If the number of dependent coordinates is n, the number of degrees of freedom of the open- and closed-chain subdivisions will be $f_1 = n - m_1$ and $f_2 = n - m_1 - m_2$, respectively.

The key point in this formulation is the fact that the matrix $\mathbf{R}^1$, which defines a basis for the nullspace of Φ_q^1, *can be directly obtained* by the procedure explained in Section 8.2 without forming and triangularizing the Jacobian matrix Φ_q^1. This leads to important savings in computational costs. Even though the matrix Φ_q^1 is never formed explicitly, the following relationships will still be satisfied:

$$\Phi_q^1 \mathbf{R}^1 = 0 \tag{8.109}$$

$$\dot{q} = \mathbf{R}^1 \dot{z}^1 \tag{8.110}$$

$$\ddot{q} = \mathbf{R}^1 \ddot{z}^1 + \mathbf{S}^1 \mathbf{c}^1 \tag{8.111}$$

where the vector z^1 is formed by the base body and relative joint coordinates of the open-chain system. Now, in the closed-chain system, these coordinates z^1 are not independent, because they are interrelated through the constraints Φ^2. The problem is that the constraints Φ^2 are not written in terms of z^1 but in terms of q. However, this problem may be easily solved as follows.

The equations of motion seen in Chapter 5 that are based on the Lagrange multipliers technique will be considered. These equations, applied to the problem at hand, become

$$
\begin{bmatrix} \mathbf{M} & \Phi_{\mathbf{q}}^{1T} & \Phi_{\mathbf{q}}^{2T} \\ \Phi_{\mathbf{q}}^{1} & 0 & 0 \\ \Phi_{\mathbf{q}}^{2} & 0 & 0 \end{bmatrix} \begin{Bmatrix} \ddot{\mathbf{q}} \\ \lambda^{1} \\ \lambda^{2} \end{Bmatrix} = \begin{Bmatrix} \mathbf{Q} \\ \mathbf{c}^{1} \\ \mathbf{c}^{2} \end{Bmatrix}
\tag{8.112}
$$

where λ^1 and λ^2 are the multipliers that correspond to the partitions $\Phi_{\mathbf{q}}^1$ and $\Phi_{\mathbf{q}}^2$, respectively. The vector $\mathbf{Q}$ includes the external as well as all the velocity dependent inertial forces.

Substituting equation (8.111) into equation (8.112) and pre-multiplying the first row of equation (8.111) by $(\mathbf{R}^1)^T$, one obtains

$$
\begin{bmatrix} \mathbf{R}^{1T}\mathbf{M}\mathbf{R}^{1} & \mathbf{R}^{1T}\Phi_{\mathbf{q}}^{1T} & \mathbf{R}^{1T}\Phi_{\mathbf{q}}^{2T} \\ \Phi_{\mathbf{q}}^{1}\mathbf{R}^{1} & 0 & 0 \\ \Phi_{\mathbf{q}}^{2}\mathbf{R}^{1} & 0 & 0 \end{bmatrix} \begin{Bmatrix} \ddot{\mathbf{z}}^{1} \\ \lambda^{1} \\ \lambda^{2} \end{Bmatrix} = \begin{Bmatrix} \mathbf{R}^{1T}\mathbf{Q} - \mathbf{R}^{1T}\mathbf{M}\mathbf{S}^{1}\mathbf{c}^{1} \\ \mathbf{c}^{1} - \Phi_{\mathbf{q}}^{1}\mathbf{S}^{1}\mathbf{c}^{1} \\ \mathbf{c}^{2} - \Phi_{\mathbf{q}}^{2}\mathbf{S}^{1}\mathbf{c}^{1} \end{Bmatrix}
\tag{8.113}
$$

However, introducing equation (8.109), the second row and column of equation (8.113) cancels out. It follows that the coefficient of λ^1 vanishes and the following equation is obtained:

$$
\begin{bmatrix} \mathbf{R}^{1T}\mathbf{M}\mathbf{R}^{1} & \mathbf{R}^{1T}\Phi_{\mathbf{q}}^{2T} \\ \Phi_{\mathbf{q}}^{2}\mathbf{R}^{1} & 0 \end{bmatrix} \begin{Bmatrix} \ddot{\mathbf{z}}^{1} \\ \lambda^{2} \end{Bmatrix} = \begin{Bmatrix} \mathbf{R}^{1T}\mathbf{Q} - \mathbf{R}^{1T}\mathbf{M}\mathbf{S}^{1}\mathbf{c}^{1} \\ \mathbf{c}^{2} - \Phi_{\mathbf{q}}^{2}\mathbf{S}^{1}\mathbf{c}^{1} \end{Bmatrix}
\tag{8.114}
$$

The new mass matrix is of size $(f_1 \times f_1)$, instead of $(n \times n)$ as was the original mass matrix $\mathbf{M}$. The new projected Jacobian matrix $(\Phi_{z1} = \Phi_{\mathbf{q}}^{2}\mathbf{R}^{1})$ is of size $(m_2 \times f_1)$. Since $m_2 << m_1$ and $f_1 << n$, this new Jacobian matrix is much smaller than the original in equation (8.112), that had a dimension of $((m_1 + m_2) \times n)$.

The matrix transformations implied in equation (8.114) may be performed in an *element-by-element* basis and thus can be parallelized in an optimal manner.

Therefore, the reduced system of equations (8.114) has a size much smaller than that of system (8.112). The equations of motion (8.114) may be solved by either one of the following methods, studied in detail in Chapter 5:

a) *Lagrange multipliers.* The system of equations (8.114) can be solved as is or including the Baumgarte stabilization terms, as explained in Section 5.1.

b) *Penalty formulation.* The application of the penalty formulation is straightforward. By introducing $\Phi_{z1} = \Phi_{\mathbf{q}}^{2}\mathbf{R}^{1}$ and $\mathbf{Q}_{z1} = \mathbf{R}^{1T}(\mathbf{Q} - \mathbf{M}\mathbf{S}^{1}\mathbf{c}^{1})$, this formulation (See equation (5.37)) leads to

$$
\left(\mathbf{R}^{1T}\mathbf{M}\mathbf{R}^{1} + \alpha\,\Phi_{z1}^{T}\Phi_{z1}\right)\ddot{\mathbf{z}}^{1} = \mathbf{Q}_{z1} - \alpha\,\Phi_{z1}^{T}\left(\dot{\Phi}_{z1}^{T}\dot{\mathbf{z}}^{1} - \dot{\Phi}_{t}(\mathbf{z}^{1}) + \right.
$$
$$
\left. + 2\Omega\mu\,\dot{\Phi}(\mathbf{z}^{1}) + \Omega^{2}\,\Phi(\mathbf{z}^{1})\right)
\tag{8.115}
$$

c) *Independent coordinates*. The kinematic part of (8.114) can be written in the following form:

$$\Phi_{z^1}\, \ddot{z}^1 = \bar{c}^2 \tag{8.116}$$

where $\Phi_{z^1} = \Phi_q^2 R^1$ and $\bar{c}^2 = c^2 - \Phi_q^2 S^1 c^1$. One can always choose a subset of independent accelerations $\ddot{z}$ from the vector $\ddot{z}^1$. Using the method of the Boolean matrix B (See Section 3.3), the following relation is satisfied:

$$B\, \ddot{z}^1 = \ddot{z} \tag{8.117}$$

By simply joining equations (8.116) and (8.117), the following equation is obtained:

$$\begin{bmatrix} \Phi_q^2\, R^1 \\ B \end{bmatrix} \{\ddot{z}^1\} = \begin{Bmatrix} \bar{c}^2 \\ \ddot{z} \end{Bmatrix} \tag{8.118}$$

If the unit values of the matrix B have been chosen according to the pivot structure of the matrix $\Phi_q^2 R^1$, the leading matrix of equation (8.118) can be inverted to yield

$$\{\ddot{z}^1\} = \begin{bmatrix} \Phi_q^2\, R^1 \\ B \end{bmatrix}^{-1} = \begin{Bmatrix} \bar{c}^2 \\ \ddot{z} \end{Bmatrix} \equiv S^2\, \bar{c}^2 + R^2\, \ddot{z} \tag{8.119}$$

Equation (8.119) leads to the definition of the matrices R^2 and $(S^2 \bar{c}^2)$. The columns of the matrix R^2 constitute a basis of the nullspace of the projected Jacobian matrix $\Phi_{z^1} = \Phi_q^2 R^1$. Substituting equation (8.119) into equation (8.114) and pre-multiplying by R^{2T}, one obtains

$$R^{2T}\, R^{1T}\, M\, R^1\, R^2\, \ddot{z} =$$
$$= R^{2T}\, (R^{1T}\, (Q - M\, S^1\, c^1) - R^{1T}\, M\, R^1\, S^2\, \bar{c}^2) \tag{8.120}$$

which is the final equation of motion in independent coordinates.

While the matrix R^1 (of size $(n \times f_1)$) corresponding to Φ_q^1, has been *directly obtained without forming and triangularizing* Φ_q^1, the matrix R^2 (of order $(f_1 \times f_2)$) has to be computed *numerically*. The size of R^2 can be significantly smaller than R^1 as may be seen in Figure 8.18, where the sizes of the matrices involved in this formulation are illustrated for a typical case.

Table 8.2 shows the scheme of a numerical algorithm applicable to closed-loop multibody systems using the method presented in this section, with the Lagrange multipliers version.

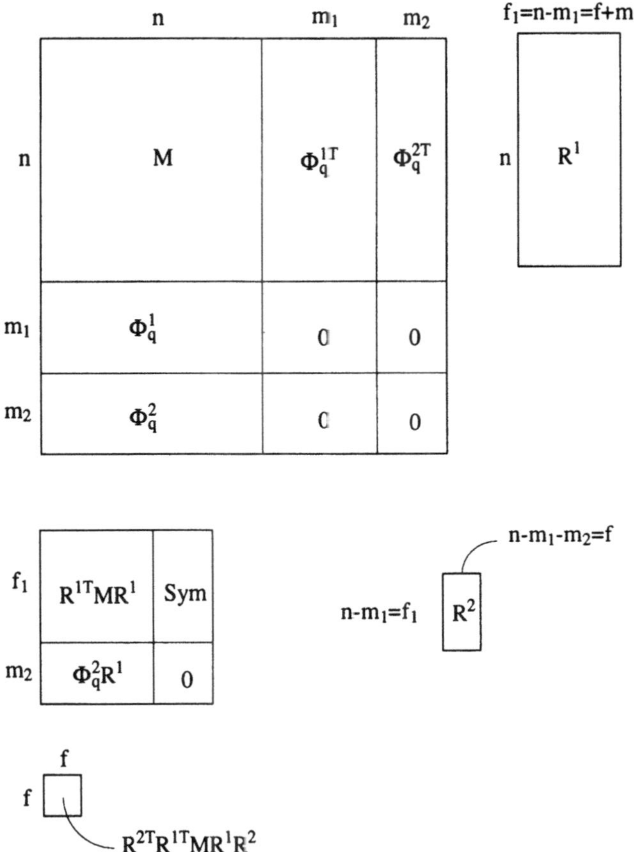

Figure 8.18. Matrix pattern for the method based on velocity transformations.

It may be seen that dealing with closed-chains does not complicate the formulation very much. Thus a good numerical efficiency and a very simple implementation can be expected.

8.4 Examples Solved by Velocity Transformations

In Sections 8.2 and 8.3, the application of velocity transformations to open- and closed-chain kinematic chains has been presented. In this section, two rather complex theoretical examples of the above methods will be described in detail. Afterwards, some numerical results will be presented to demonstrate the efficiency of these formulations.

Table 8.2. Algorithm to formulate and integrate the equations of motion of an closed-chain system, by the Lagrange multipliers method.

Step	Data	Result	Mode
1	$\mathbf{z}_1$	$\overline{\mathbf{q}}$	recursive
2	$\overline{\mathbf{q}}$	$\mathbf{R}^1$	e-by-e or rec.
3	$\dot{\mathbf{z}}_1, \overline{\mathbf{q}}$	$\dot{\mathbf{q}}$	recursive
4	$\dot{\mathbf{z}}_1, \overline{\mathbf{q}}, \dot{\mathbf{q}}$	$(\mathbf{S}^1\mathbf{c}^1)$	e-by-e
5	$\mathbf{R}^1, \mathbf{M}$	$\mathbf{R}^{1T}\,\mathbf{M}\,\mathbf{R}^1$	e-by-e
6	$\mathbf{R}^1, \mathbf{Q}$	$\mathbf{R}^{1T}\,\mathbf{Q}$	e-by-e
7	$\mathbf{R}^1, \mathbf{S}^1\mathbf{c}^1, \mathbf{M}$	$\mathbf{R}^{1T}\,\mathbf{M}\,\mathbf{S}^1\mathbf{c}^1$	e-by-e
8	$\overline{\mathbf{q}}, \dot{\mathbf{q}}$	$\Phi_q^2, \mathbf{c}^2$	e-by-e
9	$\Phi_q^2, \mathbf{R}^1$	$\Phi_q^2\,\mathbf{R}^1$	e-by-e
10		$\mathbf{c}^2 - \Phi_q^2\mathbf{S}^1\mathbf{c}^1$	e-by-e
11	linear equations	$\ddot{\mathbf{z}}_1$	global
12	$(\ddot{\mathbf{z}}_1, \dot{\mathbf{z}}_1)_t$	$(\dot{\mathbf{z}}_1, \mathbf{z}_1)_{t+h}$	global
	GO TO 1		

8.4.1 Open-Chain Example: Human Body

Figure 8.19 shows an interesting example of open kinematic chain: a dynamic model of the human body with 40 rigid bodies and 45 degrees of freedom. All joints are of the revolute type. Some of them are defined by sharing one point and one unit vector and others are defined by sharing two points.

This mechanical model of the human body has been used for some time in several studies on sport biomechanics, as the ones shown in Figures 1.5 and 1.6 which include also realistic geometric models.

Table 8.3 summarizes the resulting theoretical number of floating-point arithmetic operations both for Formulations A and B. It may be seen that important savings can be obtained with Formulation B. The most important conclusion is that, according to the number of floating-point operations of Formulation B, a 6 Mflops computer would be enough to get a function evaluation every 10 msec. This CPU performance is available in many RISC low-cost workstations.

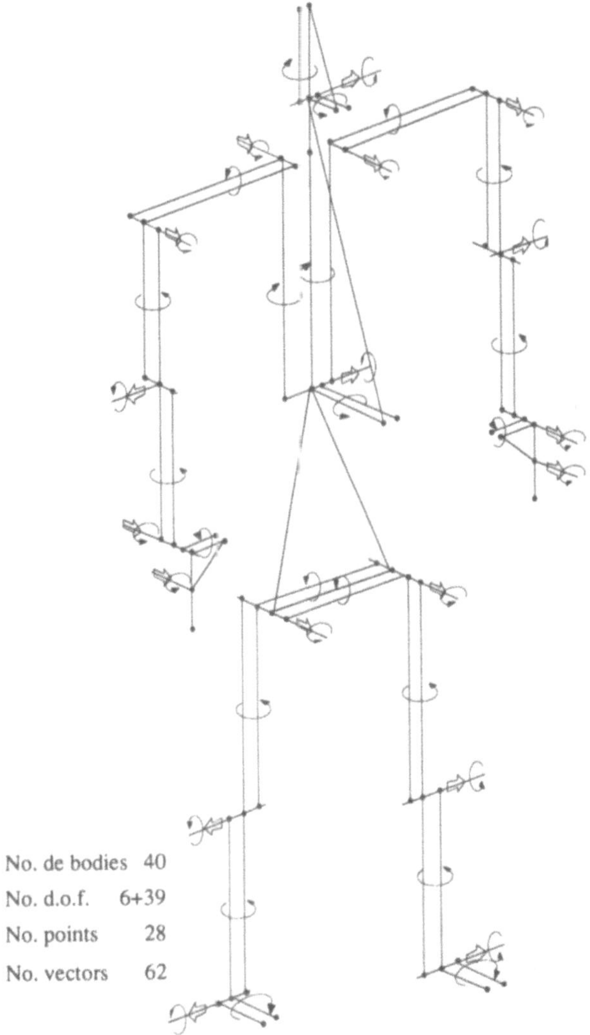

Figure 8.19. Mechanical model of the human body.

Many additional unit vectors (not represented in Figure 8.19) have been introduced in Formulation A. This was done in order for the model to have constant mass matrices with natural coordinates.

8.4.2 Closed-Chain Example: Heavy Truck

Figure 8.20 shows the scheme of a heavy truck suspension and steering system. This system consists of the chassis, the front axle, the rear axle, the leaf-springs modeled by four articulated rigid bodies, the front and rear stabilization bars, the

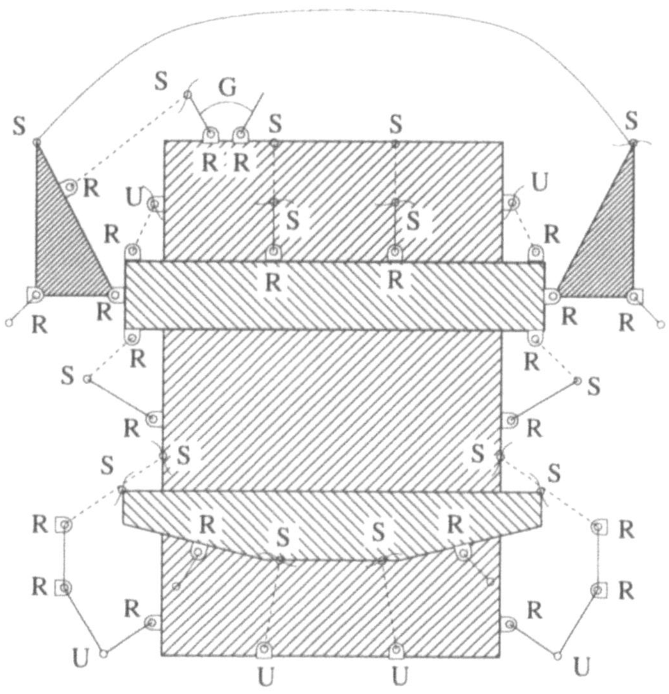

No. of rigid bodies:	33 (21 + 12)
Joints:	23R + 15S +6U + 1G
Open-loop d.o.f.:	39
No. d.o.f.:	18
Dependent coordinates:	285 (246 + 39)

Figure 8.20. Mechanical model of a heavy truck.

elements of the steering system, and the wheels. There are 33 rigid bodies and 18 degrees of freedom. With Formulation A the loops have been opened by removing the constraint equations corresponding to the elements represented by dashed lines. In this case there are 39 open-chain degrees of freedom or relative coordinates.

With formulation B, some joints have been cut to obtain an open-chain system with 48 degrees of freedom. The cut joints are shown in Figure 8.20 with a line crossing the corresponding joint. Table 8.4 presents the theoretical count of floating-point operations for both Formulations. Again important savings are obtained with Formulation B. In spite of the closed loops of this system, it needs less arithmetic operations than the human model of Figure 8.19, because the branches of the open-chain model of the truck are shorter than in the human body.

Table 8.3. Number of floating-point arithmetic operations for the human body (Formulations A and B).

Function	Mode	Formulation A	Formulation B
$z \rightarrow q$	Glob-rec.	1820m + 859a	1560m + 960a
$\dot{z} \rightarrow \dot{q}$	Glob-rec.	615m + 441a	360m + 360a
$\ddot{q}\big|_{z=0} = S\,c$	Glob-rec.	1089m + 915a	840m + 840a
M	e-by-e	--	1080m + 480a
c	e-by-e	--	600m + 360a
R	e-by-e	4716m + 3537a	1050m + 525a
$R^{T}MR$	e-by-e	25824m + 22142a	7482m + 5285a
$R^{T}MSc$	e-by-e	3060m + 2885a	1050m + 875a
$R^{T}(Q-c)$	e-by-e	3060m + 2885a	1050m + 995a
$\phi_q^2 R$	e-by-e	--	--
$c_2 - \phi_q^2 Sc$	e-by-e	--	--
Assembly	e-by-e	3918a	700a
Linear eqs.	Global	16710m + 16170a	16710m + 16170a
TOTAL		56894m + 53752a	31782m + 27550a

Table 8.4. Number of floating-point arithmetic operations for a heavy truck (Formulations A and B).

Function	Mode	Formulation A	Formulation B
$z \rightarrow q$	Glob-rec.	1500m + 1222a	1170m + 720a
$\dot{z} \rightarrow \dot{q}$	Glob-rec.	507m + 627a	585m + 360a
$\ddot{q}\big|_{z=0} = S\,c$	Glob-rec.	561m + 492a	870m + 540a
M	e-by-e	--	837m + 372a
c	e-by-e	--	465m + 279a
R	e-by-e	960m + 726a	306m + 153a
$R^{T}MR$	e-by-e	25890m + 20893a	1947m + 1394a
$R^{T}MSc$	e-by-e	4947m + 4120a	306m + 255a
$R^{T}(Q-c)$	e-by-e	2040m + 1870a	306m + 348a
$\phi_q^2 R$	e-by-e	2220m + 1850a	1224m + 1374a
$c_2 - \phi_q^2 Sc$	e-by-e	432m + 432a	108m + 108a
Assembly	e-by-e	897a	900a
Linear eqs.	Global	16000m + 16000a	16000m + 16000a
TOTAL		55084m + 49129a	24124m + 23406a

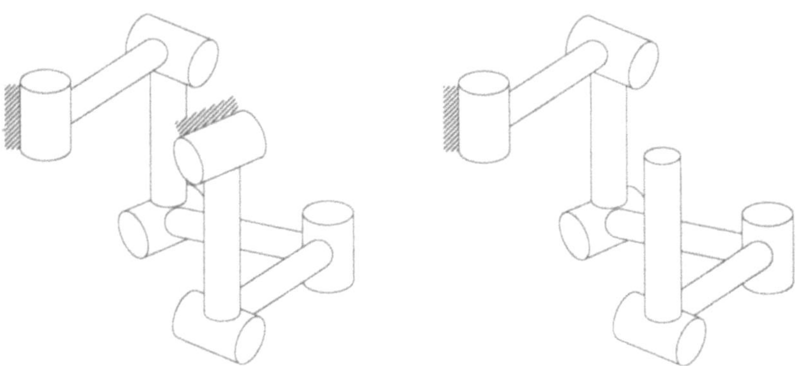

Figure 8.21. Bricard mechanism.   Figure 8.22. Five-bar pendulum.

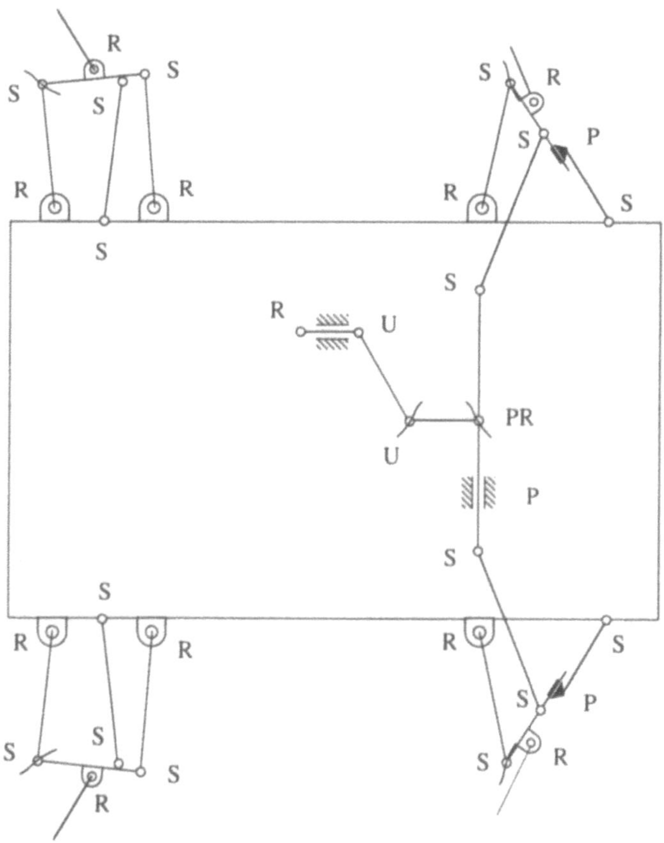

Figure 8.23. Multibody model of a car suspension and steering system.

Table 8.5. Comparative results in CPU milliseconds for function evaluation.

	Penalty method (Section 5.1.4)	Matrix R (Section 5.2.3)	Formulation B (Section 8.2.5.2)
Bricard	8,01	8,73	1,33
Five-bar pendulum	12,00	18,65	1,29
Car model	87,79	224,54	12,21
Human body	91,30	942,64	23,39

8.4.3 Numerical Results

Finally, some comparative numerical results obtained with the Formulation B, and with two other standard dynamic formulations explained in Chapter 5, will be presented. Four examples will be considered. Two of them are very simple. The other two present some degree of complexity.

Figures 8.21 and 8.22 show two simple three-dimensional multibody systems with five bodies and revolute joints only. Figure 8.21 shows the Bricard mechanism which is an over-constrained closed-chain system. According to the Grübler criterion, this mechanism has zero degrees of freedom. Because of the particular orientation of the axes in the revolute joints, it actually has one degree of freedom. Figure 8.22 illustrates a five-bar three-dimensional pendulum that can be obtained directly from the Bricard mechanism by opening the chain through the removal of one of the fixed revolute joints. This pendulum has five degrees of freedom.

Figure 8.23 shows a car model that includes the chassis, the steering and the suspension system. The front suspension is a McPherson type and the rear suspension is based on the five-point system. The complete model has 25 rigid bodies with general mass and inertial properties and 15 degrees of freedom. This car model corresponds to the more realistic model shown in Figure 1.2 of Chapter 1. The fourth example is based on the 45 bodies and 45 degrees of freedom human body model shown in Figure 8.19 which constitutes the basis of the models presented in Figures 1.5 and 1.6 of Chapter 1.

These four examples with different topology and degree of complexity have been analyzed using three dynamic formulations: a) the penalty method, based on dependent coordinates (Section 5.1.4); b) the method of independent coordinates based on the matrix **R** (Section 5.2.3); and c) the Formulation B previously explained in this chapter (Section 8.2.5.2).

Table 8.5 contains the CPU time per function evaluation (computation of accelerations from positions and velocities) expressed in milliseconds for the four examples and using the three different formulations. The results have been obtained with a SG workstation using a MIPS 3000 processor rated at 25 MHz

(3.9 DP Mflops in the Linpack test). Taking into account that there are currently newer low-cost workstations that are much more powerful, it may be concluded from the results shown in Table 8.5 that real time dynamic simulation is at hand for these systems, at least from the point of view of fast function evaluation.

These results do not provide an idea of the absolute efficiency of each method, since the total time of integration depends not only on the time required per function evaluation, but also on the type of numerical method used for the integration of the equations of motion. Standard ODE integrators, such as the DE Shampine and Gordon routine, require less function evaluations with the matrix **R** or Formulation B than with the penalty method. The penalty method leads to stiffer equations and works more efficiently in conjunction with more stable integrators (See Section 8.5). In order to get the best possible response, the concepts developed in Chapters 7 and 8 must be jointly considered along with the particular physical characteristics of the case at hand.

8.5 Special Implementations Using Dependent Natural Coordinates

So far in this chapter dynamic formulations have been considered that, although they may use dependent coordinates to define the motion of the system, try to ultimately solve the equations of motion through a minimum set of independent (in the case of open-chain systems) or dependent coordinates (in closed chain systems). In our search for formulations suitable for real time analysis, we present in this section a method proposed by Bayo et al. (1991) which is an alternative to the methods seen already in this chapter. This method is based on formulating and solving the equations of motion with respect to a full set of dependent coordinates without intermediate transformations. The constraints are considered via a modification of the penalty formulation seen in Chapter 5. The numerical integration is carried out using the trapezoidal rule (See Chapter 7) with the positions, rather than accelerations, as primary variables.

Among the possible sets of dependent coordinates, the natural coordinates (explained in Chapter 2) provide the best setting. They present important advantages over other possible sets: first, the use of elements with two basic points and two unit vectors leads to a constant mass matrix in the inertial frame and to the absence of velocity dependent (centrifugal and Coriolis) inertial forces in the formulation; and secondly, the natural coordinates originate quadratic constraint equations that yield linear Jacobian matrices. As clearly explained in Chapter 2, these Jacobian matrices can be evaluated by a number of arithmetic operations that is considerably lower than those required by other types of coordinates.

Other choices of 12 Cartesian components that define the position and orientation of the rigid body will also lead to a constant mass matrix, such as one point and three Cartesian vectors, four points, and so forth. The choice of two points and two unit vectors is due to the fact that these variables can be shared by contiguous elements; thus leading to a lesser number of dependent coordinates that will represent the multibody system.

8.5.1 Differential Equations of Motion in the Natural Coordinates

A dynamic system shall be considered whose configuration is characterized by a vector $\mathbf{q}$ of n fully Cartesian coordinates that satisfy m constraint conditions $\Phi=0$. Let $L=T-V$ be the *Lagrangian* function of the system, where T and V are the kinetic and potential energies, respectively.

In order to introduce the constraint conditions, a penalty formulation slightly different from that explained in Section 5.2 will be used, that will also lead to a set of ordinary differential equations (ODE), and will guarantee the fulfillment of the constraint equations. In order to introduce these constraints, a fictitious potential is added to the expression of the Hamilton's principle:

$$V^* = \frac{1}{2}\,\Phi^T\,\alpha\,\Phi \tag{8.121}$$

and a set of Rayleigh's dissipative forces

$$G^* = -\,\alpha\,\mu\,\dot{\Phi} \tag{8.122}$$

The penalty factors α_k, $k= 1,2,...\ m$ (with each constraint having a different factor) are large real numbers that represent the spring values of the physical system associated with the constraint $\Phi_k = 0$, and $(\alpha_k\mu_k)$ represents the damping characteristics. Similar to the development already carried out in Chapter 5, the application of the Lagrange's equations directly leads to

$$\mathbf{M}\,\ddot{\mathbf{q}} + \Phi_q^T\,\alpha\left(\Phi + \mu\,\dot{\Phi}\right) = \mathbf{Q} \tag{8.123}$$

where Φ_q is the *(m×n)* Jacobian matrix of the constraint equations, $\mathbf{M}$ the mass matrix, and $\mathbf{Q}$ the external forces plus those coming from a potential acting on the system. The matrices α and μ are *(m×m)* diagonal matrices that contain the values of the penalty numbers and damping coefficients. If the same values are used for all the constraints these matrices become identity matrices multiplied by constant factors. Remember that the *natural coordinates* described above yield a constant mass matrix. In addition, neither Coriolis nor centrifugal forces are present in the vector $\mathbf{Q}$. Consequently, the only nonlinear components in (8.123) except for position-dependent forces are the penalty terms which physically represent the forces necessary to maintain the constraint conditions. The product $\alpha(\Phi+\mu\dot{\Phi})$ represents an approximation to the Lagrange multipliers that arise in the classical formulation.

The nonlinear system (8.123) without the velocity constraints and the penalty system only composed of springs is merely stable. Depending on the type of integrator used, the numerical integration may lead to numerical instabilities in long simulations when using large time steps. These problems disappear when the velocity constraints consisting of fictitious dissipative forces are included in the formulation as done in (8.123), or when the numerical integrator supplies artificial numerical damping (for example the α-*method* of Hilber, Hughes, and Taylor seen in Chapter 7).

Augmented Lagrangian formulation. By using the integration procedure described below and penalty factors equal to 10^7 in double precision arithmetic, the solution of (8.123) can be achieved with an excellent satisfaction of the constraints. In case a wider range of penalty values is desired, a correcting scheme can be introduced by means of an augmented Lagrangian formulation also seen in Chapter 5. These methods assure convergence within the desired constraint tolerances without the need for very large penalty factors, thereby avoiding ill-conditioning problems.

The expression of the equations of motion can be augmented by adding the Lagrange multipliers as follows:

$$\mathbf{M}\,\ddot{\mathbf{q}} + \Phi_q^T \alpha \left(\Phi + \mu\,\dot{\Phi} \right) + \Phi_q^T \lambda^* = \mathbf{Q} \tag{8.124}$$

It may be seen by comparing equation (8.124) with the classical Lagrange multipliers technique (equation (5.8)) that

$$\lambda = \lambda^* + \alpha \left(\Phi + \mu\,\dot{\Phi} \right) \tag{8.125}$$

where λ are the true multipliers. Since the values of λ^* are unknown *a priori*, an iterative procedure is necessary to solve equation (8.125). In each iteration a new value of λ^* is computed as follows:

$$\lambda_{i+1}^* = \lambda_i^* + \alpha \left(\Phi_i + \mu\,\dot{\Phi}_i \right) \tag{8.126}$$

with $\lambda_0^*=0$ for the first iteration. In the limit $\lambda_0^*=\lambda$, however, enough accuracy is obtained in one or two iterations. Equation (8.126) physically represents the introduction at step $i+1$ of forces that tend to compensate the fact that the constraints are not exactly zero. A consequence of this iteration is that small deviations in the constraints resulting from either the integration process or small penalty factors will be eliminated by the Lagrange multiplier terms of (8.124). The additional computational effort that it requires is not significant when compared to that necessary to solve the system of nonlinear differential equations.

8.5.2 Integration Procedure

For real time integration, it is necessary to use a fixed integration formula that will yield the same computational time in each integration step which has to be smaller than the step size. This condition imposes severe limitations to integration methods. They must be computationally inexpensive with few function evaluations and iterations in each step and must allow large step sizes without introducing excessive loss of accuracy and –most importantly– without becoming unstable. Because of the ease of implementation, the trapezoidal rule has been chosen for the integration of the equations of motion in real time. This method is implicit, A-stable, and second order. It was shown in the numerical example of Chapter 7 how this rule performs, when it is directly combined with the equations of motion taking the positions as primary variables. It will be shown here how this integration scheme fits quite well into the framework provided by the fully Cartesian coordinates and the penalty formulation. Other recently proposed methods based on the midpoint rule that preserve energy and angular momentum (Simo and Wong (1991)) could also be used within this context.

The trapezoidal rule can be written as:

$$\mathbf{q}_{n+1} = \mathbf{q}_n + \frac{h}{2}\left(\dot{\mathbf{q}}_n + \dot{\mathbf{q}}_{n+1} \right)$$
$$\dot{\mathbf{q}}_{n+1} = \dot{\mathbf{q}}_n + \frac{h}{2}\left(\ddot{\mathbf{q}}_n + \ddot{\mathbf{q}}_{n+1} \right) \tag{8.127}$$

where h is the time step. These finite difference expressions can be rewritten, considering $\mathbf{q}_{n+1}$ as the primary variable and solving for the velocities and accelerations at step $(n+1)$. Consequently,

$$\dot{\mathbf{q}}_{n+1} = \frac{2}{h}\,\mathbf{q}_{n+1} - \hat{\dot{\mathbf{q}}}_{n+1}$$
$$\ddot{\mathbf{q}}_{n+1} = \frac{4}{h^2}\,\mathbf{q}_{n+1} - \hat{\ddot{\mathbf{q}}}_{n+1} \tag{8.128}$$

Vectors $\hat{\dot{\mathbf{q}}}_{n+1}$ and $\hat{\ddot{\mathbf{q}}}_{n+1}$ depend on the positions, velocities, and displacements at step n and can be written as:

$$\hat{\dot{\mathbf{q}}}_{n+1} = \dot{\mathbf{q}}_n + \frac{2}{h}\,\mathbf{q}_n$$
$$\hat{\ddot{\mathbf{q}}}_{n+1} = \ddot{\mathbf{q}}_n + \frac{4}{h}\,\dot{\mathbf{q}}_n + \frac{4}{h^2}\,\mathbf{q}_n \tag{8.129}$$

The setting up of the difference equations as done in (8.128) adds numerical efficiency to the computer implementation of the algorithm. Knowing that $(\dot{\Phi} = \Phi_{\mathbf{q}}\dot{\mathbf{q}} + \Phi_t)$, the substitution of (8.128) into the equations of motion (8.124) yields the following expression:

$$\frac{4}{h^2}\,\mathbf{M}\,\mathbf{q}_{n+1} + \Phi_{\mathbf{q}}^T\,\alpha\left(\Phi + \Phi_{\mathbf{q}}\left(\frac{2}{h}\,\mu\,\mathbf{q}_{n+1} - \mu\,\hat{\dot{\mathbf{q}}}_{n+1}\right) + \mu\,\Phi_t\right) =$$
$$= \mathbf{Q} + \mathbf{M}\,\hat{\ddot{\mathbf{q}}}_{n+1} \tag{8.130}$$

which constitutes a set of nonlinear algebraic equations with $\mathbf{q}_{n+1}$ as the only unknowns. The use of Newton-Raphson iteration leads to the following iteration process:

$$\Delta\mathbf{q}_{i+1} = -\left(\frac{\partial \mathbf{f}}{\partial \mathbf{q}}\right)_i^{-1}\,\mathbf{f}(\mathbf{q}_i) \tag{8.131}$$

where i represents the iteration number. The function $\mathbf{f}$ is defined as

$$\mathbf{f} = \mathbf{M}\,\ddot{\mathbf{q}}_{n+1} + \Phi_{\mathbf{q}}^{\mathrm{T}}\,\alpha\left(\Phi + \mu\,\dot{\Phi}\right) - \mathbf{Q} \tag{8.132}$$

and the tangent matrix as

$$\frac{\partial \mathbf{f}}{\partial \mathbf{q}} = \frac{4}{h^2}\mathbf{M} + \Phi_{\mathbf{q}}^{\mathrm{T}}\,\beta\,\Phi_{\mathbf{q}} + \Phi_{\mathbf{qq}}^{\mathrm{T}}\,\alpha\left(\Phi + \mu\,\dot{\Phi}\right) + \Phi_{\mathbf{q}}^{\mathrm{T}}\,\alpha\,\mu\,\Phi_{\mathbf{qq}}\,\dot{\mathbf{q}} - \mathbf{Q}_{\mathbf{q}} \tag{8.133}$$

where $\beta = \alpha\,(1 + 2\mu/h)$. $\mathbf{Q}_{\mathbf{q}}$ is the contribution to the tangent matrix coming from the nonlinear generalized forces (such as springs). A close examination at the tangent matrix reveals that the second term of the RHS of equation (8.133) is always much larger than the third and the fourth. This is so: first, by virtue of the fully Cartesian coordinates, $\Phi_{\mathbf{qq}}$ is a very sparse third order tensor in which the only non-zero terms are constants of value 2 (these correspond to the exponents of the quadratic terms); and secondly, the values of Φ and $\mu\,\dot{\Phi}$ are always negligible compared with the values of $\Phi_{\mathbf{q}}$. Furthermore, the parameter μ, which is an order of magnitude smaller than h, will make the fourth term negligible. This will be shown in the next section. Neglecting these terms may no longer assure the quadratic convergence of the Newton-Raphson iteration. However, accuracy and quasi-second order behavior are still satisfied. As a consequence, a quasi-tangent matrix can be defined as

$$\frac{\partial \mathbf{f}}{\partial \mathbf{q}} \cong \frac{4}{h^2}\mathbf{M} + \Phi_{\mathbf{q}}^{\mathrm{T}}\,\beta\,\Phi_{\mathbf{q}} - \mathbf{Q}_{\mathbf{q}} \tag{8.134}$$

where the diagonal matrix β becomes a constant times the identity matrix, when all the constraints are assigned the same penalty values.

8.5.3 Numerical Considerations

Improving convergence. The iteration process defined by (8.131) is carried out until $\|\Delta\mathbf{q}\|$ is smaller than a prescribed tolerance. For real time simulation, the iteration is stopped after a fixed number of iterations, that is, one or at most two per time step. Convergence can be accelerated at each time step if the iteration is started not from the solution at the previous time step, but from the solution given by a predictor. A good second order predictor is the *modified trapezoidal rule* or *Heun method* which gives the following explicit coordinate interpolation:

$$\widetilde{\mathbf{q}}_{n+1} = \mathbf{q}_n + h\,\dot{\mathbf{q}}_n + \frac{h^2}{2}\,\ddot{\mathbf{q}}_n \tag{8.135}$$

Once the solution has been obtained at step n, the iteration process of (8.131) can be started at step $n+1$ with $\widetilde{\mathbf{q}}_{n+1}$ rather than $\mathbf{q}_n$. Since no function evaluation is required in equation (8.135), the use of this predictor adds an insignificant amount of computations. As will be shown in an example below, it accelerates the integration process substantially.

Table 8.6. Percentage of the total time required by each algorithm phase.

	% of CPU time
Solutions of equations	35.0
Forming the tangent matrix	33.5
Forming the residual	14.6
Evaluating the Jacobian matrix and constraints	13.0
Updating velocities and accelerations	3.5
Predictor	0.4
Total	100.0

Computer implementation. Fully Cartesian coordinates yield a constant mass matrix. This obviously represents a substantial savings in numerical computations at the time of forming the quasi-tangent matrix of equation (8.134). The major burden in computing this matrix lies in the product $\Phi_q^T \beta \Phi_q$. Note that Φ_q is a sparse matrix and that due to the type of coordinates chosen the only non-zero terms are linear. The formation of this linear Jacobian matrix is rather inexpensive computationally. Furthermore, the product of the Jacobian matrices can be optimized by the use of sparse matrix operations. Note that the coordinates can be numbered so that a minimum profile of the final matrix is obtained.

This way of introducing the constraints through a penalty method leads to a tangent matrix that is dominated by the terms in the main diagonal. Consequently, the triangularization process does not require pivoting. Although the number of arithmetic operations is problem dependent, Table 8.6 gives an indication of the percentage time that each of the different phases of the solution process takes as a fraction of the total time. This table corresponds to an average of many simulations performed by the authors. The most time consuming parts correspond to the formation of the tangent matrix and the solution of the resulting equations. In all the cases studied the use of the *Heun method* as a predictor eliminates the need for an extra iteration. Yet it only takes 0.4% of the time taken to complete one iteration. For real time applications, the time step size should be chosen so that at most two iterations are performed per time step.

Choosing the velocity constraint factor. The characteristic equation corresponding to the constraint condition $\mu \dot{\Phi} + \Phi = 0$ is $\mu A + A = 0$. and its root is $\lambda = -1/\mu$. The region of absolute stability for the trapezoidal rule is the negative half-plane. The product $h\lambda$ must lie in the negative real axis. Thus, $h/m > 0$. At this stage, there are a series of possibilities. If both constraints Φ and $\dot{\Phi}$ are wanted to be satisfied within the same accuracy, then both should be penalized with the same factor. Therefore μ should be equal to one. This value, however, tends to introduce damping in the system's response, producing a gradual decrease in the energy of the system. If the intention is to just eliminate the possible instabilities during the integration process, then h/μ can be chosen away from the area of the stability region where the response of the multi-body system lies. In the examples shown below, values of h/μ were chosen between 30 and 80 to eliminate the high frequency response in the acceleration. This yielded excellent results. Very small values of the parameter μ may not provide sufficient damping to eliminate the numerical instability. On the other hand, large values of μ will eliminate the instability but may introduce artificial damping in the response of the multibody system.

The method explained above is systematic and general, and shows very good convergence characteristics, even for large time steps. A numerical simulation is shown next which demonstrates its capabilities.

Example   8.6

Five-link open-chain multibody system. The multibody system shown in Figure 8.21 is composed of five links interconnected by revolute joints and has a total of 30 coordinates, 25 constraint conditions, and five degrees of freedom. Each link has a unit mass which is equally lumped at the joints to yield a stronger motion. The motion is due to its self-weight. The simulation is carried out using the algorithms described above for 20 seconds with a time step h=0.008 seconds.

The multibody system is analyzed with $\mu=h/80$ and $\alpha=10^7$. The response is not sensitive to the value of μ. Similar responses are obtained with $h/40 < \mu < h/120$ with slightly more damping for larger values of μ. The time history of the vertical tip acceleration is depicted in Figure 8.24. This time history gives a clear indication that the multibody system undergoes a very strong motion with peak accelerations of the order of 180 m/sec^2. This again illustrates the excellent convergence characteristics of the algorithm. In fact, the analysis was stopped after 10 minutes of simulation without any appearance of convergence problems. Each time step requires two iterations with a tolerance in the positions of $5*10^{-7}$, and each iteration takes 6.5 milliseconds of CPU time of a DECstation 3100.

The time history of the energy is shown in Figure 8.25. This time history shows how the energy is well preserved for this strong motion, even with a relatively large time step (the maximum error is 1.3%). Figure 8.26 illustrates the time histories of the constraint errors for a penalty factor of: 10^7 (solid line), 10^6 (short dashed curve), and 10^6 with 1 iteration of the augmented Lagrangian formulation. It can be seen how the maximum constraint error using the penalty value of 10^7 is 10^{-5}. With the penalty value of 10^6 the maximum constraint error is about 10^{-4}, but the use of one iteration of the augmented Lagrangian method brings it down an order of magnitude.

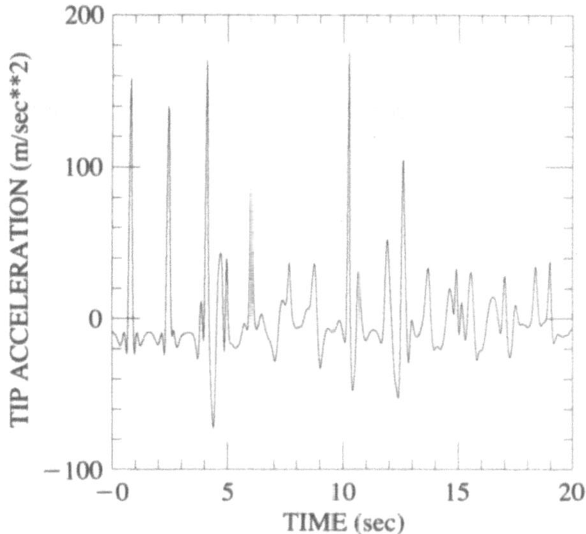

Figure 8.24. Vertical tip acceleration of the open-chain multibody system.

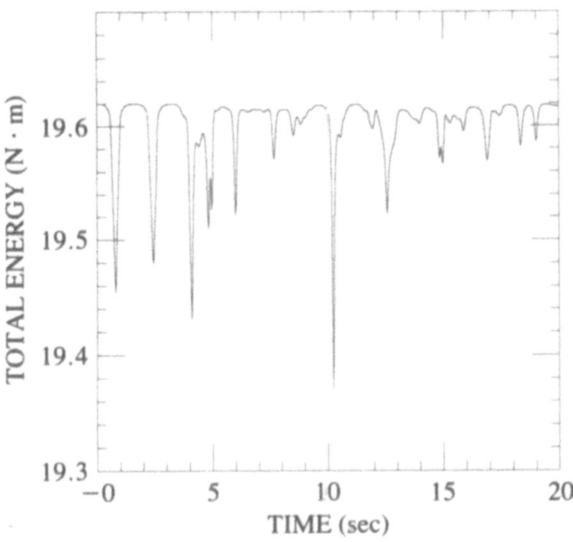

Figure 8.25. Total energy time history of the open-chain multibody system.

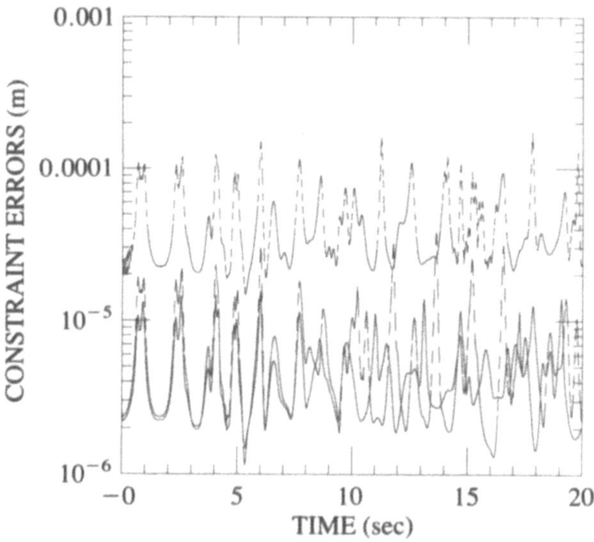

Figure 8.26. Maximum errors in the constraints using a penalty value of 10^7 (solid line), 10^6 (short dashed line), and 10^6 with augmented Lagrangian formulation (long dashed line).

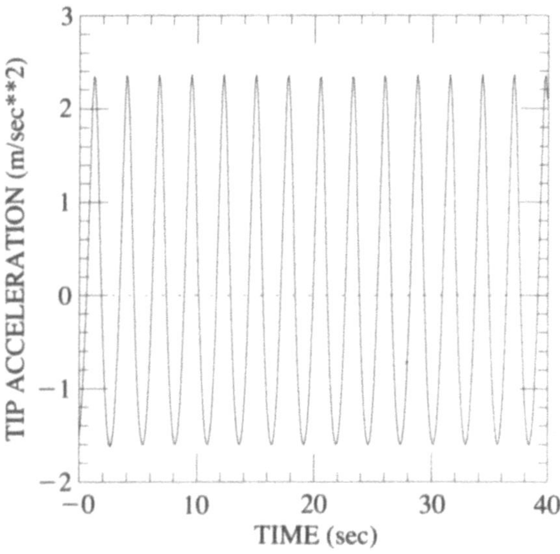

Figure 8.27. Vertical acceleration at the middle link of the Bricard mechanism.

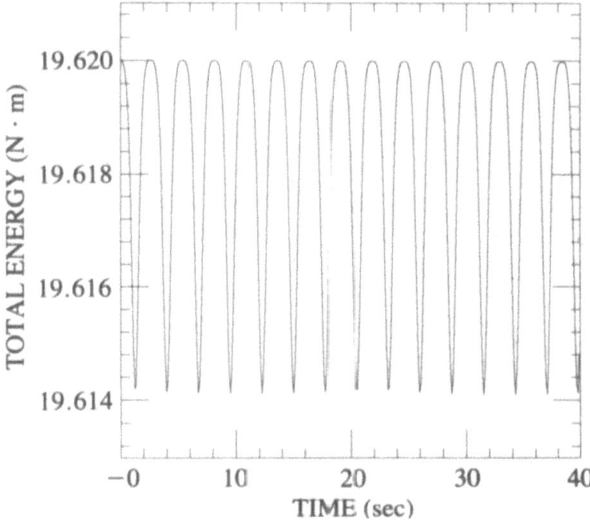

Figure 8.28. Total energy time history of the Bricard mechanism.

Example 8.7

Five-link closed-chain multibody system. The multibody system shown in Figure 8.21 is transformed into the closed-chain Bricard mechanism shown in Figure 8.22. The simulation is carried out for 40 seconds with a time step of 0.03 seconds, a penalty value of 10^7, and $\mu=h/60$. Figure 8.27 illustrates the acceleration time history of the middle link which shows that this multibody system undergoes a smoother motion than the previous one. Each time step requires two iterations with a tolerance in the positions of $5*10^{-7}$. Each iteration takes 5.0 milliseconds of CPU time. The total CPU time required to complete 40 seconds simulation is about 13 seconds. The time history of the energy is depicted in Figure 8.28 which again shows how well the energy is preserved. The maximum error is 0.03%.

References

Argyris, J., "An Excursion into Large Rotations", *Computer Methods in Applied Mechanics and Engineering*, Vol. 32, pp. 85-155, (1982).

Armstrong, W.W., "Recursive Solution to the Equations of Motion of an N-Link Manipulator", Proc. 5th World Congress on Theory of Machines and Mechanisms, Vol. 2, pp. 1343-1346, Montreal, (1979).

Avello, A., Jiménez, J.M., Bayo, E., and García de Jalón, J., "A Simple and Highly Parallelizable Method for Real-Time Dynamic Simulation Based on Velocity Transformations", to appear in *Computer Methods in Applied Mechanics and Engineering*, (1993).

Bae, D.-S. and Haug, E.J., "A Recursive Formulation for Constrained Mechanical System Dynamics. Part I: Open-Loop Systems", *Mechanics of Structures and Machines*, Vol. 15, pp. 359-382, (1987).

Bae, D.-S. and Haug, E.J., "A Recursive Formulation for Constrained Mechanical System Dynamics. Part II: Closed-Loop Systems", *Mechanics of Structures and Machines*, Vol. 15, pp. 481-506, (1987-88).

Bae, D.-S., Hwang, R.S., and Haug, E.J., "A Recursive Formulation for Real-Time Dynamic Formulation", *1988 Advances in Design Automation*, ed. by S.S. Rao, ASME, pp. 499-508, (1988).

Bae, D.-S. and Won, Y.S., "A Hamiltonian Equation of Motion for Real Time Vehicle Simulation", *1990 Advances in Design Automation*, ed. by B. Ravani, ASME, pp. 151-157, (1990).

Bayo, E., García de Jalón, J., Avello, A., and Cuadrado, J., "An Efficient Computational Method for Real Time Multibody Dynamic Simulation in Fully Cartesian Coordinates", *Computer Methods in Applied Mechanics and Engineering*, Vol. 92, pp. 377-395, (1991).

Featherstone, R., "The Calculation of Robot Dynamics Using Articulated Body Inertias", *The Int. Journal of Robotic Research*, Vol. 2, pp. 13-30, (1983).

Featherstone, R., *"Robot Dynamics Algorithms"*, Kluwer, (1987).

García de Jalón, J., Jiménez, J.M., Avello, A., Martín, F., and Cuadrado, J., "Real Time Simulation of Complex 3-D Multibody Systems with Realistic Graphics", *Advanced Research Workshop on Real Time Integration Methods for Mechanical System Simulation*, NATO, *Snowbird*, UTAH, August, (1989).

Haug, E.J., *Computer-Aided Kinematics and Dynamics of Mechanical Systems. Volume I: Basic Methods*, Allyn and Bacon, (1989).

Jerkovsky, W., "The Structure of Multibody Dynamic Equations", *Journal of Guidance and Control*, Vol. 1, pp. 173-182, (1978).

Jiménez, J.M., "Formulaciones Cinemáticas y Dinámicas para la Simulación en Tiempo Real de Sistemas de Sólidos Rígidos", Ph.D. Thesis, University of Navarre, San Sebastián, (1993).

Kim, S.S. and Vanderploeg, M.J., "A General and Efficient Method for Dynamic Analysis of Mechanical Systems Using Velocity Transformations", *ASME Journal of Mechanisms, Transmissions, and Automation in Design*, Vol. 108, 176-182, (1986).

Luh, J.Y.S., Walker, M.W., and Paul, R.P.C., "On Line Computational Scheme for Mechanical Manipulators", *Journal of Dynamic Systems, Measurements, and Control*, ASME, Vol. 102, pp. 69-76, (1980).

Nikravesh, P.E., *Computer-Aided Analysis of Mechanical Systems*, Prentice-Hall, (1988).

Rodriguez, G., Jain, A., and Kreutz, K., "A Spatial Operator Algebra for Manipulator Modeling and Control", *International Journal of Robotics Research*, Vol. 10, pp. 371-381, (1991).

Simo, J.C. and Wong, S., "Unconditionally Stable Algorithms for Rigid Body Dynamics that Exactly Preserve Energy and Momentum", *International Journal for Numerical Methods in Engineering*, Vol. 31, pp. 19-52, (1991).

Walker, M.W. and Orin, D.E., "Efficient Dynamic Computer Simulation of Robotic Mechanisms", Journal of Dynamic Systems, Measurements, and Control, ASME, Vol. 104, pp. 205-211, (1982).

9
Linearized Dynamic Analysis

Several ways of formulating the differential equations of motion of a multibody system have been presented in Chapter 5. These equations are fully nonlinear in either the independent or dependent coordinates. The solution of these non linear equations is required in order to simulate the dynamic behavior of multibody systems that undergo large displacements and rotations. However, many systems work mostly on the proximity of a fixed or constant dynamic equilibrium configuration. It is very convenient to linearize the equations of motion about this equilibrium configuration, so as to take advantage of the linear analysis tools: fast computation of linear response, eigenvalue analysis, control design by pole placement, or other linear techniques, that are not available or at least are more complicated for the fully nonlinear models.

This chapter deals with several techniques that linearize the most commonly used forms of the equations of motion. In particular, closed-form and numerical computation of the derivatives of the equations of motion will be considered. In addition, linearization methods of these equations expressed in terms of dependent (using the penalty formulation) and independent coordinates as well as in the canonical form will be explained. This chapter will end with a short review of the available methods used to compute the response of the linear system and its frequencies and mode shapes.

9.1 Linearization of the Differential Equations of Motion

In this section a constant or fixed dynamic configuration will be considered, such that position $\mathbf{y}$, velocity $\dot{\mathbf{y}}$, acceleration $\ddot{\mathbf{y}}$, and external forces $\mathbf{Q}$ that satisfy the equations of motion will be symbolically expressed in the form:

$$H(\mathbf{y}, \dot{\mathbf{y}}, \ddot{\mathbf{y}}, \mathbf{Q}) = 0 \tag{9.1}$$

The dynamic equilibrium configuration will be denoted by the subscript $(_0)$. Hence, the equations of motion at the equilibrium configuration can be written as

$$H_0 \equiv H(y_0, \dot{y}_0, \ddot{y}_0, Q_0) = 0 \tag{9.2}$$

Small perturbations of the equilibrium configuration variables can now be considered. The equations (9.1) will be linearized by replacing them with the first two terms of the Taylor's series expansion about the dynamic equilibrium configuration,

$$H(y_0 + \Delta y_0, \dot{y}_0 + \Delta \dot{y}_0, \ddot{y}_0 + \Delta \ddot{y}_0, Q_0 + \Delta Q_0) \cong H_0 + \Delta H_0 =$$
$$= H(y_0, \dot{y}_0, \ddot{y}_0, Q_0) + H_y \, \Delta y + H_{\dot{y}} \, \Delta \dot{y} + H_{\ddot{y}} \, \Delta \ddot{y} + H_Q \, \Delta Q = 0 \tag{9.3}$$

According to equation (9.2),

$$H_y \, \Delta y + H_{\dot{y}} \, \Delta \dot{y} + H_{\ddot{y}} \, \Delta \ddot{y} + H_Q \, \Delta Q = 0 \tag{9.4}$$

which constitutes the linearized set of equations of motion. The main issue is now to compute the partial derivatives that appear in equation (9.4). The following subsections deal with this problem for those cases in which y represents either the independent coordinates z, or the dependent ones q, and for the case when H is represented in canonical form.

9.1.1 Independent Coordinates

According to equation (5.67), the equations of motion in terms of independent coordinates take the form:

$$H(z, \dot{z}, \ddot{z}, Q) \equiv R(z)^T M(z) R(z) \ddot{z} -$$
$$- R(z)^T (Q(z, \dot{z}, f) - M(z) Sc(z, \dot{z})) = 0 \tag{9.5}$$

where M is the inertial or mass matrix, R is the velocity projection matrix whose columns span the nullspace of the Jacobian matrix Φ_q, (Sc) is the term that accounts for the velocity-dependent accelerations, and Q are the generalized forces that depend on the inertia and applied external forces f. In expression (9.5), the dependence of each term or factor with respect to the configuration variables has been introduced explicitly. Remember (See equations (5.64) and (5.65)) that matrix R and vector (Sc) came from the expressions

$$\begin{bmatrix} \Phi_q \\ B \end{bmatrix} \ddot{q} = \begin{Bmatrix} c \\ \ddot{z} \end{Bmatrix} \tag{9.6}$$

$$\ddot{q} = \begin{bmatrix} \Phi_q \\ B \end{bmatrix}^{-1} \begin{Bmatrix} c \\ \ddot{z} \end{Bmatrix} = S \, c + R \, \ddot{z} \tag{9.7}$$

Then the partial derivatives of function H take the following form:

$$H_z = (R^T M R_z + R_z^T M R + R^T M_z R) \ddot{z} -$$
$$- R_z^T (Q - M \, Sc) - R^T (Q_z - M_z (Sc) - M(Sc)_z) \tag{9.8}$$

$$\mathbf{H}_i = \mathbf{R}^{\mathrm{T}} (\mathbf{Q}_i - \mathbf{M}(\mathbf{Sc})_i) \tag{9.9}$$

$$\mathbf{H}_{\dot{z}} = \mathbf{R}^{\mathrm{T}} \mathbf{M} \mathbf{R} \tag{9.10}$$

$$\mathbf{H}_f = \mathbf{R}^{\mathrm{T}} \mathbf{Q}_f \tag{9.11}$$

All these derivatives must be evaluated at the equilibrium configuration. If the equilibrium configuration is static ($\dot{z}_0 = \ddot{z}_0 = 0$), equations (9.8)-(9.11) are very much simplified, because all terms depending on velocities and accelerations vanish.

In the sequel, ways of evaluating some of the partial derivatives that appear in the RHS of equations (9.8)-(9.11) will be considered.

i) Computation of $\mathbf{R}_z$. There are several possible ways to compute the partial derivative of matrix $\mathbf{R}$ with respect to the vector of independent coordinates z. Here, a method based on the acceleration analysis will be presented.

If the system is scleronomic, matrix $\mathbf{R}$ relates dependent and independent velocities in the form:

$$\dot{\mathbf{q}} = \mathbf{R} \, \dot{z} = \sum_{i=1}^{f} \mathbf{r}^i \, \dot{z}_i \tag{9.12}$$

where $\mathbf{r}^i$ is the i-column of matrix $\mathbf{R}$. Differentiating this expression with respect to time and taking into account that $\mathbf{R}$ depends explicitly on the position variables only,

$$\ddot{\mathbf{q}} = (\sum_{j=1}^{f} \frac{\partial \mathbf{R}}{\partial z_j} \, \dot{z}_j) \, \dot{z} + \mathbf{R} \, \ddot{z} = \sum_{i=1}^{f} \sum_{j=1}^{f} \frac{\partial \mathbf{r}^i}{\partial z_j} \, \dot{z}_i \, \dot{z}_j + \mathbf{R} \, \ddot{z} \tag{9.13}$$

This expression offers a simple way to compute the derivatives of $\mathbf{R}$. Remember that this matrix is of order ($n{\times}f$). Its derivative $\partial \mathbf{R}/\partial z_i$ is also a matrix of size ($n{\times}f$), but the derivative $\partial \mathbf{R}/\partial z$ is a hyper-matrix of size ($n{\times}f{\times}f$). Equation (9.13) suggests that the derivatives of the columns of $\mathbf{R}$ can be computed through an acceleration analysis. For instance, by making

a) $\ddot{z}_i = 0 \ (i=1,...,f); \ \dot{z}_i = 1, \dot{z}_j = \delta_{ij}$

$$\ddot{\mathbf{q}} = \frac{\partial \mathbf{r}^i}{\partial z_i} \tag{9.14}$$

b) $\ddot{z}_i = 0 \ (i=1,...,f); \ \dot{z}_i = 1, \dot{z}_j = 1, \dot{z}_k = 0 \ (k \neq i, j)$

$$\ddot{\mathbf{q}} = \frac{\partial \mathbf{r}^i}{\partial z_i} + \frac{\partial \mathbf{r}^j}{\partial z_j} + \frac{1}{2}(\frac{\partial \mathbf{r}^i}{\partial z_j} + \frac{\partial \mathbf{r}^j}{\partial z_i}) \tag{9.15}$$

This expression yields

$$\frac{\partial \mathbf{r}^i}{\partial z_j} = \frac{\partial \mathbf{r}^j}{\partial z_i} = \frac{1}{2}(\ddot{\mathbf{q}} - \frac{\partial \mathbf{r}^i}{\partial z_i} - \frac{\partial \mathbf{r}^j}{\partial z_j}) \tag{9.16}$$

There are a number of $(f^2 + f)/2$ acceleration analyses. They can be done rather inexpensively, because all of them use the same LU factorization of the Jacobian matrix. Therefore, only one forward reduction and back substitution is needed per acceleration analysis. If the multibody system is open-loop, these evaluations can be made even more cheaply.

ii) Computation of $\mathbf{M_z}$. The computation of the derivatives of the inertia matrix with respect to the position variables is strongly formulation dependent.

If one considers *fully Cartesian coordinates* and bodies with four non-coplanar points and/or unit vectors, the mass matrices are constant (See Section 4.2.2). Consequently, their derivatives are zero. If the bodies do not have constant inertial matrices (See Section 4.2.2.3), it is still possible to consider a virtual power transformation that allows one to compute the rigid body inertial matrix in the form:

$$\mathbf{M}_{new} = \mathbf{V}^T \mathbf{M} \mathbf{V} \tag{9.17}$$

and the additional velocity dependent inertial forces in the form:

$$\mathbf{Q}_{inertia} = \mathbf{V}^T \mathbf{M} \dot{\mathbf{V}} \dot{\mathbf{q}} \tag{9.18}$$

where $\mathbf{M}$ is constant and $\mathbf{V}$ is position dependent. Thus,

$$(\mathbf{M}_{new})_z = \mathbf{V}^T \mathbf{M} \mathbf{V}_z + \mathbf{V}_z^T \mathbf{M} \mathbf{V} \tag{9.19}$$

According to equation (4.72), matrix $\mathbf{V}$ has the expression:

$$\mathbf{V} = \begin{bmatrix} \mathbf{I}_3 & 0 & 0 \\ 0 & \mathbf{I}_3 & 0 \\ 0 & 0 & \mathbf{I}_3 \\ -c\,\tilde{\mathbf{u}} & c\,\tilde{\mathbf{u}} & c\,\tilde{\mathbf{r}}_{ij} \end{bmatrix} \tag{9.20}$$

This matrix depends explicitly on the dependent coordinates vector $\mathbf{q}$. Hence, using the chain differentiation rule,

$$\mathbf{V}_z = \mathbf{V}_q \frac{\partial \mathbf{q}}{\partial \mathbf{z}} + \mathbf{V} \mathbf{R} \tag{9.21}$$

where $\mathbf{V_q}$ is a hyper-matrix with most of its components equal to zero.

If *reference point coordinates* are used, the inertial matrix of a rigid body takes the form (See equation (8.15)):

$$\mathbf{M}_i = \begin{bmatrix} m_i \mathbf{I}_3 & 0 \\ 0 & \mathbf{A}_i \mathbf{J}_i \mathbf{A}_i^T \end{bmatrix} \tag{9.22}$$

where $\mathbf{A}_i$ is the rotation matrix of this element. This rotation matrix introduces the position dependency in matrix $\mathbf{M}_i$. Thus, it is possible to write:

$$M_{i,z} = \begin{bmatrix} 0 & 0 \\ 0 & A_{i,z} J_i A_i^T + A J_i A_{i,z}^T \end{bmatrix} \tag{9.23}$$

The derivative $A_{i,z}$ can be computed using again the chain rule of differentiation:

$$A_{i,z} = A_{i,q} \frac{\partial q}{\partial z} + A_{i,q} R \tag{9.24}$$

where $A_{i,q}$ can be computed easily in terms of the variables used to define angular orientation, that is, Euler angles or Euler parameters.

iii) Computation of $(Sc)_z$. In order to compute this derivative, it may be useful to return again to equation (5.65) written in the form:

$$\dot{q} = R(z)\,\dot{z} + S c \tag{9.25}$$

Taking time derivatives and assuming scleronomic constraints yields

$$\ddot{q} = R\,\ddot{z} + \dot{R}\,\dot{z} + (Sc)_z\,\dot{z} + (Sc)_{\dot{z}}\,\ddot{z} \tag{9.26}$$

This equation offers the opportunity of computing the derivative $(Sc)_z$ by means of a jerk analysis. This is not significantly more complicated or expensive than a velocity or acceleration analysis, because it uses again the same LU factorization of the Jacobian matrix Φ_q.

By making $\ddot{z} = \dddot{z} = 0$, $\dot{z}_i = 1$, and $\dot{z}_j = 0$, for $j \neq i$ in equation (9.26), one can obtain

$$\dddot{q} = \frac{\partial(Sc)}{\partial z_i} \tag{9.27}$$

iv) Computation of $(Sc)_{\dot{z}}$. This derivative can also be computed from equation (9.26) in two steps:
- with $\ddot{z} = \dddot{z} = 0$ and the true velocities $\dot{z}$, a jerk analysis based on equation (9.26) yields

$$\dddot{q} = (Sc)_z\,\dot{z} \tag{9.28}$$

- now, assuming that $\dot{R}$ is known and making $\dddot{z} = 0$, the following expression is obtained:

$$\dddot{q} = (\ddot{r}^i + (Sc)_z\,\dot{z}) + \frac{\partial(Sc)}{\partial \dot{z}_i} \tag{9.29}$$

from which the desired derivative can be obtained.

There is another way to compute this derivative. Equation (9.7) states that matrix S depends on the position but not on the velocity vector $\dot{z}$. Thus, taking into account that c is given by the expression:

$$c = -\dot{\Phi}_t - \dot{\Phi}_q\,\dot{q} \tag{9.30}$$

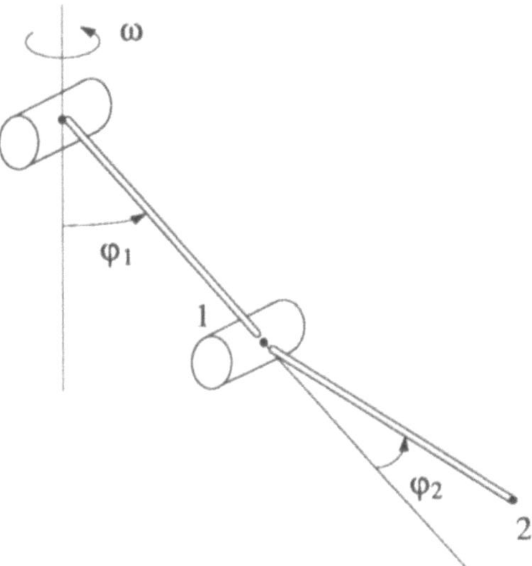

Figure 9.1. Double pendulum rotating about a vertical axis.

which contains terms that depend explicitly on $\dot{z}$. The derivative can be found as

$$(S\ c)_{\dot{z}} = S\ c_{\dot{z}} = S\ c_{\dot{q}}\ \frac{\partial \dot{q}}{\partial \dot{z}} = S\ c_{\dot{q}}\ R \qquad (9.31)$$

The main disadvantage of equation (9.31) is that it requires the explicit evaluation of matrix S. That is not necessary when computing this derivative using the first way.

v) Computation of Q_z, $Q_{\dot{z}}$, *and* Q_f. These derivatives are strongly case dependent. It is not expected that one will have any particular difficulty in their evaluation.

Example   9.1

Consider the double pendulum of Figure 9.1. Each link has a length equal to $2l$ and a lumped mass $m = 1$ at both ends. The double pendulum is rotating about the vertical axis with a constant angular velocity ω. Due to the gravity effects, the pendulum reaches the equilibrium position at angles φ_1 and φ_2. Considering the dependent Cartesian coordinates of the ends 1 and 2 $\{x_1, y_1, x_2, y_2\}$, the mass matrix M becomes

$$M = \begin{bmatrix} 2 & 0 & 0 & 0 \\ 0 & 2 & 0 & 0 \\ 0 & 0 & 1 & 0 \\ 0 & 0 & 0 & 1 \end{bmatrix}$$

The vector of generalized forces Q composed of the gravity forces and the centrifugal forces corresponding to the rotation about the vertical axis is

$$Q = \left\{ \begin{array}{c} 4l\,\omega^2 sin\,\varphi_1 \\[4pt] -2g \\[4pt] 2l\,\omega^2\left(sin\,\varphi_1 + sin(\varphi_1+\varphi_2)\right) \\[4pt] -g \end{array} \right\}$$

If the two angles φ_1 and φ_2 shown in the figure are taken as independent coordinates, the matrix R becomes

$$R = \left[\begin{array}{cc} 2l\,cos\,\varphi_1 & 0 \\[4pt] 2l\,sin\,\varphi_1 & 0 \\[4pt] 2l\left(cos\,\varphi_1 + cos(\varphi_1+\varphi_2)\right) & 2l\,cos(\varphi_1+\varphi_2) \\[4pt] 2l\left(sin\,\varphi_1 + sin(\varphi_1+\varphi_2)\right) & 2l\,sin(\varphi_1+\varphi_2) \end{array} \right]$$

and the product $(\dot{R}\dot{z})$ is

$$\dot{R}\dot{z} = \left\{ \begin{array}{c} -2l\,sin\,\varphi_1\,\dot{\varphi}_1^2 \\[4pt] 2l\,cos\,\varphi_1\,\dot{\varphi}_1^2 \\[4pt] -2l\,sin\,\varphi_1\,\dot{\varphi}_1^2 - 2l\,sin(\varphi_1+\varphi_2)\left(\dot{\varphi}_1^2+\dot{\varphi}_2^2\right) \\[4pt] 2l\,cos\,\varphi_1\,\dot{\varphi}_1^2 + 2l\,cos(\varphi_1+\varphi_2)\left(\dot{\varphi}_1^2+\dot{\varphi}_2^2\right) \end{array} \right\}$$

We desire to linearize the equations of motion about the equilibrium position given by φ_1, φ_2, $\dot{\varphi}_1 = \dot{\varphi}_2 = 0$, and $\ddot{\varphi}_1 = \ddot{\varphi}_2 = 0$. After a tedious but straightforward manipulation of expressions, one can verify that the linearized equations of motion are:

$$A\,\Delta\ddot{z} + B\,\Delta z = 0$$

where

$$A = 4l^2 \left[\begin{array}{cc} 2\left(2+sin\,\varphi_2\right) & \left(1+sin\,\varphi_2\right) \\[4pt] \left(1+sin\,\varphi_2\right) & 1 \end{array} \right]$$

and

$$B = -2\,g\,l \left[\begin{array}{cc} cos\,(\varphi_1+\varphi_2) + 3\,cos\,\varphi_1 & cos\,(\varphi_1+\varphi_2) \\[4pt] cos\,(\varphi_1+\varphi_2) & cos\,(\varphi_1+\varphi_2) \end{array} \right] +$$

$$+\,4l^2\omega^2 \left[\begin{array}{cc} 2cos\,(2\varphi_1+\varphi_2) + cos\,(2\varphi_1+2\varphi_2) + 3cos\,2\varphi_1 & cos\,(2\varphi_1+2\varphi_2) + cos\,(2\varphi_1+\varphi_2) \\[10pt] cos\,(2\varphi_1+2\varphi_2) + cos\,(2\varphi_1+\varphi_2) & cos\,(2\varphi_1+2\varphi_2) - sin\,(\varphi_1+\varphi_2)\,sin\,\varphi_1 \end{array} \right]$$

9.1.2 Dependent Coordinates

It is implied in the linearization of a set of nonlinear equations that the differentiation takes place with respect to independent variables. The linearization of the equations of motion formulated as a function of the dependent coordinates q is not entirely meaningful when the Lagrange multipliers method is used to formulate the equations of motion (See Section 5.1.1.). If the penalty formulation is used (See Section 5.1.4.), the elements of the vector q are considered as mathematically independent or unconstrained. Therefore, the partial derivatives of the equations of motion with respect to q and with respect to its time derivatives $\dot{q}$ and $\ddot{q}$ have a full mathematical meaning. In addition, it can be seen that the resulting equations are simpler than those resulting from the use of the independent coordinates z as seen in the previous section.

Using the penalty formulation with holonomic constraints (See Section 5.1.4.), the equations of motion (9.1) take the following form:

$$H(q, \dot{q}, \ddot{q}, Q) \equiv M(q)\ddot{q} + \Phi_q^T(q)\,\alpha\,(\ddot{\Phi}(q, \dot{q}, \ddot{q}) +$$
$$+\, 2\,\mu\,\Omega\,\dot{\Phi}(q, \dot{q}) + \Omega^2\,\Phi(q)) - Q(q, \dot{q}, f) = 0 \tag{9.32}$$

where α is the diagonal matrix of penalty factors, and μ and Ω also are constant diagonal matrices that control the frequency and the damping of the constraint violations. In equation (9.32), the dependencies of each term or factor with respect to the configuration variables have been made explicit again.

Assuming that the constraints are scleronomic, the first and second time derivatives of the constraint equations can be written in the form:

$$\dot{\Phi} = \Phi_q(q)\,\dot{q} \tag{9.33}$$

$$\ddot{\Phi} = \Phi_q(q)\,\ddot{q} + \dot{\Phi}_q(q, \dot{q})\,\dot{q} \tag{9.34}$$

Then the partial derivatives of the equations of motion (9.32) take the form:

$$H_q = M_q\,\ddot{q} + \Phi_{qq}^T\,\alpha\,(\ddot{\Phi} + 2\,\mu\,\Omega\,\dot{\Phi} + \Omega^2\,\Phi) +$$
$$+\,\Phi_q^T\,\alpha\,(\Phi_{qq}\,\ddot{q} + \dot{\Phi}_{qq}\,\dot{q} + 2\,\mu\,\Omega\,\Phi_{qq}\,\dot{q} + \Omega^2\,\Phi_q) - Q_q \tag{9.35}$$

$$H_{\dot{q}} = \Phi_q^T\,\alpha\,(\dot{\Phi}_{q\dot{q}}\,\dot{q} + \dot{\Phi}_q + 2\,\mu\,\Omega\,\Phi_q) - Q_{\dot{q}} \tag{9.36}$$

$$H_{\ddot{q}} = M + \Phi_q^T\,\alpha\,\Phi_q \tag{9.37}$$

$$H_f = Q_f \tag{9.38}$$

These derivatives are particularly simple to evaluate if natural (or mixed) coordinates are used. If only Cartesian coordinates are used, the constraint equations are quadratic, and then Φ_{qq} is a constant and very sparse hyper-matrix of dimension $(n{\times}n{\times}n)$. In addition, the terms $\dot{\Phi}_{qq}$ and $\dot{\Phi}_{q\dot{q}}$ will be zero. If a few relative or joint coordinates are used (mixed coordinates), these terms will no longer be zero, but only a few of their elements need to be computed. Expressions (9.35)-(9.38) become even simpler when considering a static equilibrium configuration at which $\dot{q}_0 = \ddot{q}_0 = 0$.

Example   9.2

Repeat Example 9.1 using natural coordinates and the dependent coordinates formulation.

Considering the set Cartesian coordinates of the two lumped masses at the end of each link $\{x_1, y_1, x_2, y_2\}$, the mass matrix takes the following form:

$$\mathbf{M} = \begin{bmatrix} 2 & 0 & 0 & 0 \\ 0 & 2 & 0 & 0 \\ 0 & 0 & 1 & 0 \\ 0 & 0 & 0 & 1 \end{bmatrix}$$

The constraint equations based on the constant distance between the masses is:

$$\Phi = \left\{ \begin{array}{c} \dfrac{1}{2}(x_1^2 + y_1^2 - (2l)^2 = 0) \\[2mm] \dfrac{1}{2}(x_2 - x_1)^2 + (y_2 - y_1)^2 - (2l)^2 = 0) \end{array} \right\}$$

The Jacobian matrix of the constraints becomes

$$\Phi_{\mathbf{q}} = \begin{bmatrix} x_1 & y_1 & 0 & 0 \\ (x_1 - x_2) & (y_1 - y_2) & (x_2 - x_1) & (y_2 - y_1) \end{bmatrix}$$

The force vector is directly obtained as a combination of gravity forces in the y direction and centrifugal forces in the x direction:

$$\mathbf{Q}^{\mathrm{T}} = \begin{bmatrix} 2\,\omega^2 x_1 & -2\,g & \omega^2(x_1 + x_2) & -g \end{bmatrix}$$

where g is the acceleration of gravity, and ω the constant angular velocity. At the generic position $\{x_1, y_1, x_2, y_2\}$ where the equations are to be linearized, the following conditions are met: $\Phi = 0$, $\dot{\Phi} = 0$, and $\ddot{\Phi} = 0$. Since the mass matrix is constant, $\mathbf{M}_{\mathbf{q}} = 0$, $\mathbf{Q}_{\mathbf{f}} = 0$, and $\mathbf{Q}_{\dot{\mathbf{q}}} = 0$. Finally, $\mathbf{Q}_{\mathbf{q}}$ is

$$\mathbf{Q}_{\mathbf{q}} = \begin{bmatrix} 2\omega^2 & 0 & \omega^2 & 0 \\ 0 & 0 & 0 & 0 \\ \omega^2 & 0 & \omega^2 & 0 \\ 0 & 0 & 0 & 0 \end{bmatrix}$$

Applying equations (9.35), (9.36), and (9.37) the following results are obtained:

$$\mathbf{H}_{\mathbf{q}} = \Phi_{\mathbf{q}}^{\mathrm{T}} \alpha\, \Omega^2 \Phi_{\mathbf{q}} - \mathbf{Q}_{\mathbf{q}}$$

$$\mathbf{H}_{\dot{\mathbf{q}}} = \Phi_{\mathbf{q}}^{\mathrm{T}} \alpha\, 2\, \mu\, \Omega\, \Phi_{\mathbf{q}}$$

$$\mathbf{H}_{\ddot{\mathbf{q}}} = \mathbf{M} + \Phi_{\mathbf{q}}^{\mathrm{T}} \alpha\, \Phi_{\mathbf{q}}$$

The final set of linearized equations becomes:

$$(\mathbf{M} + \Phi_{\mathbf{q}}^{\mathrm{T}} \alpha\, \Phi_{\mathbf{q}})\Delta\ddot{\mathbf{q}} + (\Phi_{\mathbf{q}}^{\mathrm{T}} \alpha\, 2\, \mu\, \Omega\, \Phi_{\mathbf{q}})\Delta\dot{\mathbf{q}} + (\Phi_{\mathbf{q}}^{\mathrm{T}} \alpha\, \Omega^2 \Phi_{\mathbf{q}} - \mathbf{Q}_{\mathbf{q}})\Delta\mathbf{q} = 0$$

where all the matrices have been previously defined.

This approach is much less involved than that resulting from the use of independent coordinates.

9.1.3 Canonical Equations

Another route to the linearization process can be taken through the canonical formulation of the equations of motion previously seen in Chapter 5. The aim of this section is to show in a simple manner another possible approach to the linearized dynamic analysis. The use of canonical equations will be limited to the case of dependent coordinates with the penalty formulation for the introduction of the constraint conditions.

The canonical approach as seen in Section 5.4 leads to $2n$ first order differential equations in terms of the momenta $\mathbf{p}$ and coordinates $\mathbf{q}$

$$\mathbf{H}_1 = \dot{\mathbf{p}} - \dot{\boldsymbol{\Phi}}_{\mathbf{q}}^{\mathsf{T}} \, \alpha \, \dot{\boldsymbol{\Phi}} + \boldsymbol{\Phi}_{\mathbf{q}}^{\mathsf{T}} \, \alpha \, (\Omega^2 \boldsymbol{\Phi} + 2\mu\Omega\dot{\boldsymbol{\Phi}}) - \mathbf{Q} = 0 \tag{9.39}$$

$$\mathbf{H}_2 = \left[\mathbf{M} + \boldsymbol{\Phi}_{\mathbf{q}}^{\mathsf{T}} \, \alpha \, \boldsymbol{\Phi}_{\mathbf{q}}\right] \dot{\mathbf{q}} - \mathbf{p} = 0 \tag{9.40}$$

where it has been assumed that the constraints are scleronomic. The vector $\mathbf{y}$ is now composed of $2n$ entries, $\mathbf{p}$ and $\mathbf{q}$. The partial derivatives of $\mathbf{H}_1$ and $\mathbf{H}_2$ become:

$$(\mathbf{H}_1)_{\mathbf{p}} = 0 \tag{9.41}$$

$$\begin{aligned}
(\mathbf{H}_1)_{\mathbf{q}} = \boldsymbol{\Phi}_{\mathbf{qq}}^{\mathsf{T}} \, \alpha \, (\Omega^2 \boldsymbol{\Phi} + 2\mu\Omega\dot{\boldsymbol{\Phi}}) + \\
+ \boldsymbol{\Phi}_{\mathbf{q}}^{\mathsf{T}} \, \alpha \, (\Omega^2 \boldsymbol{\Phi}_{\mathbf{q}} + 2\mu\Omega\dot{\boldsymbol{\Phi}}_{\mathbf{q}}) - \dot{\boldsymbol{\Phi}}_{\mathbf{qq}} \, \alpha \, \dot{\boldsymbol{\Phi}} - \dot{\boldsymbol{\Phi}}_{\mathbf{q}}^{\mathsf{T}} \, \alpha \, \dot{\boldsymbol{\Phi}}_{\mathbf{q}} - \mathbf{Q}_{\mathbf{q}}
\end{aligned} \tag{9.42}$$

$$(\mathbf{H}_2)_{\mathbf{p}} = -\mathbf{I} \tag{9.43}$$

$$(\mathbf{H}_2)_{\mathbf{q}} = \left[\mathbf{M}_{\mathbf{q}} + \boldsymbol{\Phi}_{\mathbf{qq}}^{\mathsf{T}} \, \alpha \, \boldsymbol{\Phi}_{\mathbf{q}} + \boldsymbol{\Phi}_{\mathbf{q}}^{\mathsf{T}} \, \alpha \, \boldsymbol{\Phi}_{\mathbf{qq}}\right] \dot{\mathbf{q}} \tag{9.44}$$

$$(\mathbf{H}_1)_{\dot{\mathbf{p}}} = \mathbf{I} \tag{9.45}$$

$$(\mathbf{H}_1)_{\dot{\mathbf{q}}} = -\dot{\boldsymbol{\Phi}}_{\mathbf{q}\dot{\mathbf{q}}}^{\mathsf{T}} \, \alpha \, \dot{\boldsymbol{\Phi}} - \dot{\boldsymbol{\Phi}}_{\mathbf{q}}^{\mathsf{T}} \, \alpha \, \boldsymbol{\Phi}_{\mathbf{q}} + \boldsymbol{\Phi}_{\mathbf{q}}^{\mathsf{T}} \, 2\mu \, \alpha \, \Omega \, \boldsymbol{\Phi}_{\mathbf{q}} - \mathbf{Q}_{\dot{\mathbf{q}}} \tag{9.46}$$

$$(\mathbf{H}_2)_{\dot{\mathbf{p}}} = 0 \tag{9.47}$$

$$(\mathbf{H}_2)_{\dot{\mathbf{q}}} = \mathbf{M} + \boldsymbol{\Phi}_{\mathbf{q}}^{\mathsf{T}} \, \alpha \, \boldsymbol{\Phi}_{\mathbf{q}} \tag{9.48}$$

$$(\mathbf{H}_1)_{\mathbf{f}} = -\mathbf{Q}_{\mathbf{f}} \tag{9.49}$$

$$(\mathbf{H}_2)_{\mathbf{f}} = 0 \tag{9.50}$$

Similar to the acceleration-based formulation, some of these partial derivatives cancel out when using the fully Cartesian coordinates, since $\boldsymbol{\Phi}_{\mathbf{qq}}$ is a constant tensor. The linearized set of equations in *phase space* takes the following form:

$$\Delta H_1 \equiv (H_1)_{\dot{p}}\, \Delta\dot{p} + (H_1)_{\dot{q}}\, \Delta\dot{q} + (H_1)_p\, \Delta p + (H_1)_q\, \Delta q + (H_1)_Q\, \Delta Q = 0 \quad (9.51)$$

$$\Delta H_2 \equiv (H_2)_{\dot{p}}\, \Delta\dot{p} + (H_2)_{\dot{q}}\, \Delta\dot{q} + (H_2)_p\, \Delta p + (H_2)_q\, \Delta q + (H_2)_Q\, \Delta Q = 0 \quad (9.52)$$

that expressed in matrix form gives the final result

$$\begin{bmatrix} I & (H_1)_{\dot{q}} \\ 0 & (H_2)_{\dot{q}} \end{bmatrix} \begin{Bmatrix} \Delta\dot{p} \\ \Delta\dot{q} \end{Bmatrix} + \begin{bmatrix} 0 & (H_1)_q \\ -I & (H_2)_q \end{bmatrix} \begin{Bmatrix} \Delta p \\ \Delta q \end{Bmatrix} + \begin{Bmatrix} (H_1)_Q\, \Delta Q \\ 0 \end{Bmatrix} = 0 \qquad (9.53)$$

9.2 Numerical Computation of Derivatives

The expressions for the partial derivatives in the linearized dynamic equations (9.3), both using independent and dependent coordinates, have been found in the previous section. Some of these derivatives are straightforward, but others are more complicated. As a whole, the computer implementation of these linearized dynamic equations may end up being tedious and cumbersome. An alternative way will be presented in this section to find these derivatives, that can be very interesting in practice because of its simple theoretical formulation and computer implementation. This alternative way relies on the numerical computation of the partial derivatives of the dynamic equilibrium equation (9.1).

The linearization of the differential equations of motion is a task that is normally performed only once in each program execution. It differs from the direct and/or inverse dynamics formulations that are applied again and again in each step of the numerical integration process. The linearization of dynamic equations, like the solution of the static equilibrium position problem, is a task that can be carried out in the preprocessing phase, so its numerical efficiency is a second order of importance factor. In most practical cases, it is more convenient to be able to implement it in an easier way.

Formulas for numerical computation of derivatives can be found in many text books on numerical analysis (Burden and Faires (1985), Smith (1986), and Chapra and Canale (1988)). If choice is restricted to symmetric formulas that are the most accurate for the same number of function evaluations, one can include the following expressions to evaluate the first derivative of a function:

– Three-point $O(h^2)$ formula,

$$f'(x) = \frac{1}{2h}\left[f(x+h) - f(x-h)\right] - \frac{h^2}{6} f^{(3)}(\xi) \qquad (9.54)$$

– Five-point $O(h^4)$ formula,

$$f'(x) = \frac{1}{12h}\left[f(x-2h) - 8f(x-h) + 8f(x+h) - f(x+2h)\right] - \frac{h^4}{30} f^{(5)}(\xi) \quad (9.55)$$

If formula (9.54) is applied to the dynamic equilibrium equation (9.1), it can be written (assuming independent coordinates z)

$$\frac{\partial \mathbf{H}}{\partial z_i} = \frac{1}{2h}\left[\mathbf{H}(\mathbf{z}+\mathbf{h}_i, \dot{\mathbf{z}}, \ddot{\mathbf{z}}, \mathbf{Q}) - \mathbf{H}(\mathbf{z}-\mathbf{h}_i, \dot{\mathbf{z}}, \ddot{\mathbf{z}}, \mathbf{Q})\right] \qquad (9.56)$$

where $\mathbf{h}_i^T = \{0, 0, ...,0, h, 0, ...,0\}$. Analogous expressions can be used for the remaining partial derivatives, with respect to $\dot{\mathbf{z}}$, $\ddot{\mathbf{z}}$, and $\mathbf{Q}$.

Equation (9.54) requires $2f$ function evaluations to compute the partial derivative with respect to z, and it is second order accurate. On the other hand, equation (9.55) requires $4f$ function evaluations, but it is fourth order accurate. Equation (9.55) should be preferred in most practical cases.

The choice of an adequate size for the increment h is problem dependent. It can be different for positions, velocities, accelerations, and forces and even for each component of these vectors. The use of $h = 10^{-2}$ to 10^{-5} should provide good results in most practical cases.

9.3 Numerical Evaluation of the Dynamic Response

Once the partial derivatives are obtained, the resulting linear equations of motion (9.4) or (9.53) can be integrated using the methods explained in Chapter 7 for the integration of differential equations. In addition, standard techniques widely applied in linear structural analysis may be used as well. Refer to Craig (1982) and Meirovitch (1980) for a detailed description of these methods.

Of particular importance is the evaluation of the linearized natural frequencies and mode shapes around a particular configuration. These are helpful not only for the dynamic response but also for design purposes and vibration control.

Similar to equation (9.53), the matrix differential equations of motion in accelerations (9.4) may also be written as a set of $2n$ equations in *state space* form:

$$\begin{bmatrix} \mathbf{H}_{\ddot{y}} & 0 \\ 0 & \mathbf{I} \end{bmatrix}\begin{Bmatrix} \Delta\ddot{\mathbf{y}} \\ \Delta\dot{\mathbf{y}} \end{Bmatrix} + \begin{bmatrix} \mathbf{H}_{\dot{y}} & \mathbf{H}_{y} \\ -\mathbf{I} & 0 \end{bmatrix}\begin{Bmatrix} \Delta\dot{\mathbf{y}} \\ \Delta\mathbf{y} \end{Bmatrix} = \begin{Bmatrix} -\mathbf{H}_Q\,\Delta\mathbf{Q} \\ 0 \end{Bmatrix} \qquad (9.57)$$

Whether in phase space form (9.53) or state space form (9.57), the linearized equations of motion can be written in simplified notation as

$$\mathbf{A}\,\dot{\mathbf{u}} + \mathbf{B}\,\mathbf{u} = \mathbf{U}(t) \qquad (9.58)$$

where u and U represent the linearized response and forcing terms, respectively. Notice that for dependent coordinates with the penalty formulation,

$$\mathbf{H}_{\ddot{y}} = (\mathbf{H}_2)_{\ddot{q}} = \mathbf{M} + \mathbf{\Phi}_q^T\,\alpha\,\mathbf{\Phi}_q \qquad (9.59)$$

Therefore, $\mathbf{A}$ is a non-singular matrix, with $\mathbf{B}$ being a non-symmetric matrix. The eigenvalue problem associated with (9.58) has the following form:

$$\lambda\,\mathbf{A}\,\mathbf{v} + \mathbf{B}\,\mathbf{v} = 0 \qquad (9.60)$$

where λ is the eigenvalue and v the eigenvector. Equation (9.60) may also be written as

$$\lambda\, \mathbf{v} = \mathbf{D}\, \mathbf{v} \tag{9.61}$$

where $\mathbf{D} = -\,\mathbf{A}^{-1}\,\mathbf{B}$. Equation (9.61) represents a standard eigenvalue problem, but with $\mathbf{D}$ being non-symmetric. Its solution will lead to $2n$ eigenvalues and eigenvectors that are real or complex in conjugate pairs. Jennings (1979) affirms that the reduction of equation (9.61) to its upper Hessenberg form and subsequent use of the QR method for the eigen-solution leads to a very efficient algorithm for matrices whose order does not exceed 100. These algorithms are available through standard mathematical subroutine packages that run in a diversity of computers.

The real part of the eigenvalue λ represents the decay of the amplitude of the natural mode. The imaginary part represents the damped frequency. The eigenvectors $\mathbf{v}$ can also be used to uncouple the linearized equations of motion (9.58) since they obey the orthogonality conditions:

$$\mathbf{v}_i^T\,\mathbf{A}\,\mathbf{v}_i = a_i \tag{9.62}$$

$$\mathbf{v}_i^T\,\mathbf{B}\,\mathbf{v}_i = b_i \tag{9.63}$$

Introducing the linear transformation

$$\mathbf{u} = \sum_1^{2n} \mathbf{v}_i\, r_i \tag{9.64}$$

into equation (9.58) and pre-multiplying the same equation by $\mathbf{v}_i^T$ one obtains the following set of uncoupled first order equations with complex entries:

$$\mathbf{v}_i^T\,\mathbf{A}\,\mathbf{v}_i\,\dot{r}_i + \mathbf{v}_i^T\,\mathbf{B}\,\mathbf{v}_i\, r_i = \mathbf{v}_i^T\,\mathbf{U}(t) \tag{9.65}$$

or

$$a_i\,\dot{r}_i + b_i\, r_i = \mathbf{v}_i^T\,\mathbf{u}(t) \tag{9.66}$$

which can be solved using the standard methods explained in Chapter 7.

References

Burden, R.L. and Faires, J.D., *Numerical Analysis*, 3rd edition, PWS-Kent, (1985).

Craig, R., *Structural Dynamics. An Introduction to Computer Methods*, Wiley, (1982).

Chapra, S.C. and Canale, R.P., *Numerical Methods for Engineers*, 2nd edition, McGraw-Hill, (1988).

Jennings, A., *Matrix Computations for Engineers and Scientists*, Wiley, (1979).

Meirovitch, L., *Computational Methods in Structural Dynamics*, Sijthoff and Noordhoff, (1980).

Smith, W.A., *Elementary Numerical Analysis*, Prentice-Hall, (1986).

10
Special Topics

This chapter deals with several techniques to solve some problems of particular interest in multibody simulation that have not been considered in other chapters. These techniques are neither very sophisticated nor trivial. However, they may be very useful at the time of solving practical or real problems.

The first problem to be considered in this chapter is a way to model *Coulomb friction* in a dynamic simulation. This approach to friction forces and energy dissipation is more accurate in many practical cases than the viscous friction model, but it is also far more difficult to implement. The second topic is to be *impact forces*, that is, very large forces that act on a very short period of time. There are many practical cases where impact forces play a very important role, and a simple and efficient way to model the process becomes necessary. One of these cases is the *backlash* or *clearance* in joints. This becomes the subject of another section of this chapter. *Kinematic synthesis*, which entails the finding of the best possible dimensions for a multibody system, and *sensitivity analysis*, so useful for determining the tendencies of the optimal objective function with respect to design variations, are also dealt with in this chapter. Finally, some ways to deal with *singular positions* of multibody systems will be described.

10.1 Coulomb Friction

Coulomb, or dry modeling for friction, seems to be more accurate than viscous friction for joints with small relative velocity. Coulomb friction is also more difficult to introduce in a general purpose program because it is highly nonlinear and can involve switching between sliding and stiction conditions. A consistent consideration of the Coulomb friction model can be found in Bagci (1975) and more importantly, in Haug et al. (1986). Figure 10.1a represents the friction force dependency on relative velocity, according to the Coulomb model. In order to avoid the discontinuity at zero, some authors (Threlfall (1978) and Rooney and Deravi (1982)) have introduced more continuous dependency laws similar to the one shown in Figure 10.1b. In this section, we will be consistent with the Coulomb friction model.

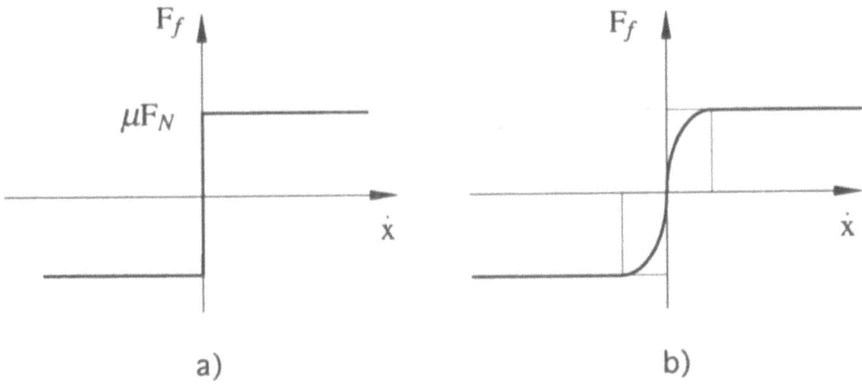

Figure 10.1. Coulomb friction models: a) standard. b) modified.

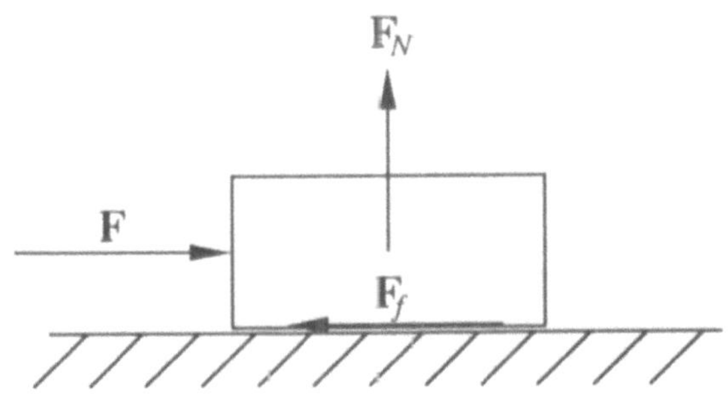

Figure 10.2. Block sliding on a plane with Coulomb friction.

10.1.1 Review of the Coulomb Friction Hypothesis

Consider a block on a flat surface as seen in Figure 10.2. The block is a body of mass m subjected to gravity forces of vertical descending direction. These forces are in equilibrium with the ground reaction force $\mathbf{F}_N$. There is some experimental evidence that if a rather small horizontal force $\mathbf{F}$ is applied to the block, no motion is obtained. This means that a horizontal reaction force $\mathbf{F}_f$ has appeared. If the external force $\mathbf{F}$ is increased little by little, it can be observed that when it reaches a particular value the block starts to move, sliding on the ground. This critical value of $\mathbf{F}$ depends on the nature of the ground and block contact surfaces and on the normal force $\mathbf{F}_N$. It also may be observed that during the relative sliding, the horizontal reaction force $\mathbf{F}_f$ (the *friction force*) depends on the normal force $\mathbf{F}_N$, but it does not depend on the velocity and/or acceleration. The mathematical model for this mechanical behavior is call *Coulomb friction* after the French scientist of the *XVIII* Century or *dry friction*, because it models reasonably well the friction forces between non-lubricated contact surfaces.

The Coulomb friction model assumes that between the block and the ground there is a normal reaction force $\mathbf{F}_N$ and a horizontal friction force $\mathbf{F}_f$.

With respect to the relative motion between the surfaces in contact, two different cases or situations are possible: *sliding* and *stiction* (or lock up)

  –   Sliding: $$\mathbf{F}_f = \mu_d \, \mathbf{F}_N \tag{10.1}$$

  –   Stiction: $$\mathbf{F}_f \le \mu_s \, \mathbf{F}_N \tag{10.2}$$

where μ_d and μ_s are the coefficients of *dynamic* and *static* friction, respectively. Two coefficients have been introduced because there is some experimental evidence that the external force $\mathbf{F}$ necessary to start the motion is different (higher) that the force necessary to maintain the motion with constant relative velocity. Both μ_d and μ_s are constant scalar coefficients that depend on the nature of the contact surfaces (material, finishing state, etc.) but not on the external forces or motion variables. Very often both coefficients are considered as equivalent, with a single value μ.

In the *sliding* condition, the friction force $\mathbf{F}_f$ is known and the motion acceleration becomes the unknown. In the *stiction* condition, the friction force $\mathbf{F}_f$ is unknown, but there is no relative acceleration.

It is very important to set the conditions for switching between the two possible states of sliding and stiction. If the block is initially at rest, the motion will start when the external force $\mathbf{F}$ reaches the critical value, that is, maximum value for the friction force $\mathbf{F}_f$. In the stiction condition, the friction force is unknown. It shall be obtained from the horizontal equilibrium equation. The motion will start when a friction force is obtained such that

  –   Stiction to sliding: $\qquad \mathbf{F}_f > \mu_s \, \mathbf{F}_N$

If there is sliding but the external forces are changing, it is possible that a state is reached in which the relative sliding velocity changes its sign. In this case stiction will occur. The mathematical condition becomes:

  –   Sliding to stiction: Sliding velocity changes sign.

In summary, when there is *stiction*, the relative velocity is zero and the friction force shall be computed and checked. When it goes over its maximum value, it is necessary to switch to the *sliding* condition. In the *sliding* condition, the friction force is known, but the relative motion shall be computed and checked. When the relative velocity changes its sign, it is necessary to switch to the *stiction* condition.

The *Coulomb* friction model is quite different from *viscous* friction model, in which friction forces depend (often linearly) on velocities. Viscous dampers and the motion of a body inside a fluid environment are examples of viscous friction. Its mathematical model is simpler than the Coulomb one, mainly because there is neither force dependency on reaction forces nor switching conditions between different states governed by different sets of differential equations.

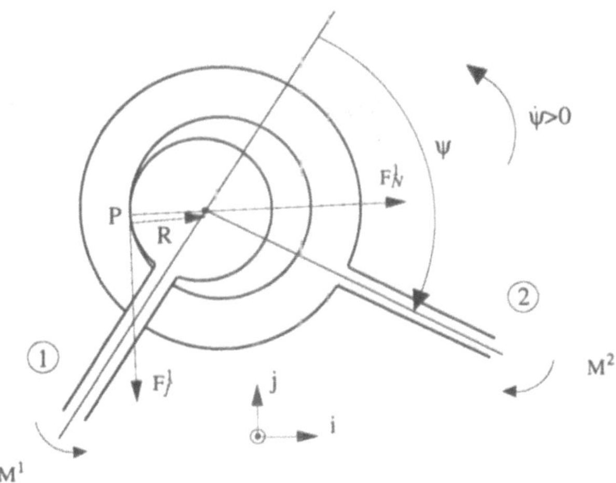

Figure 10.3. Friction forces in a planar revolute joint.

In the following sections, the Coulomb friction model will be extended for applications in complex multibody systems.

10.1.2 Coulomb Friction in Multibody Systems: Sliding Condition

In a complex multibody system, Coulomb friction may likely appear in joints where there are contacting surfaces belonging to different bodies that have relative sliding motion. The simpler and more likely to occur sliding condition will be considered first.

If there is sliding, the friction force in a joint is a known linear function of the normal reaction force in this joint. The normal reaction force can be computed from the Lagrange multipliers associated with the joint constraint equations (See Chapter 6). The corresponding equations of motion take the form (See equation (5.10)):

$$\begin{bmatrix} \mathbf{M} & \Phi_\mathbf{q}^T \\ \Phi_\mathbf{q} & 0 \end{bmatrix} \begin{Bmatrix} \ddot{\mathbf{q}} \\ \lambda \end{Bmatrix} = \begin{Bmatrix} \mathbf{Q} \\ \mathbf{c} \end{Bmatrix} \tag{10.3}$$

where the term ($\Phi_\mathbf{q}^T \lambda$) represents the joint constraint forces. For any kind of joint, the normal forces $\mathbf{F}_N$ can be expressed as a function of the constraint forces in the form:

$$\mathbf{F}_N = \mathbf{E}(\mathbf{q}) \, \Phi_\mathbf{q}^T \lambda \tag{10.4}$$

where $\mathbf{E}(\mathbf{q})$ is a term that depends on the joint geometry and sometimes on the position as well. The generalized friction forces $\mathbf{Q}_f$ are proportional to $\mathbf{F}_N$ and can be written as

$$Q_f = \mu \cdot \mathbf{u(q)} \, \mathbf{E(q)} \, \Phi_q^T \lambda \tag{10.5}$$

where f is the friction coefficient and $\mathbf{u(q)}$ another function characteristic of each joint type and geometry.

Example 10.1

Consider the planar revolute joint with Coulomb friction, shown in Figure 10.3. The backlash is assumed to be small, although large enough to assure a single point of contact between the inner (1) and outer (2) circles. Force $\mathbf{F}_N^1$ is considered to be the normal contact force that body 2 exerts on body 1. This normal constraint force can be computed by solving the inverse dynamics problem, as explained in Chapter 6. Also assume that that the joint relative angle ψ has been introduced as a dependent coordinate. This will lead to an easier definition of the friction torques on the two contacting bodies.

In order to use vector expressions, the relative velocity of body 2 respect to body 1 is defined as

$$\omega_r^{2,1} = -\,\dot{\psi}\,\mathbf{k} \tag{i}$$

where $\mathbf{k} = \mathbf{i} \wedge \mathbf{j}$ is the unit vector normal to the plane of the mechanism. The minus sign (–) is necessary for full consistency of expression (i). The tangent friction force $\mathbf{F}_f^1$ is computed as

$$\mathbf{F}_f^1 = \mu \, \text{sign}(\dot{\psi}) \, \mathbf{F}_N^1 \wedge \mathbf{k} \tag{ii}$$

The friction force $\mathbf{F}_f^1$ shall be applied to body 1 at the joint center which is a basic point. It is then necessary to apply the opposite force on body 2. However, it is also necessary to apply the friction torques on both bodies. These torques can be computed by means of the following equation:

$$\mathbf{M}_f = M_f\,\mathbf{k} = \mathbf{F}_f^1 \wedge \frac{\mathbf{F}_N^1}{|\mathbf{F}_N^1|} R = \mu\,\text{sign}(\dot{\psi})\,R\,(\mathbf{F}_N^1 \wedge \mathbf{k}) \wedge \mathbf{F}_N^1 \tag{iii}$$

The scalar value of this torque M_f is the force variable conjugated with the relative angle ψ. Therefore, it can be applied on both bodies at the same time (See Section 4.3.1).

The equations of motion with Coulomb friction become

$$\begin{bmatrix} \mathbf{M} & \Phi_q^T \\ \Phi_q & 0 \end{bmatrix} \begin{Bmatrix} \ddot{\mathbf{q}} \\ \lambda \end{Bmatrix} = \begin{Bmatrix} \mathbf{Q} + Q_f(\lambda) \\ \mathbf{c} \end{Bmatrix} \tag{10.6}$$

It may be seen that the vector of Lagrange multipliers λ appears on both sides of the equation. Therefore it must be solved iteratively. A possible algorithm, based on fixed point iteration, is the following one:

Algorithm 10-1

1. Estimate λ^0. Compute $Q_f(\lambda^0)$. Set $i = 0$

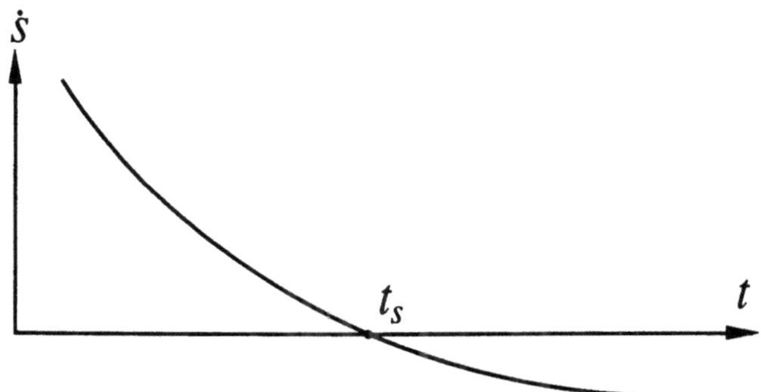

Figure 10.4. Change of sign of sliding velocity.

2. Calculate λ^{i+1} from (10.6) adjusted for fixed point iteration

$$\begin{bmatrix} \mathbf{M} & \Phi_q^T \\ \Phi_q & 0 \end{bmatrix} \left\{ \begin{array}{c} \ddot{\mathbf{q}} \\ \lambda^{i+1} \end{array} \right\} = \left\{ \begin{array}{c} \mathbf{Q} + \mathbf{Q}_f(\lambda^i) \\ \mathbf{c} \end{array} \right\}$$

3. If abs($\lambda^{i+1} - \lambda^i$) < *tolerance*, stop. Otherwise, go to step 4.
4. Set $\lambda^{i+1} = \lambda^i$, i=i+1, and go to step 2.

Remember that the friction force $\mathbf{F}_f$, which may be obtained from an equation analogous to (10.5), needs be checked in relation with $\mathbf{F}_N$ in order to see if the assumed *sliding* condition remains valid.

10.1.3 Coulomb Friction in Multibody Systems: Stiction Condition

The sliding relative velocity must be monitored during the sliding condition. At time t_s, if a change in the relative velocity sign is detected (See Figure 10.4), the joint becomes locked, and it is then necessary to switch to the stiction condition.

The joint lock-up can be represented mathematically by a new constraint equation. In a revolute joint, for instance, the relative angle must be kept constant. The stiction constraint equation can be written in the general form:

$$\Phi^r = 0 \qquad (10.7)$$

This new equation is appended to the remaining constraints, leading to a system of dynamic equations similar to (10.6)

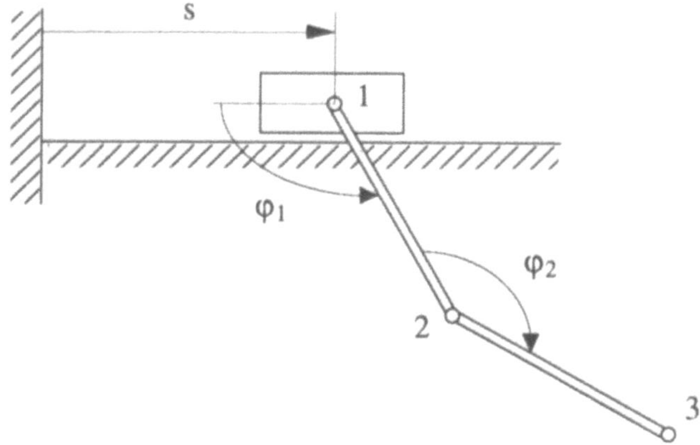

Figure 10.5. Planar mechanism with Coulomb friction in the joints.

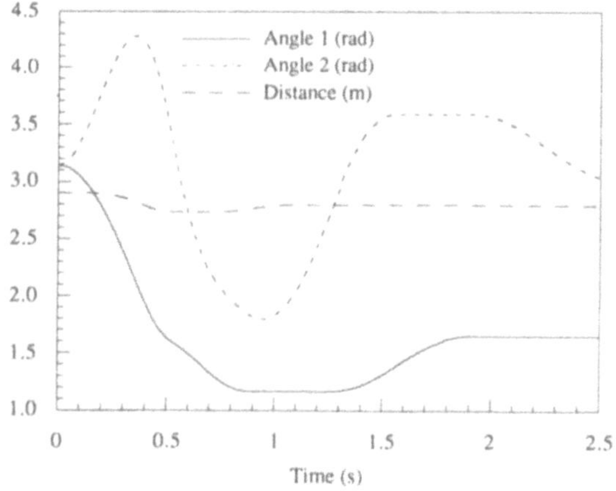

Figure 10.6. Time variation of the joint coordinates.

$$
\begin{bmatrix}
\mathbf{M} & \boldsymbol{\Phi}_\mathbf{q}^{\mathrm{T}} & \boldsymbol{\Phi}_\mathbf{q}^{r\mathrm{T}} \\
\boldsymbol{\Phi}_\mathbf{q} & 0 & 0 \\
\boldsymbol{\Phi}_\mathbf{q}^{r} & 0 & 0
\end{bmatrix}
\left\{
\begin{matrix}
\ddot{\mathbf{q}} \\
\lambda \\
\lambda^{r}
\end{matrix}
\right\}
=
\left\{
\begin{matrix}
\mathbf{Q} \\
\mathbf{c} \\
\mathbf{c}^{r}
\end{matrix}
\right\}
\tag{10.8}
$$

At this point, the stiction force $\mathbf{F}_f$ is to be computed and monitored. It can be computed from the force associated with the stiction constraint (10.7) in the form:

$$\mathbf{F}_f = \mathbf{D}(\mathbf{q})\,\mathbf{\Phi_q^{rT}}\,\lambda^r \tag{10.9}$$

where $\mathbf{D}(\mathbf{q})$ is again a term that depends on the joint type, geometry, and position. This friction or stiction force shall be checked so as to be sure that it is below its maximum value. This value is the normal force multiplied by the coefficient of friction. When it is found that

$$\mathbf{F}_f > \mu\,\mathbf{F}_N \tag{10.10}$$

it is necessary to switch to the sliding condition, releasing the previously added constraint equation (10.7).

The mathematical model for the Coulomb friction is not easy to implement in general purpose codes, even for the simplest cases. It is not currently implemented with generality in any commercial simulation package to the authors' knowledge. The difficulties stem from the switching between sliding and stiction states involving a change in the number of the system degrees of freedom and the need to iterate, in the sliding state, for each acceleration evaluation with equation (10.6).

Example   10.2

Figure 10.5 shows a planar multibody system consisting on a double pendulum joined to the fixed element through a prismatic joint that allows a horizontal translation. The physical characteristics of the problem are:

$L_{12} = L_{23} = 1$ m.
$m_1 = m_2 = 10$ kg.
$I_2 = I_3 = 0$
$\mu_P = 0,25 \qquad\qquad \mu_1 = 0,14 \qquad\qquad \mu_2 = 0,11$

where μ_P is the friction coefficient at the prismatic joint, and μ_1, and μ_2 are the friction coefficients at the revolute joints 1 and 2. The pendulum starts moving from the horizontal position and falls under gravity effects. Figure 10.6 shows the time history of the three relative coordinates. It may be seen how these curves remain flat during some time intervals. This means that during those time intervals the corresponding joint is locked due to friction.

10.2 Impacts and Collisions

Impacts are due to large impulsive forces acting over infinitesimal periods of time. The mathematical representation of impacts can be done with the *unit impulse* or *Dirac delta function* $\delta(t-a)$. This function can be seen as the limit of a rectangle function of unit area centered at time a, when the width ε tends to zero as can be seen in Figure 10.7.

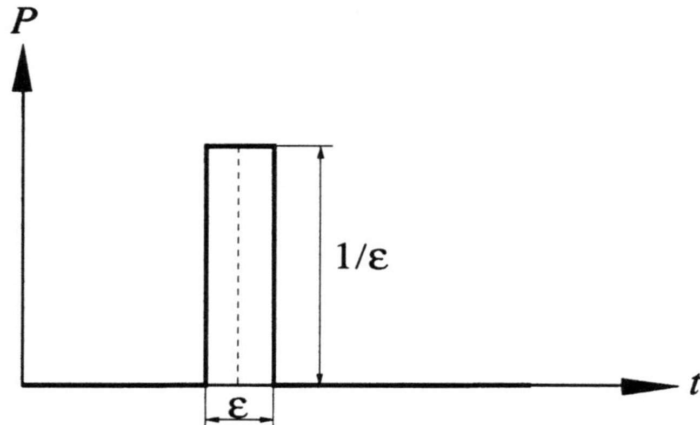

Figure 10.7. Rectangular function of unit area.

As ε goes to zero, the function value becomes infinite, but the area under the function remains equal to the unit value. The product of the Dirac delta function by an arbitrary function $f(t)$ satisfies the following property:

$$\int_{-\infty}^{\infty} f(t)\, \delta(t-a)\, dt = f(a) \tag{10.11}$$

that is used sometimes as an alternative definition for function $\delta(t-a)$.

A general discussion on impact forces in multibody systems was presented by Haug et al. (1986).

10.2.1 Known Impact Forces

It is customary in mechanics to compute the effect of impacts on bodies assuming that during the impact all the remaining finite forces can be neglected. It is also assumed that the system position does not change, because the impact time is very small. As with the Dirac function, the equations of motion under impulsive forces can be considered in an integral form. For instance, the equations of motion (10.3) with known impulse forces $\mathbf{Q}^i$ acting at time t_i become

$$\begin{bmatrix} \mathbf{M} & \mathbf{\Phi}_{\mathbf{q}}^{\mathrm{T}} \\ \mathbf{\Phi}_{\mathbf{q}} & 0 \end{bmatrix} \begin{Bmatrix} \ddot{\mathbf{q}} \\ \lambda \end{Bmatrix} = \begin{Bmatrix} \mathbf{Q} + \mathbf{Q}^i \\ \mathbf{c} \end{Bmatrix} \tag{10.12}$$

Integrating from time t_i^- to t_i^+:

$$\int_{t_i^-}^{t_i^+} (\mathbf{M}\ddot{\mathbf{q}} + \mathbf{\Phi}_{\mathbf{q}}^{\mathrm{T}}\lambda)\, dt = \int_{t_i^-}^{t_i^+} (\mathbf{Q}+\mathbf{Q}^i)\, dt \tag{10.13}$$

$$\int_{t_i^-}^{t_i^+} \boldsymbol{\Phi}_q \ddot{\mathbf{q}} \, dt = \int_{t_i^-}^{t_i^+} \mathbf{c} \, dt \tag{10.14}$$

Using incremental notation, one obtains:

$$\mathbf{M} \, \Delta\dot{\mathbf{q}} + \boldsymbol{\Phi}_q^T \, \boldsymbol{\lambda}_p = \mathbf{P}^i \tag{10.15}$$

$$\boldsymbol{\Phi}_q \, \Delta\dot{\mathbf{q}} = 0 \tag{10.16}$$

where $\boldsymbol{\lambda}_p$ are the Lagrange multipliers that are related to the internal impact forces (impact reactions), and $\mathbf{P}^i$ is the *integral effect* of the impact forces. Equation (10.15) represents the *conservation of momentum*, and equation (10.16) states that the velocity increment shall fulfill the homogeneous velocity constraint equations. In equation (10.15), $\mathbf{P}^i$ is

$$\mathbf{P}^i = \int_{t_i^-}^{t_i^+} \mathbf{Q}^i \, dt \tag{10.17}$$

Writing equations (10.15) and (10.16) jointly in matrix form, one can obtain

$$\begin{bmatrix} \mathbf{M} & \boldsymbol{\Phi}_q^T \\ \boldsymbol{\Phi}_q & 0 \end{bmatrix} \begin{Bmatrix} \Delta\dot{\mathbf{q}} \\ \boldsymbol{\lambda}_p \end{Bmatrix} = \begin{Bmatrix} \mathbf{P}^i \\ 0 \end{Bmatrix} \tag{10.18}$$

It may be seen from equation (10.18), that the mechanical effect of a known impulse is an instantaneous increment in the system velocities $\dot{\mathbf{q}}$. If the impact forces $\mathbf{P}^i$ are known, the dynamic simulation can proceed according to the following algorithm:

Algorithm   10-2

1. Integrate the equations of motion from $t = 0$ to $t = t_i$ using the equations

$$\begin{bmatrix} \mathbf{M} & \boldsymbol{\Phi}_q^T \\ \boldsymbol{\Phi}_q & 0 \end{bmatrix} \begin{Bmatrix} \ddot{\mathbf{q}} \\ \boldsymbol{\lambda} \end{Bmatrix} = \begin{Bmatrix} \mathbf{Q} \\ \mathbf{c} \end{Bmatrix} \tag{10.19}$$

2. At $t = t_i$, during the actuation of the impact force, use the equation

$$\begin{bmatrix} \mathbf{M} & \boldsymbol{\Phi}_q^T \\ \boldsymbol{\Phi}_q & 0 \end{bmatrix} \begin{Bmatrix} \Delta\dot{\mathbf{q}} \\ \boldsymbol{\lambda}_p \end{Bmatrix} = \begin{Bmatrix} \mathbf{P}^i \\ 0 \end{Bmatrix} \tag{10.20}$$

to compute jump discontinuity in velocities and find the velocities after the impact as

$$\dot{\mathbf{q}}(t_i^+) = \dot{\mathbf{q}}(t_i^-) + \Delta\dot{\mathbf{q}} \tag{10.21}$$

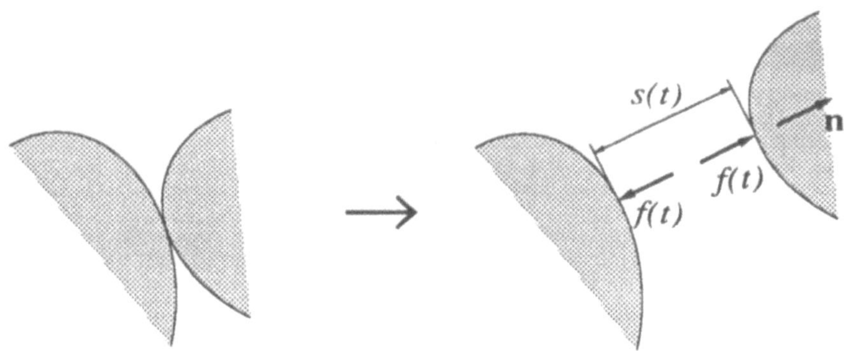

Figure 10.8. Impact between two bodies.

3. Use the new velocities $\dot{\mathbf{q}}$ resulting from (10.21) to restart the numerical integration.

10.2.2 Impacts Between Bodies

When two bodies impact against each other, an unknown impulsive force acts between them. In this case, equation (10.20) cannot be used because the impact $\mathbf{P}^i$ is not known. According to the physical characteristics of the bodies, one normally knows if the impact is perfectly *elastic* (because there is a perfect rebound), perfectly *plastic* (where there is no rebound at all), or something in between. One can use this information to compute the velocities after the impact.

In Figure 10.8, two bodies can be seen impacting and rebounding. Vector $\mathbf{n}$ is a unit vector normal to the body surfaces in the contact point; s is the distance between the contact points, and $f(t)$ is the impact force.

The virtual work produced by $f(t)$ is

$$\delta W = f \,\delta s = f \left(\frac{\partial s}{\partial \mathbf{q}}\right)^{\mathrm{T}} \delta \mathbf{q} \tag{10.22}$$

and the generalized impulse $\mathbf{P}^i$ is given by the integral of the generalized force between t_i^- and t_i^+ leading to

$$\mathbf{P}^i = \frac{\partial s}{\partial \mathbf{q}}\, p \tag{10.23}$$

where

$$p = \int_{t_i^-}^{t_i^+} f(t)\,dt \tag{10.24}$$

Since the magnitude p of the impact is unknown, the additional equation that arises from the impact characteristics of the bodies is needed. From experimental

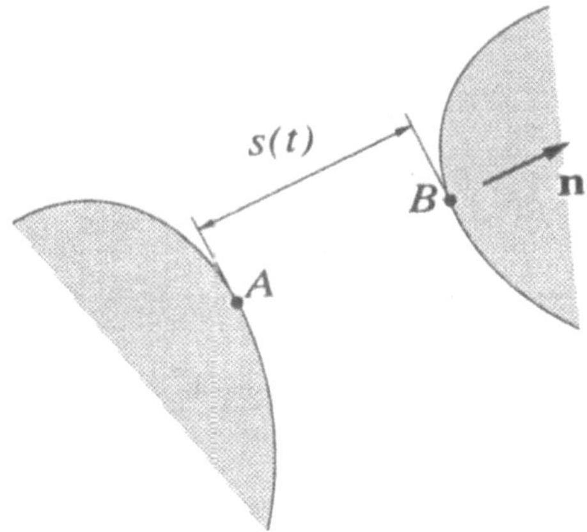

Figure 10.9. Points in contact during the impact.

testing, it is known that the normal relative velocities of the contact points before and after the impact are related by

$$\dot{s}(t_i^+) = - e\, \dot{s}(t_i^-) \tag{10.25}$$

where e is called the *coefficient of restitution* or Newton's coefficient. If the impact is perfectly *elastic* ($e=1$), the normal relative velocity changes its sign but keeps the magnitude. On the other hand, if the impact is perfectly *plastic* ($e=0$), there is no rebound or normal relative velocity after the impact. Notice that this formulation assumes that there is no friction or tangent impact forces. Impact with friction is a very specialized and difficult subject and will not be considered here.

Next the equations for the impact will be developed. Using the chain rule of differentiation for the normal relative velocity,

$$\dot{s} = \left(\frac{\partial s}{\partial \mathbf{q}}\right)^{\mathrm{T}} \dot{\mathbf{q}} \tag{10.26}$$

Substituting this in equation (10.25) yields

$$\left(\frac{\partial s}{\partial \mathbf{q}}\right)^{\mathrm{T}} \dot{\mathbf{q}}(t_i^+) = - e\left(\frac{\partial s}{\partial \mathbf{q}}\right)^{\mathrm{T}} \dot{\mathbf{q}}(t_i^-) \tag{10.27}$$

Subtracting $\dfrac{\partial s}{\partial \mathbf{q}}\, \dot{\mathbf{q}}(t_i^-)$ from both sides of equation (10.27), one obtains

$$\left(\frac{\partial s}{\partial \mathbf{q}}\right)^{\mathrm{T}} \Delta\dot{\mathbf{q}} = - (1+e)\left(\frac{\partial s}{\partial \mathbf{q}}\right)^{\mathrm{T}} \dot{\mathbf{q}}(t_i^-) \tag{10.28}$$

This last equation can be written together with the impulse equations (10.18), yielding

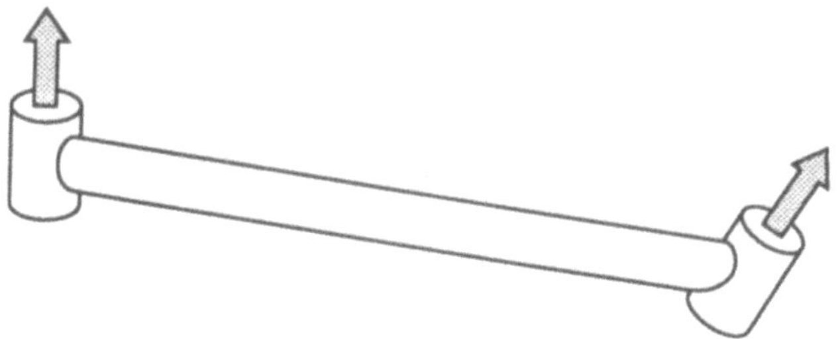

Figure 10.10. Element defined with two points and two unit vectors.

$$
\begin{bmatrix}
\mathbf{M} & \boldsymbol{\Phi}_q^T & \dfrac{\partial s}{\partial q} \\[2ex]
\boldsymbol{\Phi}_q & 0 & 0 \\[2ex]
\left(\dfrac{\partial s}{\partial q}\right)^T & 0 & 0
\end{bmatrix}
\begin{Bmatrix}
\Delta \dot{q} \\[1ex]
\lambda_p \\[1ex]
-p
\end{Bmatrix}
=
\begin{Bmatrix}
0 \\[2ex]
0 \\[2ex]
-(1+e)\left(\dfrac{\partial s}{\partial q}\right)^T \dot{q}(t_i^-)
\end{Bmatrix}
\tag{10.29}
$$

The derivative $\partial s/\partial q$ needs to be computed next. The two points that will be in contact during the impact are called A and B, as can be seen in Figure 10.9.

According to the expressions developed in Chapter 4, the position vectors $\mathbf{r}_A$ and $\mathbf{r}_B$ can be written as linear combinations of the elements of the dependent coordinates vector $\mathbf{q}$ (See equations (4.50) or (4.90)) in the form

$$
\mathbf{r}_A = \mathbf{C}_A\, \mathbf{q} \tag{10.30}
$$

$$
\mathbf{r}_B = \mathbf{C}_B\, \mathbf{q} \tag{10.31}
$$

For an element defined with two points and two vectors as the one in figure 10.10, $\mathbf{C}_A$ and $\mathbf{C}_B$ are constant matrices.

Thus, the distance s can be expressed as

$$
s^2 = \left(\mathbf{r}_A - \mathbf{r}_B\right)^T \left(\mathbf{r}_A - \mathbf{r}_B\right) \tag{10.32}
$$

Differentiating with respect to $\mathbf{q}$ yields

$$
\frac{\partial s}{\partial q} = \frac{\partial \left(\mathbf{r}_A - \mathbf{r}_B\right)^T}{\partial q}\, \frac{\mathbf{r}_A - \mathbf{r}_B}{s} = (\mathbf{C}_A^T - \mathbf{C}_B^T)\,\mathbf{n} = \mathbf{C}_{AB}^T\,\mathbf{n} \tag{10.33}
$$

Substituting this result in equation (10.29), one obtains

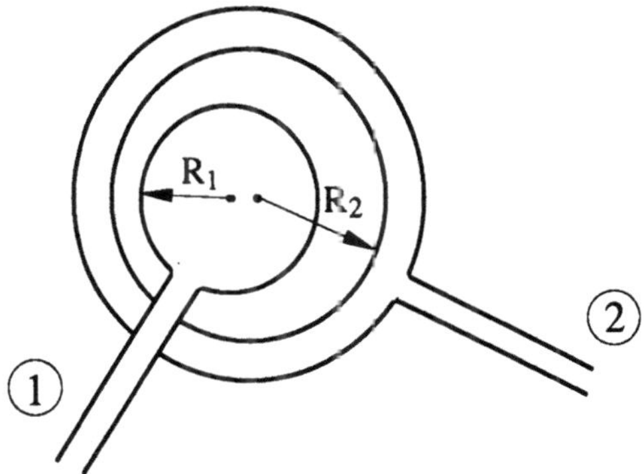

Figure 10.11. Planar revolute joint with backlash.

$$
\begin{bmatrix}
\mathbf{M} & \boldsymbol{\Phi}_\mathbf{q}^\mathsf{T} & \mathbf{C}_{AB}\,\mathbf{n}^\mathsf{T} \\
\boldsymbol{\Phi}_\mathbf{q} & 0 & 0 \\
\mathbf{C}_{AB}^\mathsf{T}\,\mathbf{n} & 0 & 0
\end{bmatrix}
\begin{Bmatrix}
\Delta\dot{\mathbf{q}} \\
\lambda_\mathbf{p} \\
-p
\end{Bmatrix}
=
\begin{Bmatrix}
0 \\
0 \\
-(1+e)\,\mathbf{C}_{AB}^\mathsf{T}\,\mathbf{n}\,\dot{\mathbf{q}}(t_i^-)
\end{Bmatrix}
\tag{10.34}
$$

This system of linear equations allows the computation of all the unknown parameters. The new velocities after the impact are

$$
\dot{\mathbf{q}}(t_i^+) = \dot{\mathbf{q}}(t_i^-) + \Delta\dot{\mathbf{q}}
\tag{10.35}
$$

and one also obtains the magnitude of the impact p and the internal reaction or constraint impacts ($\boldsymbol{\Phi}_\mathbf{q}^\mathsf{T}\,\lambda_\mathbf{p}$).

Remark. Equation (10.34) can be modified so as to eliminate the internal impact forces by expressing it in terms of the increment of the independent velocities with the use of the matrix $\mathbf{R}$ introduced in Section 3.5. After some algebraic manipulations, one can obtain

$$
\begin{bmatrix}
\mathbf{R}^\mathsf{T}\mathbf{M}\,\mathbf{R} & \mathbf{R}^\mathsf{T}\mathbf{n}^\mathsf{T}\mathbf{C}_{AB} \\
\mathbf{C}_{AB}^\mathsf{T}\,\mathbf{n}\,\mathbf{R} & 0
\end{bmatrix}
\begin{Bmatrix}
\Delta\dot{\mathbf{z}} \\
-p
\end{Bmatrix}
=
\begin{Bmatrix}
0 \\
-(1+e)\,\mathbf{C}_{AB}^\mathsf{T}\,\mathbf{n}\,\dot{\mathbf{q}}(t_i^-)
\end{Bmatrix}
\tag{10.36}
$$

This concludes the formulation of impacts between bodies.

10.3 Backlash

The problem of *backlash* or existing *clearances* in joints is an important practical problem in many applications that may produce noise, vibrations, severe damage and other difficulties in normal operation of real multibody systems. Some results on this subject for complex mechanical systems can be found in Dubowsky et al. (1987).

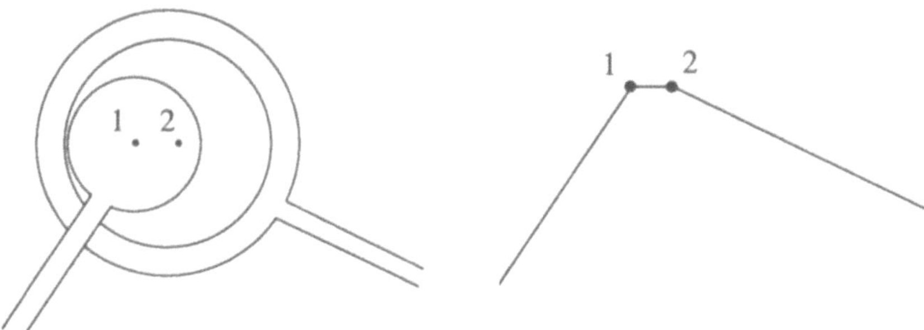

Figure 10.12. Contact condition in a revolute joint with backlash.

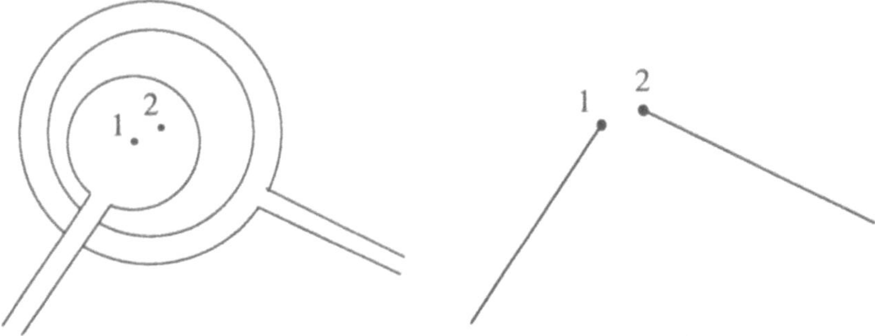

Figure 10.13. Non-contact condition in a revolute joint with backlash.

Backlash is a difficult problem to model because it depends on poorly known data, as for instance the clearance magnitude itself. Backlash is also a problem that strongly depends on the particular geometry of the joint. In a 3-D revolute joint with backlash, there are many possible ways to set the physical contact between the inner and outer cylinders, and a general formulation for backlash largely exceeds the aims and scope of this section. This text will be limited to introducing some backlash concepts using the planar revolute and prismatic joints.

10.3.1 Planar Revolute Joint

Figure 10.11 shows a planar revolute joint where the backlash has been made very large in order to make it clearly visible. This joint is materialized by two circles, the inner of radius R_1 and belonging to body 1, and the outer of radius R_2 and belonging to body 2. One can assume that points 1 and 2 are the centers of the two circles. In this case, the revolute joint shall be considered with two points and not with a single shared point, as in Chapter 2.

It is easy to see that there are two possible scenarios:

a) *The two circles are in contact.* The mathematical condition for this case is that the distance between the circle centers should be equal to the difference of radius (clearance):

$$(x_1 - x_2)^2 + (y_1 - y_2)^2 - (R_2 - R_1)^2 = 0 \qquad (10.37)$$

As long as the circles are in contact, the centers move as if they were connected by a rigid bar 1-2, of length $d_{12}^2 = (R_2 - R_1)^2$. Thus the joint can be replaced by an equivalent mechanism with an additional bar as indicated in Figure 10.12.

b) *The two circles are not in contact.* In this case, elements 1 and 2 move independently, and there is not any mathematical constraint equation between them as may be seen in Figure 10.13. The following inequality must be fulfilled:

$$(x_1 - x_2)^2 + (y_1 - y_2)^2 - (R_2 - R_1)^2 < 0 \qquad (10.38)$$

It remains to explain when the joint switches from condition (a) to condition (b), and vice versa.

Assume that the system is under condition (a). In this case, the constraint equation (10.37) must be added to the system constraint equations $\Phi(q)=0$ as an additional dynamic constraint $\phi_{m+1}(q)=0$. If equation (10.37) is added to the dynamics through a Lagrange multiplier λ_{m+1}, this multiplier represents the axial force exerted by the fictitious bar that joins the centers of the circles. This force also represents the contact force between both circles. The value of λ_{m+1} may only represent a tensional force, or, in other words, the fictitious bar can avoid a separation between points 1 and 2, but it is not able to actuate if the points tend to join each other. Therefore, the value of λ_{m+1} shall be checked in each integration step to make sure that it is positive. If so, the joint is under condition (a), and this value remains positive. If λ_{m+1} changes its sign, the constraint equation (10.37) must be removed and one should switch to case (b).

Consider now that the system is under condition (b).

In this case, bodies 1 and 2 move independently; consequently, the equations of motion are integrated without any constraint equation for the joint where the backlash is located. It is only necessary to check the inequality (10.38).

If condition (10.38) is not fulfilled, a contact between the two circles occurs. This represents a real impact between bodies 1 and 2. We can assume that because lubricant in the joint and small relative velocities, the impact is perfectly plastic with no rebound. In order to solve for the new dependent velocities after the impact $\dot{q}(t^+) = \dot{q}(t^-) + \Delta\dot{q}$, one can use equation (10.34) by taking into account that points A and B are points 1 and 2, and thus matrices C_A and C_B can be taken as unit matrices. Equation (10.34) also provides the impact force and the remaining internal impact reaction forces. At time (t^+), the numerical integration is restarted including equation (10.37), because one is now in case (a). If in addition to backlash there is also Coulomb friction in the joint, the formula-

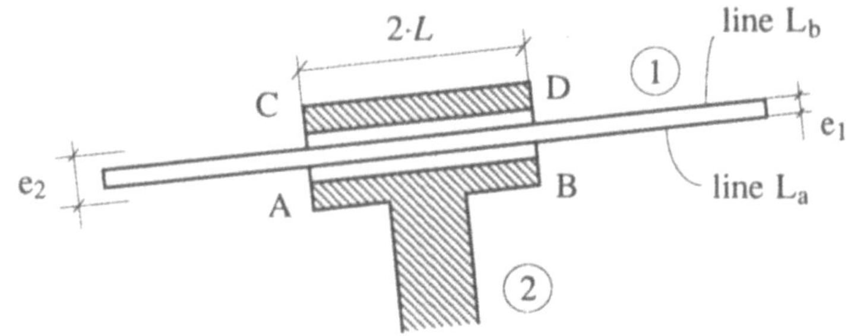

Figure 10.14. Planar prismatic joint with backlash.

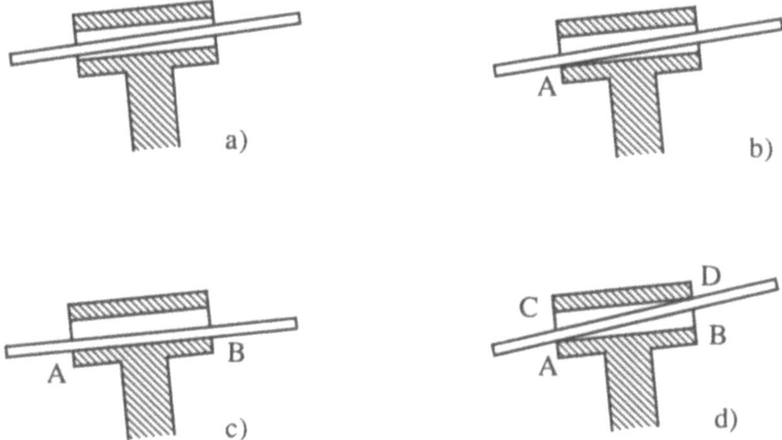

Figure 10.15. Four contact possibilities for a prismatic joint with backlash.

tion becomes more involved and will not be discussed here. It is necessary again to watch the normal force between the circles in order to check the contact condition, but now there is also a tangent friction force that can be introduced as an external force. As this external force changes the value of the normal force, it is necessary to iterate inside each integration step, as explained in Section 10.1.

10.3.2 Planar Prismatic Joint

The formulation of backlash for this kind of joint is more complicated than that for the revolute one. Figure 10.14 illustrates a possible geometry for a prismatic joint with backlash.

The possible scenarios for the backlash in a prismatic joint are illustrated in Figure 10.15, and consist of:

a) No contact (Figure 10.15a).
b) Contact on a single point (Figure 10.15b).

c) Contact on two points on the same side (Figure 10.15c).
d) Contact on two opposed points (Figure 10.15d).

Consider four points (A, B, C, and D) defined on element 2 and two lines (L and L') defined on element 1. The equations corresponding to segments L and L' can be expressed as a function of the dependent coordinates of body 1 ($\mathbf{q}^1$) and the coordinates of points A, B, C, and D, which may be expressed as functions of the dependent coordinates of body 2 ($\mathbf{q}^2$)

$$\text{Segment L:} \qquad y - m(\mathbf{q}^1)x - Y(\mathbf{q}^1) = 0 \tag{10.39}$$

$$\text{Segment L':} \qquad y - m'(\mathbf{q}^1)x - Y'(\mathbf{q}^1) = 0 \tag{10.40}$$

$$\text{Point A:} \qquad x_A = a_x(\mathbf{q}^2) \tag{10.41}$$

$$y_A = a_y(\mathbf{q}^2) \tag{10.42}$$

$$\text{Point B:} \qquad x_B = b_x(\mathbf{q}^2) \tag{10.43}$$

$$y_B = b_y(\mathbf{q}^2) \tag{10.44}$$

$$\text{Point C:} \qquad x_C = c_x(\mathbf{q}^2) \tag{10.45}$$

$$y_C = c_y(\mathbf{q}^2) \tag{10.46}$$

$$\text{Point D:} \qquad x_D = d_x(\mathbf{q}^2) \tag{10.47}$$

$$y_D = d_y(\mathbf{q}^2) \tag{10.48}$$

where m, m', Y, Y', a_x, a_y, b_x, b_y, c_x, c_y, d_x, d_y are known functions of the dependent coordinates vector of the corresponding element.

Using these expressions, it is possible to set the constraint equations corresponding to the four cases in Figure 10.15. This is done in the following way:

a) No constraint equations are necessary.
b) Point A shall be on line L. Substituting equation (10.41) on equation (10.39):

$$a_y(\mathbf{q}^2) - m(\mathbf{q}^1) \cdot a_x(\mathbf{q}^2) - Y(\mathbf{q}^1) = 0 \tag{10.49}$$

c) Points A and B shall be on line L. The corresponding equations are:

$$a_y(\mathbf{q}^2) - m(\mathbf{q}^1) \cdot a_x(\mathbf{q}^2) - Y(\mathbf{q}^1) = 0 \tag{10.50}$$

$$b_y(\mathbf{q}^2) - m(\mathbf{q}^1) \cdot b_x(\mathbf{q}^2) - Y(\mathbf{q}^1) = 0 \tag{10.51}$$

d) Points A and D shall be on lines L and L', respectively. The constraint equations are

$$a_y(\mathbf{q}^2) - m(\mathbf{q}^1) \cdot a_x(\mathbf{q}^2) - Y(\mathbf{q}^1) = 0 \qquad (10.52)$$

$$d_y(\mathbf{q}^2) - m'(\mathbf{q}^1) \cdot d_x(\mathbf{q}^2) - Y'(\mathbf{q}^1) = 0 \qquad (10.53)$$

Once again, the conditions for switching from one case to a different one depend on the dynamics. In particular they depend on the constraint forces that correspond to the constraint equations (10.49)-(10.53). If the constraint equations are introduced into the dynamics through the Lagrange multipliers or augmented Lagrangian formulations (See Chapter 5), the multipliers directly represent the contact forces (it will be assumed again that there is no friction in this case). Contrary to the fictitious bars in revolute joints, in direct contact only compressive contact forces are allowed. If the sign of a contact force changes, the corresponding constraint must be removed.

When the joint is in either case (a) or (b), there is also the possibility of going into an impact condition. This condition can be detected by checking the position of point D with respect to segment L'. If an impact is detected, equation (10.34) will have to be applied to obtain a new velocity distribution. The numerical integration will have to be restarted again with the new constrain equation included.

Backlash may have additional difficulties because of the extremely small space and time scales in which losses of contact, impacts, and so forth take place. It is necessary, therefore, to use very small time steps and some interpolation techniques to capture very precisely the time of occurrence of those events. When there are many joints with backlash within the same system, the computations become exceedingly expensive.

10.4 Kinematic Synthesis

The method presented in this section is a contribution coming from Alvarez and Jiménez (1992).

In the previous chapters and sections, the most important formulations for the kinematic and dynamic *analysis* of multibody systems have been presented. In all these problems, it has been assumed that the system was *perfectly known* either because it is an existing system or it has been previously designed. When wishing to design a new system which must comply with certain specifications and only analysis tools available, one must proceed in a iterative *trial-and-error* manner by means of *re-analysis*. A preliminary design is carried out and the system is analyzed. Once the results of the analysis have been obtained and are not entirely satisfactory, the design is then modified. Another analysis is performed, and the same mode is proceeded with until the desired effect is attained.

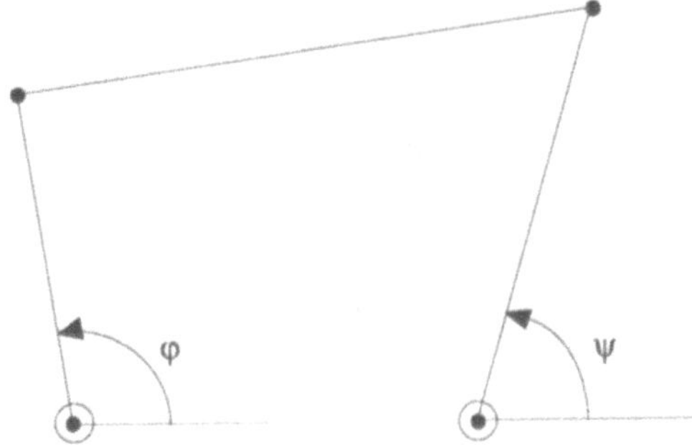

Figure 10.16. Function generation kinematic synthesis.

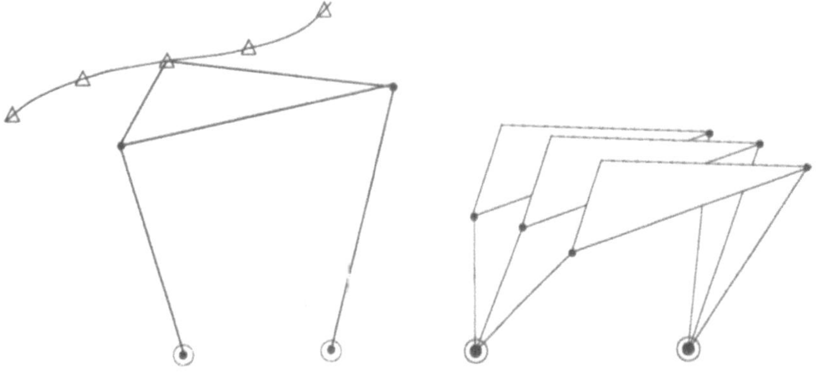

Figure 10.17. Path generation kinematic synthesis.

Figure 10.18. Rigid body guidance kinematic synthesis.

This *trial-and-error* process may be slow and quite dependent upon the experience of the designer.

Synthesis or *design* methods help overcome this difficulty, or at least lessen it. These methods lead directly without the intervention of an analyst, to a design which complies with the given specifications or which is the best one available from a certain point of view. The design of a multibody system can also be carried out from a more general perspective by taking dynamic factors into account. Two different problems will be studied: pure kinematic design, also called *kinematic synthesis*, and the more general *sensitivity analysis* for optimal dynamic problems.

Kinematic synthesis of mechanisms is mainly a geometric problem about which much has been written in the last half of the past century and in the first half of the present one (Angeles (1982), Erdman and Sandor (1978), and Suh and Radcliffe (1978)). During this time, many methods were developed. The majority of these methods were focused on the planar four-bar mechanism, with most of them graphic and containing a notable amount of ingenuity and originality.

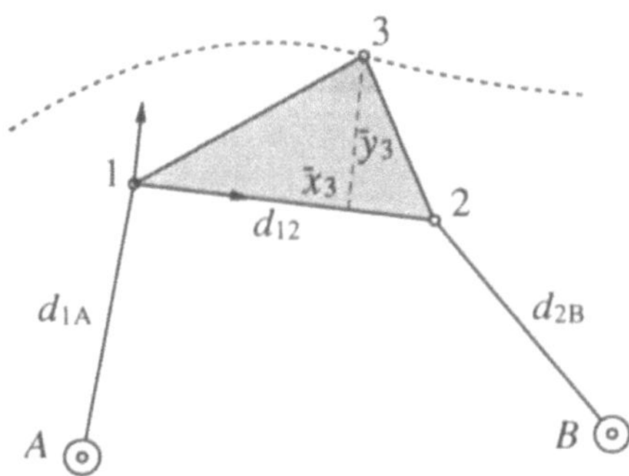

Figure 10.19. Path generated by a point of the coupling bar.

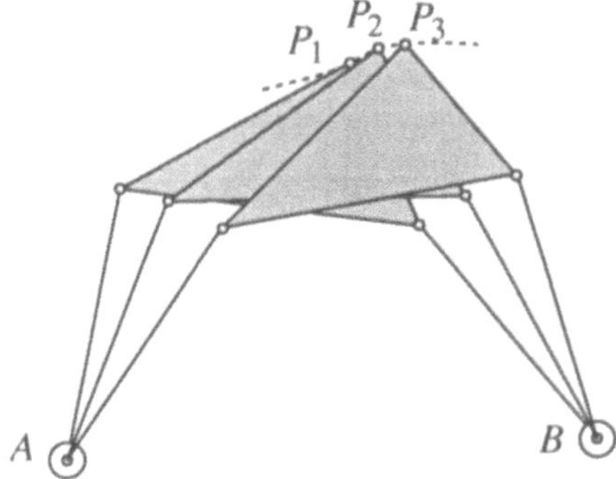

Figure 10.20. Design points for a path generation synthesis.

The traditional problems of *dimensional* synthesis are grouped together in three families: *function generation* synthesis, *path generation* synthesis, and *rigid body guidance* synthesis. An example of the function generation problem is shown in Figure 10.16. The purpose of this type of synthesis is to achieve an output angle ψ that is as close as possible to the desired nominal angle φ.

Figure 10.17 graphically illustrates a path generation problem. This basically consists of designing a four-bar mechanism so that a specific point of the coupler draws a trajectory that passes through a series of predefined points, or, at least, comes as close to them as possible.

The rigid body guidance problem can finally be seen in Figure 10.18. In this case, the design requirement is to obtain a four-bar mechanism in which a certain specific reference frame linked to the coupler passes through (or comes as close as possible to) a series of pre-established positions.

Graphic methods of kinematic synthesis are limited to simple mechanisms. They tend to be too specific and at times difficult to use. In recent years, more general programs for *optimal synthesis* have been developed. They are applicable to many different types of planar and three-dimensional multibody systems and include many different design conditions or specifications. Normally, these methods are based on numerical methods for optimization that seek the optimal solution with a minimum degree of error.

In this section, a simple and general numerical method will be described that is an improved version of the method presented by Avilés et al. (1985) for the optimal kinematic synthesis of linkages. Although this method will be described with a path generation problem for a four-bar example, it may be easily generalized for nearly any planar or three-dimensional linkage.

In order to carry out the optimum design of a multibody system for a defined set of design specifications, three steps shall be considered:

a) Choose the multibody system *topology*
b) Select the *design variables*
c) Define and minimize the *objective function*

Two kinds of constraint equations will be described: *geometric* constraints and *functional* constraints. The geometric constraints come from the multibody system topology (step (a)), and are the constraints that have been considered in Chapters 2 and 3 of this book. The functional constraints come from the specific design requirements that the multibody systems must fulfill.

Example   10.3

As a particular example, one can consider the path generated by point 3 belonging to the coupler of the four-bar mechanism in Figure 10.19.

Consider that points A and B cannot be moved; thus the design variables are the elements of the following vector:

$$\mathbf{b}^T = \{d_{1A},\ d_{12},\ d_{2B},\ \bar{x}_3,\ \bar{y}_3\} \tag{i}$$

The vector of dependent coordinates is

$$\mathbf{q}^T = \{x_1,\ y_1,\ x_2,\ y_2,\ x_3,\ y_3\} \tag{ii}$$

In this example, the geometric constraints are the constraints that correspond to the particular system being considered. In this case, the system is a four-bar mechanism with three points in the coupler, whose geometric constraints are:

$$\phi_1 \equiv (x_1 - x_A)^2 + (y_1 - y_A)^2 - d_{1A}^2 = 0 \tag{iii}$$

$$\phi_2 \equiv (x_1 - x_2)^2 + (y_1 - y_2)^2 - d_{12}^2 = 0 \tag{iv}$$

$$\phi_3 \equiv (x_2 - x_3)^2 + (y_2 - y_B)^2 - d_{2B}^2 = 0 \tag{v}$$

$$\phi_4 \equiv x_3 - x_1 + \frac{(x_2-x_1)}{d_{12}}\,\bar{x}_3 - \frac{(y_2-y_1)}{d_{12}}\,\bar{y}_3 = 0 \tag{vi}$$

$$\phi_5 \equiv y_3 - y_1 + \frac{(y_2-y_1)}{d_{12}}\,\bar{x}_3 + \frac{(x_2-x_1)}{d_{12}}\,\bar{y}_3 = 0 \tag{vii}$$

In addition to the geometric constraints, the designer also specifies the functional constraints. For the particular example being considered, we will impose the conditions of the trajectory of point 3 passing as close as possible to a finite set of design points $(P_1, P_2, P_3, ..., P_N)$, as shown in Figure 10.20.

It is clear that each design point corresponds to a different value of the dependent coordinates vector $\mathbf{q}$. These values will be called $(\mathbf{q}^1, \mathbf{q}^2, ..., \mathbf{q}^N)$. The functional constraints are now imposed for each design point. For a generic point i:

$$x_3^i - x_{P_i} = 0 \tag{viii}$$

$$y_3^i - y_{P_i} = 0 \tag{ix}$$

where $i = 1, 2, ..., N$.

In the general case, if $\mathbf{q}$ and $\mathbf{b}$ are the vectors of dependent coordinates and design variables, the geometric constraints equations can be expressed in vector form as

$$\Phi(\mathbf{q}, \mathbf{b}) = 0 \tag{10.54}$$

Using natural coordinates, the constraints equations are very simple, and the design variables $\mathbf{b}$ appear explicitly in Φ. The constraint equations (10.54) differ from the ones considered in previous chapters because the parameters in $\mathbf{b}$ are not constant as before. They are true variables, because one is in the process of finding their optimum values.

The whole set of constraints for the design point i (geometric and functional) can be written as

$$\Phi^i(\mathbf{q}^i, \mathbf{b}) = 0 \qquad i = 1, 2, ..., N \tag{10.55}$$

The *objective function* can now be introduced. Point 3 of the four-bar example should go exactly through the design points P_i. If it is not possible, a four-bar mechanism should be obtained whose dimensions guarantee that the error in getting these design points is minimum in some sense. Since exact solutions for the design problem may not exist, one must look for the optimal solution in the least square sense. One can define an objective function of the form:

$$\Psi(\mathbf{q}^1, \mathbf{q}^2, ..., \mathbf{q}^N, \mathbf{b}) = \frac{1}{2} \sum_{i=1}^{N} \Phi^{iT}(\mathbf{q}^i, \mathbf{b}) \, \Phi^i(\mathbf{q}^i, \mathbf{b}) \tag{10.56}$$

or in a more compact form,

$$\Psi(\bar{\mathbf{q}}, \mathbf{b}) = \frac{1}{2} \, \bar{\Phi}(\bar{\mathbf{q}}, \mathbf{b})^T \bar{\Phi}(\bar{\mathbf{q}}, \mathbf{b}) \tag{10.57}$$

where $\bar{\mathbf{q}}$ is the vector $\bar{\mathbf{q}}^T = \{\mathbf{q}^{1T}, \mathbf{q}^{2T}, ..., \mathbf{q}^{NT}\}$ and $\bar{\Phi}$ is a vector that contains all the geometry and functional constraints. The optimum design problem consists in minimizing the objective function Ψ with respect to vectors $\bar{\mathbf{q}}$ and $\mathbf{b}$, that is,

$$\min_{\overline{q},\,b} \Psi(\overline{q},b) = \frac{1}{2}\,\Phi(\overline{q},b)^T\,\Phi(\overline{q},b) \tag{10.58}$$

Differentiating with respect to $\overline{q}$ and b and equating to zero, the following system of nonlinear equations is obtained:

$$J(\overline{q},b)\,\overline{\Phi}(\overline{q},b) = 0 \tag{10.59}$$

where

$$J(\overline{q},b) = \begin{bmatrix} \dfrac{\partial\overline{\Phi}(\overline{q},b)}{\partial\overline{q}} \\[2ex] \dfrac{\partial\overline{\Phi}(\overline{q},b)}{\partial b} \end{bmatrix} \tag{10.60}$$

The system of nonlinear equations (10.59) may now be solved by a quasi-Newton method. Expanding $\overline{\Phi}(\overline{q},b)$ in Taylor's series,

$$\overline{\Phi}(\overline{q}+\Delta\overline{q},\ b+\Delta b) = \overline{\Phi}(\overline{q},b) + J^T\left\{\begin{matrix}\Delta\overline{q}\\ \Delta b\end{matrix}\right\} + \cdots \tag{10.61}$$

and substituting in equation (10.59), on can obtain

$$J(\overline{q},b)\,\overline{\Phi}(\overline{q},b) + J(\overline{q},b)\,J^T(\overline{q},b)\left\{\begin{matrix}\Delta\overline{q}\\ \Delta b\end{matrix}\right\} = 0 \tag{10.62}$$

from which the following iterative expression can be obtained:

$$\left\{\begin{matrix}\overline{q}\\ b\end{matrix}\right\}_{k+1} = \left\{\begin{matrix}\overline{q}\\ b\end{matrix}\right\}_{k} - \left[J(\overline{q},b)\,J^T(\overline{q},b)\right]_k^{-1} J(\overline{q},b)_k\,\overline{\Phi}(\overline{q},b)_k \tag{10.63}$$

This method is sufficiently simple and general to be applied to nearly any system topology such as planar and three-dimensional, open- and closed-chains, with any number and kind of joints and bodies. This method can accommodate any kind of functional constraints, even a mixed set.

Example 10.4

The complete set of constraint equations will now be found for the four-bar mechanism of Example 10.1 considering five design points. Particularizing equations (iii)-(vii) and (viii)-(ix) for the generic design point P_i:

$$(x_1^i - x_A)^2 + (y_1^i - y_A)^2 - d_{1A}^2 = 0 \tag{i}$$

$$(x_1^i - x_2^i)^2 + (y_1^i - y_2^i)^2 - d_{12}^2 = 0 \tag{ii}$$

$$(x_2^i - x_B)^2 + (y_2^i - y_B)^2 - d_{2B}^2 = 0 \tag{iii}$$

$$-(1 + \overline{x}_3/d_{12})\,x_1^i + (\overline{y}_3/d_{12})\,y_1^i + (\overline{x}_3/d_{12})\,x_2^i - (\overline{y}_3/d_{12})\,y_2^i + x_3^i = 0 \tag{iv}$$

$$-(\overline{y}_3/d_{12})\,x_1^i - (1 + \overline{x}_3/d_{12})\,y_1^i + (\overline{y}_3/d_{12})\,x_2^i + (\overline{x}_3/d_{12})\,y_2^i + y_3^i = 0 \tag{v}$$

$$x_3^i - x_{P_i} = 0 \qquad\qquad\text{(vi)}$$

$$y_3^i - y_{P_i} = 0 \qquad\qquad\text{(vii)}$$

for $i = 1, 2, ..., 5$. There are 35 constraint equations. The number of unknowns is also 35: five values of the six-element dependent coordinates vector q^i plus the five elements of the design variables vector b. With five design points, it is possible to get a mechanism that exactly satisfies the functional constraints. If there are more than five design points, only an optimal solution in the least square sense can be obtained.

10.5 Sensitivity Analysis and Optimization

The subject of the previous sections in this book has been the study of methods for analysis that constitute the basis simulation programs. These simulate the behavior of a multibody system once all of its geometric and dynamic characteristic have been defined. The analysis programs are certainly very useful. At the present time they are the only general purpose tools available for the largest number of applications. Currently, *design programs* are becoming more important. These programs will not only perform system analyses but also modify automatically its parameters so as to obtain an optimal behavior. An intermediate step between the analysis and optimal design programs are the *sensitivity analyses* which determine the variation of the response of the system in relation to each of the *design variables*.

The optimal design of a multibody system is started by defining an *objective function* which will optimize the system performance. The solution to the problem will be the configuration that minimizes the objective function in relation to the design variables. The problem may or may not have *design constraint equations*, that is, equalities or inequalities that should comply with certain specific functions of the design variables. The constraint equations mathematically introduce certain physical design limitations into the problem. For example, there cannot be any elements with negative mass or length, geometric limitations in the workspace, and so forth. The objective functions are defined depending on the application. Since the dynamics is a process that takes place over a period of time, the objective function is often defined as the integral of a specific function over a period of time or as a series of conditions that the multibody system must satisfy within certain intervals of time or at specific moments. The objective function depends on the design variables not only directly but also through the results of the dynamic analysis such as: positions, velocities, accelerations, stresses, and reactions.

Several optimization methods that minimize the objective function have been proposed (Gottfried and Weisman (1973), Haug and Arora (1979), Reklaitis et al. (1983)) . Almost all of the methods are based on the knowledge of the derivatives of the objective function with respect to the design variables. The determination of these derivatives is known as *sensitivity analysis* and is the first phase in the optimization process which may also be considered separately. Sensitivity analysis, which determines the tendencies of the objective function with respect to design variations, is also very useful in a non-automatic interactive design process. The derivatives of the objective function with respect to the design variables are calculated by means of the derivatives of the dynamic response with respect to positions, velocities, accelerations, stresses, and reactions.

The optimal design of multibody systems is undoubtedly an important problem, although it has not yet attained the level required for a general commercial implementation. Still, the calculation times are excessively high for even cases of average complexity, and there are important formulation and implementation problems that remain to be solved. For the time being, an interactive design based on analysis methods and sensitivity studies seems to be the most suitable alternative for general use. In this sensitivity analysis, the general ideas of Chang and Nikravesh (1985) will be followed.

First the sensitivities of the kinematic equations will be considered with respect to variations of the design variables $\delta\mathbf{b}$. As introduced in Chapter 2, the constraint equations are expressed as

$$\Phi \equiv \Phi(\mathbf{q}, t) = 0 \tag{10.64}$$

By simple differentiation, the velocity and acceleration constraint conditions can be found to be:

$$\dot{\Phi} \equiv \Phi_\mathbf{q}\,\dot{\mathbf{q}} + \Phi_t = 0 \tag{10.65}$$

$$\ddot{\Phi} \equiv \Phi_\mathbf{q}\,\ddot{\mathbf{q}} - \gamma = 0 \tag{10.66}$$

$$\gamma = -\,(\Phi_\mathbf{q}\dot{\mathbf{q}})_\mathbf{q} - 2\,\Phi_{\mathbf{q}t}\,\dot{\mathbf{q}} - \Phi_{tt} \tag{10.67}$$

The vector of dependent coordinates $\mathbf{q}$ will now be considered as a function of time t and also of the design variables $\mathbf{b}$. If a variation $\delta\mathbf{b}$ of the design variables is allowed, the following series expansion can be written:

$$\mathbf{q}(t, \mathbf{b} + \delta\mathbf{b}) = \mathbf{q}(t, \mathbf{b}) + \mathbf{q}_\mathbf{b}\,\delta\mathbf{b} + \ldots \tag{10.68}$$

and as a consequence the first variation of the positions, velocities, and accelerations become:

$$\delta\mathbf{q} = \mathbf{q}_\mathbf{b}\,\delta\mathbf{b} \tag{10.69}$$

$$\delta\dot{\mathbf{q}} = \dot{\mathbf{q}}_\mathbf{b}\,\delta\mathbf{b} \tag{10.70}$$

$$\delta\ddot{\mathbf{q}} = \ddot{\mathbf{q}}_\mathbf{b}\,\delta\mathbf{b} \tag{10.71}$$

Taking into account equation (10.69), the first variation of the constraint equations become

$$\Phi_q \, q_b \, \delta b + \Phi_b \, \delta b = (\Phi_q \, q_b + \Phi_b) \, \delta b = 0 \qquad (10.72)$$

Since the variation δb is arbitrary, it can be concluded that

$$\Phi_q \, q_b + \Phi_b = 0 \qquad (10.73)$$

Expression (10.73) is the equation that models the variation of the position vector q with respect to the design variables b. The term q_b represents a matrix of partial derivatives. Similarly, the first variation of the velocity and acceleration constraint equations becomes:

$$\Phi_q \, \dot{q}_b + \Phi_{qq} \, q_b \, \dot{q} + \Phi_{qb} \, \dot{q} + \Phi_{tq} \, q_b + \Phi_{tb} = 0 \qquad (10.74)$$

$$\begin{aligned}
&\Phi_q \, \ddot{q}_b + \Phi_{qq} \, q_b \, \ddot{q} + \Phi_{qb} \, \ddot{q} + \dot{\Phi}_{qq} \, q_b \, \dot{q} + \\
&+ \dot{\Phi}_{qb} \, \dot{q} + \dot{\Phi}_q \, \dot{q}_b + \dot{\Phi}_{tq} \, q_b + \Phi_{tb} = 0
\end{aligned} \qquad (10.75)$$

Equations (10.73) to (10.75) allow one to compute the sensitivities of the kinematic constraint equations. Turn now to the variation of the equations of motion. To this end one can consider the Lagrange multiplier version of the equations of motion (equation (5.10)):

$$M \, \ddot{q} + \Phi_q^T \, \lambda = Q \qquad (10.76)$$

Taking into account that the first variation of the Lagrange multipliers is

$$\delta \lambda = \lambda_b \, \delta b \qquad (10.77)$$

it becomes easy to see that the variation of (10.76) with respect to b is

$$\begin{aligned}
&M \, \ddot{q}_b + \Phi_q^T \, \lambda_b = \\
&= Q_b + Q_q \, q_b + Q_{\dot{q}} \, \dot{q}_b - M_b \, \ddot{q} - \Phi_{qq}^T \, q_b \, \lambda - \Phi_{qb}^T \, \lambda
\end{aligned} \qquad (10.78)$$

Equations (10.75) and (10.78) can be expressed jointly in the following form:

$$\begin{bmatrix} M & \Phi_q^T \\ \Phi_q & 0 \end{bmatrix} \begin{Bmatrix} \ddot{q}_b \\ \lambda_b \end{Bmatrix} = \begin{Bmatrix} \overline{Q}_b \\ \overline{c}_b \end{Bmatrix} \qquad (10.79)$$

where

$$\overline{Q}_b = Q_b + Q_q \, q_b + Q_{\dot{q}} \, \dot{q}_b - M_b \, \ddot{q} - \Phi_{qq}^T \, q_b \, \lambda - \Phi_{qb}^T \, \lambda \qquad (10.80)$$

$$\begin{aligned}
\overline{c}_b = &- \Phi_{qq} \, q_b \, \ddot{q} - \Phi_{qb} \, \ddot{q} - \dot{\Phi}_{qq} \, q_b \, \dot{q} - \\
&- \dot{\Phi}_{qb} \, \dot{q} - \dot{\Phi}_q \, \dot{q}_b - \dot{\Phi}_{tq} \, q_b - \Phi_{tb}
\end{aligned} \qquad (10.81)$$

Equation (10.79) closely resembles the general form of the equations of motion (5.10):

$$\begin{bmatrix} \mathbf{M} & \boldsymbol{\Phi}_{\mathbf{q}}^{\mathrm{T}} \\ \boldsymbol{\Phi}_{\mathbf{q}} & \mathbf{0} \end{bmatrix} \begin{Bmatrix} \ddot{\mathbf{q}} \\ \boldsymbol{\lambda} \end{Bmatrix} = \begin{Bmatrix} \mathbf{Q} \\ -\dot{\boldsymbol{\Phi}}_t - \dot{\boldsymbol{\Phi}}_{\mathbf{q}}\,\dot{\mathbf{q}} \end{Bmatrix} \tag{10.82}$$

The set of linear equations (10.79) and (10.82) have a common matrix of co-efficients, and therefore only one Gauss decomposition is necessary. Equation (10.82) must be solved first to obtain $\mathbf{q}$ and λ that appear in the RHS of (10.79). This last equation has as many unknowns as design variables. The position and velocity sensitivities may be obtained from those of the accelerations through numerical integration. The possible instabilities arising from the integration process may be eliminated by solving the system of mixed differential and algebraic equations (See Chapter 7) or by using the Baumgarte stabilization (See Chapter 5).

A third way to avoid the instabilities is through the use of independent variables, as explained in Section 5.2. It is possible to define a matrix of independent sensitivities that according to equations (5.64) and (5.61) must satisfy the following relation:

$$\dot{\mathbf{z}}_b = \mathbf{B}\,\dot{\mathbf{q}}_b \tag{10.83}$$

where

$$\dot{\mathbf{q}}_b = \begin{bmatrix} \boldsymbol{\Phi}_{\mathbf{q}} \\ \mathbf{B} \end{bmatrix}^{-1} \begin{Bmatrix} \gamma_b \\ \dot{\mathbf{z}}_b \end{Bmatrix} \equiv \begin{bmatrix} \mathbf{S} & \mathbf{R} \end{bmatrix} \begin{Bmatrix} \gamma_b \\ \dot{\mathbf{z}}_b \end{Bmatrix} = \mathbf{S}\,\gamma_b + \mathbf{R}\,\dot{\mathbf{z}}_b \tag{10.84}$$

$$\gamma_b = -\boldsymbol{\Phi}_{\mathbf{qq}}\,\mathbf{q}_b\,\dot{\mathbf{q}} - \boldsymbol{\Phi}_{\mathbf{qb}}\,\dot{\mathbf{q}} - \boldsymbol{\Phi}_{t\mathbf{q}}\,\mathbf{q}_b - \boldsymbol{\Phi}_{tb} \tag{10.85}$$

Taking into account equation (10.81), the sensitivities for accelerations become:

$$\ddot{\mathbf{z}}_b = \mathbf{B}\,\ddot{\mathbf{q}}_b \tag{10.86}$$

$$\ddot{\mathbf{q}}_b = \begin{bmatrix} \boldsymbol{\Phi}_{\mathbf{q}} \\ \mathbf{B} \end{bmatrix}^{-1} \begin{Bmatrix} \bar{\mathbf{c}}_b \\ \ddot{\mathbf{z}}_b \end{Bmatrix} \equiv \begin{bmatrix} \mathbf{S} & \mathbf{R} \end{bmatrix} \begin{Bmatrix} \bar{\mathbf{c}}_b \\ \ddot{\mathbf{z}}_b \end{Bmatrix} = \mathbf{S}\,\bar{\mathbf{c}}_b + \mathbf{R}\,\ddot{\mathbf{z}}_b \tag{10.87}$$

Introducing (10.87) into (10.79), pre-multiplying the result by $\mathbf{R}^{\mathrm{T}}$, and taking into account that the columns of $\mathbf{R}$ are orthogonal to the rows of $\boldsymbol{\Phi}_{\mathbf{q}}$, one can get

$$\mathbf{R}^{\mathrm{T}}\mathbf{M}\,\mathbf{R}\,\ddot{\mathbf{z}}_b = \mathbf{R}^{\mathrm{T}}(\overline{\mathbf{Q}}_b - \mathbf{M}\,\mathbf{S}\,\bar{\mathbf{c}}_b) \tag{10.88}$$

All the improved dynamic techniques described in Chapter 8 can be applied to the sensitivity equations (10.79) and (10.88). The most efficient formulations for the dynamic analysis may also be the basis of a very efficient computation of the sensitivities of the position, velocity, and acceleration vectors.

10.6 Singular Positions

The methods and results presented in this section are contributions from Bayo and Avello (1993).

A *singular position* is encountered when the multibody reaches a kinematic configuration in which there is a sudden change in the number of degrees of freedom. For instance, a slider-crank mechanism, as the one shown in Figure 10.21, reaches a singular position when the two links are in vertical position. In that configuration, both links are coincident, and the mechanism has not one but two degrees of freedom. These two degrees of freedom correspond to the two possible motions (bifurcations) that the mechanism can undergo (illustrated in Figure 10.22). Figure 10.22a shows the first possible motion that corresponds to a slider-crank mechanism. Figure 10.22b shows the second motion corresponding to a rotating bar (in fact two coincident rotating bars). Here, a singular position implies a *bifurcation* point, in which the mechanism can theoretically undergo different paths.

The existence of a singular position with both the classical Lagrange multipliers approach and the use of independent coordinates is invariably detected when the Jacobian matrix of the constraints becomes rank-deficient. These formulations are based on the decomposition of the Jacobian matrix. Since its rank suddenly falls at a singular position, the decomposition fails and therefore no solution can be found. The simulation then may crash not because of the physics of the problem but because of the inability of the dynamic formulation to overcome the sudden change in the rank of the Jacobian matrix.

Equation (5.10) is the key equation for the solution of the dynamics using the Lagrange multipliers method. This equation is again written as

$$
\begin{bmatrix} \mathbf{M} & \mathbf{\Phi}_q^T \\ \mathbf{\Phi}_q & \mathbf{0} \end{bmatrix} \begin{Bmatrix} \ddot{\mathbf{q}} \\ \lambda \end{Bmatrix} = \begin{Bmatrix} \mathbf{Q} \\ \mathbf{c} \end{Bmatrix}
\tag{10.89}
$$

Assuming that all the constraints are independent, that is if $(m=n-f)$, the rank of the leading matrix in this equations is $(n+m)$. Since the Jacobian matrix becomes rank-deficient in singular positions, this matrix becomes singular. The accelerations cannot be computed, and the dynamic simulation may either crash or introduce large errors at this point. Equation (5.67) is the alternate key equation for the independent coordinate method, which can be rewritten again as

$$
\mathbf{R}^T \mathbf{M} \mathbf{R}\, \ddot{\mathbf{z}} = \mathbf{R}^T \mathbf{Q} - \mathbf{R}^T \mathbf{M}\, \mathbf{S}\, \mathbf{c}
\tag{10.90}
$$

When a singular position is reached, the computation of the matrix $\mathbf{R}$ fails, because the Jacobian matrix becomes rank-deficient. Thus, it's decomposition can no longer be carried out.

If a singular position is not exactly reached, the leading matrix of both methods will not be strictly singular but quasi-singular with a very high condition number. If this situation is not correctly tracked, the integration and round-off errors will be amplified, and the resulting solutions may be totally erroneous. A partial solution to the problem of singular positions was provided by Park and Haug (1988) with a method that detects the ill-conditioning of the Jacobian matrix so that the integrator can step over it. Ider and Amirouche (1989) and Amirouche and Chin-Wei (1990) proposed a *regularization method* to cope with singularities. Roughly, the idea consists in substituting the linearly dependent rows of the Jacobian matrix by their derivatives, which turns the Jacobian matrix non-singular.

The penalty-augmented Lagrangian formulation provides a nice solution to this problem. A more detailed exposition of the following analysis is presented by Bayo and Avello (1993). The key equation for the augmented Lagrangian formulation (5.48) can be expressed as

$$(M + \Phi_q^T \alpha \Phi_q) \, \ddot{q} =$$
$$= Q - \Phi_q^T \alpha (\dot{\Phi}_q \dot{q} + \dot{\Phi}_t + 2 \Omega \mu \dot{\Phi} + \Omega^2 \Phi) - \Phi_q^T \lambda^* \quad (10.91)$$

It is important to note that there is a very important difference between equation (10.91) and those corresponding to the Lagrange multiplier and independent coordinates (10.89) and (10.90), respectively. The leading matrices of equations (10.89) and (10.90) become singular in singular positions. Although the mass matrix M is generally positive semi-definite, it is always strictly positive definite in the nullspace of the Jacobian matrix. A look at equation (10.91) reveals that its leading matrix $(M + \Phi_q^T \alpha \Phi_q)$ is always positive definite. It can always be factored, even in singular positions.

Remark. It is important at this stage to emphasize the difference between a singular Jacobian matrix and a singular position. While a singular position always implies a singular Jacobian matrix, the converse is not always true. A Jacobian matrix can become singular when redundant constraints are present, a dead-lock position is reached, or when the coordinate partitioning between dependent and independent coordinates is not made properly or has not been updated for a while. Contrary to the case of a singular position, these singularities can be avoided and the simulation may proceed smoothly. The difference between avoidable and unavoidable singular Jacobian matrices can be better understood by partitioning the columns of the Jacobian matrix Φ_q into two submatrices Φ_q^d and Φ_q^i, corresponding to the dependent and independent coordinates, respectively. This partition is made so that Φ_q^d has full row rank. When Φ_q^d is rank-deficient but Φ_q has full row rank, the singularity is avoidable, since the full rank of Φ_q^d can be recovered by a new suitable choice of independent coordinates. However, when Φ_q loses rank, the singularity is unavoidable. It is a physical singularity, and the rank of Φ_q^d cannot be modified by any choice of the independent coordinates.

Example 10.5

To better understand the application of the augmented Lagrangian formulation in singular and non-singular positions, consider the slider-crank mechanism shown

Table 10.1. Convergence rate with $\alpha=10^4$.

Iteration #	Error
1	6.5792 10^{-4}
2	4.3705 10^{-8}
3	2.9206 10^{-12}
4	1.6107 10^{-14}

in Figure 10.21. Both links are of length $l=1$ m with a uniformly distributed mass of $m=1$ Kg. Take as position coordinates $\mathbf{q}$, the x and y coordinates of the crank end, and the x coordinate of the slider. Thus $\mathbf{q}^T=\{x_1, y_1, x_2\}$. One can consider the gravity force with a value $g=-9.81$ m/s^2 acting in the Y direction. The (3×3) mass matrix corresponding to these variables is

$$\mathbf{M} = \frac{1}{6}\begin{bmatrix} 4 & 0 & 1 \\ 0 & 4 & 0 \\ 1 & 0 & 2 \end{bmatrix}$$

This mechanism has one degree of freedom only. Therefore there are two geometrical constraints that correspond to the constant distance conditions:

$$\Phi = \left\{ \frac{1}{2}\left(x_1^2+y_1^2- 1\right) \quad \frac{1}{2}\left[\left(x_2-x_1\right)^2+y_1^2- 1\right]\right\}$$

When the crank forms an angle of $\pi/2$ radians with the horizontal, the coupler is coincident with the crank. The crank axis is also coincident with the slider. In this position the mechanism has two instantaneous degrees of freedom, since it can undergo either the motion of a slider-crank or the motion of two superimposed rotating bars. The augmented Lagrangian formulation (Algorithm 5-3) can now be applied for the instantaneous solution of the accelerations for both a nonsingular position and a singular position.

Nonsingular Position. Consider the mechanism in an initial position in which the crank forms an angle of $\pi/4$ with the horizontal and in which the slider has a velocity $\dot{x}_2 = - 2$ m/s. The exact acceleration has been computed first with the classical Lagrange multiplier method of equation (10.89). The accelerations have been calculated with the Algorithm 5-3, using equation (10.91) iteratively with a value $\alpha=10^4$. Table 10.1 shows the norm of the difference between the exact acceleration and the one obtained with augmented Lagrangian formulation.

Table 10.1 also shows that the convergence rate of the iterative algorithm is considerably fast. A higher penalty value gives a faster convergence rate but a lower precision. For instance, a penalty value of $\alpha=10^7$ yields an error on the order of 10^{-12} in one iteration. Further iterations are unable to improve the solution, since some precision is lost in floating-point arithmetic operations between numbers with exponents of significantly different values.

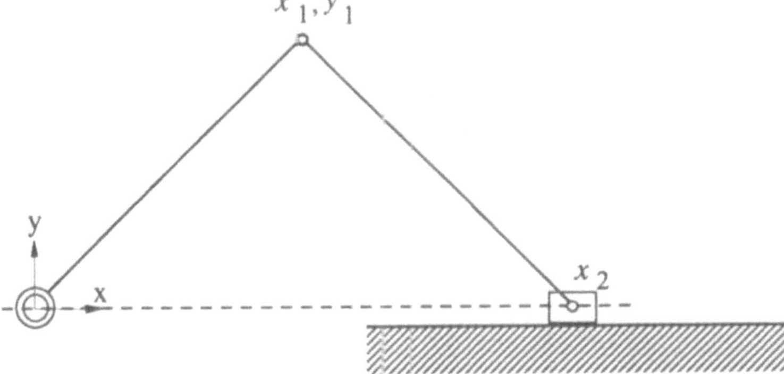

Figure 10.21. Slider-crank mechanism. Reprinted by permission of Kluwer Academic Publishers.

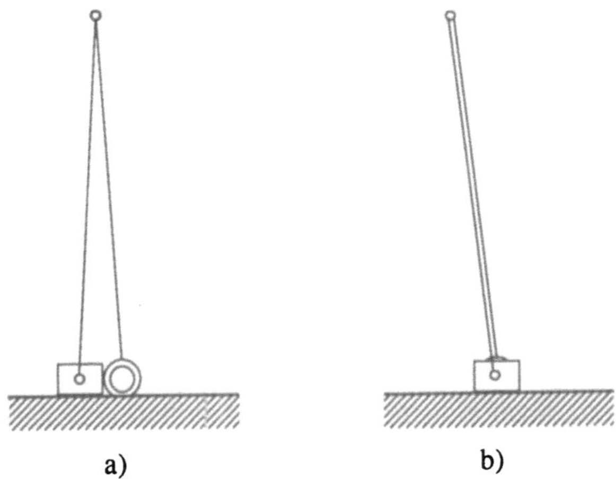

Figure 10.22. Possible bifurcations: (a) slider-crank mechanism motion, (b) rotating bar motion. Reprinted by permission of Kluwer Academic Publishers.

Singular Position. Consider the crank in a vertical position, forming an angle of $\pi/2$ radians with the horizontal. As was done in the nonsingular case, a slider velocity takes the value $\dot{x}_2 = -2$ m/s. Since the mechanism is in a singular position with two instantaneous degrees of freedom, the horizontal velocity of the crank end also has to be specified. It can be easily shown that, theoretically, the crank end can have any velocity value $\dot{x}_1 = v$. However, the slider-crank motion must satisfy the condition $x_1 = x_2/2$ over all its motion. Therefore the velocity $\dot{x}_1 = -1$ seems the obvious choice. In this example, the choice for the crank-end velocity is being made explicitly, but during a dynamic simulation, the numerical integrator will provide this value. Since the integrator assumes a continuous variation of the variables, this condition will be automatically guaranteed.

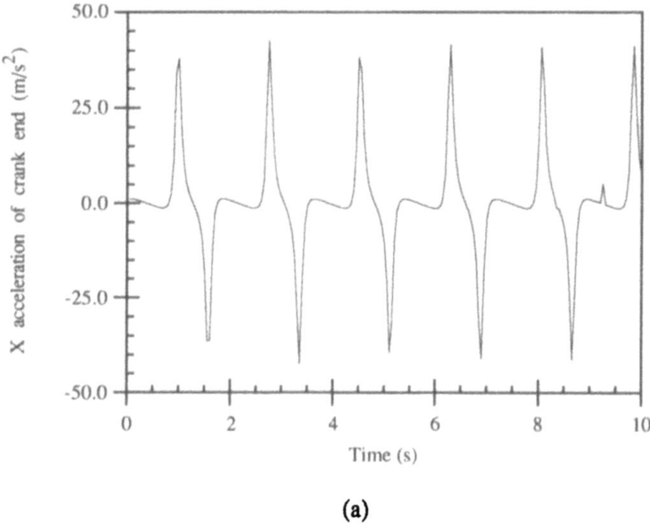

(a)

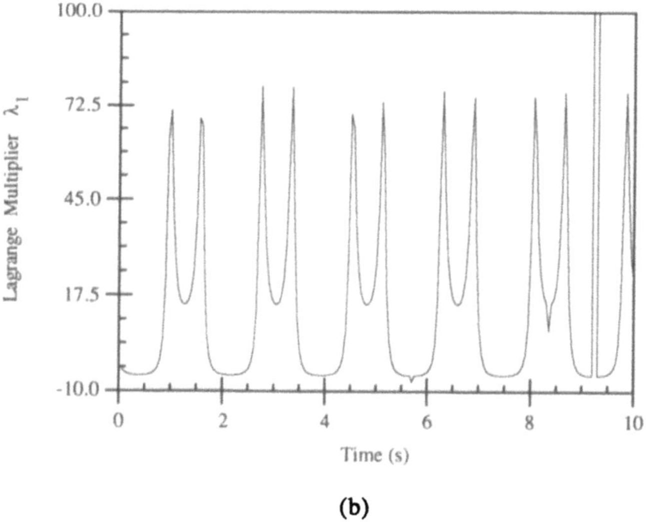

(b)

Figure 10.23. Acceleration-based augmented Lagrangian formulation: (a) acceleration of the crank-end; (b) Lagrange multiplier. Reprinted by permission of Kluwer Academic Publishers.

In this case, the exact acceleration value cannot be computed with equation (10.89), because the leading matrix is singular. However, the application of equation (10.91) with a value of $\alpha=10^4$ leads to

$$\begin{bmatrix} \dfrac{2}{3} & 0 & \dfrac{1}{6} \\[2mm] 0 & 2\left(10^4+\dfrac{1}{3}\right) & 0 \\[2mm] \dfrac{1}{6} & 0 & \dfrac{1}{3} \end{bmatrix} \begin{Bmatrix} \ddot{x}_1 \\ \ddot{y}_1 \\ \ddot{x}_2 \end{Bmatrix} = \begin{Bmatrix} 0 \\ -(2\cdot10^4+9.81) \\ 0 \end{Bmatrix}$$

which can be inverted and leads to the solution (0, –1.000473825436244272, 0). After three iterations, the result is (0,–1,0) which is accurate to 14 digits.

This example clearly and simply illustrates that the augmented Lagrangian formulation works in singular positions; whereas the classical formulations such as the Lagrange multipliers method or the reduction to independent coordinates fail.

An interesting finding is the augmented Lagrangian formulation in its canonical form defined by equation (5.119):

$$\left(\mathbf{M} + \boldsymbol{\Phi}_\mathbf{q}^\mathrm{T}\,\boldsymbol{\alpha}\,\boldsymbol{\Phi}_\mathbf{q}\right)\dot{\mathbf{q}} =$$
$$= \mathbf{p} - \boldsymbol{\Phi}_\mathbf{q}^\mathrm{T}\,\boldsymbol{\alpha}\left(\boldsymbol{\Phi}_t + 2\mu\,\Omega\,\boldsymbol{\Phi} + \Omega^2\int_{t_o}^{t}\boldsymbol{\Phi}\,d\tau\right) - \boldsymbol{\Phi}_\mathbf{q}^\mathrm{T}\,\boldsymbol{\sigma}^* \tag{10.92}$$

This formulation is more effective and reliable than its acceleration-based counterpart (equation 10.91) under singular positions. In fact, while both formulations require the triangularization of the same leading matrix for each function evaluation, there are advantages in the use of the canonical as compared to the acceleration. The kinematic constraint conditions are differentiated only once with the canonical procedure but twice in the acceleration-based formulation. This will lead to lesser violations of the constraints. The next example illustrates how this factor may become detrimental for the acceleration-based formulation under repetitive singular positions; whereas the canonical approach leads to a much better performance.

Example 10.6

Dynamic simulation of the slider-crank mechanism. Consider again the same slider-crank mechanism in an initial position so that the crank forms an angle of $\pi/4$ radians with the X axis and that the slider's velocity is $\dot{x}_2 = -4$ m/s. By integrating the equations of motion for a total of 10 seconds, a dynamic simulation is performed using a conditionally stable variable step and order integrator based on predictor-corrector multistep formulae (Shampine and Gordon (1975)). The error tolerance is set to 10^{-5} and penalty parameters $\alpha=10^7$, $\Omega=10$, and $\mu=1$ are chosen. During the simulation, the mechanism goes through the singular position eleven times following a periodical response.

First, the simulation was carried out with the acceleration-based Algorithm 5-3 (equation (10.91)). Figure 10.24a shows the X acceleration of the crank-end over the time period of 10 seconds. Figure 10.24b shows the value of the Lagrange multiplier λ_1 corresponding to the constant distance constraint condition between the crank axis and the crank end. A very interesting point can be brought up from

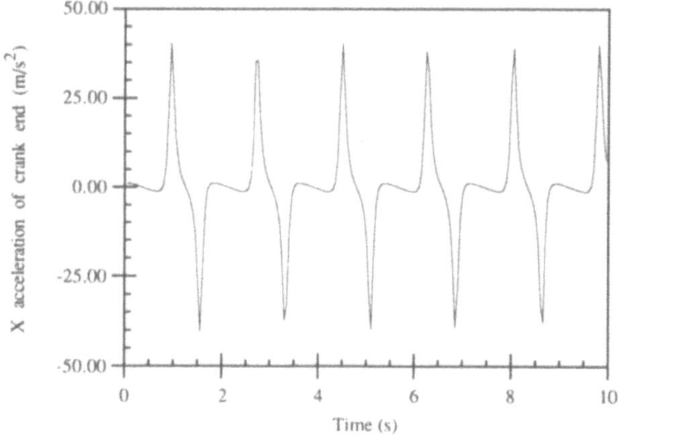

(a)

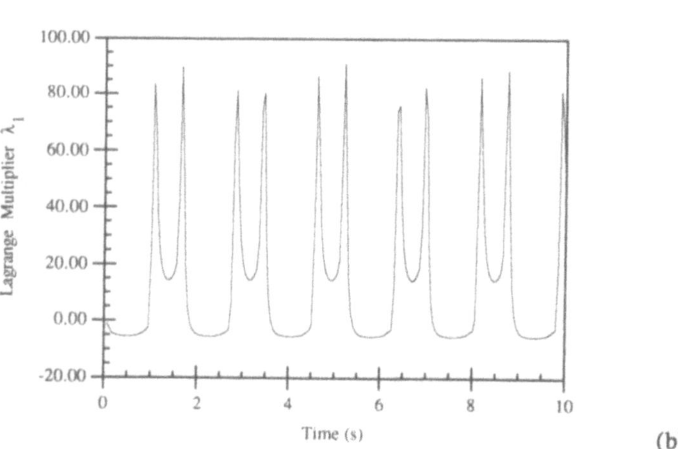

(b)

Figure 10.24. Canonical-based augmented Lagrangian formulation: (a) acceleration of the crank-end; (b) Lagrange multipler. Reprinted by permission of Kluwer Academic Publishers.

these figures. The value of the acceleration of the crank-end and λ_l presents spikes around t=9.25s. The cause of this phenomenon is the violation of the constraints around the singular position, due to the combination of the errors produced by the numerical integration routine and by the round-off errors produced by augmented Lagrangian procedure. These errors are more critical in the acceleration-based algorithm because the constraint equations are differentiated twice.

The simulation is repeated, using the Algorithm 5-8 (canonical formulation) with the same error tolerance and values for the penalty parameters. This time, the values of λ_I and the crank-end acceleration, illustrated in Figures 24a and 24b, no longer show the spikes shown in Figure 10.23. The accumulation of small constraint violations in the neighborhood of the singular position is the cause for the sudden peaks and jumps in the constraint forces and accelerations produced in Figure 10.23. The better results obtained with the canonical formulation are due to its better constraint stabilization properties that make this algorithm better suited for problems in which the constraint violation must be effectively stabilized, as is the case of the analysis of singular positions.

The way the acceleration-based augmented Lagrangian formulation may be improved, if it is to be used in repetitive singular positions, is by setting tighter error tolerances and raising the value of the parameter Ω. However, this will introduce numerical stiffness in the problem and will increase the computational effort. In this example, the value of $\omega = 20$ solves the problem satisfactorily at the cost of a lengthier and more costly integration.

References

Alvarez, G. and Jiménez, J.M., "A Simple and General Method for Kinematic Synthesis of Spatial Mechanisms", to be published, (1993)

Amirouche, F.M.L. and Chin-Wei, T., "Regularization and Stability of the Constraints in the Dynamics of Multibody Systems", *Nonlinear Dynamics*, Vol. 1, pp. 459-475, (1990).

Angeles, J., *Spatial Kinematic Chains*, Springer-Verlag, (1982).

Avilés, R., Ajuria, M.B. and García de Jalón, J., "A Fairly General Method for Optimum Synthesis of Planar Mechanisms", *Mechanism and Machine Theory*, Vol. 20, pp. 321-328, (1985).

Bagci, C., "Dynamic Motion Analysis of Plane Mechanisms with Coulomb and Viscous Damping Via the Joint Force Analysis", *ASME Journal of Engineering for Industry*, pp. 551-559, (1975).

Bayo, E. and Avello, A., "Singularity Free Augmented Lagrangian Algorithms for Constraint Multibody Dynamics", to appear in the *Journal of Nonlinear Dynamics*, (1993).

Chang, C.O. and Nikravesh, P.E., "Optimal Design of Mechanical Systems with Constraint Violation Stabilization Method", *ASME Journal of Mechanisms, Transmissions, and Automation in Design*, Vol. 107, pp. 493-498, (1985).

Dubowsky, S., Deck, J.F., and Costello, H., "The Dynamic Modeling of Flexible Spatial Machine Systems with Clearance Connections", *ASME Journal of Mechanisms, Transmissions, and Automation in Design*, Vol. 109, pp. 87-94, (1987).

Erdman, A.G. and Sandor, G.N., *Mechanism Design: Analysis and Synthesis*, Volumes 1 and 2, Prentice-Hall, (1984).

Gottfried, B.S. and Weisman, J., *Introduction to Optimization Theory*, Prentice-Hall, (1973).

Haug, E.J. and Arora, J.S., *Applied Optimal Design. Mechanical and Structural Systems*, Wiley, (1979).

Haug, E.J., Wu, S.C., and Yang, S.M., "Dynamics of Mechanical Systems with Coulomb Friction, Stiction, Impact, and Constraint Addition-Deletion. I-Theory, II-Planar Systems, and III-Spatial Systems", *Mechanism and Machine Theory*, Vol. 21, pp. 401-425, (1986).

Ider, S.K. and Amirouche, F.M.L., "Numerical Stability of the Constraints Near Singular Positions in the Dynamics of Multibody Systems", *Computers and Structures*, Vol. 33, pp. 129-137, (1989).

Park, T. and Haug, E.J., "Ill-Conditioned Equations in Kinematics and Dynamics of Machines", *Int. Journal for Numerical Methods in Engineering*, Vol. 26, pp. 217-230, (1988).

Reklaitis, G.V., Ravindran, A., and Ragsdell, K.M., *Engineering Optimization. Methods and Applications*, Wiley, (1983).

Rooney, G.T. and Deravi, P., "Coulomb Friction in Mechanism Sliding Joints", *Mechanism and Machine Theory*, Vol. 17, pp. 207-211, (1982).

Shampine, L.F. and Gordon, M., *Computer Solution of Ordinary Differential Equations: the Initial Value Problem*, W.J. Freeman, San Francisco, (1975).

Suh, C.H. and Radcliffe, C.W., *Kinematics and Mechanism Design*, Wiley, (1978).

Threlfall, D.C., "The Inclusion of Coulomb Friction in Mechanisms Programs with Particular Reference to DRAM", *Mechanism and Machine Theory*, Vol. 13, pp. 475-483, (1978).

11
Forward Dynamics of Flexible Multibody Systems

So far, several approaches to the solution of the kinematics and dynamics of multibody systems have been presented. It has been assumed in these approaches that all the bodies satisfy the rigid body condition. A body is assumed to be rigid if any pair of its material points do not present relative displacements. In practice, bodies suffer some degree of deformation; so this assumption does not hold in the strict sense. However, in the majority of the cases the relative displacements are so small that they do not affect the system's behavior. Therefore, they can be neglected without committing an appreciable error.

There are some important cases, however, in which deformation plays an important role. This is the case of lightweight spatial structures and manipulators or high-speed machinery. The dynamics of those systems is influenced by the deformation; thus the formulation of the preceding chapters cannot be applied. The complexity of the equations of motion considering deformation grows considerably. So does its size, since all the variables defining the deformation must also be considered.

In this chapter some of the methods that have been presented in the literature for the dynamics of flexible multibodies will be reviewed. Next a general method based on the moving frame approach will be described with natural coordinates that can be used when the elastic displacements are small. A formulation for beam-like elements based on the large displacement theory will be presented, and expressions for a nonlinear finite element that uses the same kind of Cartesian variables such as coordinates of points and components of unit vectors used in the previous chapters will be developed. Finally, some practical examples will be shown.

11.1 An Overview

In this section a quick overview will be given on some of the methods presented in the literature for the analysis of flexible multibody systems. Some of the work in the field was aimed at developing formulations suitable for particular mechanisms such as the four-bar linkage or the crank and rocker mechanism. These approaches are not reviewed here, but the reader is referred to the papers by Lowen and Jandrasits (1972), Lowen and Chassapis (1986), Erdman and Sandor (1972), and Erdman and Sung (1986).

All the methods currently available may be divided into three main groups: a) the simplified methods based on elasto-dynamics, b) the methods based on defining the deformation with respect to a moving reference frame, and c) the methods based on defining the overall motion plus deformation with respect to an inertial frame.

In the simplified elasto-dynamic method the deformation is considered uncoupled from the rigid body motion which is considered known by means of rigid body dynamics and is called the *nominal motion*. The main assumption is that the nominal motion induces deformations which are considered small, but that the deformations do not affect the nominal motion. This approach originally proposed by Winfrey (1971) was later expanded by Midha et al. (1978) and Sunada and Dubowsky (1981) to include inertial and centrifugal effects in the elastic equations. Naganathan and Soni (1987) proposed, for the case of open-chain flexible manipulators with independent coordinates (no constraint conditions), the use of elasto-dynamics with an iterative procedure that couples the elastic deformations with the nominal motion. For more general applications, the simplified approaches based on elasto-dynamics cannot be accepted since the coupling terms may strongly influence the solution. In such cases, one of the other two families of methods must be used.

The methods in the second group include all the nonlinear coupling terms in the formulation. Two kinds of variables are considered: first, the *rigid body variables*, that express the large nonlinear overall motion and characterize the moving frame of each body; second, the *deformation variables*, that express the state of deformation with respect to the moving frames. Both the relative displacements and the gradient of the displacements are assumed to be small, in order that the linear theory of elasticity holds. Some authors take as deformation variables the nodal variables resulting from a finite element discretization of the flexible body (Song and Haug (1980) and Serna and Bayo (1989)). Since this may lead to a large number of unknowns, one way of reducing the size of the problem consists in assuming that during the motion only a few modes will be excited and in taking the amplitude of such modes as unknowns. Shabana and Wehage (1983) used a popular substructuring technique called *component mode synthesis* (Hurty (1965)) to reduce the number of unknowns in each body. Other ways of selecting the most convenient assumed modes may be found in Craig (1981). Other authors (Book (1980), Kim and Haug (1988 and 1989), Changizi and Shabana (1988)) have developed recursive formulations that are based on the same approach for the definition of the deformation. A major advantage of the moving frame approach is that it makes use of the classical linear finite element theory to introduce either the nodal variables or the assumed mode shapes. Since there are a large number of reliable finite element codes well-known by engineers, this method has a special attractiveness. Some of the limitations of this method have been pointed out by Kane et al. (1987), who showed that the moving frame approach with linear elasticity fails to consider the rotational stiffening effects that appear at very fast speeds of operation and which become important. As pointed out by Simo and Vu-Quoc (1987), second-order strain measures are necessary to capture these centrifugal stiffening effects through the geometric stiffness.

The third group encompasses a series of more recent methods, introduced first by Simo and Vu-Quoc (1986), that are based on the large rotation theory. Its main purpose is to develop nonlinear finite elements to be embedded in the multibody formalism. There is only one kind of variables, which are the global positions and orientations of the nodes referred either to an initial undeformed state (total Lagrangian formulation) or to a previously known state of deformation (updated Lagrangian formulation). These variables define at the same time the large translations, rotations, and deformations of the body. This method allows for the existence of arbitrarily large relative displacements and displacement gradients. However, since elastic constitutive relations are most commonly used, the assumption of small strains is often made. Unlike the moving frame approach, this method incorporates automatically the correct rotational stiffening terms and is well suited to study instabilities and buckling. Its main drawback is that the size of the problem cannot be reduced as in the moving frame approach. Therefore it is usually large. Furthermore, this formulation is limited to flexible bodies that can be modeled using beam and shell elements.

After considering this overview of all the methods available so far for the dynamic analysis of flexible multibodies, we are going to concentrate in this chapter: first, in the formulation of the classical approach of the moving frame with natural coordinates (Section 11.2) and secondly on a new formulation for beam-like elements based on the large displacement theory that is also based on the use of natural coordinates (Section 11.3). Note that both formulations are non-exclusive in the sense that one may be preferred over the other depending on the type of applications. Since both are based on the natural coordinates, they can perfectly coexist with the rigid-body formulation of the previous chapters, in a general purpose simulation package. In those cases in which the geometry is complicated and only small deformations are expected, the moving frame approach with assumed modes will be the best choice. Conversely, with simplified geometry and appearance of nonlinear effects the second approach will be the way to go.

11.2 The Classical Moving Frame Approach

In this section the moving frame method using the *natural coordinates* will be described (Vukasovic et al. (1993)). A complete formulation of the moving frame approach within the setting provided by the reference point coordinates has been described by Shabana (1989)). By means of the classical formulation, we use the natural coordinates of the body (or element) to unequivocally define the

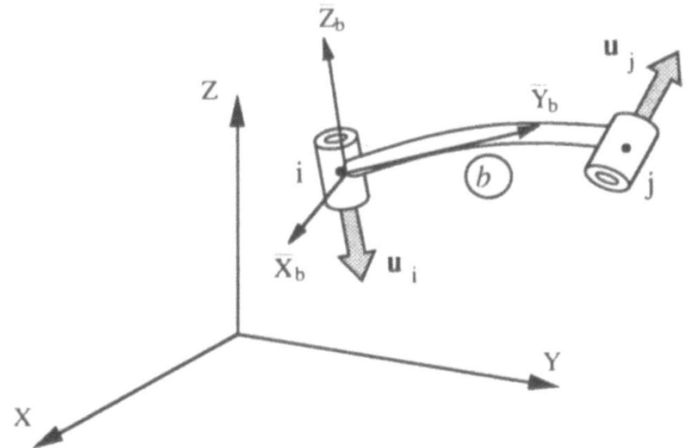

Figure 11.1. Flexible body with natural coordinates and the moving frame.

moving frame that moves with the large overall rigid body motion and to which the elastic deformation variables are referred. The natural coordinates of the body do not include relative translations or rotations and are subjected to the corresponding rigid body constraints. The formulation of the joint constraints is different than in the rigid body, because now points and vectors cannot be shared at the joints and the elastic deformations at those points need to be included.

11.2.1 *Kinematics of a Flexible Body*

Figure 11.1 shows a flexible body which will be denoted as *b* and which is defined by the two points *i* and *j* and by the two vectors $\mathbf{u}_i$ and $\mathbf{u}_j$. Note that points *i* and *j* are not material points but just two points chosen for the definition of the moving frame. We assume that $\mathbf{r}_j{-}\mathbf{r}_i$, $\mathbf{u}_i$, and $\mathbf{u}_j$ are not coplanar. If a body has more than two points and two vectors, the formulation can readily be modified to accommodate the additional coordinates by simply adding additional rigid body constraints (See Chapter 2).

Consider the moving reference frame $(\overline{X}_b,\ \overline{Y}_b,\ \overline{Z}_b)$ attached to the body. Let $\mathbf{A}^b$ be the orthogonal rotation matrix that relates the inertial frame (X, Y, Z) to the body moving frame. We can write

$$\mathbf{X}^b = \mathbf{A}^b\,\overline{\mathbf{X}}^b \tag{11.1}$$

where $\mathbf{X}^b$ represents a (3×3) matrix whose columns are, respectively, $\mathbf{r}_j{-}\mathbf{r}_i$, $\mathbf{u}_i$, and $\mathbf{u}_j$. The upper bar denotes vectors referred to the moving frame. Similarly (See Chapter 4), $\overline{\mathbf{X}}^b$ is a (3×3) constant matrix, whose columns are $\overline{\mathbf{r}}_j{-}\overline{\mathbf{r}}_i$, $\overline{\mathbf{u}}_i$, and $\overline{\mathbf{u}}_j$. Matrix $\mathbf{A}^b$ can be obtained from equation (11.1) by simple inversion:

$$\mathbf{A}^b = \mathbf{X}^b\,(\overline{\mathbf{X}}^b)^{-1} \tag{11.2}$$

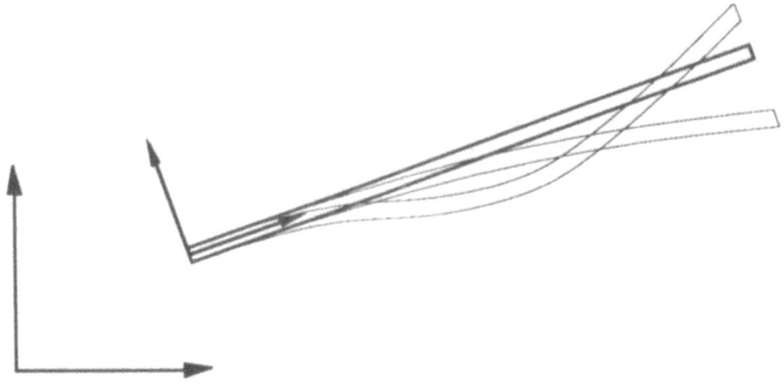

Figure 11.2. Cantilever modes of a beam-like flexible body.

Consider the body b and in it a material point P. Consider also a material segment through P, with its direction defined by a unit vector $\mathbf{u}_P$. The deformed position of P and $\mathbf{u}_P$ can be written as:

$$\mathbf{r}_P = \mathbf{r}_i + \mathbf{A}^b\,(\bar{\mathbf{r}}_P - \bar{\mathbf{r}}_i) + \mathbf{A}^b\,\delta\,\bar{\mathbf{r}}_P \tag{11.3}$$

$$\mathbf{u}_P = \mathbf{A}^b\,(\,\bar{\mathbf{u}}_P + \delta\,\bar{\mathbf{u}}_P) \tag{11.4}$$

where $\delta\,\bar{\mathbf{r}}_F$ represents the elastic displacement P, and $\delta\,\bar{\mathbf{u}}_P$ the elastic incremental rotation of $\mathbf{u}_P$, both expressed in the moving reference frame.

We can now proceed with the spatial discretization of the elastic displacement by defining a set of N_R Ritz vectors, such as finite elements or assumed modes for body b, namely $\bar{\phi}^b_{Pk}$, $k = 1, \ldots, N_R$. These vectors are functions of the material coordinates $\bar{\mathbf{r}}_P$ of the point. The set $\bar{\phi}^b_{Pk}$ contains the assumed displacement field corresponding to the assumed modes or finite elements with the rotations defined by the derivatives $\bar{\phi}'^b_{Pk}$. Using this set of Ritz vectors, the displacements and rotations at point P can be expressed as:

$$\delta\,\bar{\mathbf{r}}_P = \sum_{k=1}^{N_R} \eta^b_k(t)\,\bar{\phi}^b_{Pk} = \bar{\Phi}^b_P\,\eta^b \tag{11.5}$$

$$\delta\,\bar{\mathbf{u}}_P = \sum_{k=1}^{N_R} \eta^b_k(t)\,\bar{\phi}'^b_{Pk} = \bar{\Phi}'^b_P\,\eta^b \tag{11.6}$$

where $\eta^b_k(t)$ are the time-dependent amplitude factors of the Ritz vectors (assumed modes or finite elements).

At this point, the analyst has two choices: a) consider a finite element model from which one can extract a reduced set of assumed modes using, for instance, component mode synthesis; or b) obtain a set of such modes experimentally through a vibration analyzer. Note that $\bar{\phi}^b_{Pk}$ does not depend on time; thus it will not be differentiated. That the rigid body modes must also be eliminated from the

Ritz vectors, since the rigid body motion is already taken into account by the natural coordinates of the moving frame. One of several possible ways of imposing this condition is to select all the Ritz vectors with a clamped end at the origin of the moving frame. In this case, the moving frame attached to the elastic modes will be defined by (r_j-r_i) and u_i. Note that r_j is not a material point of the elastic body. Figure 11.2 shows the two first cantilever modes for a beam-like body.

11.2.2 Derivation of the Kinetic Energy

In order to obtain the expression for the inertial forces we first derive the expression for the kinetic energy of the body b in the form

$$T^b = \frac{1}{2} \int_{v_b} \dot{r}_P^T \dot{r}_P \, dm \tag{11.7}$$

Then, substituting equation (11.5) in (11.3) one can obtain

$$r_P = r_i + A^b (\bar{r}_P - \bar{r}_i) + A^b \bar{\Phi}_P^b \eta^b) \tag{11.8}$$

Substituting equations (4.49) and (4.50) into (11.8) one obtains

$$r_P = C_P^b q^b + A^b \bar{\Phi}_P^b \eta^b \tag{11.9}$$

where (See Chapter 4) C_P^b is a time invariant matrix that depends on the location of P, and q^b is the vector that contains the Cartesian coordinates of both the points i and j and the unit vectors u_i and u_j. The velocity of P is obtained by the differentiation of equation (11.9):

$$\dot{r}_P = C_P^b \dot{q}^b + \dot{A}^b \bar{\Phi}_P^b \eta^b + A^b \bar{\Phi}_P^b \dot{\eta}^b \tag{11.10}$$

Matrix $\dot{A}^b$ may be expressed in terms of $\dot{q}^b$ in the following manner: first, differentiate (11.2) to obtain

$$\dot{A}^b = \dot{X}^b (\bar{X}^b)^{-1} \tag{11.11}$$

then, substitute this result into (11.10) to obtain

$$\dot{r}_P = C_P^b \dot{q}^b + \dot{X}^b (\bar{X}^b)^{-1} \bar{\Phi}_P^b \eta^b + A^b \bar{\Phi}_P^b \dot{\eta}^b \tag{11.12}$$

Now the second term on the RHS of (11.12) can be modified by defining a modal transformation matrix $\bar{\Psi}_P^b$, such that

$$\bar{\Psi}_P^b = (\bar{X}^b)^{-1} \bar{\Phi}_P^b \tag{11.13}$$

The mentioned term may be expressed as

$$\dot{\mathbf{X}}^{b}\, (\overline{\mathbf{X}}^{b})^{-1}\, \overline{\Phi}_{P}^{\,b}\, \eta^{b} = \dot{\mathbf{X}}^{b}\, \overline{\Psi}_{P}^{\,b}\, \eta^{b} = \dot{\mathbf{X}}^{b} \sum_{k=1}^{NR} \overline{\mathbf{t}}_{Pk}^{\,b}\, \eta_{k}^{b} \qquad (11.14)$$

where $\overline{\mathbf{t}}_{Pk}^{\,b}$ are the transformed modal vectors. The product of $\dot{\mathbf{X}}^{b}\, \overline{\mathbf{t}}_{Pk}^{\,b}$ may be expressed as:

$$\dot{\mathbf{X}}^{b}\, \overline{\mathbf{t}}_{Pk}^{\,b} = \left[\dot{\mathbf{r}}_{j} - \dot{\mathbf{r}}_{i} \mid \dot{\mathbf{u}}_{i} \mid \dot{\mathbf{u}}_{j} \right]^{b} \left\{ \begin{matrix} \bar{t}_{1} \\ \bar{t}_{2} \\ \bar{t}_{3} \end{matrix} \right\}_{Pk}^{b}$$

$$= \left[-\bar{t}_{1}\,\mathbf{I}_{3} \quad \bar{t}_{1}\,\mathbf{I}_{3} \quad \bar{t}_{2}\,\mathbf{I}_{3} \quad \bar{t}_{3}\,\mathbf{I}_{3} \right]_{Pk}^{b} \left\{ \begin{matrix} \dot{\mathbf{r}}_{i} \\ \dot{\mathbf{r}}_{j} \\ \dot{\mathbf{u}}_{i} \\ \dot{\mathbf{u}}_{j} \end{matrix} \right\}^{b} \equiv \mathbf{T}_{Pk}^{b}\, \dot{\mathbf{q}}^{\,b} \qquad (11.15)$$

where $\mathbf{T}_{Pk}^{b}$ is a matrix that plays the same role of $\mathbf{C}_{P}^{b}$ but now is applied to the coordinates of the point P. Substituting (11.14) into (11.10) a final expression for the velocity of P is obtained:

$$\dot{\mathbf{r}}_{P} = \mathbf{C}_{P}^{b}\, \dot{\mathbf{q}}^{\,b} + \sum_{k=1}^{NR} \eta_{k}^{b}\, \mathbf{T}_{Pk}^{b}\, \dot{\mathbf{q}}^{\,b} + \mathbf{A}^{b}\, \overline{\Phi}_{P}^{\,b}\, \dot{\eta}^{\,b} \qquad (11.16)$$

Finally, by substituting (11.18) into the expression of the kinetic energy (11.7), we arrive at the following final expression:

$$T^{b} = \frac{1}{2} \left\{ \dot{\mathbf{q}}^{\,bT}\ \dot{\eta}^{\,T} \right\} \begin{bmatrix} \mathbf{M}_{rr}^{b} & \mathbf{M}_{rf}^{b} \\ \mathbf{M}_{fr}^{b} & \mathbf{M}_{ff}^{b} \end{bmatrix} \left\{ \begin{matrix} \dot{\mathbf{q}}^{\,b} \\ \dot{\eta} \end{matrix} \right\} \qquad (11.17)$$

where

$$\mathbf{M}_{rr}^{b} = \int_{Vol} \mathbf{C}_{P}^{\,bT}\, \mathbf{C}_{P}^{\,b}\, dm + \sum_{k=1}^{NR} \eta_{k}^{b} \int_{Vol} (\mathbf{C}_{P}^{\,bT}\, \mathbf{T}_{Pk} + \mathbf{T}_{Pk}^{\,T}\, \mathbf{C}_{P}^{\,b})\, dm$$

$$+ \sum_{k=1}^{NR} \sum_{l=1}^{NR} \eta_{k}^{b}\, \eta_{l}^{b} \int_{Vol} \mathbf{T}_{Pk}^{\,bT}\, \mathbf{T}_{Pl}^{\,b}\, dm \qquad (11.18)$$

$$\mathbf{M}_{rf}^{b} = \int_{Vol} \mathbf{C}_{P}^{\,bT}\, \mathbf{A}^{b}\, \overline{\Phi}_{P}^{\,b}\, dm + \sum_{k=1}^{NR} \eta_{k}^{b} \int_{Vol} \mathbf{T}_{Pk}^{\,T}\, \mathbf{A}^{b}\, \overline{\Phi}_{P}^{\,b}\, dm \qquad (11.19)$$

$$\mathbf{M}_{ff}^{b} = \int_{Vol} \overline{\Phi}_{P}^{\,bT}\, \mathbf{A}^{bT}\, \mathbf{A}^{b}\, \overline{\Phi}_{P}^{\,b}\, dm = \int_{Vol} \overline{\Phi}_{P}^{\,bT}\, \overline{\Phi}_{P}^{\,b}\, dm \qquad (11.20)$$

The subindexes $(-)_{r}$ and $(-)_{t}$ have been used to differentiate the terms corresponding to the rigid body motion characterized by $\dot{\mathbf{q}}^{\,b}$ from those that correspond to elastic deformations characterized by $\dot{\eta}$. It may be observed from those expressions how both $\dot{\mathbf{q}}^b$ and $\dot{\eta}$ are coupled in two different ways: a) through the coupling matrices $\mathbf{M}_{rf}^b$ y $\mathbf{M}_{fr}^b$; b) and by means of the η dependent terms that are included in $\mathbf{M}_{rr}^b$ and which appear in equation (11.20). The first term on the RHS of equation (11.20) coincides with the mass matrix of the rigid element as developed in Section 4.2. The second term contains a summation term that is linear in the elastic deformations η. The last term contains a double summation that depends on the square of the elastic deformations. If one is consistent with the assumption of small deformations, these square terms may be neglected for all practical purposes. However, the second linear term in η may not be neglected as a general rule. Only after a careful comparison of the magnitude of these terms with those corresponding to the rigid case, may they be neglected. It may be finally observed that $\mathbf{M}_{ff}^b$ does not depend either on the deformation or the rigid body coordinates. It is the constant mass matrix usually considered in structural dynamics (Craig (1981)).

Equation (11.18) contains integrals that are independent of both position as well as time. These can be computed only once prior to the numerical integration of the equations of motion. The integral of the first term of this equation was seen in detail in Section 4.2. In the following exercise, it will be shown how to compute in an efficient manner the second term of this equation.

Example 11.1

Assuming that the Ritz vectors have been computed using finite elements, develop a procedure to integrate

$$\int_{V_b} (\mathbf{C}_P^{bT}\,\mathbf{T}_{Pk} + \mathbf{T}_{Pk}^{T}\,\mathbf{C}_P^b)\,dm \tag{i}$$

Solution: Recall that the expressions for $\mathbf{C}_P^b$ and $\mathbf{T}_{Pk}^b$ are:

$$\mathbf{T}_{Pk}^b = [\ -\bar{t}_1\,\mathbf{I}_3\ |\ \bar{t}_1\,\mathbf{I}_3\ |\ \bar{t}_2\,\mathbf{I}_3\ |\ \bar{t}_3\,\mathbf{I}_3\ ] \tag{ii}$$

$$\mathbf{C}_P^b = [(1-\bar{c}_1)\,\mathbf{I}_3\ |\ \bar{c}_1\,\mathbf{I}_3\ |\ \bar{c}_2\,\mathbf{I}_3\ |\ \bar{c}_3\,\mathbf{I}_3\ ] \tag{iii}$$

where the vectors $\vec{t}_{Pk}^b$ y $\mathbf{c}_P$ have been defined in equations (11.16) and (4.55), respectively, as:

$$\bar{\mathbf{t}}_{Pk}^b = \left[\bar{\mathbf{X}}^b\right]^{-1} \bar{\phi}_{Pk}^b \tag{iv}$$

$$\mathbf{c}_P = \left[\bar{\mathbf{X}}^b\right]^{-1} (\bar{\mathbf{r}}_P - \bar{\mathbf{r}}_i) \tag{v}$$

It becomes obvious that one does not need to integrate the matrix products as they appear in (i). It is sufficient to compute the following integral

$$\int_{V_b} \mathbf{c}_P^b\,\mathbf{t}_{Pk}^{bT}\,dm = \int_{V_b} \begin{Bmatrix} c_1 \\ c_2 \\ c_3 \end{Bmatrix}_P^b \{t_1\ t_2\ t_3\}_{Pk}^b\,dm \tag{vi}$$

Equation (vi) contains all the terms necessary to build equation (i). If finite elements are used to compute (vi), one may proceed in the following manner. First, substitute in (vi) the results of (iv) and (v):

$$\int_{V_b} \mathbf{c}_P^b \, \mathbf{t}_{Pk}^{bT} \, dm = \left[\overline{\mathbf{X}}^b\right]^{-1} \int_{V_b} (\overline{\mathbf{r}}_P - \overline{\mathbf{r}}_i) \, \boldsymbol{\phi}_{Pk}^{bT} \, dm \left[\overline{\mathbf{X}}^b\right]^{-T} \qquad \text{(vii)}$$

Now, interpolate the spatial variables (geometry as well as deformed shapes) appearing in the integrand of (vii) using the finite element deformed shapes:

$$(\overline{\mathbf{r}}_P - \overline{\mathbf{r}}_i) = \mathbf{N}^b(\overline{\mathbf{r}}_P - \overline{\mathbf{r}}_i)\, \overline{\mathbf{x}}^b \qquad \text{(viii)}$$

$$\boldsymbol{\phi}_{Pk}^b = \mathbf{N}^b(\overline{\mathbf{r}}_P - \overline{\mathbf{r}}_i)\, \mathbf{p}_k^b \qquad \text{(ix)}$$

where $\mathbf{N}^b(\overline{\mathbf{r}}_P - \overline{\mathbf{r}}_i)$ are the finite element functions that are used to interpolate both the geometry as well as the global deformed shapes of the body. The vector $\overline{\mathbf{x}}^b$ represents the coordinates of the finite element mesh, and $\mathbf{p}_k^b$ are the values of k mode in the nodes of the finite element mesh. Substituting the equations (viii) and (ix) in the integral (vii), one can obtain

$$\int_{V_b} \mathbf{c}_P^b \, \mathbf{t}_{Pk}^{bT} \, dm = \left[\overline{\mathbf{X}}^b\right]^{-1} \int_{V_b} \mathbf{N}^b(\overline{\mathbf{r}}_P)\, (\mathbf{x}^b \, \mathbf{p}_k^{bT}) \left[\mathbf{N}^b(\overline{\mathbf{r}}_P)\right]^T \, dm \left[\overline{\mathbf{X}}^b\right]^{-T} \qquad \text{(x)}$$

Only interpolation functions are part of the integrand, since the expression inside the parenthesis in the middle of the integral does not depend on the spatial coordinates. The integral (x) may be calculated in a body-by-body basis.

The final step is to express the kinetic energy of the whole multibody system as an addition of the energies of each individual body:

$$T = \sum_b T_b = \frac{1}{2}\, \dot{\mathbf{q}}^T \, \mathbf{M} \, \dot{\mathbf{q}} \qquad (11.21)$$

where $\mathbf{M}$ and $\dot{\mathbf{q}}$ have been obtained by assembling the submatrices $\mathbf{M}^b$, and vectors $\dot{\mathbf{q}}^b$ and $\dot{\boldsymbol{\eta}}^b$, respectively.

11.2.3 Derivation of the Elastic Potential Energy

The expression for the elastic potential energy takes a very simple form with the moving frame approach, since it is only due to the contribution of the elastic displacement. The overall rigid body motion does not contribute to the potential energy. Consequently, the potential energy of a body is given by

$$\Pi_b = \frac{1}{2}\, \boldsymbol{\eta}^{bT} \, \mathbf{K}_{ff}^b \, \boldsymbol{\eta}^b \qquad (11.22)$$

where $\mathbf{K}_s^b = \mathbf{K}_{ff}^b$, if the stiffness has been obtained by using finite elements or

$$\mathbf{K}_{ff}^b = \overline{\boldsymbol{\Phi}}^{bT} \, \overline{\mathbf{K}}_s^b \, \overline{\boldsymbol{\Phi}}^b \qquad (11.23)$$

by using assumed modes defined in the local frame.

The elastic potential energy of the multibody system is obtained by assembling the potential energies of each of its bodies as:

$$\Pi = \sum_b \Pi_b = \frac{1}{2} \sum_b \left\{ \mathbf{q}^{bT} \; \boldsymbol{\eta}^{bT} \right\} \begin{bmatrix} 0 & 0 \\ 0 & \mathbf{K}_{ff}^{b} \end{bmatrix} \left\{ \begin{matrix} \mathbf{q}^b \\ \boldsymbol{\eta}^b \end{matrix} \right\} = \frac{1}{2} \mathbf{q}^T \mathbf{K} \, \mathbf{q} \qquad (11.24)$$

where $\mathbf{K}$ is the resulting global stiffness matrix that only affects the subset of $\mathbf{q}$ that corresponds to the elastic displacements $\boldsymbol{\eta}$.

11.2.4 Potential of External Forces

Only the case of a concentrated external force $\mathbf{f}$ that is applied at a point P of the body b will be considered in this section. Assuming that the force $\mathbf{f}_p$ is defined in global coordinates and making use of the equation (11.9), the virtual work of this force may be written as:

$$\delta W = \delta \mathbf{r}_P^T \, \mathbf{f}_P = (\delta \mathbf{q}^{bT} \mathbf{C}_P^{bT} + \delta \boldsymbol{\eta}^{bT} \overline{\boldsymbol{\Phi}}_P^{bT} \mathbf{A}^{bT}) \, \mathbf{f}_P =$$

$$\left\{ \delta \mathbf{q}^{bT} \; \delta \boldsymbol{\eta}^{bT} \right\} \begin{bmatrix} \mathbf{C}_P^{bT} \\ \overline{\boldsymbol{\Phi}}_P^{bT} \mathbf{A}^{bT} \end{bmatrix} \mathbf{f}_P \qquad (11.25)$$

from which one may find the generalized force

$$\mathbf{Q}_P^b = \begin{bmatrix} \mathbf{C}_P^{bT} \\ \overline{\boldsymbol{\Phi}}_P^{bT} \mathbf{A}^{bT} \end{bmatrix} \mathbf{f}_P \qquad (11.26)$$

Consequently, the potential may be easily calculated as

$$V = \int_{\mathbf{q}_0}^{\mathbf{q}} d\mathbf{q}^{bT} \mathbf{Q}_P^b \qquad (11.27)$$

The same procedure is used to calculate the generalized force and potential of any other type of loading.

11.2.5 Constraint Equations

The constraint equations for flexible bodies modeled with natural coordinates also come from two different sources, namely, rigid body constraints and joint constraints. The rigid body constraints now correspond to the definition of the moving frame and are derived and formulated as described in Chapter 2. However, the joint constraints must be modified, since these joints also include elastic deformations. As a consequence, variables such as points and unit vectors can no longer be shared at the joints.

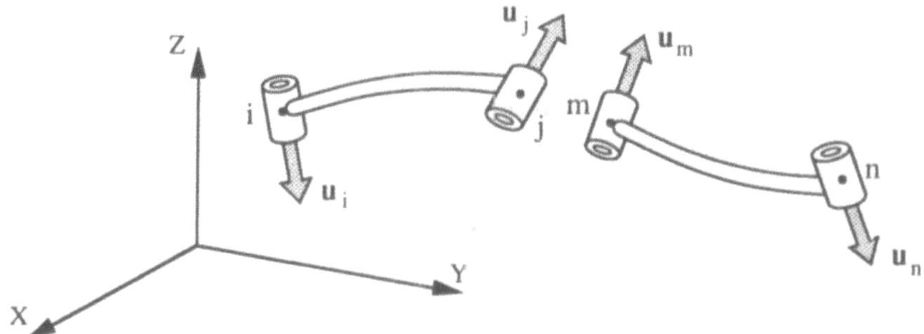

Figure 11.3. Revolute joint between two flexible bodies.

The reason why the variables cannot be shared at the joints is explained with a simple example. Consider the two contiguous bodies ij and mn shown in figure 11.3, with each of them defined by two points and two vectors. Consider for the moment that the flexibility is modeled by taking for each of them the first cantilever mode. In addition, assume that there is a revolute joint between the two bodies. Now consider that instead of taking two points j and m, only one point, say i , that is shared between the two bodies is taken. Further consider that there is only one vector in the joint, say u_j, which is shared. Note that in the flexible case the natural coordinates do not necessarily coincide with material points or directions, but they are only a mathematical tool to describe the overall motion. Since a joint constraint must be imposed between material points and directions, the sharing of variables does not in this case enforce the revolute joint constraint. Rather, the condition that the deformed end of the body ij is coincident with the deformed origin of the body mn must be imposed. Since mn is clamped at the origin, the previous condition means that the body ij cannot deform. This obviously is unacceptable. This reasoning can be extended to more than one mode, but the conclusion is always that the sharing of variables limits the deformation modes in an unacceptable way.

After this consideration, one can now formulate the constraint equations for the revolute joint shown in Figure 11.3. First, the deformed positions of j and m must coincide. Similarly, the deformed unit vectors u_j and u_m must also coincide. Those conditions can be written as:

$$\left(r_j + \delta\, r_j\right) - \left(r_m + \delta\, r_m\right) = 0 \tag{11.28}$$

$$\left(u_j + \delta\, u_j\right) - \left(u_m + \delta\, u_m\right) = 0 \tag{11.29}$$

Using expressions (11.5) and (11.6), the displacements δr_j, δr_m, δu_j, and δu_m can be expressed as a linear combination of the Ritz vectors as:

$$\delta \, \mathbf{r}_j = \mathbf{A}^{b} \, \delta \, \bar{\mathbf{r}}_j = \sum_{k=1}^{NR} \eta_k^{b}(t) \, \mathbf{A}^{b} \, \bar{\phi}_{jk}^{b} = \mathbf{A}^{b} \, \bar{\Phi}_j^{b} \, \eta^{b} \tag{11.30}$$

$$\delta \, \mathbf{u}_j = \mathbf{A}^{b} \, \delta \, \bar{\mathbf{u}}_P = \sum_{k=1}^{NR} \eta_k^{b}(t) \, \mathbf{A}^{b} \, \bar{\phi}_{Pk}^{'b} = \mathbf{A}^{b} \, \bar{\Phi}_P^{'b} \, \eta^{b} \tag{11.31}$$

and analogously $\delta \mathbf{r}_m$ and $\delta \mathbf{u}_m$.

The constraint equations for the whole multibody system that can include the definition of relative coordinates at the joints are obtained by putting together all the rigid body and joint constraints in the form

$$\Phi(\mathbf{q}, t) = 0 \tag{11.32}$$

11.2.6 Governing Equations of Motion

Several methods for deriving the equations of motion have been presented in Chapter 4 for rigid body dynamics. The equations of motion in the flexible case are derived in an analogous way, and, therefore, the final form is identical. Having developed expressions for the potential and kinetic energies, the most reasonable way of obtaining the equations of motion is through the Lagrange's equations, which leads to the following result:

$$\mathbf{M} \, \ddot{\mathbf{q}} + \mathbf{K} \, \mathbf{q} + \Phi_{\mathbf{q}}^{T} \lambda = \mathbf{Q}_{ex} - \dot{\mathbf{M}} \, \dot{\mathbf{q}} + T_{\mathbf{q}} \tag{11.33}$$

The last two terms on the RHS of this equation are velocity-dependent in both the rigid and the elastic coordinates. If the second derivative of the constraints is appended, the final form of the equations is

$$\begin{bmatrix} \mathbf{M} & \Phi_{\mathbf{q}}^{T} \\ \Phi_{\mathbf{q}} & 0 \end{bmatrix} \begin{Bmatrix} \ddot{\mathbf{q}} \\ \lambda \end{Bmatrix} = \begin{Bmatrix} \mathbf{Q}_{ex} - \dot{\mathbf{M}} \, \dot{\mathbf{q}} + T_{\mathbf{q}} - \mathbf{K} \, \mathbf{q} \\ -\dot{\Phi}_{\mathbf{q}} \, \dot{\mathbf{q}} - \dot{\Phi}_t \end{Bmatrix} \tag{11.34}$$

However, the Lagrange multiplier approach is generally not the best way of integrating these equations of motion. This is due mainly to two reasons: First, the number of constraints and therefore Lagrange multipliers is now much larger than in the rigid body case, since there is not sharing of variables at the joints. Secondly, the elastic terms in the RHS of (11.22) induce a considerable amount of numerical stiffness to the integration process, particularly if high frequency modes of vibrations such as axial modes are present in the formulation.

The first problem can be remedied by the use of the penalty formulation as done in Bayo and Serna (1989) which eliminates the multipliers from the equations of motion, thus reducing considerably the size of the system of equations. Other approaches, described in Chapters 5 and 8 to formulate the equations of motion in independent coordinates using the projection matrix $\mathbf{R}$, can also be applied for flexible multibody dynamics.

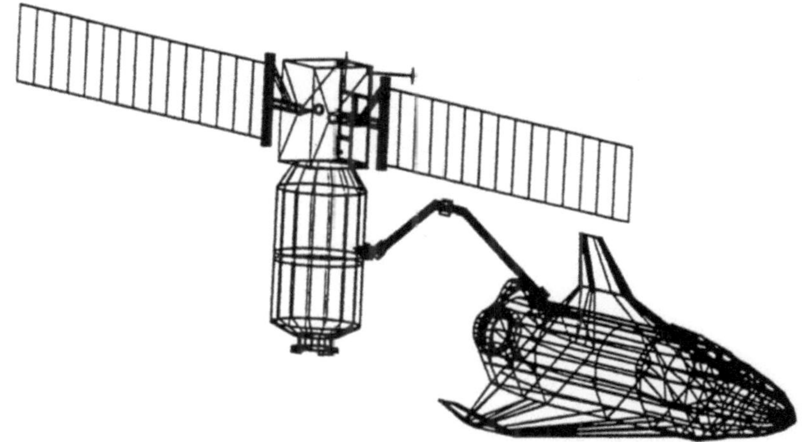

Figure 11.4 Flexible space robot in reorientation maneuver. Vukasovic et al., 1990. Reprinted by permission of ASME.

It is easy to verify that using the penalty formulation, the equations of motion become:

$$(\mathbf{M} + \Phi_q^T \alpha \Phi_q)\, \ddot{\mathbf{q}} + \mathbf{Kq} =$$
$$= \mathbf{Q}_{ex} - \dot{\mathbf{M}}\dot{\mathbf{q}} + T_q - \Phi_q^T \alpha\, (\dot{\Phi}_q\dot{\mathbf{q}} + 2\mu\Omega\,\dot{\Phi} + \Omega^2\Phi) \qquad (11.35)$$

The second problem, which is related to the stiffness of the resulting equations, may be solved by using the A-stable numerical algorithms presented in Chapter 7. The most appropriate implementation of these algorithms for the case at hand is that explained in Chapter 8, Section 8.5. In this explanation, the difference equations of the integrator are substituted into the equations of motion. The resulting set of nonlinear equations is solved using Newton-Raphson iteration.

11.2.7 Numerical Example

Using the method presented above, a satellite deployment maneuver has been simulated, where a flexible robot turns and repositions a satellite. Figure 11.4 shows the complete system, and Figure 11.5 shows the set of points and vectors used to define the mathematical model. The main links of the robot (bodies 3 and 4) are assumed to be flexible and have been modeled using six beam elements, while the other links of the robot (end-effector, wrist) are assumed rigid. The main body of the satellite is supposed rigid, while the solar arrays are flexible and modeled with eleven beam elements of equivalent stiffness and mass. An inertial reference frame is located at the base of the manipulator on the shuttle bay.

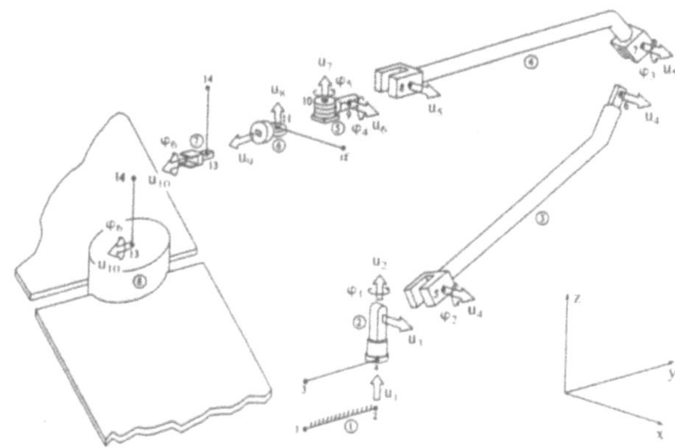

Figure 11.5. Natural coordinates model of a flexible robot. Vukasovic et al., 1990. Reprinted by permission of ASME.

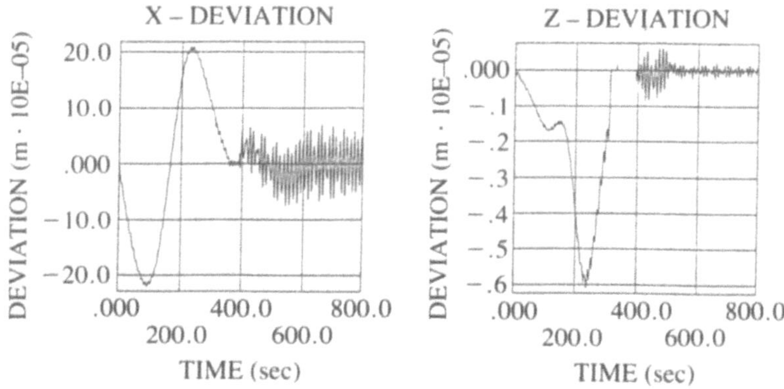

Figure 11.6 Deviation of the x-coordinate of the robot end-effector. Vukasovic et al., 1990. Reprinted by permission of ASME.

The input to the system is a known variation of the angles driving the manipulator that leads to the desired motion of the system. This motion is a 90 degrees rotation around the Z axis, and, simultaneously, a 180 degrees rotation around the Y axis, in order to re-orientate the satellite. Using this input, the dynamic response of the system has been calculated. Figure 11.6 shows the deviation of the position of the robot end-effector (x-coordinate) relative to the rigid body motion that represents the elastic vibration response superimposed on the large rigid body motion. Once the driving input is finished at time 600 sec., a residual vibration remains in the system due to the absence of damping. Calculations have been carried out on a Silicon Graphics Power Iris 4D/240 computer, using only one processor. The required CPU time has been about 2000 sec. for a maneuver that lasts 800 sec. in real time.

11.3 Global Method Based on Large Rotation Theory

The methods in this section are contributions from Avello (1990).

As mentioned in the overview of the different methods, the classical moving frame approach is based on the assumption of small displacements and equilibrium in the undeformed configuration. Kane, Ryan, and Banerjee (1987) showed that these assumptions lead to a spurious loss of stiffness, when the rotational velocities are large. Moreover, the method seen in the previous section cannot handle larger displacements than those for which the linear finite element method yields accurate results.

When both the elastic displacements are small and the stiffening effects are not important, the classical method yields sufficiently accurate results. It is usually preferred because of the reduced number of equations and the possibility of using either assumed or experimentally found modes of vibration. When the stiffening effects become important and/or displacements become finite, the global or absolute method described in this section can be applied. It is called global or absolute because the entire motion of the body (finite rotation plus deformation) is all referred to a fixed frame. This produces a shifting of nonlinearity from the inertial terms in the moving frame approach to the deformation terms in this new approach. A formulation of this type was first presented by Simo and Vu-Quoc (1986 and 1988) for multibodies modeled as planar and three-dimensional beams, respectively.

In this section it will be assumed that the flexible bodies are long and slender and that they can be correctly modeled as beams. Timoshenko's beam theory will be used, under the basic assumption that plane sections initially normal to the centroidal line remain plane after global deformation has taken place. With these basic assumptions, one will derive expressions for a simple nonlinear finite element method that can be used to model flexible bodies in a multibody formalism. The most attractive features of this formulation are its simplicity and the compatibility with the natural coordinates so far used in this book, since the nodal variables of the new beam element are also Cartesian points and unit vectors.

11.3.1 Kinematics of the Beam

Figure 11.7 shows an initially straight prismatic beam of length L and constant cross section A. One can define a fixed reference frame (X_1, X_2, X_3), with the X_1 axis coincident with the centroidal line, and axes X_2 and X_3 coincident with the principal axes of inertia. Any cross section of the beam can be described in this initial state by the coordinates $(X_1, 0, 0)$ of the intersection point between the cross section and the centroidal line, and by two mutually orthogonal vectors $\mathbf{M}$ and $\mathbf{N}$ parallel to the X_2 and X_3 axes. Vectors $\mathbf{M}$ and $\mathbf{N}$ can be considered as *co-rotational vectors* that move rigidly attached to the cross section to which they belong.

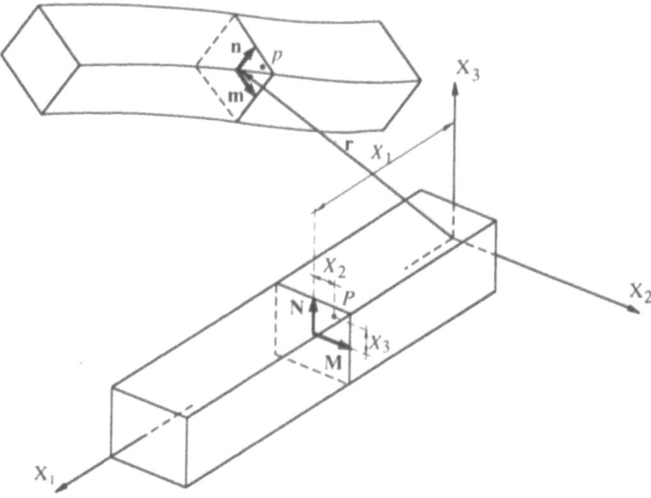

Figure 11.7. Deformed and undeformed prismatic 3-D beam.

After the beam has undergone finite displacements, the position of its cross sections can be defined with the coordinates of the intersection point **r** and with the components of the co-rotational vectors **m** and **n**, as shown in Figure 11.7. Upper-case letters will be used for the undeformed positions (*material coordinates*) and lower-case letters for deformed positions (*spatial coordinates*). If a Lagrangian formulation is used, one can write the deformed positions as a function of the undeformed ones. Since the initially straight prismatic beam is characterized in its undeformed position by just the X_1 coordinate, vectors **r**, **m**, and **n** can be written as a function of X_1, and the time t, as $\mathbf{r}=\mathbf{r}(X_1, t)$, $\mathbf{m}=\mathbf{m}(X_1, t)$, and $\mathbf{n}=\mathbf{n}(X_1, t)$. The deformed coordinates $\mathbf{x}=(x_1, x_2, x_3)$ of a particle whose material coordinates are $\mathbf{X}=(X_1, X_2, X_3)$ can be written as

$$\mathbf{x}(\mathbf{X}, t) = \mathbf{r}\left(X_1, t\right) + X_2\, \mathbf{m}\left(X_1, t\right) + X_3\, \mathbf{n}\left(X_1, t\right) \tag{11.36}$$

where X_1 is not a function of time.

11.3.2 A Non-Linear Beam Finite Element Formulation

In the finite element method, a proper inter-element continuity for the interpolated function and its derivatives must be assured by the shape functions. Typical Timoshenko beam elements require continuity only in the displacements and rotations but not in their derivatives (C^0 continuity). This is achieved by interpolating independently the displacements and rotations inside each element. In this section, an independent interpolation will be assumed for the nodal variables. However, the nodal variables that will be used in this section are different, in na-

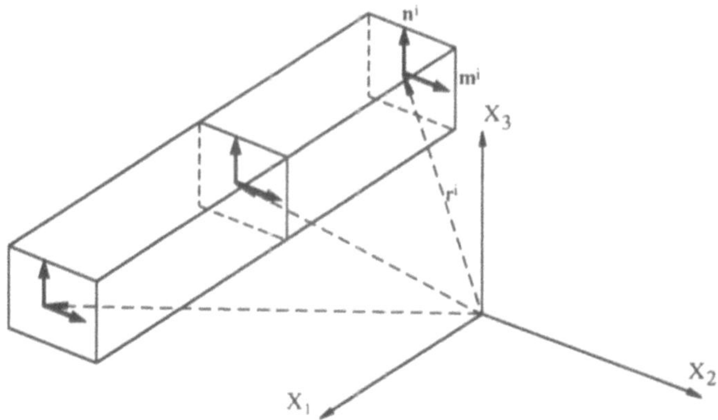

Figure 11.8. Cartesian dependent coordinates for a beam section.

ture and number, from the classical nodal variables used in linear beam elements. The nodal variables in the classical beam elements are composed of three displacements u_i and three rotations θ_i. Instead, the *nodal variables* used here are composed of the three coordinates of the position $\mathbf{r}^i$ and the six components of the two orthogonal unit vectors $\mathbf{m}^i$ and $\mathbf{n}^i$, as shown in Figure 11.8.

The nine nodal variables $(\mathbf{r}^i, \mathbf{m}^i, \mathbf{n}^i)$ are redundant, because only three of the six components of $\mathbf{m}^i$ and $\mathbf{n}^i$ are independent. In fact, there are three constraint equations that $\mathbf{m}^i$ and $\mathbf{n}^i$ must satisfy, two corresponding to the unit norm condition and the third corresponding to the orthogonality condition between them. Redundant variables have been extensively used in the kinematic and dynamic analysis of multibody systems, as has been seen throughout this book, but seldom in the finite element method. The main advantage of using redundant variables is that the overall complexity of the formulation is reduced. The degree of non-linearity of the problem is reduced as the number of variables is increased. The cost that one has to pay is the introduction of constraint equations to enforce the satisfaction of the constraints at the nodes.

Let $(\mathbf{r}^i, \mathbf{m}^i, \mathbf{n}^i)$, $i = 1, \ldots, p^e$ be the values of $(\mathbf{r}, \mathbf{m}, \mathbf{n})$ in the p^e nodes that belong to the finite element e. The values of $(\mathbf{r}, \mathbf{m}, \mathbf{n})$ inside each finite element are obtained through the following interpolation scheme:

$$\mathbf{r}^e = \sum_{i=1}^{p^e} N_i \mathbf{r}^i, \qquad \mathbf{m}^e = \sum_{i=1}^{p^e} N_i \mathbf{m}^i, \qquad \mathbf{n}^e = \sum_{i=1}^{p^e} N_i \mathbf{n}^i \tag{11.37}$$

where N_i are the shape functions that can be found in any standard book in the finite element method (See Bathe (1982)).

Note that in expression (11.37) unit vectors are being interpolated. Since the shape functions are not required to preserve the norm, vectors $\mathbf{m}^e$ and $\mathbf{n}^e$ have no longer a unit module. In the same way, the interpolated values $\mathbf{m}^e$ and $\mathbf{n}^e$ are not orthogonal. This interpolation inconsistency adds a new source of numerical error that is added to the global error of the finite element method. A full discussion on how this error affects the accuracy of the solution goes beyond the scope of this chapter, but the following points may give more insight.

- First, the magnitude of the interpolation errors depends on the relative rotation among the nodes of a single finite element. This means that small relative rotations imply small interpolation errors. For example, one can interpolate the two vectors:

$$\mathbf{m}^1 = \begin{Bmatrix} 0 \\ 1 \end{Bmatrix} \qquad \mathbf{m}^2 = \begin{Bmatrix} \cos\varphi \\ \sin\varphi \end{Bmatrix} \tag{11.38}$$

with the linear shape functions:

$$N_1 = \frac{L-x}{L} \qquad N_1 = \frac{x}{L} \tag{11.39}$$

The resulting interpolated vector inside the element can be obtained as

$$\mathbf{m}^e = N_1\,\mathbf{m}^1 + N_2\,\mathbf{m}^2 \tag{11.40}$$

The module of this vector can be easily calculated as:

$$\mathbf{m}^{eT}\mathbf{m}^e = N_1^2 + N_2^2 + 2N_1N_2\,\mathbf{m}^{1T}\mathbf{m}^2 = 2\left(\frac{x^2}{L^2} - \frac{x}{L}\right)(1 - \cos\varphi) + 1 \tag{11.41}$$

The maximum constraint violation is obtained in the middle of the element. Take $\varphi=10$ degrees and compute the module of $\mathbf{m}^e$ by taking the square root of equation (11.41). The resulting value is $|\mathbf{m}^e|=0.996195$, which represents an error below the 0.4%. Since one does not expect to handle rotations larger than 10 degrees among the nodes of the same finite element, the approximation seems quite reasonable.
- Secondly, the convergence of the finite element method is guaranteed, because as the number of elements increases the error due to the interpolation decreases. In the limit, no error is obtained.
- Finally, the results obtained with this formulation are similar to the ones obtained with other nonlinear formulations.

11.3.3 Derivation of the Kinetic Energy

In order to obtain the inertial forces, one must first develop the expression for the kinetic energy, which can be obtained from the integral:

$$T^e = \frac{1}{2}\int_{V^e} \dot{\mathbf{x}}^{eT}\dot{\mathbf{x}}^e \, dm \tag{11.42}$$

The velocity of a material point $\dot{\mathbf{x}}^e$ is obtained by differentiating expression (11.36) particularized for element e, and by substituting the interpolation scheme given in (11.37). This leads to

$$\dot{\mathbf{x}}^e = \sum_{i=1}^{p^e} N_i \left(\dot{\mathbf{r}}^i + X_2 \, \dot{\mathbf{m}}^i + X_3 \, \dot{\mathbf{n}}^i \right) \tag{11.43}$$

Substituting equation (11.43) into (11.42) yields

$$T^e = \frac{1}{2} \int_{V^e} \sum_{i=1}^{p^e} \sum_{j=1}^{p^e} N_i N_j \left(\dot{\mathbf{r}}^{iT} \dot{\mathbf{r}}^j + X_2^2 \, \dot{\mathbf{m}}^{iT} \dot{\mathbf{m}}^j + X_3^2 \, \dot{\mathbf{n}}^{iT} \dot{\mathbf{n}}^j \right.$$

$$\left. + 2\, X_2 \, \dot{\mathbf{r}}^{iT} \dot{\mathbf{m}}^j + 2\, X_3 \, \dot{\mathbf{r}}^{iT} \dot{\mathbf{n}}^j + X_2 X_3 \, \dot{\mathbf{m}}^{iT} \dot{\mathbf{n}}^j \right) dm \tag{11.44}$$

where the only terms that depend on the variables of the integral are X_2 and X_3. Since X_2 and X_3 are principal axes of inertia and recalling that X_1 coincides with the center of gravity of the cross section, the three last terms in the integral vanish, because they represent two static moments of first order and an inertial product. After reordering equation (11.44), the standard form of the kinetic energy is obtained as

$$T^e = \frac{1}{2} \dot{\mathbf{q}}^{eT} \mathbf{M}^e \dot{\mathbf{q}}^e \tag{11.45}$$

where $\mathbf{q}^e$ is a vector that contains the nodal variables of element e as

$$\mathbf{q}^{eT} = \left\{ \mathbf{r}^{1T} \, \mathbf{m}^{1T} \, \mathbf{n}^{1T} \; \cdots \; (\mathbf{r}^{p^e})^T \, (\mathbf{m}^{p^e})^T \, (\mathbf{n}^{p^e})^T \right\} \tag{11.46}$$

The matrix $\mathbf{M}^e$ is constant, symmetric, and is composed of sparse submatrices $\mathbf{M}_{ij}$ of size (9x9). In an homogeneous beam, $\mathbf{M}^e$ takes the form:

$$\mathbf{M}^e = \begin{bmatrix} \mathbf{M}_{11} & \mathbf{M}_{12} & \cdots & \mathbf{M}_{1p^e} \\ \mathbf{M}_{21} & \mathbf{M}_{22} & \cdots & \mathbf{M}_{2p^e} \\ \cdots & \cdots & \cdots & \cdots \\ \mathbf{M}_{p^e 1} & \mathbf{M}_{p^e 2} & \cdots & \mathbf{M}_{p^e p^e} \end{bmatrix} \tag{11.47}$$

$$\mathbf{M}_{ij} = \rho \begin{bmatrix} A\, c_{ij}\, \mathbf{I}_3 & \mathbf{0}_3 & \mathbf{0}_3 \\ \mathbf{0}_3 & I_2\, c_{ij}\, \mathbf{I}_3 & \mathbf{0}_3 \\ \mathbf{0}_3 & \mathbf{0}_3 & I_3\, c_{ij}\, \mathbf{I}_3 \end{bmatrix} \tag{11.48}$$

where ρ is the volumetric density, $\mathbf{I}_3$ the (3×3) unit matrix, and c_{ij} the integral over the length of the element of the product of shape functions ($N_i N_j$).

Compare this simple and constant expression for the mass matrix with the highly nonlinear matrix obtained in Section 11.2.3. Although the mass matrix is simpler, the elastic potential energy in the next section is more complicated than with the moving frame method.

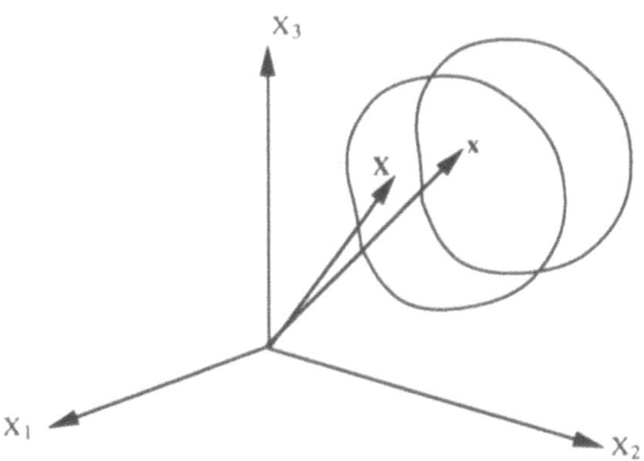

Figure 11.9. Undeformed and deformed position vectors of a point.

11.3.4 Derivation of the Elastic Potential Energy

One of the basic assumptions often made in structural theory is that displacements and displacement gradients are small. When this assumption holds, the *Cauchy strain tensor* can be used and accurate results are given. However, the Cauchy strain tensor does not work for large displacements, since it does not exhibit the proper invariance under rigid body rotations of the displacement field. Therefore, when large rotations are considered, a different measure of the strain must be used.

Several different kinds of strain measures have been proposed when displacements, displacement gradients, or both are finite (Malvern (1969)). These strain measures can be included in two major groups. *Eulerian formulations* formulate the problem in the deformed configuration, while *Lagrangian formulations* formulate it in the undeformed configuration. Eulerian formulations are used in applications where an undeformed or initial state does not exist or is unknown, as in fluid mechanics. In elasticity, however, it seems more useful to use a Lagrangian formulation, since an undeformed configuration is always assumed to exist and is taken as a reference state.

The *Green strain tensor* has typically been used in nonlinear elasticity to characterize the deformation field of bodies undergoing large displacements. As the displacements and displacement gradients get smaller, the Green tensor tends to the Cauchy tensor and, in the limit, they are identical.

Consider a continuous body and a fixed reference frame (X_1, X_2, X_3), as can be seen in Figure 11.9. Capital letters $\mathbf{X}=(X_1, X_2, X_3)$ will be used to refer to the coordinates of a particle in an initially undeformed position, and lower-case letters $\mathbf{x}=(x_1, x_2, x_3)$ will be used for the currently deformed position. In a Lagrangian formulation, $\mathbf{x}$ is taken as a function of $\mathbf{X}$ and time, and therefore can be written

$$\mathbf{x} = \mathbf{x}(\mathbf{X}, t) \tag{11.49}$$

The *deformation gradient* $\mathbf{F}$ is defined as the matrix that contains the partial derivatives of $\mathbf{x}$ with respect to $\mathbf{X}$. An infinitesimal vector in the deformed position $d\mathbf{x}$ can be expressed in terms of the deformation gradient and of its undeformed position $d\mathbf{X}$ as

$$d\mathbf{x} = \frac{\partial \mathbf{x}}{\partial \mathbf{X}} d\mathbf{X} = \mathbf{F}\, d\mathbf{X} \tag{11.50}$$

The *Green deformation tensor* $\mathbf{C}$ is defined as the one that gives the new squared length $(ds)^2$ of vector $d\mathbf{x}$, into which the given vector $d\mathbf{X}$ has deformed. Thus,

$$ds^2 = d\mathbf{X}^{\mathrm{T}}\, \mathbf{C}\, d\mathbf{X} \tag{11.51}$$

The *Green strain tensor* $\mathbf{E}$ gives, by definition, the change in squared length between the deformed and the undeformed state of a vector $d\mathbf{X}$

$$ds^2 - dS^2 = 2\, d\mathbf{X}^{\mathrm{T}}\, \mathbf{E}\, d\mathbf{X} \tag{11.52}$$

where $(dS)^2$ is the original length of vector $d\mathbf{X}$. Comparing equations (11.50), (11.51), and (11.52), the two following relations can easily be found:

$$\mathbf{C} = \mathbf{F}^{\mathrm{T}}\, \mathbf{F} \tag{11.53}$$

$$\mathbf{E} = \frac{\mathbf{C} - \mathbf{I}_3}{2} \tag{11.54}$$

where $\mathbf{I}_3$ is the (3×3) identity matrix.

The potential energy for a linearly elastic homogeneous material can be written in terms of the strain vector $\mathbf{E} = \{E_{11}\ E_{22}\ E_{33}\ E_{12}\ E_{13}\ E_{23}\}^{\mathrm{T}}$ as

$$V = \frac{1}{2}\int_{V^e} \mathbf{E}^{\mathrm{T}}\, \mathbf{D}\, \mathbf{E}\, dV \tag{11.55}$$

where the integral is extended to the body in the undeformed configuration. $\mathbf{D}$ represents the matrix of elastic constants, which is defined in terms of *Lame's* constants λ and G as

$$\mathbf{D} = \begin{bmatrix} \lambda + 2G & \lambda & \lambda & 0 & 0 & 0 \\ \lambda & \lambda + 2G & \lambda & 0 & 0 & 0 \\ \lambda & \lambda & \lambda + 2G & 0 & 0 & 0 \\ 0 & 0 & 0 & 2G & 0 & 0 \\ 0 & 0 & 0 & 0 & 2G & 0 \\ 0 & 0 & 0 & 0 & 0 & 2G \end{bmatrix} \tag{11.56}$$

The values of λ and G in terms of the *Young modulus* E and the *Poisson ratio* v are:

$$\lambda = \frac{E\,v}{(1+v)(1-2v)} \qquad\qquad G = \frac{E}{2(1+v)} \tag{11.57}$$

As seen in equation (11.36), the deformed coordinates of any point of the beam can be written as

$$x(\mathbf{X},\ t) = \mathbf{r}\,(X_1,\ t) + X_2\,\mathbf{m}\,(X_1,\ t) + X_3\,\mathbf{n}\,(X_1,\ t) \tag{11.58}$$

The deformation gradient $\mathbf{F}$ can be easily computed as

$$\mathbf{F} = \frac{\partial \mathbf{x}}{\partial \mathbf{X}} = \left[\ \mathbf{x}_{,1}\ |\ \mathbf{x}_{,2}\ |\ \mathbf{x}_{,3}\ \right] = \left[\ \mathbf{r}_{,1} + X_2\,\mathbf{m}_{,1} + X_3\,\mathbf{n}_{,1}\ |\ \mathbf{m}\ |\ \mathbf{n}\ \right] \tag{11.59}$$

where the vertical bars in equation indicate the separation between columns. The notation $(-)_{,i}$ is used to represent $\partial(-)/\partial X_i$. The Green strain tensor can be obtained by substituting equation (11.59) into equations (11.53) and (11.54) as

$$\mathbf{E} = \frac{1}{2}\begin{bmatrix} \mathbf{x}_{,1}^{T}\,\mathbf{x}_{,1} - 1 & \mathbf{x}_{,1}^{T}\,\mathbf{m} & \mathbf{x}_{,1}^{T}\,\mathbf{n} \\ \mathbf{x}_{,1}^{T}\,\mathbf{m} & 0 & 0 \\ \mathbf{x}_{,1}^{T}\,\mathbf{n} & 0 & 0 \end{bmatrix} \tag{11.60}$$

with

$$\mathbf{x}_{,1} = \mathbf{r}_{,1} + X_2\,\mathbf{m}_{,1} + X_3\,\mathbf{n}_{,1} \tag{11.61}$$

Substituting equation (11.61) into (11.60) and operating, the following expression is obtained for the components of the strain vector $\mathbf{E}$:

$$E_{11} = \frac{1}{2}\left(\mathbf{x}_{,1}^{T}\,\mathbf{x}_{,1} - 1\right) = \frac{1}{2}\left(\mathbf{r}_{,1}^{T}\,\mathbf{r}_{,1} + X_2^2\,\mathbf{m}_{,1}^{T}\,\mathbf{m}_{,1} + X_3^2\,\mathbf{n}_{,1}^{T}\,\mathbf{n}_{,1}\right.$$
$$\left. + 2\,X_2\,\mathbf{r}_{,1}^{T}\,\mathbf{m}_{,1} + 2\,X_3\,\mathbf{r}_{,1}^{T}\,\mathbf{n}_{,1} + 2\,X_2\,X_3\,\mathbf{m}_{,1}^{T}\,\mathbf{n}_{,1} - 1\right) \tag{11.62a}$$

$$E_{22} = E_{33} = 0 \tag{11.62b}$$

$$E_{12} = \frac{1}{2}\mathbf{x}_{,1}^{T}\,\mathbf{m} = \frac{1}{2}\left(\mathbf{r}_{,1}^{T}\,\mathbf{m} + X_2\,\mathbf{m}_{,1}^{T}\,\mathbf{m} + X_3\,\mathbf{n}_{,1}^{T}\,\mathbf{m}\right) \tag{11.62c}$$

$$E_{13} = \frac{1}{2}\mathbf{x}_{,1}^{T}\,\mathbf{n} = \frac{1}{2}\left(\mathbf{r}_{,1}^{T}\,\mathbf{n} + X_2\,\mathbf{m}_{,1}^{T}\,\mathbf{n} + X_3\,\mathbf{n}_{,1}^{T}\,\mathbf{n}\right) \tag{11.62d}$$

$$E_{23} = 0 \tag{11.62e}$$

If it is assumed that the strains are sufficiently small (note, however, that finite elastic displacements and rotations are still being considered), the products $(\mathbf{m}_{,1}^{T}\,\mathbf{m}_{,1})$, $(\mathbf{n}_{,1}^{T}\,\mathbf{n}_{,1})$, and $(\mathbf{m}_{,1}^{T}\,\mathbf{n}_{,1})$ in E_1 are second-order terms that can be neglected. Furthermore, the products $(\mathbf{m}_{,1}^{T}\,\mathbf{m})$ and $(\mathbf{n}_{,1}^{T}\,\mathbf{n})$ are zero, as can easily be seen by differentiating with respect to X_1 the two unit norm conditions $(\mathbf{m}^{T}\,\mathbf{m} - 1 = 0)$ and $(\mathbf{n}^{T}\,\mathbf{n} - 1 = 0)$, respectively. With these simplifications, the strain measures can finally be written as:

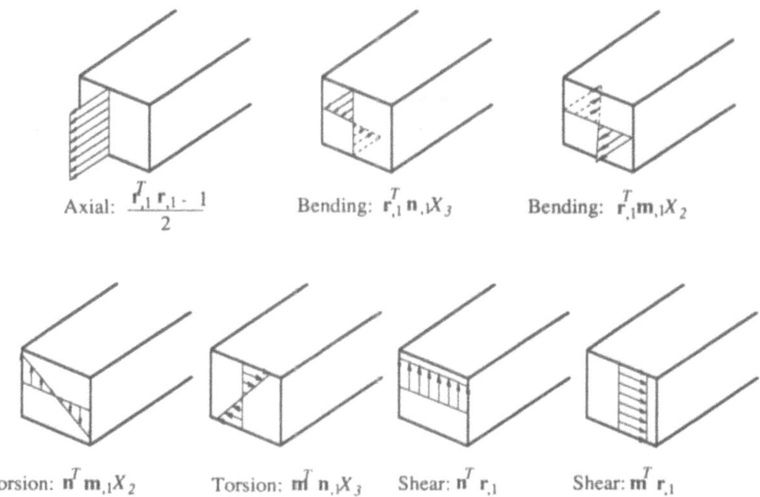

Figure 11.10. Axial, bending, torsion, and shear strains in a beam.

$$E_{11} = \frac{1}{2}\left(\mathbf{r}_{,1}^{T}\,\mathbf{r}_{,1} - 1 + 2\,X_2\,\mathbf{r}_{,1}^{T}\,\mathbf{m}_{,1} + 2\,X_3\,\mathbf{r}_{,1}^{T}\,\mathbf{n}_{,1} \right) \tag{11.63a}$$

$$E_{22} = E_{33} = E_{23} = 0 \tag{11.63b}$$

$$E_{12} = \frac{1}{2}\left(\mathbf{r}_{,1}^{T}\,\mathbf{m} - X_3\,\mathbf{n}_{,1}^{T}\,\mathbf{m} \right) \tag{11.63c}$$

$$E_{13} = \frac{1}{2}\left(\mathbf{r}_{,1}^{T}\,\mathbf{n} + X_2\,\mathbf{m}_{,1}^{T}\,\mathbf{n} \right) \tag{11.63d}$$

which is in accordance with the strain distribution predicted by the elemental
theory of strength of materials for a prismatic beam under axial, shearing, bending, and torsion loads, as illustrated in Figure 11.10. For example, the term
$(\mathbf{r}_{,1}^{T}\,\mathbf{r}_{,1} - 1)/2$ in E_{11} represents a constant strain distribution corresponding to
a pure axial load. Analogously, the term $(X_2\,\mathbf{r}_{,1}^{T}\,\mathbf{m}_{,1})$ in E_{11} represents a strain
distribution that varies linearly with X_2, with a zero value at the centroid and ex-
treme values at the edges, as corresponds to a pure bending load. The two constant shear strains predicted by the Timoshenko beam theory are known to be in-
correct, and a parabolic strain distribution should appear. This has typically been
corrected by multiplying the area of the cross section by a factor (5/6 for rectangular sections) which gives the correct shear strain energy of the beam.

The potential energy of a single element can now be written as

$$\Pi^{e} = \frac{1}{2}\int_{V}\left(E\,E_{11}^{2} + 2\,G\,E_{12}^{2} + 2\,G\,E_{13}^{2} \right)dV \tag{11.64}$$

Substituting equation (11.63) and operating, one can obtain

$$\Pi^e = \frac{1}{2} \int_V \Big[E\, A\, \Gamma_1^2 + E\, I_2\, \Gamma_2^2 + E\, I_3\, \Gamma_3^2 + G\, A_{s2}\, \Gamma_4^2$$
$$+ G\, A_{s3}\, \Gamma_5^2 + G\, I_p\, \Gamma_6^2 \Big]\, dX_1 \tag{11.65}$$

where Γ_1 represents the axial strain, Γ_2 and Γ_3 are the bending unit rotations per unit length, Γ_4 and Γ_5 are the shearing strains, and Γ_6 is the torsion rotation per unit length. Their expressions are:

$$\Gamma_1 = \frac{\mathbf{r}_{,1}^T \mathbf{r}_{,1} - 1}{2} \qquad \Gamma_2 = \mathbf{r}_{,1}^T \mathbf{n}_{,1} \qquad \Gamma_3 = \mathbf{r}_{,1}^T \mathbf{m}_{,1}$$
$$\Gamma_4 = \mathbf{r}_{,1}^T \mathbf{m} \qquad \Gamma_5 = \mathbf{r}_{,1}^T \mathbf{n} \qquad \Gamma_6 = \mathbf{n}_{,1}^T \mathbf{m} \tag{11.66}$$

where A_{s2} and A_{s3} are the equivalent shear areas, and I_2, I_3, and I_p have the following meaning:

$$I_2 = \int_A X_3^2\, dA \qquad I_3 = \int_A X_2^2\, dA \qquad I_p = \int_A (X_2^2 + X_3^2)\, dA \tag{11.67}$$

The finite element interpolation given in equation (11.37) can be introduced into equations (11.65) and (11.66). After some algebraic manipulations and rearrangements, the following expressions for the strains Γ_i can be obtained:

$$\Gamma_i = \frac{1}{2} \mathbf{q}^{eT} \mathbf{G}^i \mathbf{q}^e - \beta_i, \qquad i = 1, \ldots, 6 \tag{11.68}$$

with $\beta_1 = 1/2$, $\beta_i = 0$, $i = 2, \ldots, 6$, and where $\mathbf{q}^e$ was defined in (11.46). The matrices $\mathbf{G}^i$ are symmetric, sparse, and depend only on the shape functions and their derivatives with respect to X_1. Their expressions are as follows:

$$\mathbf{G}^1 = \begin{bmatrix} N_{i,1}\, N_{j,1}\, \mathbf{I}_3 & \mathbf{0}_3 & \mathbf{0}_3 \\ \mathbf{0}_3 & \mathbf{0}_3 & \mathbf{0}_3 \\ \mathbf{0}_3 & \mathbf{0}_3 & \mathbf{0}_3 \end{bmatrix} \tag{11.69a}$$

$$\mathbf{G}^2 = \begin{bmatrix} \mathbf{0}_3 & \mathbf{0}_3 & N_{i,1}\, N_{j,1}\, \mathbf{I}_3 \\ \mathbf{0}_3 & \mathbf{0}_3 & \mathbf{0}_3 \\ N_{i,1}\, N_{j,1}\, \mathbf{I}_3 & \mathbf{0}_3 & \mathbf{0}_3 \end{bmatrix} \tag{11.69b}$$

$$\mathbf{G}^3 = \begin{bmatrix} \mathbf{0}_3 & N_{i,1}\, N_{j,1}\, \mathbf{I}_3 & \mathbf{0}_3 \\ N_{i,1}\, N_{j,1}\, \mathbf{I}_3 & \mathbf{0}_3 & \mathbf{0}_3 \\ \mathbf{0}_3 & \mathbf{0}_3 & \mathbf{0}_3 \end{bmatrix} \tag{11.69c}$$

$$\mathbf{G}^4 = \begin{bmatrix} \mathbf{0}_3 & N_{i,1}\, N_j\, \mathbf{I}_3 & \mathbf{0}_3 \\ N_{j,1}\, N_i\, \mathbf{I}_3 & \mathbf{0}_3 & \mathbf{0}_3 \\ \mathbf{0}_3 & \mathbf{0}_3 & \mathbf{0}_3 \end{bmatrix} \tag{11.69d}$$

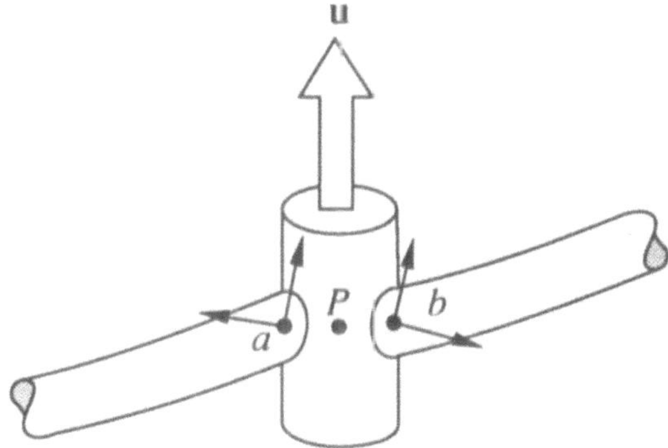

Figure 11.11. Definition of a revolute joint.

$$
\mathbf{G}^5 = \begin{bmatrix} 0_3 & 0_3 & N_{i,1}\,N_j\,\mathbf{I}_3 \\ 0_3 & 0_3 & 0_3 \\ N_{j,1}\,N_i\,\mathbf{I}_3 & 0_3 & 0_3 \end{bmatrix} \tag{11.69e}
$$

$$
\mathbf{G}^6 = \begin{bmatrix} 0_3 & 0_3 & 0_3 \\ 0_3 & 0_3 & N_{j,1}\,N_i\,\mathbf{I}_3 \\ 0_3 & N_{i,1}\,N_j\,\mathbf{I}_3 & 0_3 \end{bmatrix} \tag{11.69f}
$$

The total potential energy for the beam is obtained by adding the potential energy of all the elements given by expressions (11.65), (11.68), and (11.69) as

$$
\varPi = \sum_e \varPi^e \tag{11.70}
$$

Observe that in this beam element the potential energy is obtained as a polynomial of order 4 in the position variables because $\varPi^e$ depends on the square of $\varGamma_i$, and $\varGamma_i$ depends on the square of $\mathbf{q}^e$. It is unlike the classical moving frame formulation of Section 11.2, in which the potential energy is a quadratic function of the position variables. This complicates the implementation of the elastic forces, but the mass matrix obtained in Section 11.3.3 is constant and can be computed only once. Therefore, the complexity is transferred from the inertial forces to the elastic forces, but the overall complexity remains similar to the moving frame method's complexity.

11.3.5 Constraint Equations

Since the position variables are not independent, constraints must be introduced at the finite element nodes and at the joints. The constraints at the nodes account for the unit norm and orthogonality conditions that the unit vectors must satisfy.

The constraints at the joints restrict the relative motion of adjacent bodies to the rotations or translations allowed by the kinematic joints.

Each node introduces six constraints of the form

$$\left\{\begin{array}{c} \mathbf{l}_i^T\,\mathbf{l}_i - 1 \\ \mathbf{l}_i^T\,\mathbf{m}_i \\ \mathbf{m}_i^T\,\mathbf{m}_i - 1 \\ \mathbf{m}_i^T\,\mathbf{n}_i \\ \mathbf{n}_i^T\,\mathbf{n}_i - 1 \\ \mathbf{n}_i^T\,\mathbf{l}_i \end{array}\right\} = 0 \tag{11.71}$$

The constraint equations at the joints can be written in terms of the nodal variables of the nodes next to the joint. The constraint equations for a revolute joint are presented below as an example. Figure 11.11 shows two beam-like bodies linked at point P by a revolute joint of axis $\mathbf{u}$. Let a and b be the two nodes next to the joint, each of them belonging to one of the beams. The revolute joint constraints must enforce both the condition that point P as attached to the frame in a and as attached to the frame in b coincides, and the condition that the vector $\mathbf{u}$ as attached to a and as attached to b also coincide. Both conditions can be written through the following two vector equations equivalent to six scalar equations:

$$\boldsymbol{\phi} = \left\{\begin{array}{c} \mathbf{r}^a + \mathbf{A}^a\,{}^a\mathbf{r}^P - \mathbf{r}^b - \mathbf{A}^b\,{}^b\mathbf{r}^P \\ \mathbf{A}^a\,{}^a\mathbf{u} - \mathbf{A}^b\,{}^b\mathbf{u} \end{array}\right\} = 0 \tag{11.72}$$

where only two of the last three equations are independent. The matrices $\mathbf{A}^a$ and $\mathbf{A}^b$ are (3x3) orthogonal rotation matrices given by

$$\mathbf{A}^a = \left[\, \mathbf{m}^a \wedge \mathbf{n}^a \,\middle|\, \mathbf{m}^a \,\middle|\, \mathbf{n}^a \,\right] \quad \text{and} \quad \mathbf{A}^b = \left[\, \mathbf{m}^b \wedge \mathbf{n}^b \,\middle|\, \mathbf{m}^b \,\middle|\, \mathbf{n}^b \,\right] \tag{11.73}$$

where the vertical bars denote the separation between columns. The values of ${}^a\mathbf{r}^P$, ${}^b\mathbf{r}^P$, ${}^a\mathbf{u}$, and ${}^b\mathbf{u}$ are the coordinates of point P and the components of vector $\mathbf{u}$ expressed in frames a and b, respectively.

In the previous example, the joint is linking two beams, but the joint could also be thought of as linking a beam and a rigid body or a beam and a flexible body with assumed deformation modes. The joint constraints would be developed in the same way, but different position variables would be used for one of the bodies.

11.3.6 Governing Equations of Motion

Once more the equations of motion can be derived using any of the methods seen in Chapter 4. Here, the Lagrange multipliers method will be used again. The Lagrangian function L can be written as

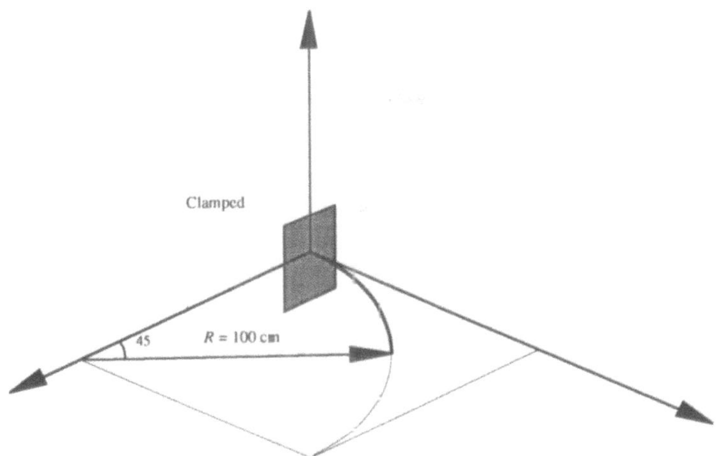

Figure 11.12. Cantilever beam 45-degree bend.

$$L = T - \Pi + \Phi^T \lambda \tag{11.74}$$

where Φ contains the constraints that arise from the unit norm and orthogonality condition that the nodal variables have to satisfy at the nodes and from the kinematic constraints imposed at the joints. Vector λ contains the Lagrange multipliers corresponding to the constraints.

The application of the Lagrange's equations leads to

$$\mathbf{M}\,\ddot{\mathbf{q}} + \Phi_{\mathbf{q}}^T\,\lambda = \mathbf{Q} - \mathbf{F} \tag{11.75}$$

where $\mathbf{M}$ is the mass matrix obtained by assembling the mass matrices $\mathbf{M}^e$ of each element, $\Phi_{\mathbf{q}}$ the Jacobian matrix of the constraint equations, $\mathbf{Q}$ the vector of generalized external forces, and $\mathbf{F}$ the elastic forces. The elastic forces are obtained by differentiating equation (11.65) with respect to $\mathbf{q}^e$, giving

$$\mathbf{F}^e = \int_0^L \left[E\,A\,\Gamma_1\,\mathbf{G}^1 + E\,I_2\,\Gamma_2\,\mathbf{G}^2 + E\,I_3\,\Gamma_3\,\mathbf{G}^3 + G\,A_{s2}\,\Gamma_4\,\mathbf{G}^4 \right.$$
$$\left. + G\,A_{s3}\,\Gamma_5\,\mathbf{G}^5 + G\,I_p\,\Gamma_6\,\mathbf{G}^6 \right] dX_1\,\mathbf{q}^e \tag{11.76}$$

The matrices $\mathbf{G}^i$ are very sparse, and consequently the multiplications by $\mathbf{q}^e$ can be carried out analytically with very few arithmetic operations.

11.3.7 Numerical Examples

In this section, the results obtained in three examples are presented in order to test both the accuracy of the present beam finite element and the numerical inte-

Table 11.1.   Tip displacement (cm) in the cantilever 45-degree bend.

	f = 300 Kg			f = 450 Kg			f = 600 Kg		
	x_1	x_2	x_3	x_1	x_2	x_3	x_1	x_2	x_3
Present method	22.14	58.66	40.65	18.23	51.84	49.31	15.26	46.48	54.54
Cardona (1989)	22.14	58.64	40.35	18.38	52.11	48.59	15.55	47.04	53.50
Simo (1986)	22.33	58.84	40.08	18.62	52.32	48.39	15.79	47.23	53.37
Bathe and B. (1979)	22.50	59.20	39.50	-	-	-	15.90	47.20	53.40
Crisfield (1990)	22.16	58.53	40.53	18.43	51.93	48.79	15.61	46.48	53.71

gration procedure. In all cases, the penalty matrix α was taken as αI, and the value of the penalty factor α was taken as 10^6 times the largest term appearing in the tangent stiffness matrix H_{qF} obtained through differentiation of equation (11.76). No attempt was made to optimize the value of the penalty factor. It was found that the iteration process converged in few iterations and that the constraint violation was kept small, roughly $|\phi| < \alpha^{-1}$. The calculations were performed in a Silicon Graphics 4D/240 using only one processor. To avoid the *shear-locking*, reduced integration has been used for the shear terms.

Example 11.2

The 45-degree bend cantilever beam shown in Figure 11.12 of radius equal to 100 *cm* shall be considered. It is located in a horizontal plane and a vertical static load f will be considered acting at the tip. The beam has a unit square cross section and $E = G = 10^7$ Kg/cm^2. It is discretized using eight linear straight elements.

In Table 11.1, the three coordinates of the tip in the deformed position are presented for three different values of the force. The values obtained with the finite element developed in this chapter are compared to the values obtained previously by Cardona (1989), Simo (1986), Bathe and Bolourchi (1979), and Crisfield (1990). The total load is applied in six equally-spaced load increments.

It is worth noting that the tip displacements are of the same order of magnitude as the length of the beam, which is 78.54 cm. Therefore, the behavior of the beam is totally nonlinear, with finite displacements which could not be studied using a mode superposition method. As the value of the load increases the solution provided by the proposed method gives larger displacements than the other formulations. This is because the interpolation in each finite element, as discussed in Section 11.3.2, does not satisfy the orthogonality condition for variables $(\mathbf{m}, \mathbf{n})$. This static example is an extreme one, and it has been presented to prove that this assumption is valid. In fact, the maximum discrepancy between the results presented in Table 11.1 for the different methods is about two per cent.

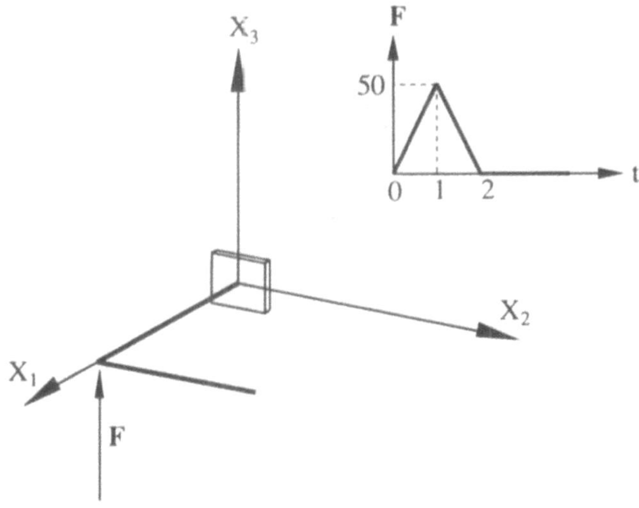

Figure 11.13. Right-angle cantilever beam.

Example   11.3

This example, a right-angle cantilever beam, was first proposed by Simo and Vu-Quoc (1988) and solved with quadratic elements. Later, it was solved by Cardona (1989) using linear elements with a very similar formulation as the previous one. The problem consists of a right-angle cantilever beam composed of two straight parts of length $L=10$, each, as shown in Figure 11.13.

The physical characteristics of the beam are not realistic, but they are useful to test the accuracy of the method in a dynamic simulation when large relative displacements appear. Their values, using the notation in Simo and Vu-Quoc (1988), are given below:

$$GA = EA = 10^6$$
$$EI_2 = EI_3 = GI_p = 10^3$$
$$A_\rho = 1$$
$$I_{\rho 1} = 2I_{\rho 2} = 2I_{\rho 3} = 20$$

There is a dynamic vertical load F acting at the elbow with a triangular variation law. The load acts for 2 sec and reaches a peak of $F_{max}=50$ at $t=1$ sec, as can be seen in Figure 11.13. The problem has been solved with two different discretizations using four and eight linear elements. The total simulation time is 30 sec. In Figures 11.14 and 11.15, the vertical displacements of the elbow and the tip obtained with four and eight elements are plotted. The agreement of this dynamic response compared to Simo and Vu-Quoc (1988) and Cardona (1989) is poor for the four elements discretization, but it is good when eight elements are used.

The results were obtained using a constant step size of 0.125 sec. The average number of iterations in the Newton-Raphson procedure was three. The *CPU* times were 20.6 sec for the four elements discretization and 44.4 sec for the eight elements discretization .

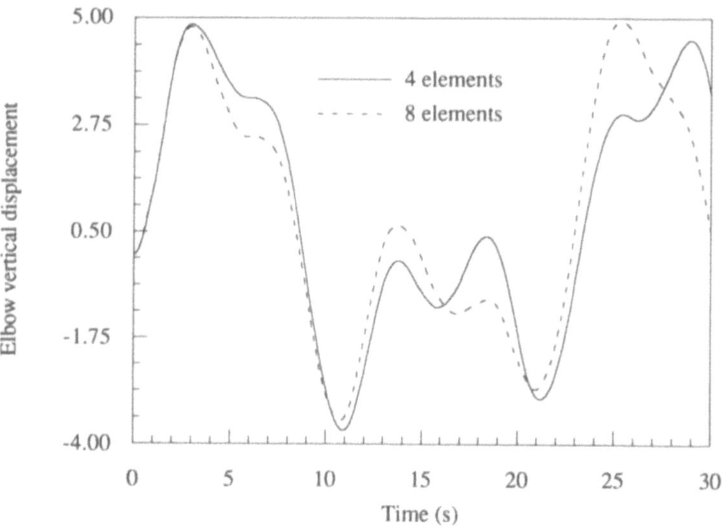

Figure 11.14. Elbow vertical displacement using four and eight finite elements.

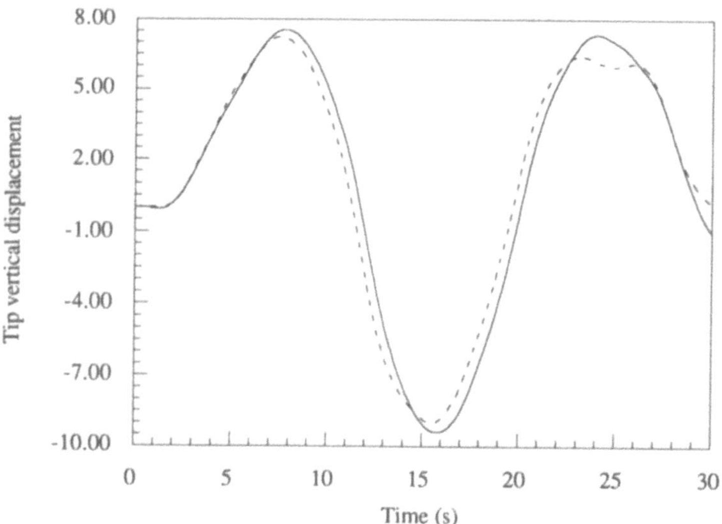

Figure 11.15. Tip vertical displacement using four and eight finite elements.

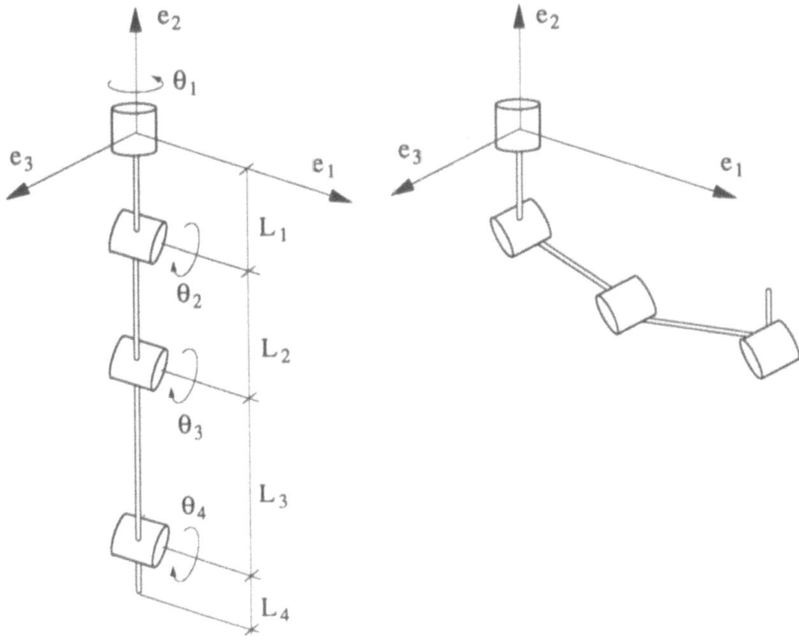

Figure 11.16. Spatial manipulator with two flexible links. Avello et al., 1991. Reprinted by permission of John Wiley & Sons, Ltd.

Example   11.4

A flexible spatial manipulator composed of two rigid and two flexible links is presented in Figure 11.16. Links 2 and 3 are flexible beams of tubular section. Each link is connected to the previous one through a revolute joint. At the midpoint of link 4 a lumped mass of 200 Kg has been attached to represent a load. The geometric and material properties of the links are:

$L_1 = 0.3$ m $\qquad$ Inner radius of the cross section for links 2 and 3.

$L_2 = 4.0$ m $\qquad\qquad\qquad\qquad$ $r_i = 0.04$ m

$L_3 = 5.0$ m

$L_4 = 0.5$ m $\qquad$ Outer radius of the cross section for links 2 and 3.

$E = 6895 \ 10^7$ N/m^2 $\qquad\qquad\qquad\qquad$ $r_o = 0.05$ m

$\rho = 2699$ Kg/m^3

Links 1 and 4 have been modeled, respectively, with a single finite element of high-elasticity modulus, and links 2 and 3 have been modeled with four elements each. The simulation that has been carried out is based on a prescribed motion in each revolute joint that moves the manipulator from the initial configuration to the final one, both shown in Figure 11.16. The prescribed motion is such that there is a rotation of 90 degrees in joints 1 and 4 and a rotation of 45 degrees in joints 2 and 3. The variation law of each joint is the following:

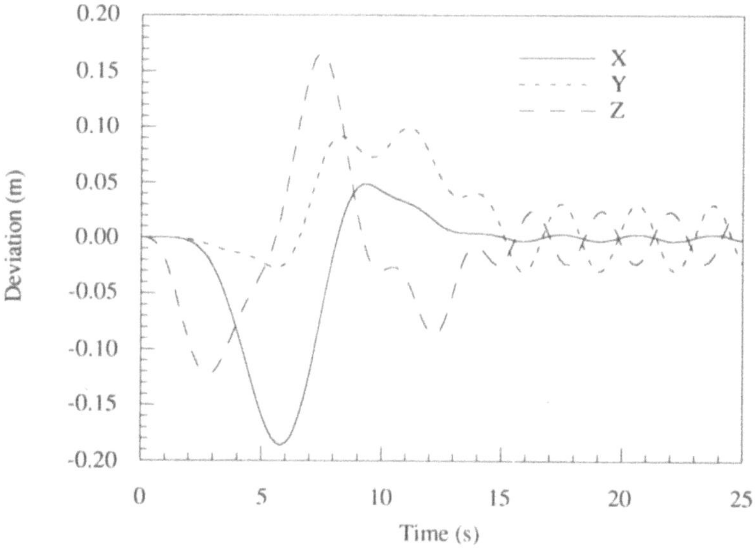

Figure 11.17. Tip deviations in the X, Y, and Z directions with respect to the rigid body trajectory.

$$
\theta_1 = \theta_4 = \begin{cases} \dfrac{\pi}{2T_s}\left(t - \dfrac{T_s}{2\pi}\,sin\!\left(\dfrac{2\pi t}{T_s}\right)\right) & 0 \le t \le T_s \\[2ex] \dfrac{\pi}{2} & t \ge T_s \end{cases}
$$

$$
\theta_2 = \theta_3 = \begin{cases} \dfrac{\pi}{4T_s}\left(t - \dfrac{T_s}{2\pi}\,sin\!\left(\dfrac{2\pi t}{T_s}\right)\right) & 0 \le t \le T_s \\[2ex] \dfrac{\pi}{4} & t \ge T_s \end{cases}
$$

The total simulation time is 25 sec, and T_S was taken to be 15 sec. Figure 11.17 illustrates the three (X,Y,Z) components of the tip deviation with respect to the nominal motion (that is, the trajectory obtained with all the links considered as rigid) as a function of time. The CPU time is 43.2 sec with a fixed step size of 0.2 sec and an average of 3.3 Newton-Raphson iterations per step.

References

Avello, A., "Dinámica de Mecanismos Flexibles con Coordenadas Cartesianas y Teoría de Grandes Deformaciones", Ph.D. Thesis, University of Navarre, San Sebastián, (1990).

Bathe, K.-J., *Finite Element Procedures in Engineering Analysis*, Prentice-Hall, (1982).

Bathe, K.-J. and Bolourchi, S., "Large Displacement Analysis of Three-Dimensional Beam Structures", *International Journal for Numerical Methods in Engineering*, Vol. 14, pp. 961-986, (1979).

Bayo, E. and Serna, M.A., "Penalty Formulations for the Dynamic Analysis of Elastic Mechanisms", *Journal of Mechanisms, Transmissions, and Automation in Design*, Vol. 111, pp.321-327, (1989).

Book, W.J., "Recursive Lagrangian Dynamics of Flexible Manipulator Arms", *International Journal of Robotics Research*, Vol. 3, pp. 87-101, (1984).

Cardona, A., "An Integrated Approach to Mechanism Analysis", Ph.D. Thesis, Université de Liége, Belgium, (1989).

Changizi, K. and Shabana, A.A., "A Recursive Formulation for the Dynamic Analysis of Open-Loop Deformable Multibody Systems", *ASME Journal of Applied Mechanics*, Vol. 55, pp. 687-693, (1988).

Craig, R.R., *Structural Dynamics*, Wiley, (1981).

Crisfield, M.A., "A Consistent Co-Rotational Formulation for Non-Linear, Three Dimensional, Beam-Elements", *Computer Methods in Applied Mechanics and Engineering*, Vol. 81, pp. 131-150, (1990).

Erdman, E.G. and Sandor, G.N., "Kineto-Elastodynamics – A Review of the State of the Art Trends", *Mechanism and Machine Theory*, Vol. 7, pp. 19-33, (1972).

Erdman, A.G. and Sung, C.K., "A Survey of Finite Element Techniques for Mechanism Design", *Mechanism and Machine Theory*, Vol. 21, pp. 351-359, (1986).

Hurty, W.C., "Dynamic Analysis of Structural Systems Using Component Modes", *AIAA Journal*, Vol. 3, pp. 678-685, (1965).

Kane T.R., Ryan R.R., and Banerjee, A.K., "Dynamics of a Cantilever Beam Attached to a Moving Base", *AIAA Journal of Guidance, Control, and Dynamics*, Vol. 10, pp. 139-151, (1987).

Kim, S.-S. and Haug, E.J., "A Recursive Formulation for Flexible Multibody Dynamics. Part I: Open-Loop Systems", *Computer Methods in Applied Mechanics and Engineering*, Vol. 71, pp. 293-314, (1988).

Kim, S.-S. and Haug, E.J., "A Recursive Formulation for Flexible Multibody Dynamics. Part II: Closed-Loop Systems", *Computer Methods in Applied Mechanics and Engineering*, Vol. 74, pp. 251-269, (1989).

Lowen, G.G. and Jandrasits, W.G., "Survey of Investigations into the Dynamic Behavior of Mechanisms Containing Links with Distributed Mass and Elasticity", *Mechanism and Machine Theory*, Vol. 7, pp. 3-17, (1972).

Lowen, G.G. and Chassapis, C., "The Elastic Behavior of Linkages: An Update", *Mechanism and Machine Theory*, Vol. 21, pp. 33-42, (1986).

Malvern, L.E., *"Introduction to the Mechanics of a Continuous Medium"*, Prentice-Hall, (1969).

Midha, A., Erdman, A.G., and Frohrib, D.A., "Finite Element Approach to Mathematical Modelling of High-Speed Elastic Linkages", *Mechanism and Machine Theory*, Vol. 13, pp. 603-618, (1978).

Naganathan, G. and Soni, A.H., "Coupling Effects of Kinematics and Flexibility in Manipulators", *International Journal of Robotics Research*, Vol. 6, pp. 75-85, (1987).

Shabana, A.A., *Dynamics of Multibody Systems*, Wiley, (1989).

Shabana, A.A. and Wehage, R.A., "A Coordinate Reduction Technique for Transient Analysis of Spatial Substructures with Large Angular Rotations", *Journal of Structural Mechanics*, Vol. 11, pp. 401-431, (1983).

Serna, M.A. and Bayo, E., "A Simple and Efficient Computational Approach for the Forward Dynamics of Elastic Robots", *Journal of Robotic Systems*, Vol. 6, pp. 363-382, (1989).

Simo, J.C., "A Three-Dimensional Finite-Strain Rod Model. Part II: Computational Aspects", *Computer Methods in Applied Mechanics and Engineering*, Vol. 58, pp. 79-116, (1986).

Simo, J.C. and Vu-Quoc L., "On the Dynamics of Flexible Beams Under Large Overall Motions – The Planar Case: Part I", *ASME Journal of Applied Mechanics*, Vol. 53, pp. 849-854, (1986).

Simo, J.C. and Vu-Quoc L., "The Role of Non-Linear Theories in Transient Dynamic Analysis of Flexible Structures", *Journal of Sound and Vibration*, Vol. 119, pp. 487-508, (1987).

Simo, J.C. and Vu-Quoc L., "On the Dynamics of Space Rods Undergoing Large Overall Motions", *Computer Methods in Applied Mechanics and Engineering*, Vol. 66, pp. 125-161, (1988).

Song, J.O. and Haug, E.J., "Dynamic Analysis of Planar Flexible Mechanisms", *Computer Methods in Applied Mechanics and Engineering*, Vol. 24, pp. 359-381, (1980).

Sunada, W. and Dubowsky, S., "The Application of Finite Element Methods to the Dynamic Analysis of Flexible Spatial and Coplanar Linkage Systems", *ASME Journal of Mechanisms, Transmissions, and Automation in Design*, Vol. 103, pp. 643-651, (1981).

Vukasovic, N., Celigüeta, J.T. ,García de Jalón, J., and Bayo, E., "Flexible Multibody Dynamics Based on a Fully Cartesian System of Support Coordinates", *ASME Journal of Mechanical Design*, Vol. 115, pp. 294-299, (1993).

Winfrey, R.C., "Elastic Link Mechanism Dynamics", *ASME Journal of Engineering for Industry*, Vol. 93, pp. 268-272, (1971).

12
Inverse Dynamics of Flexible Multibodies

Applications of artificial manipulation and robotics are steadily increasing in areas such as: microelectronics, agile space aircraft, vacuum mechatronics, satellite-mounted robots, biomedical sciences, teleoperation, assembly lines, manufacturing, and so forth. As a consequence, more demands are being placed on these systems, such as the need to design and use light and fast arms handling heavy payload with accuracy and low energy consumption. If the various links of a manipulator are to be considered rigid, they must be structurally stiff, and this leads to bulky and massive designs. If speed is not to be sacrificed, powerful and heavy actuators with high energy consumption are in turn required to move these arms. The most natural remedy is to use flexible multibodies with slender links.

The requirement that manipulators and multibody systems be flexible places new demands and challenges in their design, analysis, and control. The flexibility of their members becomes a very important factor that must be considered, so that vibrations are avoided, particularly for position control. In this chapter, concentration will be placed on the inverse dynamics of flexible multibodies. This recently introduced approach (Bayo (1987) and Bayo et al. (1989)) consists of finding the feed-forward torques that need to be applied at the joints so that the end effector can follow a desired trajectory. From the vibration control viewpoint, the inverse dynamics provides an inversion of the system dynamics. This gives the control specialist a strong tool with which he can design stable and robust control laws for the motion control of multibody systems (Paden et al. (1993)).

This chapter begins with the solution of the single-link case where the elastic Coriolis and centrifugal terms are negligible, and algorithms are proposed for the solution of the resulting time invariant system. The algorithms are then extended for the time variant case that includes the Coriolis and centrifugal terms. Solutions in the time and frequency domains are provided. It is clearly shown how the inverse solution is anticipatory, also called *non-causal*, meaning that the actuation precedes the endpoint motion and continues after it has stopped. For the sake of clarity and conciseness, we will concentrate on the planar case, and we will provide indications as to how to proceed with the 3D cases. Simulation examples are given to clarify the meaning of the inverse dynamics.

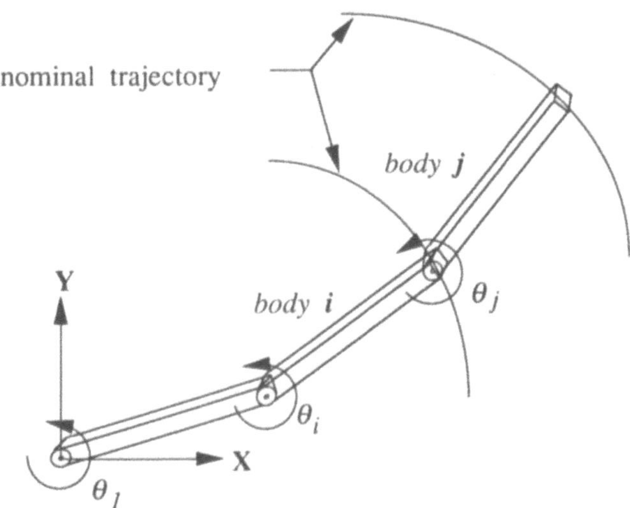

Figure 12.1. Flexible multibody system.

The techniques and methods explained in this chapter are somehow more advanced than those seen in previous chapters. *Non-causal* integration requires that the reader be familiar with the basics of Fourier and Laplace analysis. In this chapter, we deviate from the natural coordinates and use the reference point coordinates instead. This is because the reference point coordinates provide a better setting for the non-causal inversion that the inverse dynamics require.

12.1 Inverse Dynamics Equations for Planar Motion

Consider a general multi-link flexible multibody system (Figure 12.1). The objective is to move the end effector along a given trajectory without overshoot and residual elastic oscillations of the tip. These oscillations are due mainly to the transverse elastic displacements of the links. The longitudinal axial oscillations are negligible because of the much greater rigidity of the links in their axial direction. For the sake of simplicity, the equations will be derived for planar manipulators with revolute joints. The procedure, however, is also valid for general spatial manipulators.

The solution is obtained by first studying an individual link in the chain, coupling the equations of the individual links, and then setting up an iteration scheme that converges to the desired torques and corresponding joint displacements.

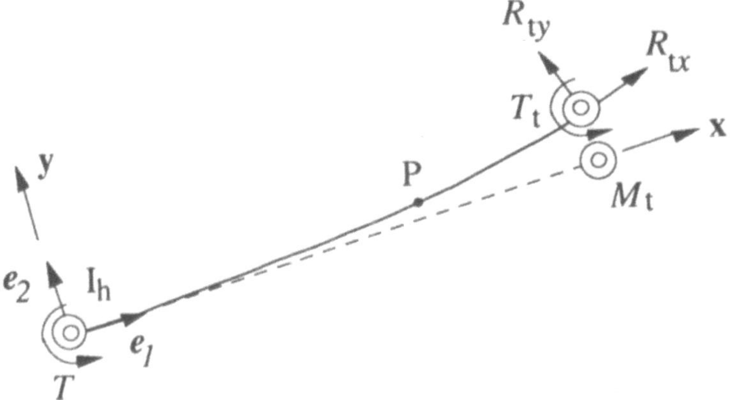

Figure 12.2. Nominal motion (dotted) and elastic deformation (solid) in a flexible body.

12.1.1 Inverse Dynamics Equations of an Individual Link

The individual flexible link depicted in Figure 12.2 forms part of a planar multi-link manipulator and has a total length L, mass per unit length $\bar{m}$, moment of inertia I, area A, Young modulus E, shear modulus G, and shear coefficient k. A tip mass of value M_t is attached at one end, and a hub of inertia I_h at the other end. The hub is attached to the actuator.

A point P at a distance x from the center of the hub has undergone elastic deflections of value u_x and u_y and rotation θ. These are defined with respect to a nominal position characterized by the moving frame (e_1, e_2) attached to the hub, that rotates at a specified (nominal) angular velocity and acceleration ω_h and $\dot{\omega}_h$, respectively. The idea of the inverse dynamics is to force the end of the link to follow the specified (desired) nominal motion. This is an important conceptual difference with the methods of Chapter 11, in which the nominal motion of the moving frame was not specified, because it was actually unknown.

As a consequence of the elastic deflections and rotating nominal motion, the point P is subjected to a total translational acceleration $\mathbf{a}_p$ and angular acceleration $\dot{\omega}_p$. Using the principles of relative motion (Greenwood (1988)), the acceleration of the point P can be set in terms of the translation and angular accelerations at the hub, $\mathbf{a}_h$ and $\dot{\omega}_h$, angular velocity ω_h at the hub, and the relative velocity $\mathbf{v}_{rel}$ and acceleration $\mathbf{a}_{rel}$ of point P. The latter are due to the elastic deflections u_x and u_y with respect to the moving frame. In vectorial notation:

$$\mathbf{a}_p = \omega_h \wedge (\omega_h \wedge \bar{\mathbf{r}}_p) + \dot{\omega}_h \wedge \bar{\mathbf{r}}_p + 2\,\omega_h \wedge \mathbf{v}_{rel} + \mathbf{a}_h + \mathbf{a}_{rel} \tag{12.1a}$$

$$\dot{\omega}_p = \dot{\omega}_h + \ddot{\theta} \tag{12.1b}$$

where $\bar{\mathbf{r}}_F = (x + u_x)\,\mathbf{e}_1(t) + u_y\,\mathbf{e}_2(t)$ is the position of P after deformation, relative to the hub.

The components of the relative velocity are $\dot{u}_x$ and $\dot{u}_y$. Those of the relative acceleration are $\ddot{u}_x$ and $\ddot{u}_y$. Performing the vectorial operations involved in (12.1) the following components of the accelerations are obtained:

$$a_x = -\omega_h^2\, u_x - \dot{\omega}_h\, u_y - 2\,\omega_h\,\dot{u}_y + \ddot{u}_x - \omega_h^2\, x + a_{hx}$$

$$a_y = -\omega_h^2\, u_y + \dot{\omega}_h\, u_x + 2\,\omega_h\,\dot{u}_x + \ddot{u}_y + \dot{\omega}_h\, x + a_{hy} \tag{12.2}$$

$$\dot{\omega}_p = \dot{\omega}_h + \ddot{\theta}$$

Using the Timoshenko beam theory which includes the effects of shear deformation and rotatory inertia, the principle of virtual displacements (Chapter 4) can be used directly to set up the equations of motion:

$$\int_0^L \left[\overline{m}\, a_x\, \delta u_x + \overline{m}\, a_y\, \delta u_y + \overline{m}\, \eta^2\, \dot{\omega}_p \delta\theta\right] dx + I_h\,(\dot{\omega}_h + \ddot{\theta}_h)\delta\theta_h + M_t\, a_t\, \delta u_t +$$

$$+ \int_0^L \left[EA\, u_x'\, \delta u_x' + EI\, \theta'\delta\theta' + GAk\,(\theta - u_x')\,\delta(\theta - u_x')\right] dx =$$

$$= T\delta\theta_h + R_{ty}\, \delta u_{ty} + R_{tx}\, \delta u_{tx} + T\delta\theta_t \tag{12.3}$$

where η is the radius of gyration of the section. The subscripts h and t indicate hub and tip, respectively. The symbol (') indicates derivative with respect to the spatial variable, and δu_x, δu_y, and $\delta\theta$ represent a set of virtual elastic displacements. T is the unknown torque to be applied at the hub, so that the prescribed tip motion is obtained. R_{ty}, R_{tx}, and T_t are the reaction forces and the torque at the tip that comes from the next link in the chain (See Figure 12.2). Note that the acceleration at the hub is decomposed into $\dot{\omega}_h$ and $\dot{\omega}_h$. The first is the nominal acceleration, and the second is the acceleration due to the elastic deflections. Also observe that the reactions at the hub do not have any effect on the total virtual work. This assumption is met by imposing the constraint that the hub move along the nominal path without any elastic deformations (See Figure 12.1). As shown later in the discussion, this condition is enforced in each of the links to compute the inverse dynamic torques. Substituting equation (12.2) into (12.3), one can obtain:

$$\int_0^L [\overline{m}\,\ddot{u}_x\delta u_x + \overline{m}\,\ddot{u}_y\delta u_y + \overline{m}\,\eta^2\,\ddot{\theta}\delta\theta]\, dx + \int_0^L [2\overline{m}\,\omega_h\,(\dot{u}_x\,\delta u_y - \dot{u}_y\,\delta u_x)]\, dx +$$

$$+ \int_0^L [\overline{m}\,\omega_h^2(-u_x\,\delta u_x - u_y\,\delta u_y) + \dot{\omega}_h\,\overline{m}\,(u_x\,\delta u_y - u_y\,\delta u_x]\, dx +$$

$$+ I_h(\dot{\omega}_h + \ddot{\theta}_h)\,\delta\theta_h + M_t\, a_t\delta u_t + \int_0^L [EI\,\theta'\delta\theta' + GAk(\theta - u_y')\,\delta(\theta - u_y') + EAu_x'\,\delta u_x']\, dx =$$

$$= T\,\delta\theta_h + R_{ty}\delta u_{ty} + R_{tx}\delta u_{tx} + T_t\delta\theta_t - \int_0^L [-\overline{m}\omega_h^2\, x\delta u_x + \overline{m}\dot{\omega}_h\, x\,\delta u_y + \overline{m}\eta^2\,\dot{\omega}_h\delta\theta]\, dx \tag{12.4}$$

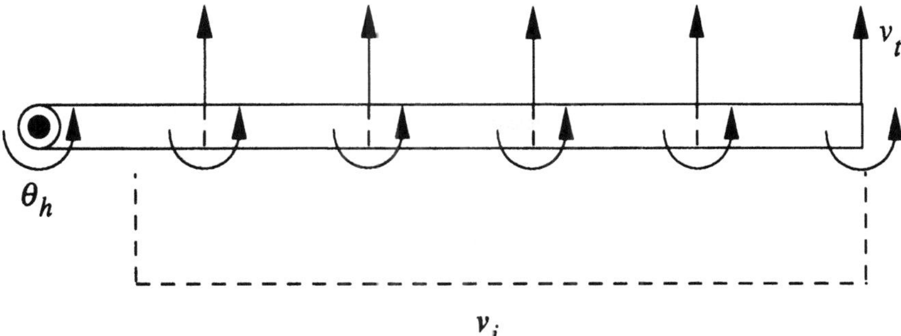

Figure 12.3. Finite element partition in the body.

The first integral on the LHS corresponds to the internal virtual work done by the inertial forces including the rotatory inertia. The second integral term represents the work done by the Coriolis forces. The third integral corresponds to the centrifugal and tangential forces due to the rotating frame. This last effect also produces additional forcing terms that are represented in the RHS of the equation. The fourth integral corresponds to the virtual work done by the internal axial and shear forces and bending moments.

The displacement field of equation (12.3) can be discretized using the finite element or assumed mode method under *pin-free* boundary conditions (Figure 12.3). A set of interpolation functions are defined within each body:

$$u_x(x,t) = \sum_1^n H_i(x)\, u_x^i(t), \quad u_y(x,t) = \sum_1^n H_i(x)\, u_y^i(t) , \quad \theta(x,t) = \sum_1^n H_i(x)\, \theta^i(t) \quad (12.5)$$

where H_i are the interpolation functions; u_x^i, u_y^i, and θ^i indicate the nodal or generalized deflections.

Substituting equations (12.5) in the virtual work expression (12.4), and following standard procedures for the formation and assemblage of element matrices (Bathe 1982), the equations of motion of the link may be expressed by a set of time varying differential equations in the form

$$\mathbf{M}\,\ddot{\mathbf{v}} + \left[\mathbf{C} + \mathbf{C}_c(\omega_h)\right]\dot{\mathbf{v}} + \left[\mathbf{K} + \mathbf{K}_c(\omega_h, \dot{\omega}_h)\right]\mathbf{v} = \mathbf{T} - \mathbf{Q}(\omega_h, \dot{\omega}_h) \quad (12.6)$$

where $\mathbf{M}$ and $\mathbf{K}$ are the conventional finite element (or assumed mode) mass and stiffness matrices, respectively. $\mathbf{C}_c$ and $\mathbf{K}_c$ are the time varying Coriolis and centrifugal stiffness matrices that depend on the nominal angular velocity ω_h and acceleration $\dot{\omega}_h$ of the link. Matrix C has been added to represent the internal viscous damping of the material. Vector $\mathbf{T}$ contains one non-zero term only, and that is the unknown torque at the hub. Finally, $\mathbf{Q}$ contains the reactions and the torque at the end of the link and the known forces produced by the rotating frame effect. A detailed description of these matrices for a Timoshenko beam finite element is given in the Appendix.

12.1.2 Solution of the Inverse Dynamics for an Individual Link

The set of equations (12.6) may be partitioned as follows:

$$\mathbf{M}\begin{bmatrix}\ddot{\theta}_h \\ \ddot{\mathbf{v}}_i \\ \ddot{v}_t\end{bmatrix} + [\mathbf{C} + \mathbf{C}(\omega_h)]\begin{bmatrix}\dot{\theta}_h \\ \dot{\mathbf{v}}_i \\ \dot{v}_t\end{bmatrix} + [\mathbf{K} + \mathbf{K}_c(\dot{\omega}_h, \omega_h)]\begin{bmatrix}\theta_h \\ \mathbf{v}_i \\ v_t\end{bmatrix} = \begin{bmatrix}1 \\ 0 \\ 0\end{bmatrix}T - \begin{bmatrix}Q_h \\ Q_i \\ Q_t\end{bmatrix} \quad (12.7)$$

where θ_h is the rotation at the hub, v_t is the elastic deflection of the end point, and $\mathbf{v}_i$ is the vector containing all the other internal finite element degrees of freedom such as displacements and rotations of the nodal points, as shown in Figure 12.3. The force vector $\mathbf{Q}$ is partitioned in the same manner.

The solution of equations (12.7) for a specified nominal motion, defined by ω_h and $\dot{\omega}_h$, and a set of external forces applied to the link, constitutes the forward dynamic problem. This simply requires the numerical integration of a set of time variant ODEs for which techniques are readily available (See Chapter 7). A problem of different and more complicated nature is the inverse dynamics. This problem is to find the torque T that will ensure that the endpoint will move according to a specified trajectory avoiding any possible oscillations and deviations from the path. The problem can be quantitatively stated as the finding of $T(t)$ in equation (12.7) under the condition that elastic normal deflection at the tip $v_t(t)$ be zero.

12.1.2.1 The Time Invariant Case

Under low nominal speeds of operations, the elastic Coriolis and centrifugal terms acting on the LHS of equation (12.7) become insignificant and are customarily neglected. An example will be shown later which clearly illustrates this assumption. Consequently, equation (12.6) becomes

$$\mathbf{M}\,\ddot{\mathbf{v}} + \mathbf{C}\,\dot{\mathbf{v}} + \mathbf{K}\,\mathbf{v} = \mathbf{T} - \mathbf{Q}(\omega_h, \dot{\omega}_h) \quad (12.8)$$

Observe that the force vector $\mathbf{Q}$ in the RHS of (12.8) contains the external Coriolis and centrifugal forcing terms due to the rigid body motion which are not neglected (See Appendix).

Direct integration in the time domain. The inverse dynamics is, in the classical sense, an ill-posed problem, because its solution does not depend continuously on the data. The intention of this section is to briefly describe that a standard integration in the time domain leads to an unbounded (thus unstable) solution and that the *unique stable solution is found to be non-causal*. Actuation is required before the endpoint has started to move as well as after the endpoint has stopped. After a first glance at equation (12.8) one may be tempted to partition the system of equations as follows:

$$\begin{bmatrix}\mathbf{M}_{11} & m_{12} \\ \mathbf{M}_{21} & \mathbf{M}_{22}\end{bmatrix}\begin{bmatrix}\ddot{\theta}_h \\ \ddot{\mathbf{v}}_i \\ \ddot{v}_t\end{bmatrix} + \begin{bmatrix}\mathbf{C}_{11} & c_{12} \\ \mathbf{C}_{21} & \mathbf{C}_{22}\end{bmatrix}\begin{bmatrix}\dot{\theta}_h \\ \dot{\mathbf{v}}_i \\ \dot{v}_t\end{bmatrix} + \begin{bmatrix}\mathbf{K}_{11} & k_{12} \\ \mathbf{K}_{21} & \mathbf{K}_{22}\end{bmatrix}\begin{bmatrix}\theta_h \\ \mathbf{v}_i \\ v_t\end{bmatrix} = \begin{bmatrix}1 \\ 0 \\ 0\end{bmatrix}T - \begin{bmatrix}Q_h(t) \\ Q_i(t) \\ Q_t(t)\end{bmatrix} \quad (12.9)$$

where for a system with n degrees of freedom (n equations): $\mathbf{M}_{11}$, $\mathbf{C}_{11}$, and $\mathbf{K}_{11}$ are $1\times(n-1)$ row vectors; m_{12}, c_{12} and k_{12} are (1×1) elements; $\mathbf{M}_{21}$, $\mathbf{C}_{21}$, and $\mathbf{K}_{21}$ are $(n-1)\times(n-1)$ matrices; and, $\mathbf{M}_{22}$, $\mathbf{C}_{22}$, and $\mathbf{K}_{22}$ are $(n-1)\times1$ column vectors. Collecting the $v_t(t)$ dependent known terms on the right-hand side, the last $n-1$ equations of (12.9) can be rewritten as:

$$\mathbf{M}_{11}\begin{bmatrix}\ddot{\theta}_h \\ \ddot{\mathbf{v}}_i\end{bmatrix} + m_{12}\ddot{v}_t + \mathbf{C}_{11}\begin{bmatrix}\dot{\theta}_h \\ \dot{\mathbf{v}}_i\end{bmatrix} + c_{12}\dot{v}_t + \mathbf{K}_{11}\begin{bmatrix}\theta_h \\ \mathbf{v}_i\end{bmatrix} + k_{12}v_t = T(t) - Q_h(t) \quad (12.10)$$

while the first equation of (12.9) is

$$\mathbf{M}_{11}\begin{bmatrix}\ddot{\theta}_h \\ \ddot{\mathbf{v}}_i\end{bmatrix} + m_{12}\ddot{v}_t + \mathbf{C}_{11}\begin{bmatrix}\dot{\theta}_h \\ \dot{\mathbf{v}}_i\end{bmatrix} + c_{12}\dot{v}_t + \mathbf{K}_{11}\begin{bmatrix}\theta_h \\ \mathbf{v}_i\end{bmatrix} + k_{12}v_t = T(t) - Q_h(t) \quad (12.11)$$

In principle, given initial conditions for $\theta_h(t=0)$ and $v_i(t=0)$, equation (12.10) may be integrated to yield $\theta_h(t)$ and $v_i(t)$ (inverse kinematics). Then, substitution of $\theta_h(t)$ and $v_i(t)$ into equation (12.11) would yield the required actuating torque $T(t)$ (inverse dynamics). However, this simple approach yields an *unbounded* and thus unacceptable solution, which does include a time delay between the actuation and the response at the endpoint.

In order to clarify this point, let the beginning of the prescribed tip motion be at time zero; so that $v_i(t)$ is zero for $t < 0$, and let the arm be initially at rest. Since the elastic waves in a solid have a finite speed of propagation, intuition dictates that in the case of a flexible arm, the torque must be applied before the tip starts moving; that is, it must be non-zero before $t=0$. This time anticipation is necessary for the actuation to reach the end of the arm. Using a term customarily used in control theory, the desired torque should be *non-causal*. As a consequence, if standard numerical integration of ODEs for initial value problems is carried out, no time anticipation would be present, and the resulting torque becomes unbounded. From a control point of view it can be said that this way of proceeding leads to a *causal* inverse which, in the case of a non-minimum phase problem such as the one at hand, is always unbounded and thus unstable. This important point is addressed in detail in Moulin (1989), and Moulin and Bayo (1991).

Allowing for this time delay, let the actuation begin at some negative time τ with the arm initially at rest and initial conditions equal to zero. Now, over $[\tau, 0]$ $v_t(t)$ is identically zero, and the unique solution of equation (12.9) with zero initial conditions and zero terms on the right-hand side will be $\theta_h(t)=0$ and $v_i(t)=0$. Substitution of $\theta_h(t)$ and $v_i(t)$ into equation (12.11) leads again to $T(t) = 0$ over $[\tau, 0]$. Hence, the torque resulting from this approach is always zero before the tip starts moving and does not lead to the desired time anticipatory non-causal inverse dynamic solution described above. The delay effect does not appear, regardless of the value of the time τ introduced, and the resulting torque becomes unbounded.

In summary, the inverse dynamic equations need to be solved by means of an integration process that will yield the time delay between actuation and response. The fact that the stable solution is *non-causal* in that it starts at negative time and extends to future time precludes standard time domain initial value ODE solvers from obtaining the proper solution. We will show in the next sections how this problem can be circumvented by solving the inverse dynamics in the *frequency domain* using the Fourier transform, or in the *time domain* using the bilateral Laplace transform.

Solution in the frequency domain. The frequency domain approach captures the desired time delay between actuation and response, because the initial and end conditions are imposed at $-\infty$ and $+\infty$, respectively. The system of equations (12.8) can be transformed by means of the fast Fourier transform (FFT) (Newland (1984)) into a set of algebraic equations with complex entries. The frequency associated with each of the Fourier pairs is equal to $\overline{w}_i = (2\pi/\Pi)\,i$, where Π is the total time interval considered for the motion of the system and i is the number of the Fourier pair. For a particular frequency $\overline{w}$, equation (2) becomes

$$\left[\mathbf{M} + \frac{1}{j\,\overline{w}}\,\mathbf{C} - \frac{1}{\overline{w}^2}\,\mathbf{K}\right]\begin{Bmatrix} \hat{\ddot{v}}_h \\ \hat{\ddot{v}}_i \\ \hat{\ddot{v}}_t \end{Bmatrix} = \begin{Bmatrix} \hat{T}(\overline{w}) \\ 0 \\ 0 \end{Bmatrix} - \begin{Bmatrix} \hat{Q}_h \\ \hat{Q}_i \\ \hat{Q}_t \end{Bmatrix} \tag{12.12}$$

where the caret stands for the Fourier transform, and $\mathbf{Q}$ represents all the known forcing terms. Equation (12.12) may be expressed in simplified notation as

$$\mathbf{H}\,\hat{\ddot{v}} = \mathbf{I}\,\hat{T}(\overline{w}) - \hat{\mathbf{Q}} \tag{12.13}$$

The transfer matrix $\mathbf{H}$ is a complex non-singular symmetric matrix except for $\overline{w} = 0$ for which it is not defined. However, the zero frequency represents the rigid body motion. Therefore, the corresponding component of the torque can be obtained by simply applying equilibrium of the moments produced by all the external forces about the origin of the link. For the Fourier pairs with $\overline{w} \neq 0$, $\hat{T}(\overline{w})$ may be obtained by solving (12.13) for each frequency as follows:

$$\begin{Bmatrix} \hat{\ddot{v}}_h \\ \hat{\ddot{v}}_i \\ \hat{\ddot{v}}_t \end{Bmatrix} = \begin{bmatrix} \mathbf{G}_{hh} & \mathbf{G}_{hi} & \mathbf{G}_{ht} \\ \mathbf{G}_{ih} & \mathbf{G}_{ii} & \mathbf{G}_{it} \\ \mathbf{G}_{th} & \mathbf{G}_{ti} & \mathbf{G}_{tt} \end{bmatrix} \left(\begin{Bmatrix} \hat{T}(\omega) \\ 0 \\ 0 \end{Bmatrix} - \begin{Bmatrix} \hat{Q}_h \\ \hat{Q}_i \\ \hat{Q}_t \end{Bmatrix} \right) \tag{12.14}$$

where $\mathbf{G}$ is the inverse of the complex transfer matrix $\mathbf{H}$. From (12.14), it is obvious that

$$\hat{\ddot{v}}_t = G_{th}\,\hat{T}(\overline{w}) - G_{th}\,\hat{Q}_h - G_{ti}\,\hat{Q}_i - G_{tt}\,\hat{Q}_t \tag{12.15}$$

Equation (12.15) may now be solved for the required torque T under the condition that $v_t = 0$:

$$\hat{T}(\overline{w}) = G_{th}^{-1}\left(G_{th}\,\hat{Q}_h + G_{ti}\,\hat{Q}_i + G_{tt}\,\hat{Q}_t\right) \tag{12.16}$$

The values of the torque $T(t)$ are obtained through the application of the inverse discrete Fourier transform. The joint angles and velocities that will yield the desired endpoint motion (inverse kinematics), and which are used for control purposes (Paden et al. (1993)), can be obtained in the time domain by a forward integration of equation (12.8). Since the forces $\mathbf{Q}$ are linear functions of the nominal accelerations, equation (12.16) also leads to an expression for the transfer function between the torque T and the endpoint acceleration. This transfer function contains poles in the right half-plane, thus characterizing this non-minimum phase system.

By using the frequency domain, the initial and final conditions are imposed at $-\infty$ and $+\infty$. This leads to an inverse dynamics torque that shows the above-mentioned time anticipation with respect to the endpoint motion (See example below); thus providing the *non-causal* inverse to this non-minimum phase problem. A formal explanation of this issue is given in Moulin (1989), and Moulin and Bayo (1991).

The following algorithm summarizes the steps necessary for the inverse dynamics and kinematics of a single-link arm in the frequency domain:

Algorithm 12-1

1. Define the rigid nominal motion, ω_h and $\dot{\omega}_h$.
2. Evaluate the forcing terms Q_h, Q_i, and Q_t which depend on ω_h and $\dot{\omega}_h$.
3. Apply the fast Fourier transform to the forcing terms.
4. Solve for the torque $\hat{T}(w)$ in the frequency domain using equation (12.16). This involves the solution of a set of algebraic linear complex equations.
5. Obtain $T(t)$ through the inverse discrete Fourier transform.
6. Perform a forward time integration of equation (12.8) to obtain the joint angles and velocities (inverse kinematics).

Stable integration in the time domain. From the above observation that in order to obtain a stable inverse that includes the delay between the actuation and response, the integration process needs to be carried out from $-\infty$ to $+\infty$. The equivalent of equation (12.16) in the time domain is

$$T(t) = \int_{-\infty}^{\infty} \left(h_{th}(t-\tau)\, Q_h(\tau) + \mathbf{h}_{ti}(t-\tau)\, Q_i(\tau) + h_{tt}(t-\tau)\, Q_t(\tau) \right) d\tau \qquad (12.17)$$

where h_{th}, $\mathbf{h}_{ti}$, and h_{tt} are the impulse response functions that correspond to the transfer functions defined in (12.16). The former are the inverse Fourier transforms of the latter.

Since there is no complex algebra involved, the integration in the time domain will in general be faster than that in the frequency domain. In addition, since the impulse response functions depend only on the physical characteristics of the link, they can be computed off-line. The only on-line computations involved consequently to obtain the torque $T(t)$ will be the evaluation of the integral (12.17).

It should be observed that equation (12.17) differs from the familiar Duhamel integral:

$$y(t) = \int_{0}^{t} h(t-\tau)\, i(\tau)\, d\tau \qquad (12.18)$$

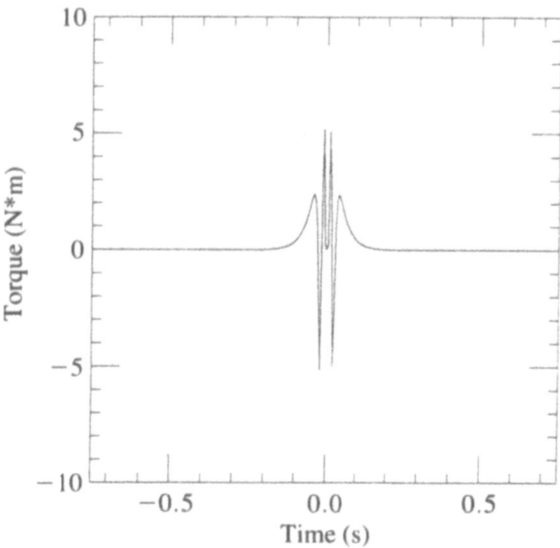

Figure 12.4. Typical tip-to-hub impulse response function of a flexible body.

relating the input $i(t)$ to the output $y(t)$ of a linear system with impulse response function $h(t)$ by the bounds of integration.

Equation (12.18), which is typical of a causal response, is valid when the input $i(\tau)$ and the response function $h(t)$ are zero for $\tau < 0$. The condition on $h(\tau)$ states that there is no output before there is an input, in other words the system is *causal*. In the inverse dynamics problem, the input is formally the tip acceleration profile, and the output is formally the actuating torque. As mentioned above, there is a delay in the forward problem between hub torque actuation and tip response. Therefore in the inverse problem, the torque must be applied before the tip starts moving. The inverse is *non-causal*. For a non-causal system, the impulse response functions in (12.17) are not identically zero for negative time. Figure 12.4 shows a typical impulse response function for inverse dynamics. The convolution integral of (12.17) cannot be reduced to that of (12.18).

The use of equation (12.17) hinges on the availability of the impulse response functions. This can be obtained in the following manner: if the input α_h is a Dirac delta function at time τ, then the output $T(t)$ is the impulse response function translated by τ. One way of obtaining $h_i(t)$ numerically is to use equation (12.16) with an approximation of a delta function for $\dot{\omega}_h$. Having computed the impulse response functions, we now turn to the evaluation of the convolution integral defined by (12.17). This equation can be evaluated at discrete time intervals of length Δt, using a suitable composite integration rule.

The following paragraph summarizes the steps involved in the inverse dynamics integration in the time domain:

Algorithm 12-2

1. Define the rigid nominal motion: ω_h and $\dot{\omega}_h$.
2. Evaluate the impulse response functions.
3. Calculate the forcing terms F_h, F_i, and F_t which depend on ω_h and $\dot{\omega}_h$.
4. Solve for the torque $T(t)$ by means of the convolution integrals (12.17).
5. Perform a forward time integration of equation (12.8) to obtain the joint angles and velocities (inverse kinematics).

This algorithm leads to a very fast computation of the inverse dynamics and allows one to obtain real time response (Bayo and Moulin (1989a)). Another efficient time domain approach consists of decomposing the transfer function into its causal and non-causal components (Kwon and Book (1990)).

12.1.2.2 The Time Varying Case

In case the manipulator undergoes a fast motion and one desires to include the elastic Coriolis and centrifugal terms, the solution for the torque $T(t)$ can still be found in either the frequency or time domains by means of an iteration procedure. In order to set up the iterative process, equation (12.6) may be restated as follows:

$$\mathbf{M}\,\ddot{\mathbf{v}} + \mathbf{C}\,\dot{\mathbf{v}} + \mathbf{K}\,\mathbf{v} = \mathbf{l}\,T - \mathbf{Q}(\dot{\omega}_h, \omega_h) - \mathbf{C}_c(\omega_h)\,\dot{\mathbf{v}} - \mathbf{K}_c(\dot{\omega}_h, \omega_h)\,\mathbf{v} \qquad (12.19)$$

where all the time invariant terms have been left in the LHS of the equation and the time varying ones have been collected on the RHS. The vector $\mathbf{l}$ contains a unit value for the degree of freedom corresponding to the hub rotation and zero for the rest, as shown in equation (12.7).

The iteration process is initiated by solving equation (12.19) for the unknown torque T, using either the frequency or time domain procedures described above. The first iteration is done in the absence of the last two terms involving $\mathbf{C}_c$ and $\mathbf{K}_c$ in the RHS and yields a displacement vector $\mathbf{v}^1(t)$ which in turn will be used to compute the last two terms in (12.19). The process is then repeated with the new force vector under the constraint that $v_t(t)=0$. The iteration procedure may be stopped, when the norm $\|\mathbf{v}^i - \mathbf{v}^{i-1}\|$ for the solution of two consecutive iterations is smaller than a prescribed tolerance. A formal proof of convergence of this algorithm is given by Moulin et al. (1992). The experience of the authors in all the cases analyzed reveals that unless the accelerations and velocities are large, the terms involving $\mathbf{K}_c$ and $\mathbf{C}_c$ are insignificant, and this iterative procedure can be neglected. Also and most importantly, when the speeds of operations are large, not only the Coriolis and centrifugal terms need be included but also the nonlinear geometric stiffening effects (Bayo and Moulin (1989b)).

Algorithm 12-3

1. Define the rigid nominal motion: ω_h and $\dot{\omega}_h$.
2. Evaluate the forcing terms on the RHS of (12.19) that depend on ω_h, $\dot{\omega}_h$, $\mathbf{v}$, and $\dot{\mathbf{v}}$.
3. Apply the fast Fourier transform to the forcing terms.
4. Solve for the torque $\widehat{T}(w)$ in the frequency or time domains.
5. Add $T(t)$ to the forces on the RHS of equation (12.19) and perform a forward integration to obtain $\mathbf{v}(t)$ (inverse kinematics).
6. Check convergence. If convergence is obtained, stop; otherwise go to step 2.

12.2 Recursive Inverse Dynamics for Open-Chain Configurations

In this section, the procedure outlined above for the inverse dynamics of the single-link is extended in a recursive manner for multi-link open-chain configurations. These cases can be decomposed into a series of individual links that can be analyzed recursively. This analysis is similar to that used in the previous section.

12.2.1 The Planar Open-Chain Case

Similar to the single-link case, the solution process for an open-chain robot is started by defining the nominal motion consisting of the inverse kinematics of the robot as if it were rigid and characterized by the θ_h, ω_h, and $\dot{\omega}_h$ of each individual link. A difference with respect to the single link arises from the fact that an intermediate link in the chain contains reaction forces at its endpoint. These forces come from the next distal link and are to be added to those arising from the moving frame effect to form the forcing vector $\mathbf{Q}$ in equation (12.19). The solution of equation (12.19) for the desired torque requires that these reaction forces contained in the force vector $\mathbf{Q}$ be known in each link.

In the open-chain case this difficulty can be easily overcome by starting the inverse dynamics with the last link, since in this case there are no link reaction forces at its end. Once the torque for the last link has been obtained under the condition that the elastic displacement at the tip is zero, the next step is to compute the reactions at the hub which will be transmitted to the previous link in the chain. These reactions may be obtained simply by equilibrium considerations. The procedure continues with the next link in the chain in the same manner as before. Reaction forces are present and therefore included. This process is conceptually similar to the recursive Newton-Euler scheme for inverse dynamics of rigid manipulators.

The method continues with the rest of the links, until the torque on the first link is determined. This way of proceeding assures that the end of each link moves along the desired nominal trajectory without oscillating from it, as shown in Figure 12.1. Once the torques T have been obtained, the motions at any point of the links specified by v_i or the angles θ_h (inverse kinematics) follow from equation (12.6) by a direct analysis. The basic steps involved in the process are summarized as follows:

Algorithm 12-4

1. Define the nominal motion consisting of inverse kinematics of the system considered rigid.
2. For each link j starting from the last in the chain:
   a) Compute the torque T^j imposing $v^j_{tip}=0$ and obtain $v(t)$ (inverse kinematics) following the single-link time-varying approach.
   b) Compute the link reaction forces R^j from equilibrium.

12.2.2 The Spatial Open-Chain Case

In the case of spatial manipulators with elastic properties in all directions, the elastic displacements contained in the plane defined by the joint axis and the tip of the link cannot be controlled by only one joint torque. These elastic displacements will influence the motions of subsequent links, introducing perturbations in their nominal motion. The nominal position of each link will be modified by the elastic displacements at the end of the previous link as follows:

$$r^i_n = r^{i-1}_n - v_{tip} \tag{12.20}$$

where r_n is a vector describing the nominal position and orientation of the link, v_{tip} indicates the elastic deformations at the end of the previous link, and "i" is the iteration number. The steps described above for the open- and closed-chain will have to be repeated introducing the displacement corrections, starting with the last link. Assuming that the elastic deformations are small compared to the overall nominal motions, the process should converge rapidly.

According to the authors' experience, the recursive procedure described in the previous section is the most suitable for open-chain configurations in which the elastic deflections on the proximal links does not affect the overall motion of the distal links. Such is the case in the planar open-chain case or in spatial manipulators that are designed with large stiffness in the plane defined by the joint axis and the endpoint of the link; so that the corresponding elastic displacements are negligible. If this is not the case, the recursive procedure demands an iteration process to account for those perturbations, and the inverse dynamics for each link needs to be repeated until convergence. Non-recursive approaches are more suitable and efficient under these circumstances, and we will see them in the next section. Other special recursive methods that avoid the need to iterate are currently being developed in this exciting area of research (Ledesma and Bayo (1992b)).

12.3 Non-Recursive Inverse Dynamics

The methods of this section are due to R. Ledesma and E. Bayo.

We presented in the previous section a recursive procedure for the inverse dynamics that relied on a pinned-free finite element or assumed mode model of a flexible beam, and the equation for the inverse dynamics torque was formulated by imposing the condition, that the transverse deformation of the free end of each link be zero throughout the motion. The recursive procedure is suitable for open-chain but not for closed-chain configurations.

In this section, we describe a non-recursive approach to solve the general planar inverse dynamics and kinematics, that has been introduced by Ledesma and Bayo (1992a). For the sake of simplicity, we will not describe herein the more general non-recursive procedure for spatial flexible multibodies which is presented in Ledesma and Bayo (1993). Compared to the recursive procedure, this non-recursive approach is more systematic and general and becomes the only choice, when closed-chain systems are encountered. The finite element model of the elastic links now has *pinned-pinned* boundary conditions. This allows one to express the end effector trajectory in terms of the rigid body coordinates only. In addition it leads to a simplified form of the inverse kinematics equations. Once these are solved, the equations of motion give an explicit expression for the inverse dynamics torque.

The starting point is the equation for the forward dynamics (11.33) that was developed in Section 11.2. In partitioned form, this equation may be written as

$$
\begin{bmatrix} m_{RR} & m_{R\theta} & m_{Rf} \\ m_{\theta R} & m_{\theta\theta} & m_{\theta f} \\ m_{fR} & m_{f\theta} & m_{ff} \end{bmatrix} \begin{Bmatrix} \ddot{R} \\ \ddot{\theta} \\ \ddot{q}_f \end{Bmatrix} + \begin{bmatrix} 0 & 0 & 0 \\ 0 & 0 & 0 \\ 0 & 0 & c_{ff} \end{bmatrix} \begin{Bmatrix} \dot{R} \\ \dot{\theta} \\ \dot{q}_f \end{Bmatrix} + \begin{bmatrix} 0 & 0 & 0 \\ 0 & 0 & 0 \\ 0 & 0 & k_{ff} \end{bmatrix} \begin{Bmatrix} R \\ \theta \\ q_f \end{Bmatrix}
$$

$$
+ \begin{bmatrix} \Phi_R^T \\ \Phi_\theta^T \\ \Phi_{q_f}^T \end{bmatrix} \lambda = \begin{bmatrix} Q_{eR} \\ Q_{e\theta} \\ Q_{ef} \end{bmatrix} + \begin{bmatrix} Q_{vR} \\ Q_{v\theta} \\ Q_{vf} \end{bmatrix}
\tag{12.21}
$$

where, as shown in Figure 12.5, R represents the Cartesian components of the origins of all the body axes with respect to the inertial frame and θ are the angles of rotation of the body axes. These are the coordinates that define the rigid body motion namely $q_r = [R, \theta]^T$. The term Q_e represents the external loads, and Q_v the quadratic velocity terms. The second set of equations in (12.21) can be rearranged to express the externally applied joint forces as

$$
Q_{e\theta} = m_{\theta R}\ddot{R} + m_{\theta\theta}\ddot{\theta} + m_{\theta f}\ddot{q}_f + \Phi_\theta^T\lambda - Q_{v\theta}
\tag{12.22}
$$

Equation (12.22) constitutes the inverse dynamics equation that yields the joint forces (torques) necessary for the endpoint or any other control point to follow a prescribed trajectory. In order to obtain $Q_{e\theta}$ the nodal acceleration vector $\ddot{q}_f$ is needed. This vector can be obtained from the third set of equations in (12.21) which may be written as

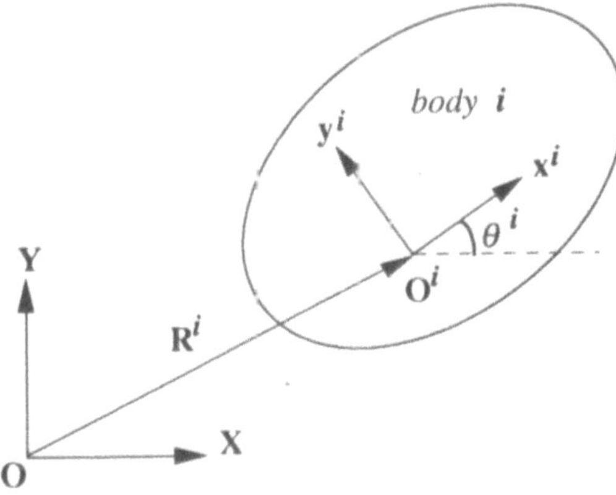

Figure 12.5. Reference frame for a planar flexible body.

$$\mathbf{m}_{ff}\ddot{\mathbf{q}}_f + \mathbf{c}_{ff}\dot{\mathbf{q}}_f + \mathbf{k}_{ff}\mathbf{q}_f = \mathbf{Q}_{ef} + \mathbf{Q}_{vf} - \mathbf{m}_{fR}\ddot{\mathbf{R}} - \mathbf{m}_{f\theta}\ddot{\theta} - \Phi_f^T\lambda \qquad (12.23)$$

The vector of applied nodal forces $\mathbf{Q}_{ef}$ can be expressed in terms of the externally applied torques through the following mapping:

$$\mathbf{Q}_{ef} = \mathbf{G}_e\,\mathbf{Q}_{e\theta} \qquad (12.24)$$

where in the planar case the matrix $\mathbf{G}_f$ is a constant Boolean matrix which maps the externally applied torques to the vector of externally applied nodal forces. For example, in the open-chain planar multibody system shown in Figure 12.6, the Boolean matrix $\mathbf{G}_f$ is constructed such that: the external moment on the node located at the base of the first link is equal to the base motor torque, the moment on the node located at the tip of the first link is the negative of the elbow motor

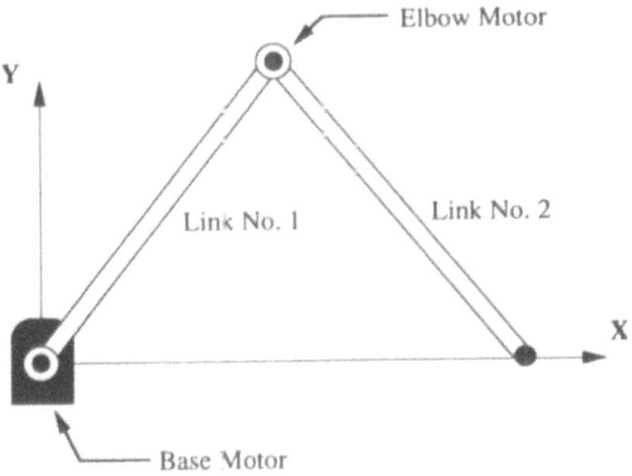

Figure 12.6. Open-chain planar multibody system.

torque, the external moment on the node located at the base of the second link is equal to the elbow motor torque, and all other external forces are zero.

Substituting equations (12.22) and (12.24) into (12.23) yields

$$\mathbf{m}_{ff}\ddot{\mathbf{q}}_f + \mathbf{c}_{ff}\dot{\mathbf{q}}_f + \mathbf{k}_{ff}\mathbf{q}_f = \mathbf{G}_f\mathbf{m}_{\theta f}\ddot{\mathbf{q}}_f + \mathbf{F}_1(\lambda, \mathbf{q}_r, \dot{\mathbf{q}}_r, \ddot{\mathbf{q}}_r, \mathbf{q}_f, \dot{\mathbf{q}}_f) \tag{12.25}$$

where $\mathbf{F}_1$ is a force vector that includes the inertial terms, reaction terms between contiguous bodies, and quadratic velocity terms.

The problem statement for the inverse kinematics is that of finding the non-causal internal states $\mathbf{q}_f$ so that the endpoint coordinates characterized by a subset of the rigid body coordinates $\mathbf{q}_r$ follow a prescribed trajectory. The inverse kinematics equations (12.25) are nonlinear in the variable $\mathbf{q}_f$. As pointed out before, the nonlinear non-causal inversion cannot be carried out by standard numerical integration of ODEs. It requires a linearization process in either the frequency domain or time domain or splitting the linearized system into its causal and anti-causal components.

The key to the linearization process for the non-recursive approach relies on decomposing the inertial coupling submatrix $\mathbf{m}_{\theta f}$ into the sum of a time-invariant matrix and a time-varying matrix:

$$\mathbf{m}_{\theta f} = \mathbf{m}_{\theta f}^{c} + \mathbf{m}_{\theta f}^{t} \tag{12.26}$$

where the first and second components in the RHS are the time-invariant and the time-varying parts of $\mathbf{m}_{\theta f}$, respectively. This decomposition is essential for the iteration process needed to obtain the non-causal solution to the nonlinear inversion problem. Substituting (12.26) into (12.25), one can obtain the inverse kinematics equation of motion for the internal nodal displacements q_f:

$$\mathbf{m}_{ff}^{*}\ddot{\mathbf{q}}_f + \mathbf{c}_{ff}\dot{\mathbf{q}}_f + \mathbf{k}_{ff}\mathbf{q}_f = \mathbf{F}(\lambda, \mathbf{q}_r, \dot{\mathbf{q}}_r, \ddot{\mathbf{q}}_r, \mathbf{q}_f, \dot{\mathbf{q}}_f) \tag{12.27}$$

where

$$\mathbf{m}_{ff}^{*} = \mathbf{m}_{ff} - \mathbf{G}_f\mathbf{m}_{\theta f}^{c} \tag{12.28}$$

The mass matrix $\mathbf{m}_{ff}^{*}$ is non-symmetric. It is precisely the non-symmetry of the mass matrix that produces internal states $\mathbf{q}_f$ (nodal deformations). These states are non-causal with respect to the endpoint motion when non-causal techniques are employed to obtain the proper inversion of the nonlinear, non-minimum problem characterized by (12.27). The nonlinear inversion can now be carried out efficiently in the frequency domain, since the leading matrices have been constructed such that they remain constant throughout the motion. It is thus solved (12.27) iteratively in the frequency domain to yield the nodal deformation vector $\mathbf{q}_f$ that is non-causal with respect to the endpoint motion. Note that this iterative procedure is similar to that used in the recursive case. Each iteration can also be carried out in the time domain through the use of an equation similar to (12.17).

Once the non-causal nodal accelerations are known, equation (12.22) can be used to explicitly compute the non-causal inverse dynamics joint efforts. The inverse dynamics torques and internal states given by equations (12.22) and (12.25), respectively, depend on the Lagrange multipliers and rigid body coordinates which in turn depend on the internal states and the applied torque. The rigid body coordinates and Lagrange multipliers are different from their normal values, when the components of the multibody system are flexible. Therefore, a forward dynamic analysis is required to obtain an improved estimate of the generalized coordinates and Lagrange multipliers. In order to ensure that the iteration process converges to obtain the joint efforts that will cause the end-effector to follow the desired trajectory, the forward dynamics analysis is carried out with the additional constraint that the coordinates of the endpoint follow the desired trajectory.

To summarize, the procedure for obtaining the inverse dynamics solution for flexible multibody systems involves the following steps:

Algorithm 12-5

1. Perform a rigid body inverse dynamic analysis to obtain the nominal values of the rigid body coordinates q_r and Lagrange multipliers λ.
2. Solve the inverse kinematics equation (12.25), either in the frequency or time domain, to obtain the *non-causal* nodal accelerations $\ddot{q}_f$
3. Compute the inverse dynamics joint efforts $Q_{e\theta}$ using equation (12.22).
4. Perform a forward dynamic analysis, using equation (12.21) to obtain new values for the generalized coordinates and Lagrange multipliers.
5. Repeat steps 2 through 4 until convergence in the inverse dynamics torques is achieved.

Compare the recursive procedure and the non-recursive Lagrangian procedure for the inverse dynamics of multibody systems. In the former method, each body in the multibody system is analyzed sequentially, starting from the last element in the chain. For each element, the joint torques are determined first under the assumption that the rigid body coordinates are moving according to the nominal trajectory. With the joint actuation known for this component, a forward dynamic analysis is carried out for this component to determine the nodal deformations. The reaction forces from the next element in the chain are subsequently determined from equilibrium considerations. This recursive method works very well for open-chain systems but is not suitable for closed-chains. In these cases, the reaction forces at the cuts need to be accounted for by *ad hoc* procedures (Bayo et al. (1989)). The non-recursive method avoids this problem, since the reactions between different bodies are automatically accounted for by the Lagrange multipliers and no distinction is made between open-chain and closed-chain configurations. *The non-recursive procedure, although more involved, is therefore more systematic and general.*

12.3.1 A Planar Open-Chain Example

In order to illustrate the performance of the above-mentioned algorithms we describe in this section some results. Consider the system of Figure 12.6 which consists of two flexible aluminum links and two revolute joints driven directly

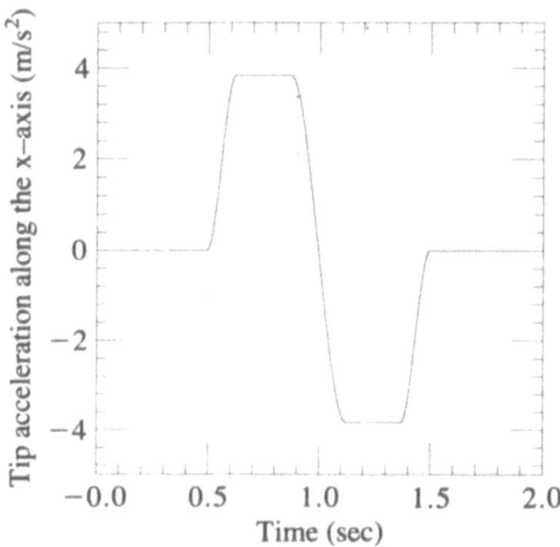

Figure 12.7. Tip acceleration profile.

by servo motors. The intention is to apply both the recursive and non-recursive approaches and compare the results to test their validity.

The links have the following characteristics:

First link: $L = 0.66$m, $A = 1.2097 \times 10^{-4}$ m^2, $I = 2.2864 \times 10^{-10}$ m^4, $M_t = 1.049$ Kg, $I_h = 0.0011823$ m^4

Second link: $L = 0.66$ m, $A = 0.4032 \times 10^{-4}$ m^2, $I = 8.4683 \times 10^{-12}$ m^4, $M_t = 0.067$ Kg, $I_h = 0.00048$ m^4

They both share the following properties: $E = 7.11 \times 10^{10}$ N/m^2, mass density $\rho = 2715$ Kg/m^3, shear coefficient $k = 5/6$, and a damping ratio of 0.002. The cross section of the links is such that the arm is rigid in the vertical direction and flexible in the horizontal direction.

A straight-line tip trajectory along the x axis is generated according to the acceleration profile shown in Figure 12.7 which corresponds to an endpoint displacement of 0.483 meters. Optimal acceleration profiles for inverse dynamics have also been proposed (Bayo and Paden (1987), and Serna and Bayo (1990)).

Figure 12.8 shows the inverse dynamics torque profile for the base motor. Both the recursive and non-recursive methods yield the same results that superimpose to each other (solid curve). The inverse dynamics torque profiles for the elbow motor computed by both methods also coincide and are superimposed to each other in Figure 12.9 (solid curve). Therefore, it is a good validation to see that the results obtained are the same regardless of the method used. The corresponding rigid body torques which are torques obtained considering that the system is rigid are also shown as dashed curves in Figures 12.8 and 12.9 to clearly illustrate the pre-actuation present in the inverse dynamics torque profiles of the flexible system.

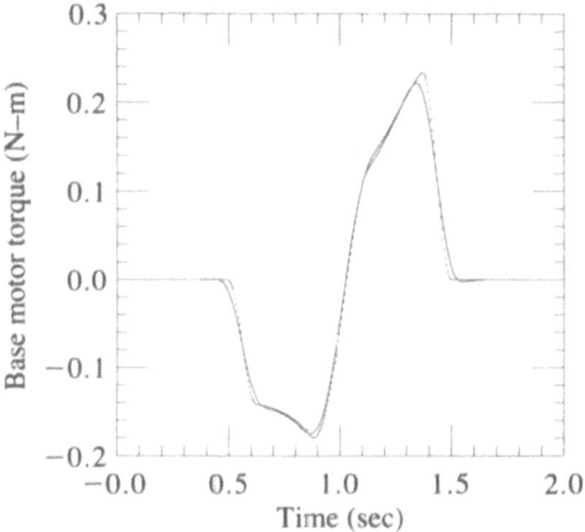

Figure 12.8. Rigid (dotted) and flexible (solid) inverse dynamics torques for first motor.

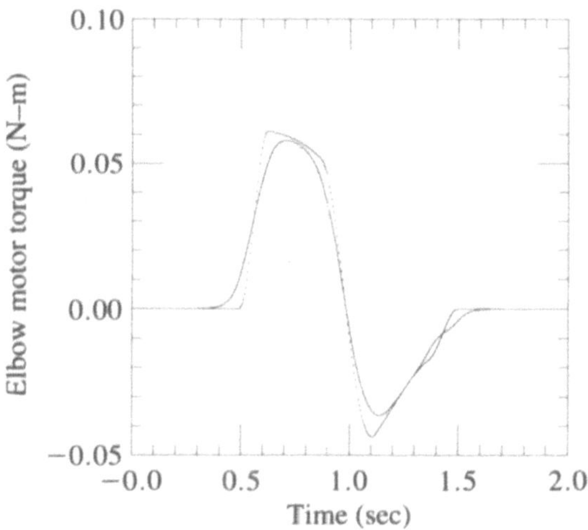

Figure 12.9. Rigid (dotted) and flexible inverse (solid) dynamics torques for second motor.

The inverse dynamics torques produce the desired tip trajectory *without overshoot or residual oscillations*. Figure 12.10 shows a comparison of the tip position error in the y direction resulting from feed forwarding the inverse dynamics torque (solid curve) and the rigid body torque (dashed curve). While the inverse dynamics torque provides an excellent tracking of the tip trajectory, the rigid torque induces a large oscillation in the tip motion. Figure 12.11 shows a com-

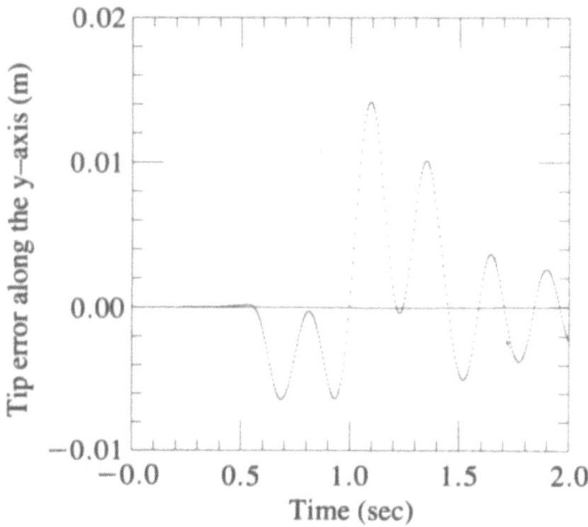

Figure 12.10. Tip error: inverse dynamic (solid) vs. rigid torques (dotted).

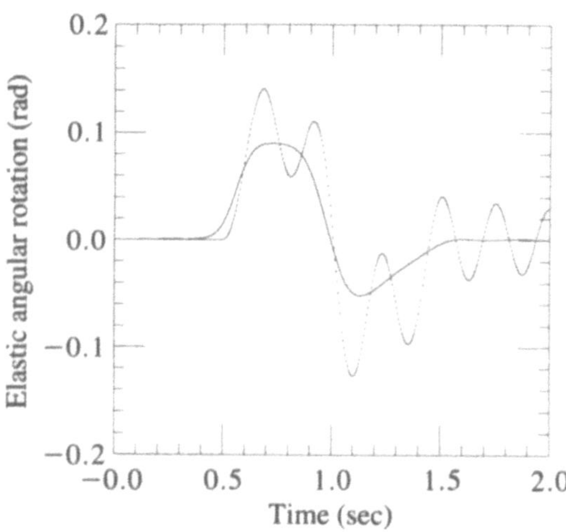

Figure 12.11. Elastic hub rotation: inverse dynamic (solid) vs. rigid torques (dotted).

parison of the elastic angular rotation at the base of the second link, obtained by a feed forward of the inverse dynamics torque (solid curve) to that obtained by a feed forward of the rigid body torque (dashed curve). One can observe that while the inverse dynamics torque does not induce residual vibration, the rigid body torque induces substantial residual oscillation.

   A very important feature of the inverse dynamics is that not only the tip trajectory is tracked but also that the vibrations are minimized; so that the actual

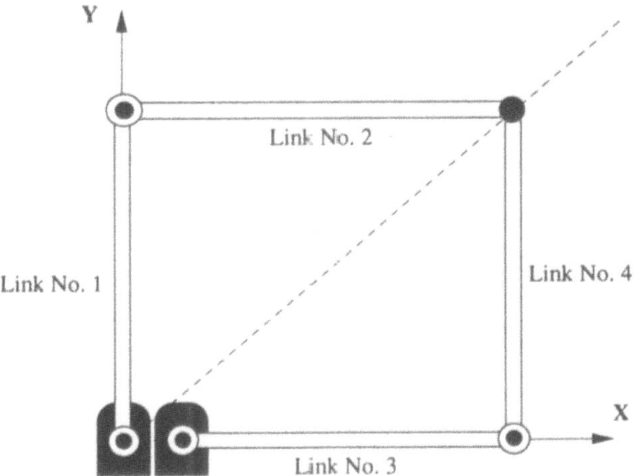

Figure 12.12. Closed-chain flexible multibody systems.

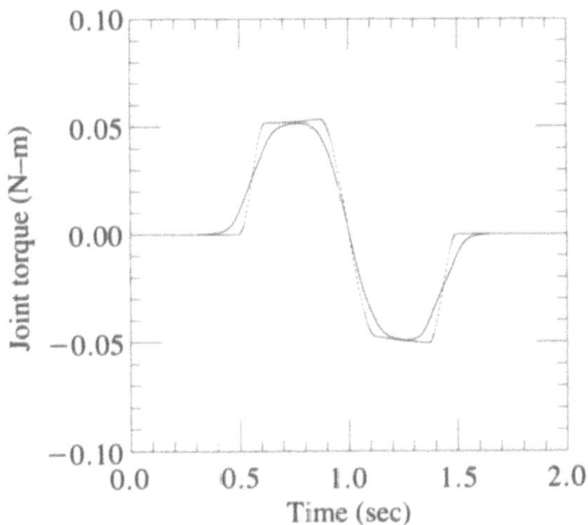

Figure 12.13. Rigid (dotted) and flexible inverse (solid) dynamics torques.

motion of the whole system resembles that of a rigid system (Bayo et al. (1988 and 1989)). An experimental validation consisting in feed forwarding the inverse dynamics torques to this flexible multibody system is presented in Bayo et al. (1989). An exponentially stable control scheme based on the inverse dynamics has been recently proposed and experimentally validated by Paden et al. (1993).

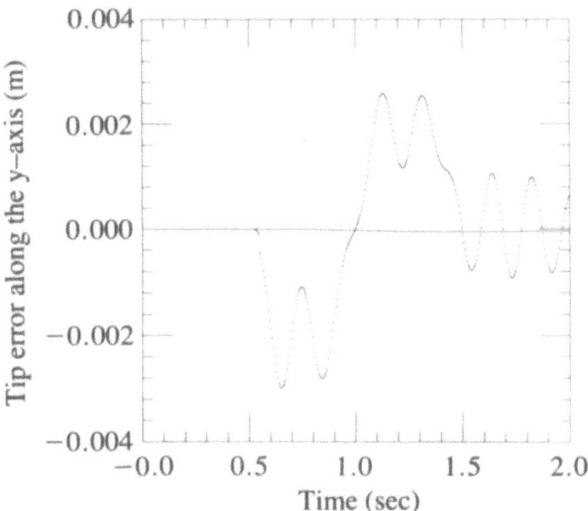

Figure 12.14. Tip error: inverse dynamic (solid) vs. rigid torques (dotted).

12.3.2 A Planar Closed-Chain Example

The non-recursive procedure can be applied to closed-chain configurations without loss of generality. For example, the inverse dynamics and kinematics are calculated for the closed-chain system depicted in the Figure 12.12. A straight-line tip trajectory along the diagonal (See Figure 12.12) is generated according to an acceleration profile similar to that of Figure 12.7. Figure 12.13 shows the inverse dynamics torque profile for one of the base motors. In this case, only the non-recursive method is applied, since the closed-loop configuration precludes the recursive procedure from a straightforward application. The corresponding rigid-body torque (torques obtained considering that the system is rigid) is also shown as a dashed curve in Figures 12.13 to again illustrate the pre-actuation present in the inverse dynamics torque profile of the flexible system.

Figure 12.14 shows a comparison of the tip position error in the y direction resulting from feed forwarding the inverse dynamics torque (solid curve) and the rigid body torque (dashed curve). While the inverse dynamics torque provides an excellent tracking of the tip trajectory, the rigid torque induces a large oscillation in the tip motion.

Appendix

We present in this Appendix the body matrices of equation (12.6) corresponding to a Timoshenko beam finite element. The axial deformation is modeled with linear element shapes defined by

$$N_1(x) = \frac{x}{L} \quad \text{and} \quad N_2(x) = 1 - \frac{x}{L}$$

and Hermite polynomials for the bending deformation are

$$H_1(x) = 1 - 3\left(\frac{x}{L}\right)^2 + 2\left(\frac{x}{L}\right)^3 \quad , \quad H_2(x) = x - 2L\left(\frac{x}{L}\right)^2 + L\left(\frac{x}{L}\right)^3$$

$$H_3(x) = 3\left(\frac{x}{L}\right)^2 - 2\left(\frac{x}{L}\right)^3 \quad , \quad H_4(x) = -L\left(\frac{x}{L}\right)^2 + L\left(\frac{x}{L}\right)^3$$

Element mass matrix

$$\frac{\overline{m}L}{420}\begin{bmatrix} 140 & 0 & 0 & 70 & 0 & 0 \\ 0 & 156+504\frac{\eta^2}{L^2} & 22L+42\frac{\eta^2}{L} & 0 & 54-504\frac{\eta^2}{L^2} & -13L+42\frac{\eta^2}{L} \\ 0 & 22L+42\frac{\eta^2}{L} & 4L^2+56\eta^2 & 0 & 13L-42\frac{\eta^2}{L} & -3L^2-14\eta^2 \\ 70 & 0 & 0 & 140 & 0 & 0 \\ 0 & 54-504\frac{\eta^2}{L^2} & 13L-42\frac{\eta^2}{L} & 0 & 156+504\frac{\eta^2}{L^2} & -22L-42\frac{\eta^2}{L} \\ 0 & -13L+42\frac{\eta^2}{L} & -3L^2-14\eta^2 & 0 & -22L-42\frac{\eta^2}{L} & 4L^2+56\eta^2 \end{bmatrix}$$

Element centrifugal matrix

$$\frac{\overline{m}L}{420}\begin{bmatrix} -140\,\omega_h^2 & -147\,\dot{\omega}_h & 21\,L\,\dot{\omega}_h & -70\,\omega_h^2 & -63\,\dot{\omega}_h & 14\,L\,\dot{\omega}_h \\ 147\,\dot{\omega}_h & -156\,\omega_h^2 & -22\,L\,\omega_h^2 & 63\,\dot{\omega}_h & -54\,\omega_h^2 & 13\,L\,\omega_h^2 \\ -21\,L\,\dot{\omega}_h & -22\,L\,\omega_h^2 & -4\,L^2\,\omega_h^2 & 14\,L\,\dot{\omega}_h & -13\,L\,\omega_h^2 & 3\,L^2\,\omega_h^2 \\ -70\,\omega_h^2 & -63\,\dot{\omega}_h & -14\,L\,\dot{\omega}_h & -140\,\omega_h^2 & -147\,\dot{\omega}_h & 21\,L\,\dot{\omega}_h \\ 63\,\dot{\omega}_h & -54\,\omega_h^2 & -13\,L\,\omega_h^2 & 147\,\dot{\omega}_h & -156\,\omega_h^2 & 22\,L\,\omega_h^2 \\ -14\,L\,\dot{\omega}_h & 13\,L\,\omega_h^2 & 3\,L^2\,\omega_h^2 & -21\,L\,\dot{\omega}_h & 22\,L\,\omega_h^2 & -4\,L^2\,\omega_h^2 \end{bmatrix}$$

The part of the element centrifugal matrix multiplying $\dot{\omega}_h$ is skew-symmetric, and the part multiplying ω_h^2 is symmetric.

Element Coriolis matrix

$$-\frac{2\,\overline{m}\,L\,\omega_h}{60}
\begin{bmatrix}
0 & 21 & 3L & 0 & 9 & -2L \\
-21 & 0 & 0 & -9 & 0 & 0 \\
-3L & 0 & 0 & -2L & 0 & 0 \\
0 & 9 & 2L & 0 & 21 & -3L \\
-9 & 0 & 0 & -21 & 0 & 0 \\
2L & 0 & 0 & 3L & 0 & 0
\end{bmatrix}$$

The element Coriolis matrix is skew-symmetric.

Element stiffness matrix

$$\begin{bmatrix}
\dfrac{EA}{L} & 0 & 0 & -\dfrac{EA}{L} & 0 & 0 \\[2ex]
0 & 12\dfrac{EI}{L^3} & 6\dfrac{EI}{L^2} & 0 & -12\dfrac{EI}{L^3} & 6\dfrac{EI}{L^2} \\[2ex]
0 & 6\dfrac{EI}{L^2} & 4\dfrac{EI}{L} & 0 & -6\dfrac{EI}{L^2} & 2\dfrac{EI}{L} \\[2ex]
-\dfrac{EA}{L} & 0 & 0 & \dfrac{EA}{L} & 0 & 0 \\[2ex]
0 & -12\dfrac{EI}{L^3} & -6\dfrac{EI}{L^2} & 0 & 12\dfrac{EI}{L^3} & -6\dfrac{EI}{L^2} \\[2ex]
0 & 6\dfrac{EI}{L^2} & 2\dfrac{EI}{L} & 0 & -6\dfrac{EI}{L^2} & 4\dfrac{EI}{L}
\end{bmatrix}$$

Element force vector

$$\frac{\overline{m}\,L^2}{60}
\begin{Bmatrix}
10\,\omega_h^2\left(1 + 3\dfrac{d}{L}\right) \\[2ex]
-\dot\omega_h\left(9 + 30\dfrac{d}{L}\right) \\[2ex]
-\dot\omega_h\left(2L + 5d\right) \\[2ex]
10\,\omega_h^2\left(2 + 3\dfrac{d}{L}\right) \\[2ex]
-\dot\omega_h\left(21 + 30\dfrac{d}{L}\right) \\[2ex]
-\dot\omega_h\left(-3L - 5d\right)
\end{Bmatrix}$$

References

Bathe, K.-J., *Finite Element Procedures in Engineering Analysis*, Prentice-Hall, (1982).

Bayo, E., "A Finite Element Approach to Control the End-Point Motion of a Single Link Flexible Robot", *Journal of Robotic Systems*, Vol. 4, pp. 63-75, (1987).

Bayo, E. and Paden, B., "On Trajectory Generation of Flexible Robots", *Journal of Robotic Systems*, Vol. 4, pp. 229-235, (1987).

Bayo, E., Movaghar, R., and Medus, M., "Inverse Dynamics of a Single-Link Flexible Robot. Analytical and Experimental Results", *International Journal of Robotics and Automation*, Vol. 3, pp. 150-157, (1988).

Bayo, E. and Moulin, H., "An Efficient Computation of the Inverse Dynamics of Flexible Manipulators in the Time Domain", *Proceedings of the 1989 IEEE Conference on Robotics and Automation*, Vol. 2, pp. 710-715, Scottsdale, Arizona, (1989a).

Bayo, E. and Moulin, H., "Stiffening Effects in the Inverse Dynamics of Flexible Manipulators", in *Robotics Research – 1989*, pp. 161-167, edited by Y. Youcef-Touni and H. Kazerooni, ASME Press, (1989b).

Bayo, E., Serna, M.A., Papadopoulos, P., and Stubbe, J.R., "Inverse Dynamics and Kinematics of Multi-Link Elastic Robots: An Iterative Frequency Domain Approach", *The International Journal of Robotics Research*, Vol. 8, pp. 49-62, (1989).

Greenwood, T.J., *Principles of Dynamics*, 2nd edition, Prentice-Hall, (1988).

Kwon, D.S. and Book, W., "An Inverse Dynamic Method Yielding Flexible Manipulator State Trajectory", *Proceedings of the 1990 ACC Conference*, Vol. 1, pp. 186-193, (1990).

Ledesma, R. and Bayo, E., "A Non-recursive Lagrangian Approach for the Non-causal Inverse Dynamics of Flexible Multibody Systems. The Planar Case", Report # UCSB-ME-92-2, Mechanical Engineering Department, University of California, Santa Barbara, CA 93106, (1992a). Also to appear in the *International Journal of Numerical Methods in Engineering*, Vol. 36, (1993).

Ledesma, R. and Bayo, E., "Inverse Dynamics of Spatial Flexible Manipulators with Lumped and Distributed Actuators", Report #UCSB-ME-92-4 Mechanical Engineering Department. University of California. Santa Barbara, CA 93106, (1992b).

Ledesma, R. and Bayo, E., "A Non-recursive Lagrangian Approach for the Non-Causal Inverse Dynamics of Flexible Multibody Systems. The Spatial Case", Report # UCSB-ME-93-2, Mechanical Engineering Department, University of California, Santa Barbara, CA 93106, (1993).

Moulin, H., "Problems in the Inverse Dynamics Solution for Flexible Manipulators", Ph.D. Dissertation, University of California, Santa Barbara, (1989).

Moulin, H. and Bayo, E., "On the End-Point Trajectory Tracking for Flexible Manipulators through Non-Causal Inverse Dynamics", *ASME Journal of Dynamic Systems, Measurement, and Control*, Vol. 113, pp. 320-324, (1991).

Moulin, H., Bayo, E., and Paden, B., "Existence and Uniqueness of Solutions of the Inverse Dynamics of Multi-Link Flexible Arms. Convergence of a Numerical Scheme", *Journal of Robotic Systems*, Vol. 10, pp. 73-102, (1993).

Newland, D.E., *An Introduction to Random Vibrations and Spectral Analysis. Second Edition*, Longman, (1984).

Paden, B., Bayo, E., Dagang, C., and Ledesma, R. "Exponentially Stable Tracking Control for Multi-Joint Flexible-Link Manipulators", *ASME Journal of Dynamic Systems, Measurement, and Control*, Vol. 115, pp. 53-60, (1993).

Serna, M.A. and Bayo, E., "Off-Line Trajectory Planning for Flexible Manipulators", in *Modelling the Innovation: Communications, Automation and Information Systems*, Elsevier Science Publishers, pp. 213-220, (1990).

Index

A

acceleration analysis 11; 81-83; 292; 328

angular velocity vector 41; 182; 284

articulated inertia 276-279

assumed modes (see modal analysis)

augmented Lagrangian (see penalty formulation)

B

backlash 351-356
    planar prismatic joint 354-356
    planar revolute joint 352-354

Baumgarte stabilization 163; 166; 188; 365

bifurcation point (see singular positions)

body
    base body 181; 182; 240; 283; 286; 292; 300
    body-by-body calculations 286; 287; 288; 297-302
    junction body 240; 280; 302

branched systems 180; 239; 279; 297

Bryant angles 39; 183

C

canonical
    equations 128; 130; 189; 334; 371
    momenta 129; 189; 193

clearances (see backlash)

closed-chain systems 7
    cut-joint method 188; 280; 304
    improved formulations 187-188; 279-281; 303-307
    inverse dynamics 422; 430

collisions (see impacts)

comparative results
    CPU times per function evaluation 313
    finite displacement problem 88
    improved dynamic formulations 308
    RSCR Jacobian computation 61

component mode synthesis 376

composite inertia 273-275

constraint equations 8; 17; 74; 157
    complexity 46; 64
    driving 71; 72; 156
    holonomic 123; 160; 165
    joint 24; 29; 41-44; 52-56; 384
    joint coordinates 57-58; 386
    loop closure 20; 38; 188; 304
    non-holonomic 107-113; 167; 195
    open-chain constraints 188
    orthogonality 39; 399
    redundant constraints 88-93; 166; 219; 231
    rigid body 18; 29; 32; 47-51; 384
    stabilization methods 162; 174
    stiction 343
    unit module 39; 399

continuation methods 77

coordinates
    coordinate partitioning 173
    dependent 8; 17
        mixed 34-35; 56-61; 64; 80; 82